# RECENT PROGRESS IN OPTICAL FIBER RESEARCH

Edited by **Moh. Yasin,**
**Sulaiman W. Harun**
and **Hamzah Arof**

**Recent Progress in Optical Fiber Research**
Edited by Moh. Yasin, Sulaiman W. Harun and Hamzah Arof

**Published by InTech**
Janeza Trdine 9, 51000 Rijeka, Croatia



**Publishing Process Manager** Martina Blecic

**Technical Editor** Teodora Smiljanic

**Cover Designer** InTech Design Team

Additional hard copies can be obtained from orders@intechopen.com

Recent Progress in Optical Fiber Research,
Edited by Moh. Yasin, Sulaiman W. Harun and Hamzah Arof
p. cm.
ISBN 978-953-307-823-6

For EU product safety concerns:
IN TECH d.o.o., Prolaz Marije Krucifikse Kozulić 3, 51000 Rijeka, Croatia,
info@intechopen.com or visit our website at intechopen.com.

# Meet the editors

Moh. Yasin received his BSc in 1990 from Airlangga University. He went on to receive MSc and PhD degrees from Gadjah Mada University in 1999 and 2010.. His research interests include fiber optic communications and sensors. Currently, he is an associate professor at the Faculty of Science and Technology, Airlangga University, Indonesia.

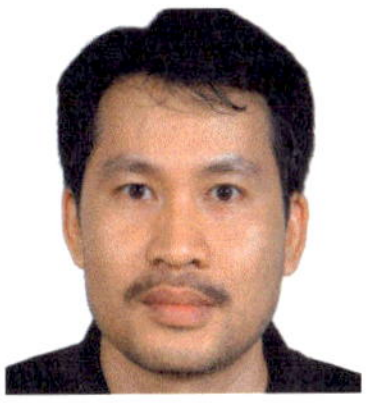

Hamzah Arof received his BSc from Michigan State University, and PhD from the University of Wales. Both degrees were in electrical engineering. His current research interests include signal processing and photonics. Currently he is affiliated with the Department of Electrical Engineering, University of Malaya, Malaysia.

Sulaiman Wadi Harun received the BE degree in Electrical and Electronics System Engineering from Nagaoka University of Technology, Japan in 1996. He received his MSc and PhD degrees in photonic technology from the University of Malaya in 2001 and 2004. He is actively working on optical amplifiers, fiber optic sensors and fiber lasers. Sulaiman Wadi Harun is also an Associate Fellow Member for both Academy of Sciences Malaysia and Malaysia Science Association. He is currently working as a full Professor in the Department of Electrical Engineering, University of Malaya, Malaysia.

# Contents

# Preface

Optical fiber refers to technology associated with the transmission of information as light pulses along a glass or plastic strand or fiber. Optical fiber carries much more information than conventional copper wire. It's immune to electromagnetic interference and is cheaper. From the early experiments by John Tyndall in the guided transmission light, through the development of light-emitting diodes and lasers and the emergence of dense wavelength-division multiplexing (DWDM), the applications for optical fiber have increased. Today, optical fiber technologies permeate a variety of industries. For instance, delivering high-definition broadcast (HDTV) at high resolutions has become possible through the deployment of fiber-to-the-home (FTTH) networks. This new book reviews current researches on optical fiber including nonlinear effects and polarization related issues in optical fibers, photonic crystal fibers (PCFs) as well as other characteristics.

This book is divided into four sections. The first section (chapters 1 to 5) discusses the nonlinear effects in optical fibers which has become an area of academic research and of great importance in the optical fiber based systems. In optical communication systems, the term nonlinearity refers to the dependence of the system on the power of the optical beams being launched into the fiber cable. The nonlinearities in optical fibers are due to the third order susceptibility ($X^3$) according to the Schrödinger equation. The real part of the equation describes self-phase modulation (SPM), cross-phase modulation (XPM) and four-wave mixing (FWM) while the imaginary part of the equation describes stimulated Brillouin scattering (SBS) and stimulated Raman scattering (SRS). Chapter 1 discusses the modeling tools based on a multimode generalized nonlinear Schrödinger equation, its simplification to experimentally relevant situations and a few selected applications. Chapter 2 reviews the physics of spontaneous nonlinear scattering processes in optical fibers. The variable-coefficient higher order nonlinear Schrödinger equation is investigated in Chapter 3. This equation describes the wave propagation in a nonlinear fiber medium with higher order effects such as third order dispersion, self-steepening and SRS. Chapter 4 describes the nonlinear effect applications in supercontinuum generation using a continuous wave (CW) pump source. Compared to the pulsed source, average powers available for the CW pump are stronger and thus increase the spectral power of the supercontinuum source. The dynamic of the supercontinuum generation is considerably different in this case and, thus, intensive numerical studies to optimize the fiber parameters are presented in this chapter. Recently, on the other hand, researchers found ways to realize slow light operating in room temperature, and solid-state materials via various techniques such as SBS, coherent population oscillations (CPO), and tunable time delays based on group velocity dispersion or conversion/dispersion. In Chapter 5 we describe some of the physical mechanisms that can be used to induce slow and fast light effects in room-temperature solids and some of the exotic propagation effects that can be observed.

The second section comprises three chapters discussing polarization losses in optical fibers, optical fiber birefringence effects and spun fibers. Chapter 6 provides details of polarization mode dispersion (PMD), chromatic dispersion and polarization dependent loss. Chapter 7 presents the basic effects, which lead to the occurrence of linear and circular birefringence in single-mode fibers. Chapter 8 discusses the theory and characterization of spun fibers which are used for compensation of PMD.

The third section consists of 5 chapters which cover some selected topics on PCF researches. The use of PCF is associated with their unique chromatic dispersion properties and nonlinear behavior, making them suitable for dispersion compensation or zero-dispersion propagation. A systematic study of dispersion properties in PCFs is presented in Chapter 9. Chapter 10 elaborates on PCF loop mirrors and their applications in several optical devices and sensors. Chapter 11 reports on a broadband super-continuum light generated using a highly nonlinear PCF in conjunction with high power picoseconds pulses for applications in ultrahigh-resolution optical coherence tomography (OCT) system for ophthalmology, dermatology and dental imaging. Chapter 12 explains the behavior of optical fiber on the basis of photonic crystal. Chapter 13 proposes and demonstrates a new design concept for microstructured fiber taper that can be produced on a traditional optical fiber draw tower with multi-pressure control.

The final section presents current research on new applications for optical fibers. This section consists of seven chapters. Chapter 14 discusses the fabrication of long period gratings (LPGs) in new generation fibers such as D-shaped fibers and PCFs as well as the novel application fields offered by the hosting fiber. Chapter 15 discusses the propagation of vortex multimode laser beams in an optical fiber for applications in mode division multiplexing and multimode self-imaging. Chapter 16 explains the use of optical fibers and some limitations in two exoplanet detection methods: nulling interferometry and the radial velocity method. Chapter 17 reviews the published work, covering the transfer of low frequency and time, and the necessary techniques for accurate optical frequency transmission. Chapter 18 presents the current applications of the fiber-optic chip platform including spectro-electro-chemical measurements, fluorescence detection of a bioassay, a broadband fiber optic light source, and Raman interrogation of molecular adsorbates. Chapter 19 discusses a theoretical framework of optical fibers in phase space. This chapter is intended to give a brief review of the phase-space analysis applied to fiber optics, stressing the use of the Wigner distribution function (WDF). Chapter 20 discusses the effect of $\gamma$-ray radiation on commercial optical fibers. Two different dose and dose rate γ-ray pulses are employed to irradiate four kinds of optical fibers and radiation-induced losses are measured by using five lasers with different wavelengths as carriers.

**Dr Moh. Yasin**
Dept. of Physics, Faculty of Science,
Airlangga Univ. Surabaya,
Indonesia

**Prof. Sulaiman W. Harun**
Dept. of Electrical Engineering, Faculty of Engineering,
Univ. of Malaya,
Malaysia

**Dr Hamzah Arof**
Dept. of Electrical Engineering, Faculty of Engineering,
Univ. of Malaya,
Malaysia

# Part 1

## Nonlinear Effects in Optical Fibers

# 1

# Multimode Nonlinear Fibre Optics: Theory and Applications

Peter Horak and Francesco Poletti
*University of Southampton*
*United Kingdom*

## 1. Introduction

Optical fibres have been developed as an ideal medium for the delivery of optical pulses ever since their inception (Kao & Hockham, 1966). Much of that development has been focused on the transmission of low-energy pulses for communication purposes and thus fibres have been optimised for singlemode guidance with minimum propagation losses only limited by the intrinsic material absorption of silica glass of about 0.2dB/km in the near infrared part of the spectrum (Miya et al., 1979). The corresponding increase in accessible transmission length simultaneously started the interest in nonlinear fibre optics, for example with early work on the stimulated Raman effect (Stolen et al., 1972) and on optical solitons (Hasegawa & Tappert, 1973). Since the advent of fibre amplifiers (Mears et al., 1987), available fibre-coupled laser powers have been increasing dramatically and, in particular, fibre lasers now exceed kW levels in continuous wave (cw) operation (Jeong et al., 2004) and MW peak powers for pulses (Galvanauskas et al., 2007) in all-fibre systems. These developments are pushing the limits of current fibre technology, demanding fibres with larger mode areas and higher damage threshold. However, it is increasingly difficult to meet these requirements with fibres supporting one single optical mode and therefore often multiple modes are guided.

Non-fibre-based laser systems are capable of delivering even larger peak powers, for example commercial Ti:sapphire fs lasers now reach the GW regime. Such extreme powers cannot be transmitted in conventional glass fibres at all without destroying them (Gaeta, 2000), but there is a range of applications for such pulses coupled into hollow-core capillaries, such as pulse compression (Sartania et al., 1997) and high-harmonic generation (Rundquist et al., 1998). For typical experimental parameters, these capillaries act as optical waveguides for a large number of spatial modes and modal interactions contribute significantly to the system dynamics.

In order to design ever more efficient fibre lasers, to optimise pulse delivery and to control nonlinear applications in the high power regime, a thorough understanding of pulse propagation and nonlinear interactions in multimode fibres and waveguides is required. The conventional tools for modelling and investigating such systems are based on beam propagation methods (Okamoto, 2006). However, these are numerically expensive and provide little insight into the dependence of fundamental nonlinear processes on specific fibre properties, e.g., on transverse mode functions, dispersion and nonlinear mode coupling. For such an interpretation a multimode equivalent of the nonlinear Schrödinger equation, the

standard and highly accurate method for describing singlemode nonlinear pulse propagation (Agrawal, 2001; Blow & Wood, 1989), is desirable.
In this chapter, we discuss the basics of such a multimode generalised nonlinear Schrödinger equation (Poletti & Horak, 2008), its simplification to experimentally relevant situations and a few select applications. We start by introducing and discussing the theoretical framework for fibres with $\chi^{(3)}$ nonlinearity in Sec. 2. The following sections are devoted to multimode nonlinear applications, presented in the order of increasing laser peak powers. A sample application in the multi-kW regime is supercontinuum generation, discussed in Sec. 3. Here we demonstrate how fibre mode symmetries and launching conditions affect intermodal power transfer and spectral broadening. For peak powers in the MW regime, self-focusing effects become significant and lead to strong mode coupling. The spatio-temporal evolution of pulses in this limit is the topic of Sec. 4. Finally, at GW peak power levels, optical pulses can only be delivered by propagation in gases. Still, intensities become so high that nonlinear effects related to ionisation must be taken into account. An extension of the multimode theory to include these extreme high power effects is presented in Sec. 5 and the significance of mode interaction is demonstrated by numerical examples pertaining to a recent experiment. Finally, we end this chapter with conclusions in Sec. 6.

## 2. The multimode generalised nonlinear Schrödinger equation

Pulse propagation in singlemode fibres is frequently modelled by a generalised nonlinear Schrödinger equation (NLSE) which describes the evolution of the electric field amplitude envelope of an optical pulse as it propagates along the fibre (Agrawal, 2001; Blow & Wood, 1989). This framework has been extremely successful in incorporating all linear and nonlinear effects usually encountered in fibres, such as second and higher order dispersion, Kerr and Raman nonlinearities and self-steepening, and its predictions have been corroborated by numerous experiments using conventional fibres, photonic crystal fibres and fibre tapers of different materials, as well as laser sources from the continuous wave regime down to few cycle pulses. Perhaps the most prominent application of the NLSE is in the description of supercontinuum generation where all the linear and nonlinear dispersion effects come together to induce spectacular spectral broadening of light, often over very short propagation distances (Dudley et al., 2006).
For very high power applications, as motivated above, a further extension of the NLSE is required to deal with the multimode aspects of large-mode area fibres. A very general multimode framework has been presented recently allowing for arbitrary mode numbers, polarisations, tight mode confinements and ultrashort pulses (Poletti & Horak, 2008). Here we describe a slightly simplified version that is more easily tractable and still is applicable to many realistic situations, e.g., for the description of high power applications as discussed in the later sections.
We start by considering a laser pulse propagating in a multimode fibre. The pulse can be written as the product of a carrier wave $\exp[i(\beta_0^{(0)}z - \omega_0 t)]$, where $\omega_0$ is the carrier angular frequency and $\beta_0^{(0)}$ is its propagation constant in the fundamental fibre mode, and an envelope function $\mathbf{E}(\mathbf{x}, t)$ in space and time. Note that throughout this chapter we adopt the notation that vectorial quantities are written in bold face and $\mathbf{x} = (x, y, z)$. For convenience, we assume $\mathbf{E}(\mathbf{x}, t)$ to be complex-valued, so that it includes the envelope phase as well as the amplitude, and we consider the pulse evolution in a reference frame moving with the group velocity of the fundamental mode, so that in the absence of dispersion a pulse would stay centred at time

$t = 0$ throughout its propagation. Finally, we use units such that $|\mathbf{E}(\mathbf{x}, t)|^2$ is the field intensity in $\mathrm{W/m^2}$. The envelope function can then be expanded into a superposition of individual modes $p = 0, 1, 2, \ldots$, each represented by a discrete transverse fibre mode profile $\mathbf{F}_p(x, y)$ and a modal envelope $A_p(z, t)$, as

$$\mathbf{E}(\mathbf{x}, t) = \sum_p \frac{\mathbf{F}_p(x, y)}{\left[\int \mathrm{d}x\, \mathrm{d}y\, |\mathbf{F}_p|^2\right]^{1/2}} A_p(z, t). \tag{1}$$

Note that $|A_p(z, t)|^2$ gives the instantaneous power propagating in mode $p$ in units of W, and that a simplified normalisation has been used compared to a more rigorous previous formulation (Poletti & Horak, 2008). The accuracy of this approximation improves as the fibre core size is increased and the core-cladding index contrast is decreased, leading to guided modes with an increasingly negligible longitudinal component of polarisation.
The multimode generalised nonlinear Schrödinger equation (MM-NLSE) is then given by the following set of coupled equations to describe the dynamics of the mode envelopes,

$$\begin{aligned}\frac{\partial A_p}{\partial z} &= \mathcal{D}\{A_p\} \\ &+ i\frac{n_2 \omega_0}{c}\left(1 + \frac{i}{\omega_0}\frac{\partial}{\partial t}\right) \sum_{l,m,n} \left\{(1 - f_R) S^K_{plmn} A_l A_m A_n^* + f_R S^R_{plmn} A_l [h * (A_m A_n^*)]\right\}.\end{aligned} \tag{2}$$

The following approximations have been applied here: (i) we have assumed that the Raman response and the pulse envelope functions vary slowly on the time scale of a single cycle of the carrier wave, so that we can neglect a rapidly oscillating term, and (ii) an additional term related to the frequency dependence of the mode functions has been omitted, assuming the variation of $S^{K,R}_{plmn}$ is slow compared to the $1/\omega_0$ self-steepening term. In Eq. (2),

$$\mathcal{D}\{A_p\} = i(\beta_0^{(p)} - \Re[\beta_0^{(0)}])A_p - (\beta_1^{(p)} - \Re[\beta_1^{(0)}])\frac{\partial A_p}{\partial t} + i\sum_{n\geq 2} \frac{\beta_n^{(p)}}{n!}\left(i\frac{\partial}{\partial t}\right)^n A_p \tag{3}$$

yields the effects of dispersion of mode $p$ with coefficients $\beta_n^{(p)} = \partial^n \beta^{(p)}/\partial\omega^n$. Here we allow for complex values of the modal propagation constants $\beta^{(p)}$ where the imaginary part describes mode and wavelength dependent losses; $\Re[..]$ denotes the real part only. The second line of (2) represents the effects of optical nonlinearity with a nonlinear refractive index $n_2$. The term $\propto \partial/\partial t$ describes self-steepening and the two terms within the sum describe Kerr and Raman nonlinearities. The Raman term contributes with a fraction $f_R$ to the overall nonlinearity ($f_R = 0.18$ for silica glass fibres) and contains the Raman mode overlap factors

$$S^R_{plmn} = \frac{\int \mathrm{d}x\, \mathrm{d}y \left[\mathbf{F}_p^* \cdot \mathbf{F}_l\right]\left[\mathbf{F}_m \cdot \mathbf{F}_n^*\right]}{\left[\int \mathrm{d}x\, \mathrm{d}y\, |\mathbf{F}_p|^2 \int \mathrm{d}x\, \mathrm{d}y\, |\mathbf{F}_l|^2 \int \mathrm{d}x\, \mathrm{d}y\, |\mathbf{F}_m|^2 \int \mathrm{d}x\, \mathrm{d}y\, |\mathbf{F}_n|^2\right]^{1/2}} \tag{4}$$

as well as a convolution of the time dependent Raman response function $h(t)$ with two mode amplitudes

$$[h * (A_m A_n^*)](z, t) = \int \mathrm{d}\tau\, h(\tau) A_m(z, t - \tau) A_n^*(z, t - \tau). \tag{5}$$

The mode overlap factors responsible for the instantaneous Kerr effect are given by

$$S^K_{plmn} = \frac{2}{3} S^R_{plmn} + \frac{1}{3} \frac{\int \mathrm{d}x\,\mathrm{d}y \left[\mathbf{F}^*_p \cdot \mathbf{F}^*_n\right]\left[\mathbf{F}_m \cdot \mathbf{F}_l\right]}{\left[\int \mathrm{d}x\,\mathrm{d}y\,|\mathbf{F}_p|^2 \int \mathrm{d}x\,\mathrm{d}y\,|\mathbf{F}_l|^2 \int \mathrm{d}x\,\mathrm{d}y\,|\mathbf{F}_m|^2 \int \mathrm{d}x\,\mathrm{d}y\,|\mathbf{F}_n|^2\right]^{1/2}}. \tag{6}$$

Numerically, the mode functions of all the modes involved in the nonlinear effects under consideration are first evaluated at $\omega_0$ and a table of overlap integrals is calculated. The number of modes and overlap integrals can be greatly reduced based on mode symmetry arguments (Poletti & Horak, 2008); all the applications discussed in the following will employ such reduced sets of modes. Next, the dispersion curves for these modes are calculated. Finally, the system of equations (2) is integrated numerically using a standard symmetrised split-step Fourier method (Agrawal, 2001), where adaptive step size control is implemented by propagating the nonlinear operator using a Runge-Kutta-Fehlberg method (Press et al., 2006). In order to avoid numerical artifacts, we also found it necessary to further limit the maximum step size to a fraction of the shortest beat length between all the modes considered. The accuracy and convergence of the results is further checked by running multiple simulations with increasingly small longitudinal step sizes.

The framework presented above still allows for modes of arbitrary polarisation. In most practical situations, however, one is interested in a subset of modes representing only a specific polarisation state which is determined by the pump laser. The two most common cases are briefly discussed in the following.

### 2.1 Circular polarisation

Under the weak guiding condition, modes fall into groups of $LP_{mn}$ modes containing either two ($m = 0$) or four ($m > 0$) degenerate modes. Within each group, the modes can be combined into modes that are either $\sigma_+$ or $\sigma_-$ circularly polarised at every point in the fibre. If the light launched into the fibre is, for example, $\sigma_+$ polarised, the form of the overlap integrals (4) and (6) guarantees that no light is coupled into the $\sigma_-$ polarised modes during propagation and those modes can therefore be eliminated entirely from the model. It is worth emphasising that this is an *exact result* within the weak guiding limit. Using the properties of circular polarisation vectors, the overlap integrals are then simplified to

$$\begin{aligned} S^R_{plmn} &= \frac{\int \mathrm{d}x\,\mathrm{d}y\, F_p F_l F_m F_n}{\left[\int \mathrm{d}x\,\mathrm{d}y\, F_p^2 \int \mathrm{d}x\,\mathrm{d}y\, F_l^2 \int \mathrm{d}x\,\mathrm{d}y\, F_m^2 \int \mathrm{d}x\,\mathrm{d}y\, F_n^2\right]^{1/2}}, \\ S^K_{plmn} &= \frac{2}{3} S^R_{plmn}, \end{aligned} \tag{7}$$

where the mode functions have been written as $\mathbf{F}_p = \mathbf{e}_+ F_p$ for $\sigma_+$ polarised modes with real-valued scalar mode functions $F_p$.

### 2.2 Linear polarisation

The situation is slightly more complicated in the case of linearly polarised pump light. In this case, nonlinear coupling between orthogonal polarisation modes is in principle allowed, leading to, for example, birefringent phase matching and vector modulation instability (Agrawal, 2001; Dupriez et al., 2007). However, for many practical situations where modes can be described as $LP_{mn}$ modes, if linearly polarised light is launched into the fibre, nonlinear coupling to orthogonal polarisation states is effectively so small that most of the pulse energy remains in its original polarisation throughout the entire pulse propagation. This allows

halving the number of modes to be considered in the model with significant computational advantage, and a simpler definition of the overlap factors (4) and (6). There are several important practical situations where this approximation can be acceptable:
(i) For degenerate modes (no birefringence), the overlap factor (6) for four-wave mixing (FWM) between modes of parallel polarisation is three times larger than that for orthogonal polarisation. Since the dispersion properties, and therefore the phase matching conditions, are the same, nonlinear gain is much higher for the same polarisation and thus will dominate the dynamics.
(ii) For few-moded fibres power transfer to orthogonal modes by FWM can be negligible if either the phase matching condition cannot be fulfilled at all, or if the phase matching condition is achieved only for widely separated wavelength bands where the difference in group velocities limits the effective interaction length due to walk-off effects.
In these situations one can therefore use an approximate theoretical description of pulse propagation by restricting the MM-NLSE to the $LP_{mn}$ modes of the fibre with the same linear polarisation everywhere. Assuming real-valued $x$-polarised mode functions $\mathbf{F}_p = \mathbf{e}_x F_p$, the overlap integrals then reduce to

$$S^R_{plmn} = S^K_{plmn} = \frac{\int \mathrm{d}x\,\mathrm{d}y\, F_p F_l F_m F_n}{\left[\int \mathrm{d}x\,\mathrm{d}y\, F_p^2 \int \mathrm{d}x\,\mathrm{d}y\, F_l^2 \int \mathrm{d}x\,\mathrm{d}y\, F_m^2 \int \mathrm{d}x\,\mathrm{d}y\, F_n^2\right]^{1/2}}. \tag{8}$$

A further simplification is also sometimes possible. If linearly polarised light is predominantly launched in an $LP_{0n}$ mode, power transfer into $LP_{mn}$ modes with $m > 0$ can only be initiated by spontaneous FWM processes. By contrast, other $LP_{0n}$ modes of the same polarisation can be excited by stimulated processes, see Sec. 3.1. Thus, if the dominant processes within the pulse propagation are stimulated ones, e.g., in the regime of high powers and relatively short propagation distances, the study can be effectively restricted to $LP_{0n}$ modes with the same polarisation.

## 3. Supercontinuum generation in multimode fibres

One of the first applications where the MM-NLSE presented in the previous section can provide deep insights is that of supercontinuum (SC) generation in multimode fibres. As already mentioned, the complex dynamic underlying SC generation in *singlemode* fibres is by now well understood. Octave spanning SC in suitably designed fibres arises as a combination of various nonlinear phenomena, including soliton compression and fission, modulation instability, parametric processes, intrapulse Raman scattering, self phase modulation (SPM) and cross phase modulation (XPM) (Dudley et al., 2006). As the fibre diameter is increased though, as required for example to increase the SC power spectral density without destroying the fibre, the fibre starts to support multiple modes. Previous theoretical models were usually restricted to two polarisation modes of a birefringent fibre (Agrawal, 2001; Coen et al., 2002; Lehtonen et al., 2003; Martins et al., 2007) or included a maximum of two spatially distinct modes (Dudley et al., 2002; Lesvigne et al., 2007; Tonello et al., 2006). Using the full MM-NLSE, however, fibres with arbitrary modal contents can be studied, for which a rich new list of *intermodal* nonlinear phenomena emerges, causing the transfer of nonlinear phase and/or power between selected combinations of modes (Poletti & Horak, 2009).
In this section, using simulations of a specific few-moded fibre as an illustrative example, we will discuss how modal symmetries and launch conditions can have a drastic influence on intermodal power transfer dynamics. For pump peak powers in the range of tens to hundreds

of kW, if the nonlinear length of the pump pulses is shorter than the walk-off length between the modes involved, significant power transfer into high-order modes with the appropriate symmetry can occur, which can be beneficial, for example, to further extend the SC spectrum to shorter wavelengths. Even if conditions for significant intermodal power transfer are not met, it is found that intermodal XPM can still play a significant role in the SC dynamics by broadening the spectrum of modes which would not otherwise present a significant spectral broadening if pumped on their own.

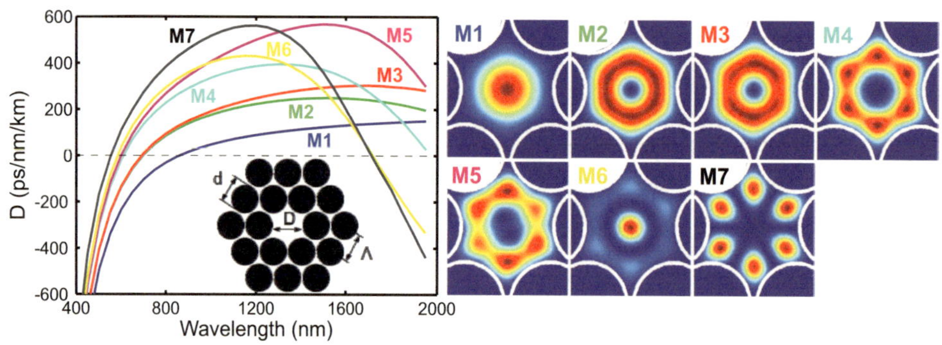


Fig. 1. GVD curves and transverse mode functions, calculated at 850nm, of the 7 circularly polarised modes guided in a HF with $\Lambda = 2.7\mu$m and $d = 2.5\mu$m .

To discuss the intermodal nonlinear dynamics leading to SC generation we focus on a *moderately* multimoded holey fibre (HF) consisting of two rings of large circular air holes with pitch $\Lambda = 2.7\mu$m and relative hole size $d/\Lambda = 0.93$, surrounding a solid core with a diameter of a few optical wavelengths ($D = 2\Lambda - d = 2.9\mu$m ), see Fig. 1. From 400nm to 2000nm the fibre supports 14 modes with effective areas ranging between 3.6 and $6.1\mu\text{m}^2$. To reduce the computational time it is possible to combine these modes into 7 pairs of circularly polarised modes and to exploit the forbidden power exchange between modes with opposite circular polarisation (see Sec. 2.1), only to focus on the 7 right-handed circularly polarised modes M1, M2,..., M7 shown in Fig. 1. The group velocity dispersion (GVD) curves of these modes are significantly different from each other, with a first zero dispersion wavelength (ZDW) ranging from $\lambda_7 = 550$nm for M7 to $\lambda_1 = 860$nm for M1.

### 3.1 Effect of modal symmetries and launch conditions on intermodal power transfer

Equation (2) shows that the transfer of power between modes is mediated by FWM terms of the form $S^K_{plmn} A_l A_m A^*_n$, with $l, m \neq n$. If only a single mode $l$ is initially excited with a narrow spectral line, the strongest power transfer to mode $p$ and therefore the first to be observed in the nonlinear process is the one controlled by *degenerate* FWM terms of the form $S^K_{plln} A_l A_l A^*_n$. If both modes $p$ and $n$ are initially empty, power transfer starts with a spontaneous FWM process and is therefore slow. If one of the generated photons is however returned into the pump $l$ by *stimulated emission*, the process becomes much faster and tends to dominate the nonlinear dynamics in the limit of high-power pulse propagation over short distances. Interestingly, these $S^K_{plll} A_l A_l A^*_l$ processes produce automatic phase-locking of mode $p$ to the pump mode $l$, similarly to what happens in non-phase matched second and third harmonic generation processes (Roppo et al., 2007). However, processes $S^K_{plll}$ require (i) that modes $p$ and $l$ belong to the same symmetry class, and (ii) that they present a large overlap. For the HF under investigation these conditions are only fulfilled for the two $LP_{0n}$ modes M1 and M6, and therefore one would expect significant power transfer only between them.

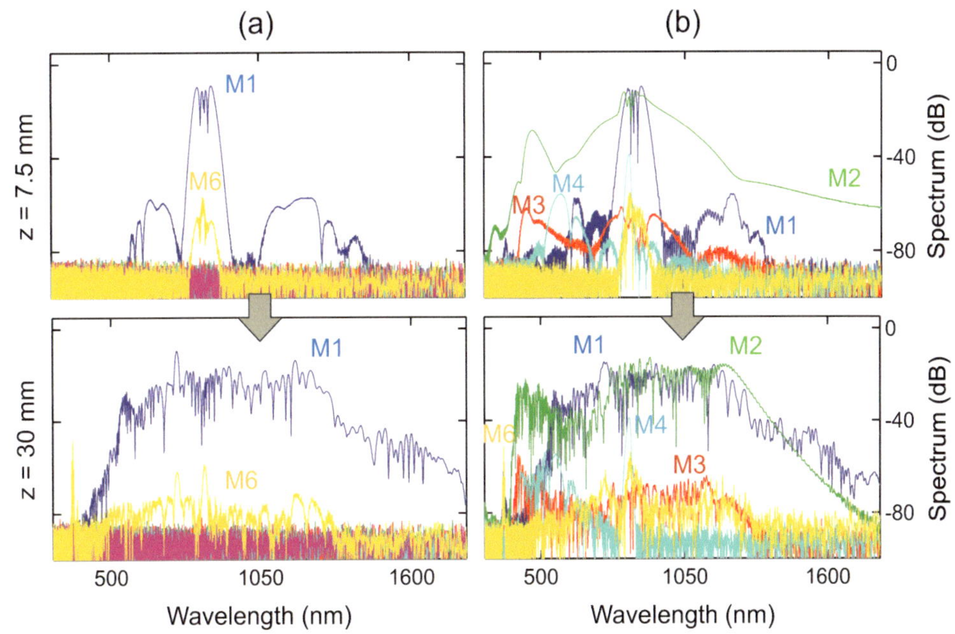


Fig. 2. Simulations of multimode nonlinear propagation in the HF of Fig. 1 after 7.5mm (top row) and 30mm (bottom row), for a 100fs sech-shaped pump centred at 850nm. (a) Only M1 is excited and (b) both M1 and M2 are excited with a 50kW peak power pulse.

This expected behaviour is indeed confirmed by the numerical simulation shown in Fig. 2(a), where a hyperbolic secant pump pulse with temporal profile $A_p(0,t) = \sqrt{P_0}\,\mathrm{sech}(t/T_0)$ with $T_0 = 100$fs (full width at half maximum 176fs) and centred at $\lambda_p = 850$nm is launched into M1 only and propagated through 30mm of the HF. Here the pulse peak power $P_0$ is set to 50kW, corresponding to a 10nJ pulse and, for mode M1, to a soliton of order $N = 166$. As one would expect from single mode SC theory (Dudley et al., 2006), besides SPM-induced spectral broadening, such a high-$N$ pulse develops sidebands which grow spontaneously from noise, through an initial modulation instability (MI) process. The characteristic distance of this phenomenon $L_{MI} \sim 16L_{NL} = 16\lambda/(3\pi n_2 S^K_{1111} P_0) = 6.9$mm correlates well with the simulation results. As expected, of all the other 6 modes only M6 is significantly amplified at wavelengths around $\lambda_p$, and subsequently develops a wide spectral expansion and an isolated peak at 360nm. Further analysis of spectrograms and phase matching conditions indicates that this peak is a dispersive wave in M6, phase matched to a soliton in M1 and slowly shifting to shorter wavelengths as the soliton red-shifts due to the effect of intrapulse Raman nonlinearity. Under these launching conditions the study can thus be restricted to the $LP_{0n}$ modes of the fibre without loss of accuracy. Simulations also show that if either M2, M3, M4, M5 or M7 are selectively launched, no power is transferred to any of the other modes, and each of them evolves as in the single mode case.

When two or more modes contain a significant amount of power, they can all act as pumps for weaker modes. Moreover, if these modes belong to different symmetry classes, additional

FWM terms come into play, giving rise to a much richer phenomenology. As an example, Fig. 2(b) shows what happens when both M1 and M2 are simultaneously excited with a $P_0 = 50$kW sech pulse. This pulse corresponds to an $N = 27$ soliton for M2, due to its much larger value of $\beta_2^{(2)}$ at the pump wavelength. As a result, the SC generated in M2 has a more temporally coherent nature, as it originates from soliton compression and fission mechanisms (the fission length $L_{\text{fiss}} = N \cdot L_{\text{NL}}$ is around 16mm). Due to a shorter ZDW than M1, the final SC in M2 also extends to much shorter wavelengths than the one in M1 (400nm versus 550nm, respectively), which can be one of the benefits of using multimode fibres for SC generation. Moreover, in addition to M6, also M3 and M4 are amplified from noise, generating a complex output spectrum, where the final relative magnitude of different modes is a strong function of wavelength. This is reminiscent of early experimental results (Delmonte et al., 2006; Price et al., 2003).

### 3.2 Non-phase matched permanent intermodal power transfer

To understand the complex dynamics of intermodal power transfer it is useful to refer to the approximate analytical theory of cw pumped parametric processes, which neglects the effects of GVD and pulse walk-off but still provides a valid reference (Stolen & Bjorkholm, 1982). Within this framework, parametric gain leading to exponential signal amplification requires the propagation constant mismatch $\Delta\beta_{plmn} = \beta^{(l)}(\omega_l) + \beta^{(m)}(\omega_m) - \beta^{(p)}(\omega_p) - \beta^{(n)}(\omega_n)$ to be smaller than a few times the *average* inverse nonlinear length $1/L_{NL} = \overline{\gamma}P_0$.

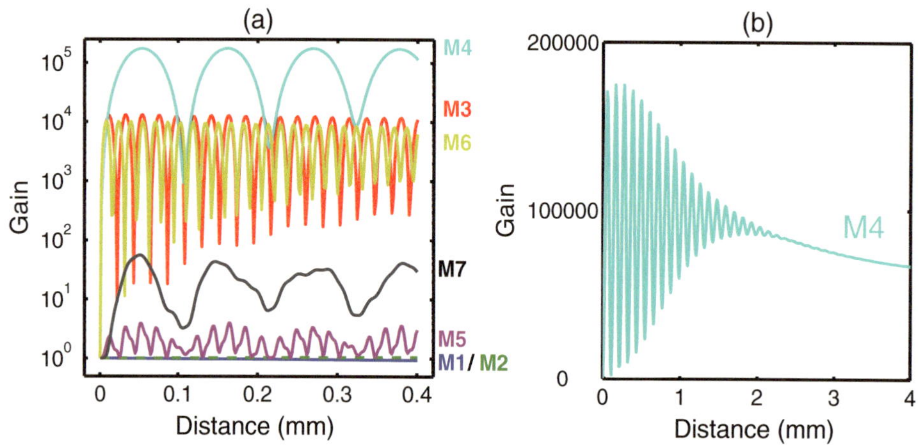


Fig. 3. (a) Dynamic gain evolution for each individual mode when M1 and M2 are simultaneously excited at launch as in Fig. 2(b), showing the oscillatory behaviour typical of non-phase matched parametric processes. (b) Permanent power transfer to M4 despite the lack of parametric phase matching due to walk-off between the pumps in M1 and M2 and the signal in M4.

For multimode processes, an estimate of $\overline{\gamma}$ can be obtained by averaging all the intermodal nonlinearities $\gamma_{plmn} = \frac{3\pi n_2}{\lambda} S^K_{plmn}$ which contribute to SPM and XPM between the relevant modes. However, in most practical situations involving SC generation in highly nonlinear multimode fibres, $\Delta\beta_{plmn} \gg \overline{\gamma}P_0$ for all the relevant FWM processes considered. Thus, no parametric gain is typically observed and each FWM term leads to an oscillatory power exchange between modes, as shown by the dynamic gain curves of high order modes when only M1 and M2 are initially pumped, reported in Fig. 3(a). The oscillation periods are given by the beat lengths $L_b \sim 2\pi/|\Delta\beta|$. For example, for the process leading to amplification of M6, $\Delta\beta_{6111} = 4.1 \cdot 10^5$ m$^{-1}$, corresponding to a value of $L_b = 15.3\mu$m in agreement with the simulation. For modes amplified by a cascade of intermodal FWM processes, such as M5 and M7 in the example, the signature of multiple beating frequencies can be clearly observed.

Despite the non-phase matched nature of most FWM processes, simulations show that after long enough propagation some power is *permanently* transferred into the weaker modes. This is shown, for example, in Fig. 3(b) extending the propagation distance of M4 from 0.4mm to 4mm. A more detailed analysis excluding XPM and Raman effects found this behaviour to be uniquely caused by the temporal walk-off between the pulses involved. The typical length scale of this permanent power transfer is therefore of the order of the walk-off length of all the pulses involved, given by $L_W^{pq} = T_0/|1/v_g^{(p)} - 1/v_g^{(q)}| = T_0/|\beta_1^{(p)} - \beta_1^{(q)}|$ for modes $p$ and $q$. For the example in Fig. 3(b), $L_W^{12} = 3$mm, $L_W^{24} = 2.4$mm and $L_W^{14} = 1.3$mm, which correlate well with the simulation.

In conclusion, nonlinear intermodal power transfer is governed by two length scales, a beat length leading to fast initial power oscillations and a walk-off length leading to permanent power transfer. In order to observe in practice intermodal nonlinear effects, the nonlinear length of the pump pulses must be shorter than the walk-off length, i.e., high peak powers are required. Otherwise, nominally multimode fibres can exhibit the same nonlinear behaviour as singlemode ones. Scaling a fixed fibre structure to larger core sizes allows for larger power throughput, but at the same time longer beat and walk-off lengths lead to much stronger mode coupling, and significant amounts of power can be transferred into higher order modes. In this case, as shown in Fig. 2, higher order modes may also serve to extend the SC spectral extension to much shorter wavelengths.

### 3.3 Effect of intermodal cross phase modulation

Intermodal power transfer mediated by FWM terms, which can permanently exchange power between modes even in the absence of proper phase matching, is not the only intermodal nonlinear effect which can occur in a multimode fibre. Intermodal XPM can also play a role in significantly broadening the spectrum of a mode which would not undergo a significant spectral expansion if propagated on its own (Chaipiboonwong et al., 2007; Schreiber et al., 2005).

To illustrate this phenomenon, we simulate the propagation of a pulse launched in M1 and/or M2 at 725nm, where M1 is in the normal dispersion region and M2 is in the anomalous region. In order to observe significant spectral expansion and intermodal effects within the distance where the pulses are temporally overlapped, we increase the input power up to a value of $P_0 = 500$kW, close to the estimated fibre damage threshold.

Figs. 4(a) and (b) show that when M1 is individually launched, only some SPM-based spectral expansion is visible, whereas if only M2 is launched, a wide MI-based SC develops. On the other hand, if the same input pulse is launched simultaneously in *both* modes as in Fig. 4(c), a much wider output spectrum is developed also in M1. Under these operating conditions the intermodal power transfer is negligible, as confirmed by nearly identical spectral results obtained when all $S^K_{plmn}$ and $S^R_{plmn}$ coefficients responsible for intermodal FWM are set to zero. Therefore, the increased spectral expansion in M1 must be generated by intermodal XPM effects alone. This is indeed confirmed by the simulation in Fig. 4(d), showing that when all intermodal XPM effects are artificially switched off, M1 and M2 produce a very similar spectrum to that of their individual propagation.

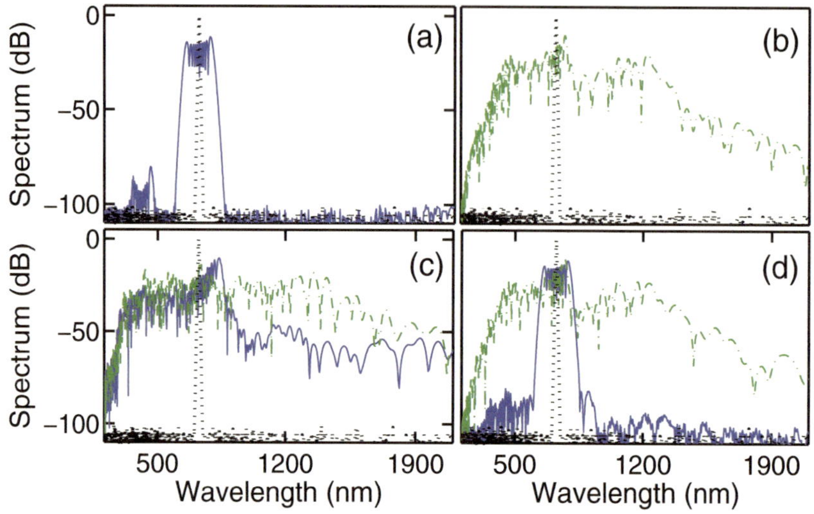


Fig. 4. Spectral output after 2mm propagation in the HF of Fig. 1 of a $T_0 = 100$fs and $P_0 = 500$kW sech pulse centred at 725nm and launched in: (a) M1 only (blue, solid line); (b) M2 only (green, dashed-dot line); (c) both M1 and M2, and (d) both M1 and M2 when all intermodal XPM coefficients are artificially set to zero. The input pulse is shown as a black dotted line.

## 4. Self-focusing in optical fibres in a modal picture

For laser powers larger than discussed in the previous section and into the MW regime, the nonlinear refractive index induced in the glass by the laser may become strong enough to introduce significant spatial reshaping of the beam in the transverse direction. The refractive index of a material is given by $n_0 + n_2 I$, including both the linear, $n_0$, and nonlinear term, $n_2$, and where $I$ is the position-dependent intensity of the laser. Thus, if the beam has a Gaussian-like transverse profile and the optical Kerr nonlinearity $n_2$ is positive, as is the case in most of the commonly used transparent materials, the induced nonlinear refractive index is maximum at the centre of the beam and decreases towards the pulse edges. Therefore, the induced index profile forms a focusing lens, acting back on the laser beam itself. This effect is known as *self-focusing* and has been studied extensively in bulk materials for nearly 50 years (Askaryan, 1962; Chiao et al., 1964). For input powers $P$ below a critical power $P_{crit}$, self-focusing is finally overcome by the beam divergence. In the case of $P > P_{crit}$, however, the pulse undergoes catastrophic collapse leading to permanent damage of the material (Gaeta, 2000). The critical power is given by

$$P_{crit} = 1.86 \frac{\lambda^2}{4\pi n_0 n_2}, \tag{9}$$

where the numerical factor slightly depends on the beam profile in a bulk material (Fibich & Gaeta, 2000). Numerically, self-focusing in bulk media is most commonly modelled by slowly-varying envelope models or, more accurately, by a nonlinear envelope equation (NEE) describing the dynamics of the transverse beam profile $\Phi(\mathbf{x}, t)$ (Brabec & Krausz, 1997;

Ranka & Gaeta, 1999),

$$\frac{\partial}{\partial z}\Phi = \mathcal{D}_{mat}\{\Phi\} + \frac{i}{2\beta_0}\left(1+\frac{i}{\omega_0}\frac{\partial}{\partial t}\right)^{-1}\nabla_\perp^2\Phi + i\frac{n_2 n_0 \omega_0}{2\pi}\left(1+\frac{i}{\omega_0}\frac{\partial}{\partial t}\right)|\Phi|^2\Phi, \quad (10)$$

where $\mathcal{D}_{mat}\{\Phi\}$ is a dispersion term similar to (3) describing the effect of material dispersion and $\nabla_\perp^2$ is the transverse Laplace operator. The NEE incorporates many features similar to the MM-NLSE (2), e.g., higher order dispersion, Kerr nonlinearity and self-steepening terms. However, even in the presence of rotational symmetry, the envelope function $\Phi$ is a two-dimensional object (radial and temporal coordinate), in contrast to the MM-NLSE which only uses a finite number of one-dimensional (temporal) envelope functions to describe the same situation. If the number of modes is small, the MM-NLSE is thus computationally significantly more efficient, both in terms of reduced memory requirements and faster dynamics simulation.

It is now well established that the same process of self-focusing occurs in optical waveguides and fibres and that the same power threshold for catastrophic collapse applies (Farrow et al., 2006; Gaeta, 2000). However, for powers below $P_{crit}$ the observed light propagation behaviour is qualitatively different from that observed in bulk media, since here the light is additionally bound by total internal reflection at the core-cladding interface, which can lead to additional spatial and temporal interference and dispersion effects, such as periodic oscillations of the beam profile or catastrophic pulse collapse even when the launched peak power is below the critical value. In this section we will discuss these effects within the framework of the MM-NLSE, which leads to an easy understanding of fibre-based self-focusing within a modal picture (Horak & Poletti, 2009; Milosevic et al., 2000). Such an interpretation is particularly useful in the context of high-power fibre lasers, which now achieve peak powers close to the critical power with pulse lengths approaching the nanosecond regime (Galvanauskas et al., 2007).

### 4.1 Continuous wave limit

We start our discussion with the case of cw propagation, which in practice is also a good approximation to the behaviour of long pulses (ps to ns regime) near the pulse peak, and use the MM-NLSE restricted to the linearly polarised $LP_{0n}$ modes, as discussed in Sec. 2.2. The MM-NLSE thus reduces to

$$\frac{\partial A_p}{\partial z} = i(\beta_0^{(p)} - \beta_0^{(0)})A_p + i\frac{n_2\omega_0}{c}\sum_{l,m,n} S_{plmn}^K A_l A_m A_n^* \quad (11)$$

with $S_{plmn}^K$ given by (8). Specifically, we assume propagation in a short piece of a step-index fibre with a pure silica core of 40$\mu$m diameter and a refractive index step of 0.02 between core and cladding. This fibre is similar to photonic crystal large-mode area fibres which are commercially available, where the index step has been increased such that the fibre supports eight $LP_{0n}$ modes. The zero-dispersion wavelength of this fibre is at 1.26$\mu$m, and we assume a pump laser operating at 1300nm wavelength. The critical power (9) for silica at this wavelength is $P_{crit}$ = 5.9MW. Note that at this power level pulses up to approximately 100ps length can be transmitted through the fibre without fibre damage (Stuart et al., 1996).

Figure 5 shows the dynamics of light propagation along this fibre when cw light is launched into the fundamental $LP_{01}$ mode with a power of 0.7$P_{crit}$=4.84MW. The curves in Fig. 5(a) show the power $|A_p|^2$ in the lowest order modes obtained by solving Eq. (11). Power

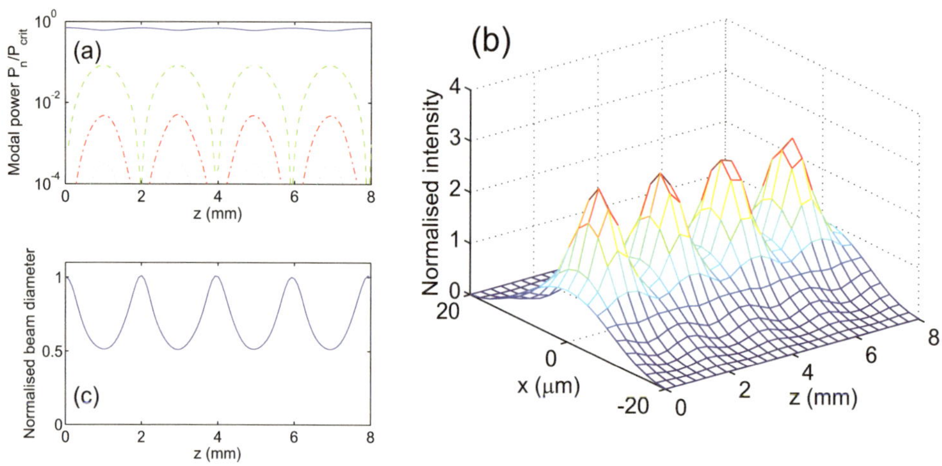


Fig. 5. Propagation of cw laser light at 1.3$\mu$m wavelength through a multimode silica step-index fibre with 40$\mu$m core diameter and core-cladding index difference of 0.02. The launched power is $0.7P_{crit}$=4.84MW in the fundamental $LP_{01}$ mode. (a) Power in the lowest four fibre modes versus propagation distance. (b) 2D (transverse and longitudinal) spatial intensity profile of the beam. (c) Dynamics of the transverse beam width (FWHM), normalised to the width of the fundamental fibre mode.

from the fundamental mode is quickly transferred over sub-mm propagation distances into higher order modes by FWM processes, most prominently by induced FWM involving three pump photons as described by terms of the form $\partial A_p/\partial z \propto iA_0^2 A_0^*$, see Sec. 3.1. However, because of the phase mismatch $\beta_0^{(p)} - \beta_0^{(0)}$ between the fundamental mode and the higher order modes the initial FWM gain is reversed after a certain propagation distance (about 1mm for the chosen parameters) and power is coherently transferred back into the pump from the higher order modes. This process is repeated subsequently leading to a periodic exchange of power between modes. The phase mismatch increases for increasing mode order and thus the maximum transferred power decreases.

In Fig. 5(b) we depict the corresponding 2D beam intensity $|\mathbf{E}(x,z)|^2$ calculated by summing the modal contributions (1), normalised to the maximum field $|\mathbf{E}(0,0)|^2$ at the fibre input. The field experiences significant periodic enhancement on the beam axis at positions where large fractions of the total power propagate inside higher order $LP_{0n}$ modes. At these positions of enhanced intensity, the full width at half maximum (FWHM) of the beam profile is strongly reduced, as shown in Fig. 5(c). The intermodal FWM processes together with the modal phase mismatch are therefore responsible for periodic beam self-focusing and defocusing in a fibre. This complements the standard interpretation of self-focusing in a bulk medium using Gaussian beam propagation, which describes the same phenomenon as focusing by a Kerr-induced lensing effect, followed by beam divergence and subsequent total internal reflection at the core-cladding interface. We finally note that a stationary solution can be obtained for the cw MM-NLSE in which the modal amplitudes and phases are locked in such a way that no oscillations occur. In the bulk interpretation this corresponds to the situation where nonlinear focusing and diffraction are perfectly balanced, thereby generating a stationary spatial soliton.

It may seem that this modal description of self-focusing is only possible in multimode fibres but breaks down in singlemode fibres, for example in large-mode area photonic crystal fibres designed for endlessly single mode operation (Mortensen et al., 2003). However, in this case the role of the higher order bound modes of a multimode fibre is taken over by the cladding modes, and it is the FWM-induced power exchange between the guided mode of a singlemode fibre and its cladding modes which provides a modal interpretation of self-focusing.

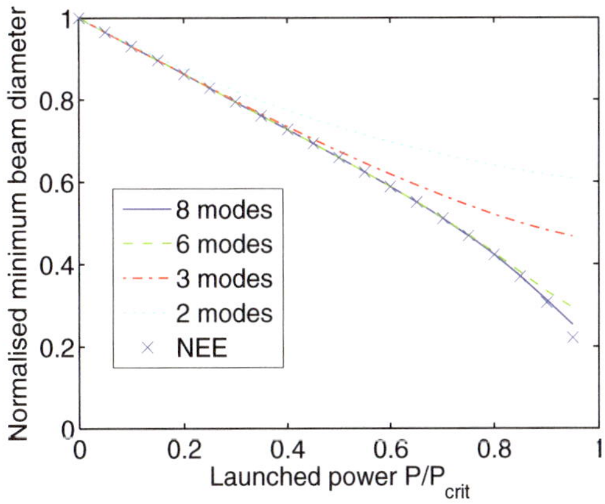


Fig. 6. Minimum beam diameter during the first period of self-focusing oscillation under cw pumping vs pump power for the same fibre parameters as in Fig. 5. The curves correspond to MM-NLSE simulations involving the lowest 2, 3, 6 modes only, and all 8 $LP_{0n}$ modes (from top to bottom). The crosses indicate simulation results using the nonlinear envelope equation.

Using only a finite number of modes in the simulation of the MM-NLSE necessarily limits the transverse spatial resolution that can be achieved by this method. For example, the $LP_{0n}$ mode function exhibits $n$ maxima and $n-1$ zeros along the radial direction within the fibre core region. With simulations using $n$ different modes one can therefore expect a maximum resolution of the order of $R/n$ where $R$ is the core radius. Simulations with pump powers approaching the critical power $P_{crit}$ will thus require a larger number of modes in order to correctly describe the increasingly small minimum beam diameter. We investigate this behaviour in Fig. 6. Here we show the minimum beam diameter achieved during the first period of self-focusing and diffraction, i.e., at approximately 1mm of propagation for the parameters of Fig. 5, when the MM-NLSE is restricted to different numbers of modes. For clarity, the beam diameter is normalised to the diameter of the launched beam ($LP_{01}$ mode). We observe that simulations with 2, 3, and 6 modes are accurate up to pump powers of approximately $0.2P_{crit}$, $0.4P_{crit}$, and $0.8P_{crit}$, respectively, compared to simulations involving all 8 bound fibre modes of this sample fibre. For comparison, we also show the results of the NEE beam propagation method (10). This confirms the accuracy of the MM-NLSE with 8 modes up to $0.95P_{crit}$ corresponding to a nearly five-fold spatial compression of the beam.

For the simulations shown in Fig. 6 we used the same 4th-5th order Runge-Kutta integration method with adaptive step size control (MATLAB R2010b by MathWorks, Inc.) for both the MM-NLSE and the NEE. Each data point required approximately 0.9s of CPU time on a standard desktop computer with the 8-mode MM-NLSE and $<$0.2s with 6 modes. In contrast, the corresponding NEE simulations with 1024 radial grid points required 101s, that is, two to three orders of magnitude slower than the MM-NLSE.

**4.2 Short pulse propagation**

Next, we consider the propagation of short pulses in the regime of peak powers close to the critical power, where in addition to transverse spatial effects the pulse may exhibit complex temporal dynamics related to intermodal and intramodal dispersion, self-steepening and nonlinear effects. As an example we consider sech-shaped pulses with a temporal FWHM of 100fs launched with a peak power of $0.8P_{crit}$ into the fundamental mode of the multimode fibre considered above. The pump wavelength is again set to $1.3\mu$m. The simulations discussed in the following used a 6-mode MM-NLSE with 2048 temporal grid points solved with a split-step Fourier method (Poletti & Horak, 2008; 2009).

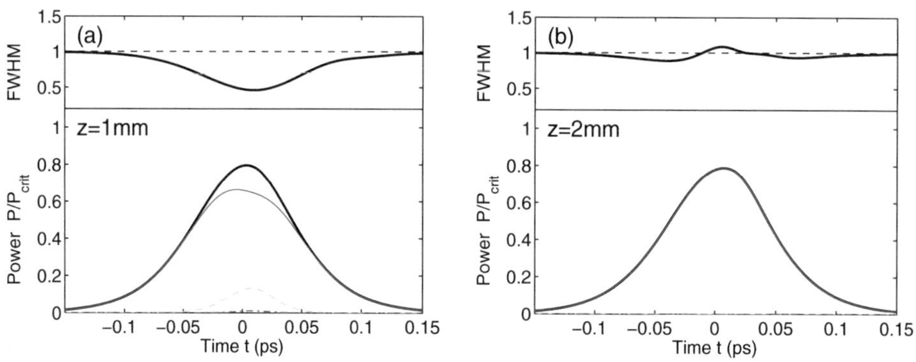


Fig. 7. Propagation of a 100fs sech-shaped pulse with $0.8P_{crit}$ peak power at $1.3\mu$m wavelength with the same fibre parameters as in Fig. 5 after (a) 1mm and (b) 2mm of propagation. The bottom part of the figure shows the overall temporal pulse profile (thick solid line) as well as its contributions from the fundamental mode (thin solid), first (dashed) and second (dash-dotted) higher order modes. The top part of the figure shows the spatial FWHM beam diameter along the pulse, normalised to the FWHM of the fundamental mode.

The initial dynamics of the pulse propagation are shown in Fig. 7. After 1mm of propagation, Fig. 7(a), a significant amount of power has been transferred from the fundamental mode into the higher order modes, leading to a transverse beam focusing to approximately 40% of the input beam width. The transverse beam size depends on the pulse power and thus varies along the pulse shape: the beam diameter is smallest near the temporal peak of the pulse, but remains unchanged in the trailing and leading edges where the power is low. Propagating further to 2mm, Fig. 7(b), most of the power has been converted back into the fundamental mode, similar to the cw case of Fig. 5. However, the transfer is not complete and is not uniform along the pulse. This is related to the walk-off of the higher order modes because of intermodal dispersion as well as a slight dependence of the beam oscillation period on power. Therefore, the spatial FWHM of the beam at 2mm propagation length is below that of the fundamental mode in some parts of the pulse while it exceeds it in other parts.

Continuing the propagation of Fig. 7, the spatial beam variations persist, but the deviations from a simple oscillation become more prominent. This is shown clearly in Fig. 8(a) in the beam properties after 7mm of propagation. At this point the initial sech-shaped temporal profile has steepened on the trailing edge and an ultrashort pulse peak is forming due to the interference of the modal contributions. In particular, the first high order mode exhibits a similar power level as the fundamental mode. Simultaneously, the beam diameter is strongly reduced. At 7.4mm of propagation, Fig. 8(b), this peak has narrowed further and reaches

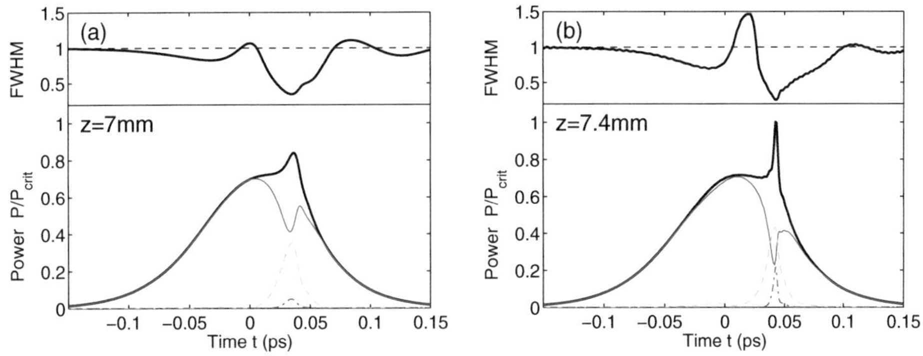


Fig. 8. Continuation of the pulse propagation of Fig. 7 to (a) 7mm and (b) 7.4mm of fibre length exhibiting simultaneous spatial and temporal collapse.

the critical power for catastrophic collapse while the beam diameter has reduced to 20% of the fundamental mode. For even longer propagation lengths the simulations show the pulse breaking up into many ultrashort high-intensity parts around this initial instability, however the MM-NLSE with 6 modes becomes invalid at this point due to its limited spatial and temporal resolution. Simulations with the MM-NLSE restricted to the fundamental mode reveal only a very small amount of pulse reshaping due to self-steepening over this propagation distance (a shift of the pulse peak by about 10fs) and exhibit none of the complex dynamics seen in Fig. 8. We therefore conclude that the simultaneous spatial and temporal collapse of the pulse observed here is a pure multimode effect, driven by FWM-based power exchange together with modal dispersion and self-steepening, in agreement with investigations based on beam propagation methods (Zharova et al., 2006).

## 5. Multimode effects in gas-filled waveguides

As discussed above, the peak power that can be transmitted in optical fibres is limited by the critical power for self-focusing and catastrophic collapse to levels of a few MW. According to Eq. (9), for a fixed laser wavelength $P_{crit}$ only depends on the material linear and nonlinear refractive index. In general, the linear refractive index does not vary much across transparent media, between 1 for vacuum and $\sim$4 for some non-silica glasses (Price et al., 2007) and semiconductors, whereas the nonlinear index $n_2$ can span many orders of magnitude. A common method for guiding extremely high power pulses is thus in hollow-core capillaries or fibres, where most of the light propagates in a gas. For example, $n_2 \approx 5 \times 10^{-23}$ m$^2$/W in air, compared to $2.5 \times 10^{-20}$ m$^2$/W in silica glass, thus pushing $P_{crit}$ into the GW regime. In contrast to solid-core fibres, gas-filled capillaries do not support strictly bound modes, but all modes are intrinsically leaky with losses scaling proportional to $\lambda^2/R^3$ where $\lambda$ is the light wavelength and $R$ is the radius of the capillary hole (Marcatili & Schmeltzer, 1964). Hence, the capillary hole must be sufficiently large in order to allow for transmission of light over long distances. For example, 800nm wavelength light propagating in the fundamental $LP_{01}$ mode of a silica glass capillary with a 75$\mu$m radius hole experiences losses of $\sim$3dB/m. For such a large hole compared to the laser wavelength, the capillary is multimoded, and this is the situation we will consider in the following. It should be noted, however, that single-mode guidance in hollow-core fibres is in principle possible using bandgap effects in photonic crystal fibres (Knight et al., 1998; Petrovich et al., 2008).

Using fs pulses at 800nm wavelength from commercial Ti:sapphire laser systems it is possible to reach peak powers large enough to observe nonlinear effects, and even self-focusing, in gases. Capillary guidance is used in this context for several high-power applications. One of these is pulse compression, where the nonlinearity of the gas in the capillary is exploited to spectrally broaden a pulse by self-phase modulation, which allows the pulse to be compressed after the capillary by purely dispersive means such as gratings or dispersive mirrors (Sartania et al., 1997). For intensities above $\sim 10^{13}\mathrm{W/cm^2}$, the electric field of the laser is large enough to start ionising the gaseous medium. The generated plasma exhibits a negative refractive index, which can counteract the self-focusing effect of the neutral gas and lead to pulse filamentation (Couairon & Mysyrowicz, 2007). In another application, ionisation and recombination effects are used for high harmonic generation of XUV and soft X-ray radiation, processes whose efficiencies can be enhanced significantly by phase matching techniques in capillaries (Rundquist et al., 1998).
In the following we will therefore discuss how the MM-NLSE can be extended to include these important effects and demonstrate a few sample effects related to the multimode nature of hollow capillaries typically used for such high-power applications.

### 5.1 Ionisation and plasma effects in the multimode nonlinear Schrödinger equation

The starting point for this derivation is the capability of high-intensity light to ionise the gas inside the capillary. Two effects contribute to the ionisation: (i) direct multiphoton ionisation, where several photons are absorbed simultaneously to eject one electron from its orbit, and (ii) tunneling ionisation, where the electric field of the laser is so strong that it deforms the electric potential of the nucleus and allows an electron to tunnel through the potential barrier. Tunneling ionisation occurs at higher field strengths than multiphoton ionisation, and is the dominant process for the effects we want to discuss here. The rate of tunneling ionisation $W$ can be calculated using Keldysh theory (Popov, 2004) as

$$W(\mathbf{x},t) = W_0\kappa^2\sqrt{\frac{3}{\pi}}C_{\kappa l}^2 2^{2n^*}F(\mathbf{x},t)^{1.5-2n^*}\exp\left(-\frac{2}{3F(\mathbf{x},t)}\right), \tag{12}$$

where $\kappa^2 = I_p/I_H$ is the ratio of the ionisation potential $I_p$ of the gas species over the ionisation potential for hydrogen $I_H = 13.6\mathrm{eV}$, $W_0 = m_e e^4/\hbar^3 = 4.13\times 10^{16}\mathrm{s}^{-1}$, $F(\mathbf{x},t) = E'(\mathbf{x},t)/(\kappa^3 E_a)$ is the reduced electric field of the laser with $E_a = 5.14\times 10^{11}\mathrm{V/m}$ the atomic unit of field intensity and $E'(\mathbf{x},t)$ the real-valued electric field in units of V/m corresponding to $\mathbf{E}(\mathbf{x},t)$, Eq. (1). The dimensionless parameters $C_{\kappa l}$ and $n^*$ are specific for the gas and can be looked up in tables (Popov, 2004). For the case of argon, which we will use as our example here, we have $I_p = 15.76\mathrm{eV}$, $C_{\kappa l} = 0.95$, and $n^* = 0.929$.
Given the modal amplitudes $A_p(z,t)$ we can calculate the electric field $\mathbf{E}(\mathbf{x},t)$ and thus the ionisation rate $W(\mathbf{x},t)$ at every point and time in the capillary. From this we obtain the fraction of neutral atoms $r_0(\mathbf{x},t)$ and the fraction of ionised atoms $r_1(\mathbf{x},t) = 1 - r_0(\mathbf{x},t)$ by solving

$$\frac{\partial r_0(\mathbf{x},t)}{\partial t} = -W(\mathbf{x},t)r_0(\mathbf{x},t). \tag{13}$$

The generated plasma modifies the refractive index of the gas to

$$n(\mathbf{x},t) = \sqrt{1-\frac{\omega_{pl}(\mathbf{x},t)^2}{\omega^2}}, \tag{14}$$

where the plasma frequency is given by

$$\omega_{pl}(\mathbf{x},t) = \sqrt{\frac{\rho r_1(\mathbf{x},t)e^2}{m_e \epsilon_0}}. \tag{15}$$

Here $\rho$ is the gas density and $e$ and $m_e$ are the electron charge and mass, respectively. The MM-NLSE thus aquires a new nonlinear term $\partial A_p(z,t)/\partial z \propto \mathcal{N}_{pl}\{A_p\}$ with

$$\mathcal{N}_{pl}\{A_p\} = -\left(1 - \frac{i}{\omega_0}\frac{\partial}{\partial t}\right)\frac{i}{2}k_0 \int \mathrm{d}x\mathrm{d}y \frac{\mathbf{F}_p(x,y)^* \cdot \mathbf{E}(\mathbf{x},t)}{\left[\int \mathrm{d}x\,\mathrm{d}y\,|\mathbf{F}_p|^2\right]^{1/2}} \frac{\omega_{pl}(\mathbf{x},t)^2}{\omega_0^2}, \tag{16}$$

which includes a self-steepening correction term and the projection of the modified laser field onto mode $p$ via a spatial overlap integral.

In addition to the effect of the plasma induced refractive index, we also have to consider the loss of energy from the propagating laser pulse due the ionisation process itself (Courtois et al., 2001). In the modal decomposition, this leads to a nonlinear loss term in the propagation of the mode envelope $A_p$ of the form

$$\mathcal{L}_{ion}\{A_p\} = -\frac{1}{2}\int \mathrm{d}x\mathrm{d}y \frac{\mathbf{F}_p(x,y)^* \cdot \mathbf{E}(\mathbf{x},t)}{\left[\int \mathrm{d}x\,\mathrm{d}y\,|\mathbf{F}_p|^2\right]^{1/2}} \frac{\rho r_0(\mathbf{x},t)W(\mathbf{x},t)I_p}{|\mathbf{E}(\mathbf{x},t)|^2}. \tag{17}$$

The full MM-NLSE in the presence of gas ionisation by tunneling in the strong-field limit thus becomes (Chapman et al., 2010)

$$\frac{\partial A_p}{\partial z} = \mathcal{D}\{A_p\} + \mathcal{N}\{A_p\} + \mathcal{N}_{pl}\{A_p\} + \mathcal{L}_{ion}\{A_p\} \tag{18}$$

where the individual terms are given by (2), (3), (16) and (17).

### 5.2 Ultrashort pulse propagation in capillaries

In the following we present simulation results of the extended MM-NLSE (18) for a specific experimental situation (Froud et al., 2009). In particular, we consider a 7cm long capillary with a 75$\mu$m radius hole filled with argon at a pressure of 80mbar in the central 3cm of the capillary; the Ar pressure tapers down over 2cm to 0mbar at the input and output. Laser pulses of 40fs length at 780nm wavelength are launched with a Gaussian waist of 40$\mu$m centred into the capillary. For the simulations, 20 linearly polarised $LP_{0n}$ modes are considered, as discussed in Sec. 2.2.

Results from two sets of simulations with different launched pulse energies, 0.5mJ and 0.7mJ, respectively, are presented in Fig. 9. The distribution of $Ar^+$ ions in the capillary is shown in Figs. 9(a) and (b). As expected, ionisation mainly occurs on axis where the laser intensity is maximum. Moreover, because the transverse beam size of the launched laser pulses is not ideally matched to the fundamental mode of the capillary, power is also coupled into the first higher order mode, which leads to mode beating and thus to the periodic ionisation pattern along the capillary length with a periodicity of $\sim$2cm, observed most clearly at lower powers, Fig. 9(a). At higher powers, the nonlinear ionisation processes become much stronger and a spate of additional radial and longitudinal structures are found in the ionisation pattern, Fig. 9(b). In Fig. 9(c) the partial $Ar^+$ pressures of (a) and (b) are averaged over the transverse cross section of the capillary. The distribution shown in this figure can be easily verified

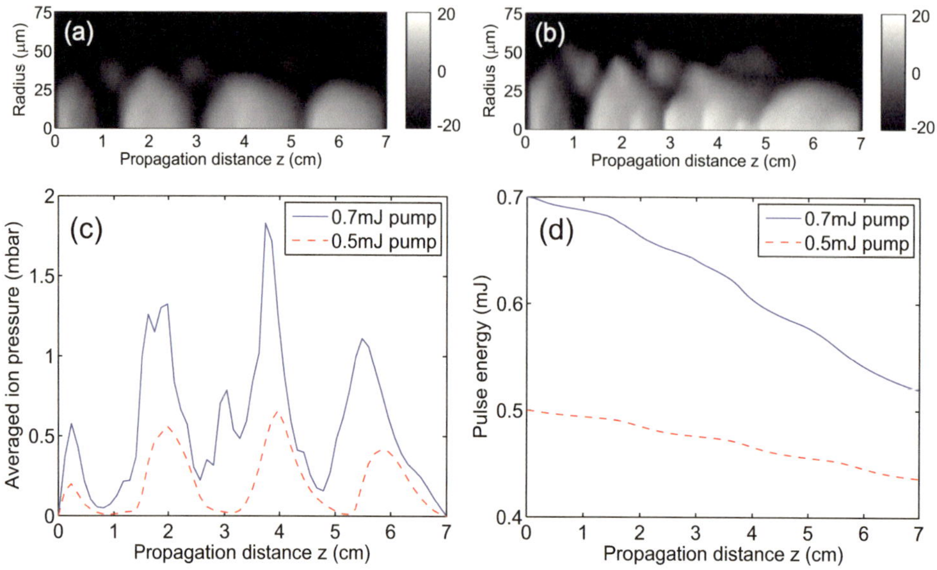


Fig. 9. Propagation of 40fs pulses at 780nm wavelength in a hollow-core capillary (length 7cm, hole radius 75$\mu$m) filled with argon with partial ionisation. (a), (b) Partial pressure of $Ar^+$ ions (in dB of mbar) vs position $z$ and radius $r$ inside the capillary for launched pulse energies of 0.5mJ and 0.7mJ, respectively. (c) $Ar^+$ pressure averaged over the capillary cross section vs $z$. (d) Corresponding integrated pulse energy vs $z$. The total gas pressure in the capillary centre is 80mbar.

experimentally as it is proportional to the intensity of the $Ar^+$ ion fluorescence observed at 488nm (Chapman et al., 2010; Froud et al., 2009). Finally, in Fig. 9(d) the pulse energy summed over all modes is presented versus the propagation distance for these two simulations. The effect of propagation losses due to ionisation, described by the term $\mathcal{L}_{ion}\{A_p\}$ in Eq. (17), is clearly visible with strong losses associated with the peaks of large ionisation in Fig. 9(c). Because of the highly nonlinear nature of tunneling ionisation, losses at slightly higher input energies (0.7mJ instead of 0.5mJ) are several times larger.

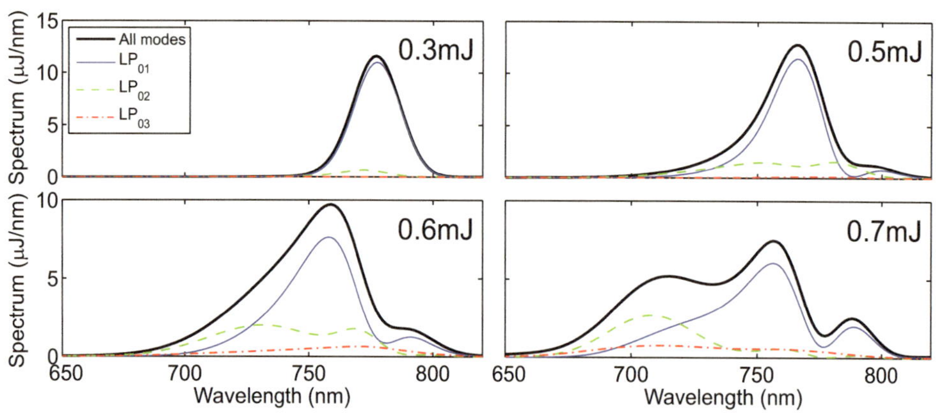


Fig. 10. Pulse spectra and modal contributions at the capillary output for launched pulse energies in the range 0.3mJ to 0.7mJ. Other parameters as in Fig. 9.

The spatial and temporal distribution of ions generated by the propagating laser pulse acts back on the pulse through its (negative) refractive index, according to the term $\mathcal{N}_{pl}\{A_p\}$ given in Eq. (16). Because of the strong localisation of the regions with high ionisation, different capillary modes are affected differently resulting in strong intermodal scattering and mode-specific spectral broadening, as is demonstrated in Fig. 10. At a relatively low pulse energy of 0.3mJ where ionisation is weak, a slight blue-shift of the spectral contribution of the excited $LP_{02}$ mode is observed, but no higher order mode excitation. Increasing the pulse energy to 0.5-0.7mJ, more and more light is scattered into higher order modes. Moreover, the spectrum first develops a small peak at the long-wavelength side of the pump (790-800nm) and then a very broad and high-intensity shoulder at short wavelengths. It is interesting to note that these short wavelength parts of the spectrum are more pronounced in the higher order modes $LP_{02}$ and $LP_{03}$ of the capillary, in fact they contain more power than the fundamental mode at these wavelengths for launched pulse energies above 0.6mJ. This finding has again been confirmed by experiments, where a strong position-dependence of the spectrum was observed in the far field beyond the capillary (Chapman et al., 2010).
These selected results demonstrate clearly that mode interference and mode coupling, i.e., transverse spatial effects, play a significant role in the propagation of high-intensity laser pulses in regimes where ionisation becomes important. This also impacts other applications of such systems, for example the angular dependence of high harmonic generation as recently observed in a capillary-based XUV source (Praeger et al., 2007).

## 6. Conclusions and outlook

To summarise, we presented an analysis of nonlinear effects of short laser pulses propagating in multimode optical fibres. We developed a general theoretical framework which is based on the modal decomposition of the propagating light and takes the form of a multimode generalised nonlinear Schrödinger equation. This approach provides new insights into the significance of fibre properties, e.g., modal dispersion and mode overlaps, for nonlinear pulse propagation, and for moderately multimode fibres and waveguides it has been shown to be numerically significantly more efficient than beam propagation methods. We subsequently discussed several applications of the model covering laser peak powers in the kW (supercontinuum generation), MW (self-focusing effects) and GW regime (ionisation and plasma nonlinearities) highlighting the importance of multimode effects throughout.
While we focused our discussion here on the high-power regime, we emphasise that there is also rapidly growing interest in the application of multimode fibres at low, W-level peak powers. A fast emerging area of interest comes, for example, from optical telecommunications, where in an attempt to increase the fibre capacity researchers are now considering the use of several fibre modes, or several cores within a single fibre, as independent channels. Intermodal nonlinear effects are expected to pose an ultimate limit to the maximum information capacity of the link, which we believe could be estimated by simulations using our model. Various sensing and imaging applications can also benefit from multimode fibres. Moreover, new sources in the mid-IR spectral region are currently being developed for spectroscopy and sensing applications that require novel waveguides such as soft glass fibres or semiconductor-based waveguides and fibres, some of which are intrinsically multimoded at near-IR pump wavelengths. We therefore expect that the multimode nonlinear Schrödinger equation discussed in this work will provide a valuable tool in the analysis and investigation of many future photonics applications.

## 7. Acknowledgements

The authors acknowledge financial support by the U.K. Engineering and Physical Sciences Research Council (EPSRC) and the Royal Society. We thank Prof. D. J. Richardson, Dr. W. S. Brocklesby, and Prof. J. G. Frey for valuable discussions.

## 8. References

Agrawal, G. P. (2001). *Nonlinear Fiber Optics*, 3rd ed., Academic Press, San Diego, USA.

Askaryan, G. A. (1962). Effects of the gradient of strong electromagnetic beam on electrons and atoms, *Sov. Phys. JETP*, Vol. 15, 1088–1090.

Blow, K. J., & Wood, D. (1989). Theoretical description of transient stimulated Raman scattering in optical fibers. *IEEE J. Quantum Electron.*, Vol. 25, No. 12, 2665–2673.

Brabec, T., & Krausz, F. (1997). Nonlinear optical pulse propagation in the single-cycle regime. *Phys. Rev. Lett.*, Vol. 78, No. 17, 3282–3285.

Chaipiboonwong, T., Horak, P., Mills, J. D., & Brocklesby, W. S. (2007). Numerical study of nonlinear interactions in a multimode waveguide. *Opt. Expr.*, Vol. 15, 9040–9047.

Chapman, R. T., Butcher, T. J., Horak, P., Poletti, F., Frey, J. G., & Brocklesby, W. S. (2010). Modal effects on pump-pulse propagation in an Ar-filled capillary. *Opt. Expr.*, Vol. 18, No. 12, 13279–13284.

Chiao, R. Y., Garmire, E., & Townes, C. H. (1964). Self-trapping of optical beams. *Phys. Rev. Lett.*, Vol. 13, No. 15, 479–482.

Coen, S., Chau, A. H. L., Leonhardt, R., Harvey, J. D., Knight, J. C., Wadsworth, W. J., & Russell, P. St. J. (2002). Supercontinuum generation by stimulated Raman scattering and parametric four-wave mixing in photonic crystal fibers. *J. Opt. Soc. Am. B*, Vol. 19, No. 4, 753–764.

Couairon, A., & Mysyrowicz, A. (2007). Femtosecond filamentation in transparent media. *Phys. Rep.*, Vol. 441, 47–189.

Courtois, C., Couairon, A., Cros, B., Marquès, J. R., & Matthieussent, G. (2001). Propagation of intense ultrashort laser pulses in a plasma filled capillary tube: Simulations and experiments. *Phys. Plas.*, Vol. 8, No. 7, 3445–3456.

Delmonte, T., Watson, M. A., O'Driscoll, E. J., Feng, X., Monro, T. M., Finazzi, V., Petropoulos, P., Price, J. H. V., Baggett, J. C., Loh, W., Richardson, D. J., & Hand, D. P. (2006). Generation of mid-IR continuum using tellurite microstructured fiber. *Conference on Lasers and Electro-Optics*, paper CTuA4, Long Beach, USA.

Dudley, J. M., Provino, L., Grossard, N., Maillotte, H., Windeler, R. S., Eggleton, B. J., & Coen, S. (2002). Supercontinuum generation in air-silica microstructured fibers with nanosecond and femtosecond pulse pumping. *J. Opt. Soc. Am. B*, Vol. 19, No. 4, 765–771.

Dudley, J. M., Genty, G., & Coen, S. (2006). Supercontinuum generation in photonic crystal fiber. *Rev. Mod. Phys.*, Vol. 78, No. 4, 1135–1184.

Dupriez, P., Poletti, F., Horak, P., Petrovich, M. N., Jeong, Y., Nilsson, J., Richardson, D. J., & Payne, D. N. (2007). Efficient white light generation in secondary cores of holey fibers. *Opt. Expr.*, Vol. 15, No. 7, 3729–3736.

Farrow, R. L., Kliner, D. A. V., Hadley, G. R., & Smith, A. V. (2006). Peak-power limits on fiber amplifiers imposed by self-focusing. *Opt. Lett.*, Vol. 31, No. 23, 3423–3425.

Fibich, G., & Gaeta, A. (2000). Critical power for self-focusing in bulk media and in hollow waveguides. *Opt. Lett.*, Vol. 25, No. 5, 335–337.

Froud, C. A., Chapman, R. T., Rogers, E. T. F., Praeger, M., Mills, B., Grant-Jacob, J., Butcher, T. J. , Stebbings, S. L., de Paula, A. M., Frey, J. G., & Brocklesby, W. S. (2009). Spatially resolved $Ar^*$ and $Ar^{+*}$ imaging as a diagnostic for capillary-based high harmonic generation. *J. Opt. A*, Vol. 11, 054011.

Gaeta, A. (2000). Catastrophic collapse of ultrashort pulses. *Phys. Rev. Lett.*, Vol. 84, No. 16, 3582–3585.

Galvanauskas, A., Cheng, M.-Y., Hou, K.-C., Liao, K.-H. (2007). High peak power pulse amplification in large-core Yb-doped fiber amplifiers. *IEEE J. Sel. Top. Quantum Electron.*, Vol. 13, No. 3, 559–566.

Hasegawa, A., & Tappert, F. (1973). Transmission of stationary nonlinear optical pulses in dispersive dielectric fibers. I. Anomalous dispersion. *Appl. Phys. Lett.*, Vol. 23, No. 3, 142–144.

Horak, P., & Poletti, F. (2009). Effects of pulse self-focusing on supercontinuum generation in multimode optical fibers. *International Conference on Transparent Optical Networks*, Ponta Delgada, Portugal, 28 June - 2 July 2009.

Jeong, Y., Sahu, J. K., Payne, D. N., & Nilsson, J. (2004). Ytterbium-doped large-core fiber laser with 1.36 kW continuous-wave output power. *Opt. Expr.*, Vol. 12, No. 25, 6088–6092.

Kao, K. C., & Hockham, G. A. (1966). Dielectric-fibre surface waveguides for optical frequencies. *Proc. IEE*, Vol. 113, No. 7, 1151–1158.

Knight, J. C., Broeng, J., Birks, T. A., & Russell, P. St. J. (1998). Photonic band gap guidance in optical fibers. *Science*, Vol. 282, 1476–1478.

Lehtonen, M., Genty, G., Ludvigsen, H., & Kaivola, M. (2003). Supercontinuum generation in a highly birefringent microstructured fiber. *Appl. Phys. Lett.*, Vol. 82, No. 14, 2197–2199.

Lesvigne, C., Couderc, V., Tonello, A., Leproux, P., Barthelemy, A., Lacroix, S., Druon, F., Blandin, P., Hanna, M., & Georges, P. (2007). Visible supercontinuum generation controlled by intermodal four-wave mixing in microstructured fiber. *Opt. Lett.*, Vol. 32, No. 15, 2173–2175.

Marcatili, E. A. J., & Schmeltzer, R. A. (1964). Hollow metallic and dielectric waveguides for long distance optical transmission and lasers. *Bell Tech. Syst. J.*, Vol. 43, 1783–1809.

Martins, E. R., Spadoti, D. H., Romero, M. A., & Borges, B.-H. V. (2007). Theoretical analysis of supercontinuum generation in a highly birefringent D-shaped microstructured optical fiber. *Opt. Expr.*, Vol. 15, No. 22, 14335–14347.

Mears, R. J., Reekie, L., Jauncey, I. M., & Payne, D. N. (1987). Low-noise Erbium-doped fiber amplifier at 1.54$\mu$m. *Electron. Lett.*, Vol. 23, No. 19, 1026–1028.

Milosevic, N., Tempea, G., & Brabec, T. (2000). Optical pulse compression: bulk media versus hollow waveguides. *Opt. Lett.*, Vol. 25, No. 9, 672–674.

Miya, T., Terunuma, Y., Hosaka, T., & Miyashita, T. (1979). Ultimate low-loss single-mode fibre at 1.55$\mu$m. *Electron. Lett.*, Vol. 15, No. 4, 106–108.

Mortensen, N. A., Nielsen, M. D., Folkenberg, J. R., Petersson, A., & Simonsen, H. R. (2003). Improved large-mode-area endlessly single-mode photonic crystal fibers. *Opt. Lett.*, Vol. 28, No. 6, 393–395.

Okamoto, K. (2006). *Fundamentals of Optical Waveguides*, 2nd ed., Academic Press, San Diego, USA.

Petrovich, M. N., Poletti, F., van Brakel, A., & Richardson, D. J. (2008). Robustly single mode hollow core photonic bandgap fiber. *Opt. Expr.*, Vol. 16, No. 6, 4337–4346.

Poletti, F., & Horak, P. (2008). Description of ultrashort pulse propagation in multimode optical fibers. *J. Opt. Soc. Am. B*, Vol. 25, No. 10, 1645–1654.

Poletti, F., & Horak, P. (2009). Dynamics of femtosecond supercontinuum generation in multimode fibers. *Opt. Expr.*, Vol. 17, No. 8, 6134–6147.

Popov, V. S. (2004). Tunnel and multiphoton ionization of atoms and ions in a strong laser field (Keldysh theory). *Physics-Uspekhi*, Vol. 47, No. 9, 855–885.

Praeger, M., de Paula, A. M., Froud, C. A., Rogers, E. T. F., Stebbings, S. L., Brocklesby, W. S., Baumberg, J. J., Hanna, D. C., & Frey, J. G. (2007). Spatially resolved soft X-ray spectrometry from single-image diffraction. *Nat. Phys.*, Vol. 3, 176–179.

Press, W. H., Teukolsky, S. A., Vetterling, W. T., & Flannery, B. P. *Numerical recipes: the art of scientific computing*, 3rd ed. (Cambridge University Press, New York, USA, 2007).

Price, J. H. V., Monro, T. M., Furusawa, K., Belardi, W., Baggett, J. C., Coyle, S., Netti, C., Baumberg, J. J., Paschotta, R., & Richardson, D. J. (2003). UV generation in a pure-silica holey fiber. *Appl. Phys. B*, Vol. 77, No. 2–3, 291–298.

Price, J. H. V., Monro, T. M., Ebendorff-Heidepriem, H., Poletti, F., Horak, P., Finazzi, V., Leong, J. Y. Y., Petropoulos, P., Flanagan, J. C., Brambilla, G., Feng, X., & Richardson, D. J. (2007). Mid-IR supercontinuum generation from nonsilica microstructured optical fibers. *J. Sel. Top. Quantum Electron.*, Vol. 13, No. 3, 738–749.

Ranka, J. K., & Gaeta, A. L. (2009). Breakdown of the slowly varying envelope approximation in the self-focusing of ultrashort pulses. *Opt. Lett.*, Vol. 23, No. 7, 534–536.

Roppo, V., Centini, M., Sibilia, C., Bertolotti, M., de Ceglia, D., Scalora, M., Akozbek, N., Bloemer, M. J., Haus, J. W., Kosareva, O. G., & Kandidov, V. P. (2007). Role of phase matching in pulsed second-harmonic generation: Walk-off and phase-locked twin pulses in negative-index media. *Phys. Rev. A*, Vol. 76, No. 3, 033829.

Rundquist, A., Durfee, C. G., Chang, Z., Herne, C., Backus, S., Murnane, M. M., & Kapteyn, H. C. (1998). Phase-matched generation of coherent soft X-rays. *Science*, Vol. 280, 1412–1415.

Sartania, S., Cheng, Z., Lenzner, M., Tempea, G., Spielmann, Ch., Krausz, F., and Ferencz, K. (1997). Generation of 0.1-TW 5-fs optical pulses at a 1-kHz repetition rate. *Opt. Lett.*, Vol. 22, No. 20, 1562–1564.

Schreiber, T., Andersen, T., Schimpf, D., Limpert, J., & Tünnermann, A. (2005). Supercontinuum generation by femtosecond single and dual wavelength pumping in photonic crystal fibers with two zero dispersion wavelengths. *Opt. Expr.*, Vol. 13, No. 23, 9556–9569.

Stolen, R. H., Ippen, E. P., & Tynes, A. R. (1972). Raman oscillation in glass optical waveguide. *Appl. Phys. Lett.*, Vol. 20, No. 2, 62–64.

Stolen, R. H., & Bjorkholm, J. B. (1982). Parametric amplification and frequency conversion in optical fibers. *IEEE J. Quantum Electron.*, Vol. 18, No. 7, 1062–1072.

Stuart, B. C., Feit, M. D., Herman, S., Rubenchik, A. M., Shore, B. W., & Perry, M. D. (1996). Nanosecond-to-femtosectond laser-induced breakdown in dielectrics. *Phys. Rev. B*, Vol. 53, No. 4, 1749–1761.

Tonello, A., Pitois, S., Wabnitz, S., Millot, G., Martynkien, T., Urbanczyk, W., Wojcik, J., Locatelli, A., Conforti, M., & De Angelis, C. (2006). Frequency tunable polarization and intermodal modulation instability in high birefringence holey fiber. *Opt. Expr.*, Vol. 14, No. 1, 397–404.

Zharova, N. A., Litvak, A. G., & Mironov, V. A. (2006). Self-focusing of wave packets and envelope shock formation in nonlinear dispersive media. *J. Exp. Theor. Phys.*, Vol. 103, No. 1, 15–22.

2

# Spontaneous Nonlinear Scattering Processes in Silica Optical Fibers

Edouard Brainis
*Université libre de Bruxelles*
*Belgium*

## 1. Introduction

When light travels in a optical fiber, a fraction of its total power is always scattered to other wavelengths (or polarization) due to material non linearity. Whether that scattering is weak or strong, desirable or not, depends on the situation. One distinguishes (i) scattering stimulated by the presence of a seed wave (at another wavelength or polarization), (ii) spontaneous scattering, and (iii) amplified spontaneous scattering. Stimulated Raman scattering (SRS), stimulated Brillouin scattering (SBS) and four-wave mixing (FWM) are examples of stimulated scatterings. Those have been thoroughly studied in the past thirty years and are well summarized in classic nonlinear fiber optics textbooks, e.g. (Agrawal, 2007). Several chapters of this book also deal with specific aspects and applications of stimulated scattering. The present chapter focuses on *spontaneous* scattering processes, cases (ii) and (iii).
The chapter also concentrates on nonlinear scattering in *silica* fibers because nowadays those are the most common and widely used types of fibers. Gas-filled hollow core fibers (Benabid et al., 2005) and ion doped fibers (Digonnet, 2001) are not considered here, and it is assumed that the fiber has not been subjected to poling (Bonfrate et al., 1999; Huy et al., 2007; Kazansky et al., 1997), so that the main non linearity is of third order. In this context the most important spontaneous nonlinear scattering processes are

1. the spontaneous Raman scattering (RS),
2. the spontaneous Brillouin scattering (BS), and
3. the spontaneous four-photon scattering (FPS).

These phenomena play an important role in many applications of optical fibers. This role can be positive as in remote optical sensing (Alahbabi et al., 2005a;b; Dakin et al., 1985; Farahani & Gogolla, 1999; Wait et al., 1997). It can also be detrimental as in fiber optics telecommunication, where spontaneous nonlinear scattering processes contribute to decrease the signal-to-noise ratio (SNR) or in supercontinuum generation, where it limits the coherence and stability of the supercontinuum (Corwin et al., 2003; Dudley et al., 2006). In the emerging field of quantum photonics, fiber optical photon-pair sources are intrinsically based on the physics of the FPS (Amans et al., 2005; Brainis, 2009; Brainis et al., 2005), while at the same time RS is the main factor that limits the SNR (Brainis et al., 2007; Dyer et al., 2008; Fan & Migdall, 2007; Lee et al., 2006; Li et al., 2004; Lin et al., 2006; 2007; Rarity et al., 2005; Takesue, 2006).
This chapter reviews the physics of spontaneous nonlinear scattering processes in optical fibers. In Sec. 2, the physical origin of RS, BS and FPS in explained. Because those are pure

quantum mechanical effects, they cannot be properly described in the framework of classical nonlinear optics. A quantum mechanical treatment is presented in Sec. 3. Finally, in Sec. 4, the coupling between different scattering processes in considered.

## 2. Physics of nonlinear scattering processes in optical fibers

### 2.1 Raman scattering

Light at frequency $\omega_p$ traveling in an optical fiber, can excite the fiber molecules from ground to excited vibrational states. In amorphous silica fiber, vibrational states have energy $\hbar|\Omega|$ with $|\Omega|/(2\pi)$ in the 0-40 THz range.These energies (about 0.05 eV) being much smaller than the photon energy $\hbar\omega_p$, no direct excitation of the vibrational states is possible. However, the states can be excited through a second order *Raman transition* involving a second photon at frequency $\omega_s$ and a virtual state as shown in Fig. 1. The spontaneous inelastic scattering that converts a $\omega_p$ photon into a $\omega_s = \omega_p - |\Omega|$ photon and a vibrational excitation at frequency $|\Omega|$ is call a spontaneous *Stokes* process. If a vibrational state at frequency $|\Omega|$ is initially populated, the complementary process in which a $\omega_p$ photon is converted into a $\omega_a = \omega_p + |\Omega|$ photon is also allowed and called a spontaneous *anti-Stokes* process, see Fig. 1.

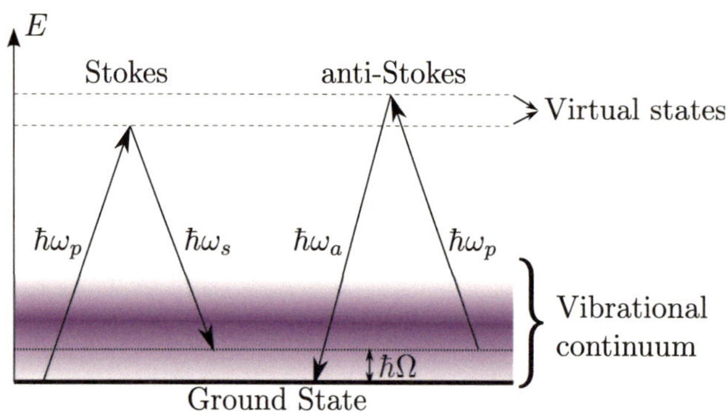


Fig. 1. Spontaneous Stokes and anti-Stokes processes in amorphous silica fibers

Molecular vibrations behave like waves (phonons). The momentum of these vibrational waves corresponds to the momentum mismatch of the pump and (anti-)Stoke waves and does not depend on $|\Omega|$. For this reason, Raman scattering has no preferential direction. It happens in the forward but also in the backward direction. The damping of a phonon wave depends on the wave number and is stronger for shorter wavelength. In fibers the damping is very strong because of the amorphous nature of silica. Therefore the molecular vibration can to a good approximation be considered as local. Yet the small difference in the forward and backward damping explains that the strengths of Raman scattering in forward and backward directions is slightly different (Bloembergen & Shen, 1964).

In addition to the Stokes and anti-Stokes processes that convert pump photons to other wavelengths, Raman scattering can also convert the Stokes and anti-Stokes photons at $\omega_s$ and $\omega_a$ back to the pump mode through *reverse* Stokes and anti-Stokes scattering. In Sec. 3.1, both direct and reverse scattering processes are taken into account to derive the basic equations governing the *net* energy transfer from the pump to Stokes and anti-Stokes waves. For a single monochromatic pump wave at $\omega_p$ the scattered spectral power density $S(z,\omega)$ obeys the following propagation equation

$$\frac{\mathrm{d}}{\mathrm{d}z}\mathcal{S}(z,\omega) = \left[\mathcal{S}(z,\omega)g(\omega_p,\Omega,\theta) + \frac{\hbar\omega}{2\pi}\,[m_{\text{th}}(|\Omega|) + \nu(\Omega)]\,|g(\omega_p,\Omega,\theta)|\right] P_p(z) \tag{1}$$

where $\Omega = \omega_p - \omega$ (positive for a Stokes process and negative for an anti-Stokes ones), $\nu(\Omega)$ is the Heaviside step function, and

$$m_{\text{th}}(|\Omega|) = \left[\exp\left(\frac{\hbar|\Omega|}{k_B T}\right) - 1\right]^{-1} \tag{2}$$

is the thermal equilibrium expectation value of the number of vibrational excitations at angular frequency $|\Omega|$. The function $g(\omega_p,\Omega,\theta)$ in Eq. (1) is the Raman *gain*. The Raman gain measures the scattering strength and is polarization dependent. For a linearly polarized pump field, the Raman gain is maximal for photon scattered with polarization parallel to the pump and minimal for photons scattered with polarization orthogonal to the pump (Stolen, 1979):

$$g(\omega_p,\Omega,\theta) = g_{\|}(\omega_p,\Omega)\cos^2(\theta) + g_{\perp}(\omega_p,\Omega)\sin^2(\theta), \tag{3}$$

where $\theta$ is the angle between the linear polarization vectors of pump and scattered photons. The parallel and orthogonal gains are $g_{\|}(\omega_p,\Omega)$ and $g_{\perp}(\omega_p,\Omega)$ are material properties that can be measured experimentally. It can be shown (see Sec. 3.1) that the ratio of the Stokes to anti-Stokes gain corresponding to the same vibrational mode $|\Omega|$ is

$$\frac{g(\omega_p,\Omega,\theta)}{g(\omega_p,-\Omega,\theta)} = -\frac{n(\omega_p+|\Omega|)}{n(\omega_p-|\Omega|}\left(\frac{\omega_p-|\Omega|}{\omega_p+|\Omega|}\right)^3. \tag{4}$$

Stokes and anti-Stokes gain have opposite signs: Stokes gain is positive while anti-Stokes gain is negative.

### 2.1.1 Spontaneous scattering

With initial condition $\mathcal{S}(0,\omega) = 0$, $\forall\omega \neq \omega_p$, Eq. (1) describes both spontaneous Raman scattering and its subsequent amplification. In the initial propagation stage, the first term in the square bracket can be neglected. This regime corresponds to pure spontaneous Raman scattering. The solution of Eq. (1) is

$$\mathcal{S}(L,\omega) = \frac{\hbar\omega}{2\pi}\,[m_{\text{th}}(|\Omega|) + \nu(\Omega)]\,|g(\omega_p,\Omega,\theta)|\,P_p\,L, \tag{5}$$

where $L$ is the propagation length. The strength of the spontaneous Raman parallel and orthogonal scattering is often measured by the parallel and orthogonal *spontaneous Raman coefficients*

$$R_{\|,\perp}(\omega_p,\Omega,T) = \frac{\hbar\omega_p}{2\pi}\,[m_{\text{th}}(|\Omega|) + \nu(\Omega)]\,|g_{\|,\perp}(\omega_p,\Omega)|. \tag{6}$$

Spontaneous Raman scattering has been observed and measured in bulk glass (Hellwarth et al., 1975; Stolen & Ippen, 1973) and in optical fibers (Stolen et al., 1984; Wardle, 1999). In optical fiber, the polarization properties are usually more difficult to measure because standard fibers do not preserve and even scramble polarization. For this reason, the *effective* spontaneous Raman coefficient is often taken to be $R = (R_{\|} + R_{\perp})/2$.

It is interesting to note that the ratio of Stoke to anti-Stokes spectral components only depend on temperature:

$$\frac{\mathcal{S}(L,\omega_p-|\Omega|)}{\mathcal{S}(L,\omega_p+|\Omega|)}=\frac{n(\omega_p+|\Omega|)}{n(\omega_p-|\Omega|}\left(\frac{\omega_p-|\Omega|}{\omega_p+|\Omega|}\right)^4\exp\left(\frac{\hbar|\Omega|}{k_BT}\right). \tag{7}$$

This is the reason why spontaneous Raman scattering is used for temperature sensing (Alahbabi et al., 2005a;b; Dakin et al., 1985; Farahani & Gogolla, 1999; Wait et al., 1997).

### 2.1.2 Amplified spontaneous scattering

According to Eq. (1), the spontaneous scattering regime ends as soon as $\mathcal{S}(z,\omega)$ becomes significant compared to $\frac{\hbar\omega}{2\pi}[m_{\text{th}}(|\Omega|)+\nu(\Omega)]$. At that point, the scattering becomes stimulated and the system enters the amplification regime. From Eq. (5), one sees that the amplification regime is reached when $g(\omega_p,\Omega,\theta)\ P_p\ L\approx 1$. For $\mathcal{S}(0,\omega)=0$, the solution of Eq. (1) is

$$\mathcal{S}(L,\omega)=\frac{\hbar\omega}{2\pi}\left[m_{\text{th}}(|\Omega|)+\nu(\Omega)\right]\left|e^{g(\omega_p,\Omega,\theta)\ P_p\ L}-1\right|. \tag{8}$$

Stokes radiation ($\Omega>0,\ g>0$) is *grows exponentially* while anti-Stokes ($\Omega>0,\ g<0$) radiation saturates at $\mathcal{S}(L,\omega)=\frac{\hbar\omega}{2\pi}m_{\text{th}}(|\Omega|)$. When losses are taken into account, the gain must overcome a threshold value to enter the amplification regime. Since the Raman gain is frequency dependent, the amplification bandwidth depends on the input power. The effective amplification threshold is usually considered to be reached when Stokes and pump intensity have the same value at the output of the fiber (Agrawal, 2007; Smith, 1972).
By measuring the grows of the Stokes wave, one can deduced the Raman gain as a function of frequency (Mahgerefteh et al., 1996; Stolen et al., 1984). Amplified spontaneous Stokes wave plays an important role in Raman fiber amplifiers (Aoki, 1988; Mochizuki et al., 1986; Olsson & Hegarty, 1986).
Fig. 2 shows the typical (forward) Raman gain $g_{\parallel}(\omega_p,\Omega)$ and the spontaneous Raman coefficient $R_{\parallel}(\omega_p,\Omega,T)$ in a silica fiber at $\lambda_p=1.5$ µm. The parallel Raman gain has a peak at $\Omega_R=13.2$ THz and a width of about 5 THz. The peak value varies for fiber to fiber. A typical value is $g_R=1.6\ \text{W}^{-1}\ \text{km}^{-1}$. The orthogonal gain $g_{\perp}(\omega_p,\Omega)$ is about 30 times smaller (Agrawal, 2007; Dougherty et al., 1995; Stolen, 1979). The parallel gain can be fit using a 10-Lorentzian model, each Lorentzian having three independent parameters : strength, central frequency, and width (Drummond & Corney, 2001). Note that spontaneous anti-Stokes scattering can be eliminated by lowering the temperature, while the spontaneous Stokes coefficient $R_{\parallel}$ is at least $\hbar\omega_p/(2\pi)\times g_{\parallel}$.

## 2.2 Brillouin scattering

Brillouin scattering is very similar to Raman scattering in the sense it couples two light modes to material vibrations. However, in the contrast with Raman scattering which couples light to molecular vibrations, Brillouin scattering couples light to vibration modes of the fiber itself, that is *sound* waves. Therefore the vibrational frequencies involved in Brillouin scattering are much lower: $|\Omega|/(2\pi)$ is usually in the 10 GHz range. BS is also polarization dependent: as long as the fiber can be considered as mechanically isotropic, their is no orthogonal BS, that is $g_{\perp}=0$ (Benedek & Fritsch, 1966; McElhenny et al., 2008; Stolen, 1979). The major difference between Raman and Brillouin scattering lies in the dispersion relation of acoustics vibrations:

$$|\Omega|=v_A\ |\mathbf{k}_A|, \tag{9}$$

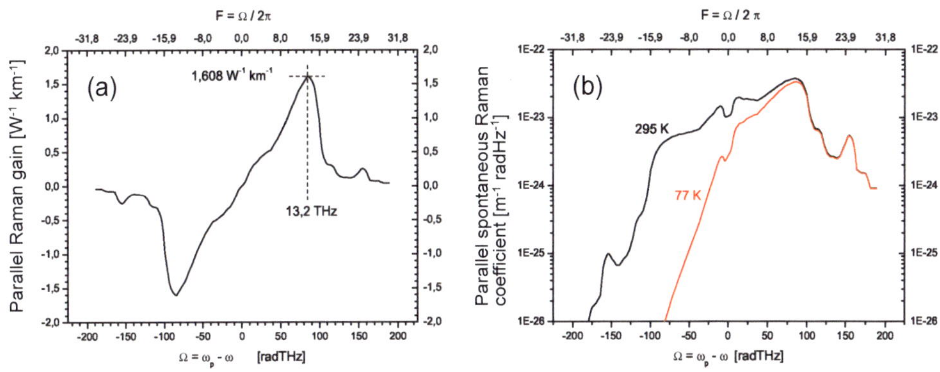


Fig. 2. (a) Raman gain $g_{\parallel}$ in a silica fiber for $\lambda_p = 1.5$ µm and forward propagation. Peak value: $g_R = 1.6$ W$^{-1}$ km$^{-1}$. Peak position: $\Omega_R = 13.2$ THz. (b) Spontaneous Raman coefficient $R_{\parallel}$ for $\lambda_p = 1.5$ µm and forward propagation at $T = 295$ K and 77 K.

where $v_A = 5.96$ km/s in silica fibers and $\mathbf{k}_A$ is the wave vector of the acoustic wave. Because of momentum conservation, $\mathbf{k}_A$ is equal to the wave vector mismatch between pump and (anti-)Stokes waves: $\mathbf{k}_A = \mathbf{k}_p - \mathbf{k}_{s,a}$. Since the energy difference between pump, Stokes and anti-Stokes waves is very small, $|\mathbf{k}_p| \approx |\mathbf{k}_{s,a}|$ and $|\mathbf{k}_A|^2 \approx 2|\mathbf{k}_p|^2\,[1-\cos(\phi)] = 4|\mathbf{k}_p|^2\sin^2(\phi/2)$, where $\phi$ is the angle between $\mathbf{k}_p$ and $\mathbf{k}_{s,a}$. Eq. (9) yields

$$|\Omega| = 2\,v_A\,|\mathbf{k}_p|\,\sin^2(\phi/2) = 4\pi\,v_A\,\frac{n(\omega_p)}{\lambda_p}\,\sin^2(\phi/2). \tag{10}$$

The maximum value of $|\Omega|$ occurs for backward propagation ($\phi = \pi$), while for forward propagation of (anti-)Stokes waves ($\phi = 0$), $|\Omega| = 0$. Therefore, forward Brillouin scattering is not observed. In the backward direction the Brillouin gain as a peak is at $\Omega_R/(2\pi) = 11.1$ GHz when $\lambda_p = 1.55$ µm. The Brillouin gain has a Lorentzian spectrum

$$g_{\parallel}(\omega_p, \Omega) = \text{sign}(\Omega)\frac{g_B(\Gamma_B/2)^2}{(|\Omega| - \Omega_B)^2 + (\Gamma_B/2)^2} \tag{11}$$

and its spectral width $\Gamma_B/(2\pi)$ is in the 10-100 MHz range. $1/\Gamma_B$ is the decay time of the sound waves. The peak value $g_B$ is usually of the order of 1000 W$^{-1}$ km$^{-1}$, one thousand times higher than the Raman gain peak $g_R$.

The Brillouin gain spectrum discussed so far corresponds to a plane acoustic wave propagating along the fiber axis. Other smaller peaks may occur due to other acoustic modes, the presence of dopants and their spatial distribution (Lee et al., 2005; Yeniay et al., 2002). Guided acoustic wave can also produce narrow and very low frequency Brillouin shifts (50 kHz to 1 GHz) and can be even observed in the forward direction (Shelby et al., 1985a;b).

Despite the differences in the Raman and Brillouin gain functions, the underlying scattering mechanism is the same. Therefore, the principle explained in Sec. 3.1 in the context of RS also apply to BS. In particular spontaneous BS exhibits the same temperature dependence as spontaneous RS. Spontaneous Brillouin scattering can be used for temperature sensing (Alahbabi et al., 2005a;b; Pi et al., 2008; Wait et al., 1997).

### 2.3 Four-photon scattering

The four-photon scattering process differs from the previous scattering processes in that it involves four photons and no material vibration. Since a silica fiber is centro-symmetric, it is the lowest order nonlinear scattering phenomenon that involves only photons in the input and output channels. As shown in Fig. 3, a FPS process consists in the conversion

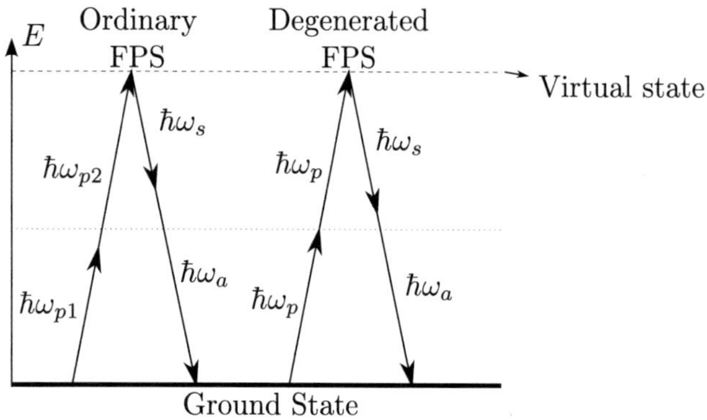


Fig. 3. Spontaneous four-photon scattering processes in silica fibers: ordinary and degenerated case.

of two pump photons at frequencies $\omega_{p1}$ and $\omega_{p2}$ into two other photons at frequencies $\omega_s$ and $\omega_a$. The photon of lower energy is called "Stokes", the one of higher energy in called "anti-Stokes" as in the RS and BS processes. The conversion process satisfies the energy and moment conservation laws

$$\omega_{p1} + \omega_{p2} = \omega_s + \omega_a, \tag{12}$$

$$\mathbf{k}_{p1} + \mathbf{k}_{p2} = \mathbf{k}_s + \mathbf{k}_a. \tag{13}$$

When $\omega_{p1} = \omega_{p2}$ the FPS process is said to *degenerated* and is by far the most studied case, both experimentally and theoretically. FPS is a non resonant process. Therefore many different resonances can contribute to it. In silica, the main contribution comes from electronic resonances. Molecular vibrations contribute to a fraction $f_R = 18\%$ of the FPS strength.

The spectrum of a spontaneous FPS process is usually very broadband. It is not limited by resonance conditions (as RS) or losses (as BS), but merely by the phase matching conditions (13). If a single mode fiber, the wave number of a optical wave has a linear part $k_L(\omega) = n(\omega)\omega/c$ that depends on the effective index $n(\omega)$ of the mode, and a nonlinear part $k_{NL}$ that depends on the power carried by the wave itself (self-phase modulation) and the power of the other waves propagation in the fiber (cross-phase modulation) (Agrawal, 2007). In a spontaneous FPS problem, Stokes and anti-Stokes waves are so faint that their contribution to self or cross-phase modulation is negligible. On the other hand, the pump wave modulates its own phase as well as the phases of the Stokes and anti-Stokes waves. If a wave carries a power $P$, self-phase modulation changes its own wave number by $k_{NL} = \gamma P$, where $\gamma$ is the nonlinear coefficient of the fiber. At the same time, that wave modifies the wave number of any other co-polarized wave by $k_{NL} = 2\gamma P$ and any other orthogonally polarized wave by $k_{NL} = (2/3)\gamma P$, through the cross-phase modulation effect. For instance, for a degenerate co-polarized FPS, the wave number mismatch is $\Delta k = k_s + k_a - 2k_p = (k_{Ls} + 2\gamma P_p) + (k_{La} + 2\gamma P_p) - 2(k_{Lp} + \gamma P_p) = \Delta k_L + 2\gamma P_p$, where $P_p$ is the pump power. Using quantum perturbation theory (Brainis, 2009), it can be shown that the spectral density of power at Stokes and anti-Stokes wavelengths is

$$S(L,\omega_s) = S(L,\omega_a) = \frac{\hbar\omega_{s,a}}{2\pi}\left(\gamma P_p L\right)^2 \mathrm{sinc}^2\left(\frac{\Delta k}{2}L\right) \tag{14}$$

in the case of degenerate co-polarized FPS. Equivalent formulas for non co-polarized degenerate FPS processes can be found in (Brainis, 2009). Whatever the FPS process (degenerated or not, co-polarized or not, ...), Stokes and anti-Stokes powers are always equal because those photons are created in pairs and the spectrum always depends on the wave number mismatch through the same sinc-function factor, see also Sec. 4.2.

It is important to note that the strength of a spontaneous FPS process scales as $(P_p L)^2$, while the strength of spontaneous RS and BS scales as $P_p L$. The spontaneous FPS spectrum is also independent on temperature. Increasing the propagation length $L$ not only increases the amount of scattered photons, but also narrows the spectrum. In contrast, raising the pump power increases scattering, but as little impact of the spectrum. Therefore, adjusting both parameters, it is possible to set the spectral width of the Stokes and anti-Stokes waves as well as their intensities. Because Stokes and anti-Stokes photons are created in pair, FPS as been extensively studied in the context of photon-pair generation for quantum optics and quantum information applications (Amans et al., 2005; Brainis, 2009; Brainis et al., 2005; 2007; Dyer et al., 2008; Fan & Migdall, 2007; Lee et al., 2006; Li et al., 2004; Lin et al., 2006; 2007; Rarity et al., 2005; Takesue, 2006).

When the scattered intensity becomes high enough ($\gamma P_p \gtrsim 1$), spontaneous scattering gets amplified. In the case of the degenerate co-polarized FPS, the growth of Stokes and anti-Stokes waves in the amplification regime in described by (Brainis, 2009; Dyer et al., 2008)

$$S(L,\omega_s) = S(L,\omega_a) = \frac{\hbar\omega_{s,a}}{2\pi}\left(\gamma P_p L\right)^2 \left|\frac{\sinh\left(g(\omega_{s,a})L\right)}{g(\omega_{s,a})L}\right|^2, \tag{15}$$

where $g(\omega_{s,a}) = \sqrt{(\gamma P_p)^2 - (\Delta k/2)^2}$ is the *parametric gain* function that appears in the classical theory of four-wave mixing (Agrawal, 2007). Amplification only occurs at those frequencies for which $g(\omega_{s,a}) \in \mathbb{R}$. Because such a condition is never satisfied in the spontaneous regime ($g(\omega_{s,a}) \xrightarrow{P\to 0} \mathrm{i}\sqrt{\Delta k_L \gamma P_p}$), it strongly modifies the FPS spectrum when amplification begins. It the amplified regime, the spectral width is determined by $g(\omega_{s,a})$ rather by the propagation length. The peak value of the parametric gain $g_P$ is larger by 70% that the Raman peak gain $g_R$.

## 3. Quantum mechanical description of nonlinear scattering

Spontaneous scattering of light cannot be understood in the framework of classical nonlinear optics. A proper description requires the quantum theory. There are two possible approaches. The most elementary one consists in (i) applying quantum perturbation theory to calculate the scattering of light by a single molecule in the first place, then (ii) extending the result to continuous media. The drawback of this method is that it gives access the scattered power density, but not to the field amplitudes. The second approach consists in using a quantum field theory of propagation of light in the fiber that is based on an effective matter/light interaction Hamiltonian.

The "perturbation theory" approach is used in Sec. 3.1 to derive, from first principles, the main formula of Sec. 2.1 for RS. Having identified the limitations of that method, the "field theory method" will be presented in Sec. 3.2.

### 3.1 Perturbation theory of Raman scattering

The simplest way to model Stokes and anti-Stokes Raman scattering from a coherent pump wave at $\omega_p$ consists in applying second order perturbation theory (Crosignani et al., 1980; Wardle, 1999) to matter/light coupling described by the interaction Hamiltonian $H = -\mathbf{d} \cdot \mathbf{E}$, where

$$\mathbf{E}(\mathbf{r}, t) = \mathrm{i}\frac{f(x,y)}{\sqrt{L}} \left[ \sqrt{\frac{\hbar\omega_p}{2\epsilon_0 n^2(\omega_p)}}\, \alpha_p\, \mathrm{e}^{\mathrm{i}(k(\omega_p)z-\omega_p t)}\, \mathbf{e}_p - c.c. + \sqrt{\frac{\hbar\omega}{2\epsilon_0 n^2(\omega)}}\, a\, \mathrm{e}^{\mathrm{i}(k(\omega)z-\omega t)}\, \mathbf{e} - h.c. \right] \tag{16}$$

is the electric field operator associated to the light travelling in the fiber and $\mathbf{d}$ is the electronic dipole moment operator of a scattering fiber molecule at position $\mathbf{r}$. In Eq. (16), the field as been reduced to a pump mode in a coherent state with amplitude $\alpha$ (treated as a strong classical field) and a signal mode representing either the Stokes or anti-Stokes wave at frequency $\omega$. The polarization of the pump and signal modes is defined by the unit vectors $\mathbf{e}_p$ and $\mathbf{e}$. In Eq. (16), the quantity $L$ is the quantization length, a formal parameter that will disappear at the end of the calculation and $f(x,y)$ is normalized so that $\iint f^2(x,y)\mathrm{d}A = 1$, where the integration is over the entire fiber cross-section. This normalization is such that

$$\hbar\omega \frac{c}{n(\omega)L} \langle a^\dagger a \rangle = P(\omega) \quad \text{and} \quad \hbar\omega_p \frac{c}{n(\omega_p)L} |\alpha_p|^2 = P_p, \tag{17}$$

with $P_p$ and $P(\omega)$ the powers in the pump mode and the signal mode, respectively.
The vibration of a molecule can be decomposed in normal modes. Assuming that only one normal mode is excited, the electronic dipole moment of the molecule can be written to first order as

$$\mathbf{d} = \mathbf{d}_0 + \mathbf{d}' Q, \tag{18}$$

where $Q$ is the normal mode coordinate of the vibration, $\mathbf{d}_0$ the dipole moment around the molecular equilibrium point and $\mathbf{d}' = \frac{\partial \mathbf{d}}{\partial Q}$.
Consider the Stokes process first ($\omega < \omega_p$). Assuming that the molecule starts in the electronic ground state $|g\rangle$ and the vibrational number state $|m\rangle$ and that the Stokes mode is in the Fock state $|n\rangle$, the transition probability amplitude to the state $|g, m+1, n+1\rangle$ after an interaction time $t$ can be calculated using *second* order perturbation theory (Wardle, 1999):

$$c(\Omega, m, n, t) = \frac{f^2(x,y)}{L}\, \mathrm{e}^{\mathrm{i}[k(\omega)-k(\omega_p)]z} \sqrt{\frac{\omega_p \omega}{8\hbar\epsilon_0^2 n^2(\omega_p) n^2(\omega) M\Omega}}\, \sqrt{m+1}\, \sqrt{n+1}\, \alpha_p \\ [\mathbf{e}_p]^\dagger \cdot R \cdot \mathbf{e}\, \frac{\mathrm{e}^{\mathrm{i}(\Omega+\omega-\omega_p)t} - 1}{\Omega + \omega - \omega_p}, \tag{19}$$

where $M$ and $\Omega$ are the effective mass and angular frequency of the molecular normal mode of vibration, while $R \approx 2\sum_e \frac{1}{\omega_{eg}} \langle e|\mathbf{d} \otimes \mathbf{d}' + \mathbf{d}' \otimes \mathbf{d}|g\rangle$, where $\otimes$ denotes the tensor product of two vectors and the sum runs over all the electronic excited states of the scattering molecule having Bohr frequencies $\omega_{eg}$ with respect to the ground state.

Around each point **r**, the material medium is made of many molecules and each molecule has several normal modes of vibration. Since all the molecules are at thermal equilibrium, their vibration have no locked phase relationship. The local field is thus simultaneously coupled to a large thermal reservoir of independent vibration modes and the contribution $c(\Omega, m, n, t)$ of each of them to the overall scattering probability can be added *incoherently*. Writing $\rho(\Omega)$ the number of vibration modes in the volume $dV$ centered on **r** with a frequency in the interval $[\Omega, \Omega + d\Omega]$ and and taking the thermal average of the number of excitations in a vibration mode $\Omega$, the *Stokes scattering rate* from the point **r** (integrated over all the possible vibration modes frequencies $\Omega$) is found to be

$$S(n;\mathbf{r}) = \frac{dV}{L^2} f^4(x,y) \frac{\pi\omega_p\omega}{4\hbar\epsilon_0^2 n^2(\omega_p) n^2(\omega)} \frac{\rho(|\Omega|)\langle(\mathbf{e}_p^\dagger \cdot R \cdot \mathbf{e})^2\rangle}{M|\Omega|} (m_{\text{th}}(|\Omega|)+1)(n+1)|\alpha_p|^2, \quad (20)$$

where $\Omega$ is not an independent variable anymore but is now *defined* as $\Omega := \omega_p - \omega$, and $m_{\text{th}}(|\Omega|)$ – given by Eq. (2) – is the Bose-Einstein expectation value of the number of vibrational excitations. The average $\langle(\mathbf{e}_p^\dagger \cdot R \cdot \mathbf{e})^2\rangle$ is taken over arbitrary molecular orientation in amorphous silica. Therefore the quantity $\langle(\mathbf{e}_p^\dagger \cdot R \cdot \mathbf{e})^2\rangle$ only depends on the angle $\theta$ between $\mathbf{e}_p$ and **e**. As a result, one can write

$$\frac{\rho(|\Omega|)\langle(\mathbf{e}_p^\dagger \cdot R \cdot \mathbf{e})^2\rangle}{M|\Omega|} = K_\parallel(|\Omega|)\cos^2(\theta) + K_\perp(|\Omega|)\sin^2(\theta) = K(|\Omega|, \theta), \quad (21)$$

where $K_\parallel(|\Omega|)$ and $K_\perp(|\Omega|)$ are material characteristics that can be determined experimentally[1]. The total scattering rate from $\omega_p$ to $\omega$ due to the Stokes process in a fiber segment dz is obtained by integration $S(n, \mathbf{r})$ over the fiber cross-section. As a consequence,

$$S_{p\to s}(n) = dz \frac{\pi\omega_p\omega}{4\hbar\epsilon_0^2 n^2(\omega_p) n^2(\omega) A_{\text{eff}} L^2} K(|\Omega|, \theta)\, (m_{\text{th}}(|\Omega|)+1)\, (n+1)\, |\alpha_p|^2, \quad (22)$$

where $A_{\text{eff}} = 1/\left(\iint f^4(x,y) dA\right)$ is the effective area of the fiber (Agrawal, 2007).
The scattering rate for the anti-Stokes process ($\omega > \omega_p$) can be computed according to the same lines: the rate is the same as in Eq. (22) with the exception that $(m_{\text{th}}(|\Omega|)+1)$ is replaced by $m_{\text{th}}(|\Omega|)$ since an vibrational excitation is destroyed in that process:

$$A_{p\to a}(n) = dz \frac{\pi\omega_p\omega}{4\hbar\epsilon_0^2 n^2(\omega_p) n^2(\omega) A_{\text{eff}} L^2} K(|\Omega|, \theta)\, m_{\text{th}}(|\Omega|)\, (n+1)\, |\alpha_p|^2. \quad (23)$$

During propagation, light is not only scattered from the pump to the Stokes and anti-Stokes modes at $\omega_p - |\Omega|$ and $\omega_p + |\Omega|$ but also *from* these mode to the pump wave. The rates associated to these *reverse* Raman processes are

$$A_{s\to p}(n) = dz \frac{\pi\omega_p\omega}{4\hbar\epsilon_0^2 n^2(\omega_p) n^2(\omega) A_{\text{eff}} L^2} K(|\Omega|, \theta)\, m_{\text{th}}(\Omega)\, n\, |\alpha_p|^2, \quad (24)$$

$$S_{a\to p}(n) = dz \frac{\pi\omega_p\omega}{4\hbar\epsilon_0^2 n^2(\omega_p) n^2(\omega) A_{\text{eff}} L^2} K(|\Omega|, \theta)\, (m_{\text{th}}(|\Omega|)+1)\, n\, |\alpha_p|^2. \quad (25)$$

[1] Note that $K_\parallel(|\Omega|)$ and $K_\perp(|\Omega|)$ are slightly different in the core and in the cladding of the fiber because of the dopants. Here, we neglect this difference.

Therefore the *net* Raman scattering rates from a coherent pump to Stokes and anti-Stokes modes (containing $n$ photons initially) are

$$S(n) = S_{p\to s}(n) - A_{s\to p}(n) = \frac{\pi\omega_p\omega \mathrm{d}z}{4\hbar\epsilon_0^2 n^2(\omega_p)n^2(\omega)A_{\mathrm{eff}}L^2}K(|\Omega|,\theta)\ \left(n + m_{\mathrm{th}}(|\Omega|)+1\right)|\alpha_p|^2, \tag{26}$$

$$A(n) = A_{p\to a}(n) - S_{a\to p}(n) = \frac{\pi\omega_p\omega \mathrm{d}z}{4\hbar\epsilon_0^2 n^2(\omega_p)n^2(\omega)A_{\mathrm{eff}}L^2}K(|\Omega|,\theta)\ \left(-n + m_{\mathrm{th}}(|\Omega|)\right)|\alpha_p|^2. \tag{27}$$

When a pump wave is launched in an optical fiber it scatters photons to many Stokes and anti-Stokes modes simultaneously. The variation in the power spectral density $\mathrm{d}\mathcal{S}(z,\omega)$ due to the scattering in the fiber slice dz is found by multiplying Eqs. (26) or (27) by the photon energy $\hbar\omega$ and summing over the contribution from all the $c/(n(\omega)L)\mathrm{d}\omega$ modes in the interval $[\omega,\omega+\mathrm{d}\omega]$. Therefore the following differential equations hold for Stokes and anti-Stokes radiation, respectively:

$$\frac{\mathrm{d}}{\mathrm{d}z}\mathcal{S}(z,\omega) = \begin{cases} \left[\mathcal{S}(z,\omega) + \frac{\hbar\omega}{2\pi}\left(m_{\mathrm{th}}(|\Omega|)+1\right)\right] g(\omega_p,\Omega,\theta)P_p(z) & \text{if } \Omega=\omega_p-\omega>0 \text{ (Stokes)} \\ \left[\mathcal{S}(z,\omega) - \frac{\hbar\omega}{2\pi}\, m_{\mathrm{th}}(|\Omega|)\right] g(\omega_p,\Omega,\theta)P_p(z) & \text{if } \Omega=\omega_p-\omega<0 \text{ (anti-Stokes)} \end{cases} \tag{28}$$

where

$$\begin{aligned} g(\omega_p,\Omega,\theta) =& \mathrm{sign}(\Omega)\frac{\pi}{4\hbar^2c^2\epsilon_0^2}\frac{\omega}{n(\omega_p)n(\omega)A_{\mathrm{eff}}}\left(K_{\parallel}(|\Omega|)\cos^2(\theta)+K_{\perp}(|\Omega|)\sin^2(\theta)\right) \\ =& g_{\parallel}(\omega_p,\Omega)\cos^2(\theta)+g_{\perp}(\omega_p,\Omega)\sin^2(\theta) \end{aligned} \tag{29}$$

is the *Raman gain*. Eq. (28) is identical to Eq. (1).

Unfortunately, BS and FPS laws cannot be established in the same manner. For FPS, one can start the analysis at the molecular scale, but fourth order perturbation theory is required. In addition, transition amplitudes must be added *coherently* to get the phase matching right (see Sec. 2.3). For BS, a molecular approach is not possible since BS couples light to the excitation of an acoustic wave involving many molecules (see Sec. 2.2).

### 3.2 Nonlinear quantum field theory

The purpose of the quantum field theory approach is to establish a quantum generalization of the nonlinear Schrödinger equation (NLSE) that governs the propagation of the optical field in a fiber, accounting for dispersion and nonlinear interaction with matter, as well as for spontaneous effects.

#### 3.2.1 Operator equation for nonlinear propagation

Kärtner et al. presented a field theory model of Raman scattering (Kärtner et al., 1994). In this model, a light field $A(z,t)$ is coupled to an harmonic field $Q(z)$, the amplitude of which depends on the position in the fiber. The light field $A(z,t)$ represents the envelope of the E-field oscillating at the carrier frequency $\omega_0$ and is assumed to travel in the fiber at group velocity $v_g$ and with no dispersion. Depending on the dispersion relationship of the field $Q(z)$, it can represent acoustical phonons (if $\omega(q) = v_A q$) or optical photons (if $\omega(q) = \Omega_R$

is independent of $q$). Coupling to acoustical and optical phonons is responsible for BS and RS, respectively. In (Kärtner et al., 1994) it is assumed that $Q(z)$ is an optical phonon field. Physically $Q(z)$ represents the coordinate of a molecular normal mode of vibration at position $z$ in the fiber. Such a field does not propagate but is nevertheless damped. In order to model the damping (with a rate $\Gamma_R$), it is assumed that $Q(z)$ itself is coupled to a large bath of harmonic oscillators at many different frequencies that are at thermal equilibrium. These harmonic oscillators represent other optical and acoustical vibration modes. After eliminating the field $Q(z)$ and the bath variables from the equations, one founds that $A(z,t)$ obeys the nonlinear field equation

$$\begin{aligned}\frac{\partial}{\partial z}A(z,t) = &-\frac{1}{v_g}\frac{\partial}{\partial t}A(z,t) + \mathrm{i}(1-f_R)\gamma A^{\dagger}(z,t)A(z,t)A(z,t)\\ &+\mathrm{i}f_R\gamma\int_{-\infty}^{t} h_R(t-t')A^{\dagger}(z,t')A(z,t')\mathrm{d}t'A(z,t) + \mathrm{i}\sqrt{f_R\gamma}N_R(z,t)A(z,t),\end{aligned} \tag{30}$$

and the commutation relationship

$$\left[A(z,t), A^{\dagger}(z,t')\right] = \hbar\omega_0\delta(t-t'), \tag{31}$$

where $f_R = 0.18$ (see Sec. 2.3), $\gamma$ is the nonlinear coefficient (see Sec. 2.3),

$$h_R(t) = \frac{\Omega_R^2}{\sqrt{\Omega_R^2 - (\Gamma_R/2)^2}} \sin\left(\sqrt{\Omega_R^2 - (\Gamma_R/2)^2}\, t\right) \exp\left(-(\Gamma_R/2)t\right)\, \nu(t) \tag{32}$$

is the Raman *response function*, and $N_R(z,t)$ is the Raman noise field (see Eqs . (35) and 36 below). In Eq. (32), $\nu(t)$ is the Heaviside step function. The Fourier transform of $h_R(t)$ is called the Raman susceptibility:

$$\chi_R^{(3)}(\Omega) = \chi_R'(\Omega) + \mathrm{i}\chi_R''(\Omega) = \int_{-\infty}^{\infty} h_R(t)\, \mathrm{e}^{-\mathrm{i}\Omega t}\mathrm{d}t. \tag{33}$$

Since $h_R(t) \in \mathbb{R}$ and is normalized such that $\int_{-\infty}^{\infty} h_R(t)\mathrm{d}t = 1$,

$$\chi_R'(-\Omega) = \chi_R'(\Omega), \quad \chi_R''(-\Omega) = -\chi_R''(\Omega), \quad \chi_R'(0) = 1, \quad \chi_R''(0) = 0. \tag{34}$$

The Raman noise operator is such that its Fourier transform

$$\tilde{N}_R(z,\Omega) = \int_{-\infty}^{\infty} N_R(z,t)\, \mathrm{e}^{-\mathrm{i}\Omega t}\, \mathrm{d}t \tag{35}$$

satisfies the following spectral correlations (Boivin et al., 1994; Drummond & Corney, 2001):

$$\langle \tilde{N}_R^{\dagger}(z,\Omega)\tilde{N}_R(z',\Omega')\rangle = \hbar(\omega_0-\Omega)\frac{|\chi_R''(\Omega)|}{\pi}\left[m_{\mathrm{th}}(|\Omega|) + \nu(\Omega)\right]\delta(z-z')\delta(\Omega-\Omega'). \tag{36}$$

The second term at the right-hand side of Eq. (30) does not come out of Kärtner's model but has been added phenomenologically to account for the $(1-f_R)$ fraction of the total non linearity that originates in the interaction of light with bound electrons rather than molecular vibrations.

The last term in Eq. (30) is the one responsible for the spontaneous Raman scattering. In order to make the connection with the description given in Secs. 2.1 and 3.1, consider the propagation of a strong continuous pump field $A_p(z,t) = \sqrt{P_p}$ together with a weak scattered field $A_{sc}(z,t)$ that is null at the input of the fiber: $A(z,t) = A_p + A_{sc}(z,t)$. Ignoring all the terms but the last one in Eq. (30) one easily finds that $A_{sc}(z,t) \approx \mathrm{i}\sqrt{f_R \gamma P_p} \int_0^L N_R(z,t)\mathrm{d}z$. Therefore, the total scattered power is

$$\int_{-\infty}^{\infty} S(L, \omega_0 - \Omega)\mathrm{d}\Omega = \langle A_{sc}^{\dagger}(z,t) A_{sc}(z,t)\rangle = f_R \gamma P_p \int_0^L \mathrm{d}z \int_0^L \mathrm{d}z' \langle N_R^{\dagger}(z,t) N_R(z',t')\rangle$$
$$= f_R \gamma P \int_{-\infty}^{\infty} \mathrm{d}\Omega \int_{-\infty}^{\infty} \mathrm{d}\Omega' \int_0^L \mathrm{d}z \int_0^L \mathrm{d}z' \langle \tilde{N}_R^{\dagger}(z,\Omega) \tilde{N}_R(z',\Omega')\rangle$$

Using Eq. (36), one finally gets

$$S(L, \omega = \omega_0 - \Omega) = \hbar(\omega_0 - \Omega) f_R \gamma \frac{|\chi_R''(\Omega)|}{\pi} \left[m_{\mathrm{th}}(|\Omega|) + \nu(\Omega)\right] P_p L. \tag{37}$$

Comparing this expression with Eq. (5), the Raman gain is found to be related to the imaginary part of the Raman susceptibility by the following relationship:

$$g_{\parallel}(\omega_p, \Omega) = -2 f_R \gamma \chi_R''(\Omega) \tag{38}$$

According to the Kärtner's model, the Raman gain would be Lorentzian in shape because

$$\chi_R''(\Omega) = \frac{\Omega \Omega_R^2 \Gamma_R}{(\Omega_R^2 - \Omega^2)^2 + \Omega^2 \Gamma^2}, \tag{39}$$

according to Eqs. (32) and (33). This would be a rough approximation of the actual Raman gain in Fig. 2a. As explained in (Drummond & Corney, 2001), the Raman gain is well fitted by a 10-Lorentzian model. Modifying the quantum field model to couple light to ten Lorentzian vibration modes is trivial: it only changes the shape of the Raman response function $h_R(t)$ in Eq. (32), which becomes a linear superposition of damped sine functions with appropriate oscillation frequencies and damping constants.

With this modification, the quantum propagation equation (30) is able to simulate the spontaneous grow of Stokes and anti-Stokes wave and their amplification. However, Eq. (30) is unable to simulate FPS despite that all the terms (second and third term of the right-hand side) responsible for photon-pair generation are included. This is because phase-matching is of crucial importance for the FPS process and Eq. (30) does not properly deal with the group velocity dispersion of the traveling waves.

### 3.2.2 Dispersion

Dispersion plays an important role in the physics of spontaneous and stimulated nonlinear effects. The exact dispersion of the fiber can be include in the quantum non linear propagation equation (30) by replacing the first term in the right-hand side

$$-\frac{1}{v_g}\frac{\partial}{\partial t} A(z,t) \tag{40}$$

by the generalized dispersion operator

$$\mathcal{D}[A(z,t)] = +\mathrm{i} \sum_{a=1}^{\infty} (\mathrm{i})^a \frac{k_a}{a!} \frac{\partial^a}{\partial t^a} A(z,t), \tag{41}$$

where

$$k_a = \frac{\mathrm{d}^a}{\mathrm{d}\omega^a} k_L(\omega)\bigg|_{\omega=\omega_0} \tag{42}$$

are the derivatives of the propagation constant $k_L(\omega)$. The dispersion operator can also be written as a convolution integral (Kärtner et al., 1994; Lin et al., 2007)

$$\mathcal{D}[A(z,t)] = \mathrm{i}\int_{-\infty}^{t} h_L(t-t')A(z,t')\mathrm{d}t', \tag{43}$$

where

$$h_L(t) = \frac{1}{2\pi}\int_{-\infty}^{\infty} [k_L(\omega_0-\Omega) - k_L(\omega_0)]\mathrm{e}^{\mathrm{i}\Omega t}\mathrm{d}\Omega \tag{44}$$

is the *linear* response function of the fiber. Using (43) Eq. (30) reads:

$$\begin{aligned}\frac{\partial}{\partial z}A(z,t) =& \mathrm{i}\int_{-\infty}^{t} h_L(t-t')A(z,t')\mathrm{d}t' + \mathrm{i}(1-f_R)\gamma A^{\dagger}(z,t)A(z,t)A(z,t) \\ &+ \mathrm{i}f_R\gamma\int_{-\infty}^{t} h_R(t-t')A^{\dagger}(z,t')A(z,t')\mathrm{d}t' A(z,t) + \mathrm{i}\sqrt{f_R\gamma}N_R(z,t)A(z,t),\end{aligned} \tag{45}$$

### 3.2.3 Brillouin and polarization effects

As mentioned in Sec. 3.2.1, the quantum propagation model couples light to non propagative phonons. Strictly speaking, such a model is unsuitable for describing BS. However, if the propagation length is long enough to consider that momentum conservation (opto-acoustical phase matching) is verified, the phonon field has a well defined oscillation frequency $\Omega_B = 4\pi\ v_A\ \frac{n(\omega_p)}{\lambda_p}$, see Eq. (10). Therefore, the Brillouin Lorentzian gain can be included as an eleventh Lorentzian (ultra-low frequency) contribution to the nonlinear Raman response $h_R(t)$ (Drummond & Corney, 2001).

Eq. (45) only takes into account nonlinear effects that involve photons with the same polarization state. One can generalize the model to take polarization into account (Brainis, 2009; Brainis et al., 2005; Lin et al., 2006; 2007).

### 3.2.4 Solving the quantum propagation equation

There are two main methods to solve the quantum nonlinear propagation equation.

The first one is using numerical integration and consists in converting Eq. 45 into a set of c-number equations with *stochastic terms* in order to solve them on a computer (Brainis et al., 2005; Kennedy & Wright, 1988). These methods have been first introduced to solve the scalar quantum equation without the Raman effect ($f_R = 0$) to study the squeezing of a quantum soliton (Carter et al., 1987; Drummond & Carter, 1987) and co-polarized FPS (Brainis et al., 2005). It has been then generalized to study different types of *non* co-polarized FPS processes (Amans et al., 2005; Brainis et al., 2005; Kennedy, 1991) and squeezing in birefringent fibers (Kennedy & Wabnitz, 1988), as well as Raman scattering noise (Drummond & Corney, 2001).

The second method consists in linearizing the quantum nonlinear equation around a classical solution such as a continuous pump wave or a soliton in order to derive linear couple mode operator equations that can be solved analytically (Brainis, 2009; Brainis et al., 2007; Lin et al., 2006; 2007). Coupled mode equations are easier to establish from the Fourier transform of Eq. (45). Defining the Fourier components of the wave as

$$\tilde{A}(z,\Omega) = \int_{-\infty}^{\infty} A(z,t)\mathrm{e}^{-\mathrm{i}\Omega t}\mathrm{d}t, \tag{46}$$

one finds that they satisfy the following equation:

$$\frac{\partial}{\partial z}\tilde{A}(z,\Omega) = \mathrm{i}[k_L(\omega_0-\Omega)-k_L(\omega_0)]\tilde{A}(z,\Omega) + \mathrm{i}\sqrt{f_R\gamma}\frac{1}{2\pi}\int_{-\infty}^{\infty} d\omega_1 N_R(z,\Omega-\omega_1)A(z,\omega_1)$$
$$+\mathrm{i}\gamma\frac{1}{(2\pi)^2}\int_{-\infty}^{\infty} d\omega_1 \int_{-\infty}^{\infty} d\omega_2 \chi(\omega_2-\omega_1)\tilde{A}(z,\omega_1)\tilde{A}(z,\omega_2)\tilde{A}(z,\Omega+\omega_1-\omega_2), \tag{47}$$

where

$$\chi(\Omega) = (1-f_R) + f_R\,\chi_R(\Omega) \tag{48}$$

is the total third order susceptibility which takes into account both electronic and vibrational non linearity. $\chi(\Omega)$ is a complex function that has the same symmetry properties as $\chi_R(\Omega)$, see Eq. (34)

$$\chi'(-\Omega) = \chi'(\Omega), \quad \chi''(-\Omega) = -\chi''(\Omega), \quad \chi'(0) = 1, \quad \chi''(0) = 0. \tag{49}$$

Using Eq. (31), one finds that the operators $\tilde{A}(z,\Omega)$ satisfy the following commutation relations

$$\left[\tilde{A}(z,\Omega), \tilde{A}^{\dagger}(z,\Omega')\right] = 2\pi\hbar\omega_0\delta(\Omega-\Omega'). \tag{50}$$

A generalization of Eqs. (47)-(50) that takes into account polarization can be found in (Lin et al., 2007). Linearized coupled mode equations are directly obtained from Eq. (47). Hereafter, the result is given for one and two pump waves. These coupled-mode equations will be used in Sec. 4 to analyze the competition between the RS process and the FPS process.

3.2.4.1 Single pump configuration

Let us assume that that a monochromatic pump wave with frequency $\omega_p = \omega_0$ and spectral amplitude $\tilde{A}(z=0,\Omega) = 2\pi\sqrt{P_p}\delta(\Omega)$ is launched in the fiber. During the propagation, the pump remains monochromatic but acquires a nonlinear phase modulation: $\tilde{A}(z,\Omega) = 2\pi A_p(z)\delta(\Omega)$. The amount of phase modulation can be derived by injecting this ansatz in Eq. (47). One finds that

$$\frac{dA_p}{dz} = \mathrm{i}\gamma A_p^{\dagger}(z)A_p(z)A_p(z). \tag{51}$$

The solution of this equation is

$$A_p(z) = \sqrt{P_p}e^{\mathrm{i}\gamma P_p z}. \tag{52}$$

However, this solution is not a stable solution of Eq. (47). Brillouin, Raman and four-photon scattering, will spontaneously scattered power from the pump to Stokes and anti-Stokes frequencies. Nevertheless, for the calculation of the Stokes and anti-Stokes amplitudes, one can make the assumption that the pump remains *undepleted*, i.e. (52) is approximately valid. Injecting the ansatz

$$\tilde{A}(z,\Omega) = \left[2\pi A_p(z)\delta(\Omega) + \tilde{A}_{sc}(z,\Omega)\right], \tag{53}$$

and retaining only the terms of highest order in $P$, one finds that the scattered field $\tilde{A}_{sc}(z,\Omega)$ satisfies

$$\frac{\partial}{\partial z}\tilde{A}_{sc}(z,\Omega) = \mathrm{i}\ \left[[k_L(\omega_p-\Omega)-k_L(\omega_p)] + B(\Omega)\gamma P_p\right]\ \tilde{A}_{sc}(z,\Omega)$$
$$+\mathrm{i}\,\chi(\Omega)\gamma P e^{\mathrm{i}2\gamma P_p z}\ \tilde{A}_{sc}^{\dagger}(z,-\Omega) + \mathrm{i}\ \sqrt{f_R\gamma P_p}e^{\mathrm{i}\gamma P_p z}\ N_R(z,\Omega), \tag{54}$$

where

$$B(\Omega) = \chi(0) + \chi(\Omega) = 1 + \chi(\Omega) = 2 - f_R[1 - \chi_R(\Omega)]. \tag{55}$$

The coefficient $B(\Omega)$ measures the relative strength of the cross-phase modulation of the scattered field by the pump and the self-phase modulation of the pump, see Eq. (52). If Raman scattering is ignored ($f_R = 0$), it takes the usual value $B = 2$. At frequencies close to the pump ($\Omega \to 0$), one also finds $B \approx 2$, because $\chi_R \approx 1 + 0\mathrm{i}$. Very far away from the pump frequency ($\Omega \to \infty$), $B \approx 1.82$ because $\chi_r \approx 0 + 0\mathrm{i}$ and $f_R = 0.18$. The third term of the Eq. (54) represents a FWM process with a complex coupling coefficient $\chi(\Omega)\gamma P_p$. In Sec. 4, it will be shown that this term is responsible for stimulated FWM, stimulated Raman and Brillouin scattering, as well as spontaneous FPS. One may notice that this term couples each spectral component at $\Omega > 0$ (Stokes) to the symmetric component at frequency $\Omega < 0$ (anti-Stokes), as required by the aforementioned processes. The last term in the right-hand side of Eq. (54) is the source of spontaneous Raman and Brillouin scattering. Since Stokes and anti-Stokes frequencies are always coupled, the coupled-mode equations

$$\frac{\partial}{\partial z}\tilde{A}_{\mathrm{sc}}(z,\Omega) = \mathrm{i}\ \left[[k_L(\omega_p - \Omega) - k_L(\omega_p)] + B(\Omega)\gamma P_p\right]\ \tilde{A}_{\mathrm{sc}}(z,\Omega) + \mathrm{i}\,\chi(\Omega)\gamma P_p \mathrm{e}^{\mathrm{i}2\gamma P_p z}\ \tilde{A}^{\dagger}_{\mathrm{sc}}(z,-\Omega) + \mathrm{i}\,\sqrt{f_R\gamma P_p}\mathrm{e}^{\mathrm{i}\gamma P_p z}\ N_R(z,\Omega), \tag{56}$$

$$\frac{\partial}{\partial z}\tilde{A}_{\mathrm{sc}}(z,-\Omega) = \mathrm{i}\ \left[[k_L(\omega_p + \Omega) - k_L(\omega_p)] + B(-\Omega)\gamma P_p\right]\ \tilde{A}_{\mathrm{sc}}(z,-\Omega) + \mathrm{i}\,\chi(-\Omega)\gamma P_p \mathrm{e}^{\mathrm{i}2\gamma P_p z}\ \tilde{A}^{\dagger}_{\mathrm{sc}}(z,\Omega) + \mathrm{i}\,\sqrt{f_R\gamma P_p}\mathrm{e}^{\mathrm{i}\gamma P_p z}\ N_R(z,-\Omega). \tag{57}$$

must be solved together to solve the propagation problem. These are linear, but inhomogeneous equations. Note that

$$\chi(-\Omega) = \chi^*(\Omega), \qquad \text{and} \qquad B(-\Omega) = B^*(\Omega). \tag{58}$$

#### 3.2.4.2 Dual pump configuration

If two pump waves at frequencies $\omega_{p1} = \omega_0 - \Omega_p$ and $\omega_{p2} = \omega_0 + \Omega_p$ are launched simultaneously in the fiber, the spectral amplitude can be written

$$\tilde{A}(z,\Omega) = 2\pi A_{p1}(z)\delta(\Omega - \Omega_p) + 2\pi A_{p2}(z)\delta(\Omega + \Omega_p) \tag{59}$$

Injecting this ansatz in Eq. 47 shows the the two pumps will interact through nonlinear effects:

$$\frac{\mathrm{d}A_{p1}}{\mathrm{d}z} = \mathrm{i}\left[[k_L(\omega_0 - \Omega_p) - k_L(\omega_0)] + \gamma|A_{p1}|^2 + B(2\Omega_p)\gamma|A_{p2}|^2\right] A_{p1}(z) \tag{60}$$

$$\frac{\mathrm{d}A_{p2}}{\mathrm{d}z} = \mathrm{i}\left[[k_L(\omega_0 + \Omega_p) - k_L(\omega_0)] + \gamma|A_{p2}|^2 + B(-2\Omega_p)\gamma|A_{p1}|^2\right] A_{p2}(z) \tag{61}$$

The third terms on the right-hand side are responsible for both cross-phase modulation and stimulated Raman scattering: Eq. (38) shows that

$$\begin{aligned} \mathrm{i}B(\pm 2\Omega_p)\gamma &= \mathrm{i}\left[2 - f_R[1 - \chi'(\pm 2\Omega_p)]\right]\gamma - f_R\gamma\chi''(\pm 2\Omega_p), \\ &= \mathrm{i}\left[2 - f_R[1 - \chi'(\pm 2\Omega_p)]\right]\gamma + \frac{g_{\parallel}(\pm 2\Omega_p)}{2}. \end{aligned} \tag{62}$$

If the propagation distance $L$ and the initial pump powers $P_{p1}$ and $P_{p2}$ are such that $g_{\|}(2\Omega_p)(P_{p1}+P_{p2})L \ll 1$, power transfer due to stimulated Raman scattering is negligible and the solution of Eqs. (60) and (61) is:

$$A_{p1}(z) = P_{p1}\,\exp\left[\mathrm{i}\left[[k_L(\omega_0-\Omega_p)-k_L(\omega_0)]+\gamma P_{p1}+\Re[B(2\Omega_p)]\gamma P_{p2}\right]z\right], \tag{63}$$

$$A_{p2}(z) = P_{p2}\,\exp\left[\mathrm{i}\left[[k_L(\omega_0+\Omega_p)-k_L(\omega_0)]+\gamma P_{p2}+\Re[B(-2\Omega_p)]\gamma P_{p1}\right]z\right]. \tag{64}$$

As in the single pump case, such a solution in unstable and light will be spontaneously scattered to other wavelength. To analyze that scattering, we introduce the ansatz

$$\tilde{A}(z,\Omega) = 2\pi A_{p1}(z)\delta(\Omega-\Omega_p)+2\pi A_{p2}(z)\delta(\Omega+\Omega_p)+\tilde{A}_{\mathrm{sc}}(z,\Omega) \tag{65}$$

in Eq. (47) and only keep the terms of highest order in $P_{p1}$ and $P_{p2}$. It is found that

$$\begin{aligned}
\frac{\partial}{\partial z}\tilde{A}_{\mathrm{sc}}(z,\Omega) = {} & \mathrm{i}\,\left[[k_L(\omega_0-\Omega)-k_L(\omega_0)]+B(\Omega-\Omega_p)\gamma P_{p1}+B(\Omega+\Omega_p)\gamma P_{p2}\right]\,\tilde{A}_{\mathrm{sc}}(z,\Omega)\\
&+\mathrm{i}\gamma\chi(\Omega-\Omega_p)A_{p1}(z)A_{p1}(z)\tilde{A}^{\dagger}_{\mathrm{sc}}(z,2\Omega_p-\Omega)\\
&+\mathrm{i}\gamma\chi(\Omega+\Omega_p)A_{p2}(z)A_{p2}(z)\tilde{A}^{\dagger}_{\mathrm{sc}}(z,-2\Omega_p-\Omega)\\
&+\mathrm{i}\gamma 2\Re\left[\chi(\Omega_p-\Omega)\right]A_{p1}(z)A_{p2}(z)\tilde{A}^{\dagger}_{\mathrm{sc}}(z,-\Omega)\\
&+\mathrm{i}\gamma\chi(2\Omega_p)A^{\dagger}_{p1}(z)A_{p2}(z)\tilde{A}_{\mathrm{sc}}(z,-2\Omega_p)+\mathrm{i}\gamma\chi(-2\Omega_p)A^{\dagger}_{p2}(z)A_{p1}(z)\tilde{A}_{\mathrm{sc}}(z,2\Omega_p)\\
&+\mathrm{i}\gamma\chi(\Omega_p-\Omega)A^{\dagger}_{p1}(z)A_{p2}(z)\tilde{A}_{\mathrm{sc}}(z,\Omega+2\Omega_p)\\
&+\mathrm{i}\gamma\chi(\Omega-\Omega_p)A^{\dagger}_{p2}(z)A_{p1}(z)\tilde{A}_{\mathrm{sc}}(z,\Omega-2\Omega_p)\\
&+\mathrm{i}\sqrt{f_R\gamma}\left[A_{p_1}(z)N_R(z,\Omega-\Omega_p)+A_{p_2}(z)N_R(z,\Omega+\Omega_p)\right].
\end{aligned} \tag{66}$$

In striking contrast with Eq. (54), the light scattered at frequency $\omega_0-\Omega$ is not only coupled to the symmetric mode $\omega_0+\Omega$ but also to six other modes at frequencies: $\omega_0-2\Omega_p-\Omega$, $\omega_0-2\Omega_p$, $\omega_0-2\Omega_p+\Omega$, $\omega_0+2\Omega_p-\Omega$ , $\omega_0+2\Omega_p$, $\omega_0+2\Omega_p+\Omega$. As a consequence, there is no way to write down a *closed* set of coupled mode equations for that problem. However, the perturbation theory technique introduced in (Brainis, 2009) can be applied to investigation the quantum regime of scattering (see Sec. 4.2).

## 4. Coupling between spontaneous scattering processes

When light propagates in an optical fiber, spontaneous RS, BS and FPS take place simultaneously. Several processes may scatter light to the same modes so that it may not be possible the decouple the processes. Hereafter, that point is illustrated in the single and dual pump configuration.

### 4.1 Single pump configuration

The field evolution in the single pump configuration is fully described by the coupled-mode equations (56) and ( 57), the solution of which is

$$\begin{aligned}
\tilde{A}_{\mathrm{sc}}(L,\Omega) = {} & \mu_1(L,\Omega)\tilde{A}_{\mathrm{sc}}(0,\Omega)+\mu_2(L,\Omega)\tilde{A}^{\dagger}_{\mathrm{sc}}(0,-\Omega)\\
&+\mathrm{i}\sqrt{f_R\gamma P}\int_0^L N_R(z,\Omega)\left(\mu_1(L-z,\Omega)-\mu_2(L-z,\Omega)\right)\mathrm{d}z
\end{aligned} \tag{67}$$

$$\tilde{A}_{sc}(L,-\Omega) = \mu_1(L,-\Omega)\tilde{A}_{sc}(0,\Omega) + \mu_2(L,-\Omega)\tilde{A}^{\dagger}_{sc}(0,\Omega) + i\sqrt{f_R\gamma P}\int_0^L N_R(z,-\Omega)\left(\mu_1(L-z,-\Omega) - \mu_2(L-z,-\Omega)\right)dz. \quad (68)$$

If $\Omega > 0$, $\tilde{A}_{sc}(L,\Omega)$ corresponds to the Stokes part of the spectrum and $\tilde{A}_{sc}(L,-\Omega)$ to the anti-Stokes part. The functions $\mu_1(z,\Omega)$ and $\mu_2(z,\Omega)$ are defined as

$$\mu_1(z,\Omega) = \left[\cosh(g(\Omega)L) + i\frac{\Delta k(\Omega)}{2g(\Omega)}\sinh(g(\Omega)L)\right]\exp\left(i\frac{k_L(\Omega)-k_L(-\Omega)}{2}L\right) \quad (69)$$

$$\mu_2(z,\Omega) = i\frac{\gamma\chi(\Omega)P_p e^{i2\gamma P_p L}}{g(\Omega)}\sinh(g(\Omega)L)\exp\left(i\frac{k_L(\Omega)-k_L(-\Omega)}{2}L\right) \quad (70)$$

where

$$\Delta k(\Omega) = k_L(\omega_p-\Omega) + k_L(\omega_p+\Omega) - 2k_L(\omega_p) + 2\gamma P_p[B(\Omega)-1] = \Delta k_L(\Omega) + 2\gamma P_p\chi(\Omega) \quad (71)$$

is the total phase mismatch and

$$g(\Omega) = \sqrt{(\chi(\Omega)\gamma P_p)^2 - (\Delta k(\Omega)/2)^2} = \sqrt{-\Delta k_L(\Omega)\gamma P_p\chi(\Omega) - (\Delta k_L(\Omega)/2)^2} \quad (72)$$

is the parametric gain. The square-root in Eq. (72) is chosen such that $\Re(g) > 0$. Comparing with the results of Sec. 2.3, both the phase mismatch parameter and the parametric gain have a modified value due to the *simultaneous* action of FPS and RS. This modification impacts the stimulated FWM (Golovchenko et al., 1990; Vanholsbeeck et al., 2003) as well as spontaneous FPS regime (Brainis et al., 2007; Lin et al., 2006).

The spontaneous regime corresponds to the initial conditions $\langle\tilde{A}^{\dagger}_{sc}(0,\Omega)\tilde{A}_{sc}(0,\Omega)\rangle = 0$, for any value of $\Omega$. The spectral power density $S(L,\omega_p-\Omega)$ at the fiber output can be calculated as follows (Brainis et al., 2005; 2007):

$$S(L,\omega_p-\Omega) = \lim_{\epsilon\to 0}\frac{1}{2\pi\epsilon}\int_{\Omega-\epsilon/2}^{\Omega+\epsilon/2}\int_{\Omega-\epsilon/2}^{\Omega+\epsilon/2}\langle\tilde{A}^{\dagger}_{sc}(L,\Omega_1)\tilde{A}_{sc}(L,\Omega_2)\rangle d\Omega_1 d\Omega_2. \quad (73)$$

Using Eqs. (67) and (68), one finds (Brainis et al., 2007)

$$\frac{S(L,\omega_p-\Omega)}{\hbar(\omega_p-\Omega)} = \frac{1}{2\pi}\left|\chi(\Omega)\gamma P_p L\right|^2\left|\frac{\sinh(g(\Omega)L)}{g(\Omega)L}\right|^2 + \frac{|\Im[\chi(\Omega)]|\gamma P_p}{\pi}\rho(L,\Omega)\left(m_{th}(|\Omega|) + \nu(\Omega)\right), \quad (74)$$

where

$$\rho(L,\Omega) = \int_0^L\left|\cosh(g(\Omega)z) + i\,\mathrm{sign}(\Omega)\frac{\Delta k(\Omega)}{2g(\Omega)}\sinh(g(\Omega)z)\right|^2 dz. \quad (75)$$

The first and second terms in the right-hand side of Eq. (74) represent the photons scattered through the four-photon and Raman processes, respectively. Note that $|\Im[\chi(\Omega)]|\gamma = f_R\gamma|\chi''_R(\Omega)| = |g_{\parallel}(\omega_p,\Omega)|/2$, see Eq. (38).

In the spontaneous regime ($|g(\Omega)|P_p \to 0$), Eq. (74) reduces to

$$\frac{S(L,\omega_p-\Omega)}{\hbar(\omega_p-\Omega)} = \frac{(|\chi(\Omega)|P_p L)^2}{2\pi}\mathrm{sinc}^2\left(\frac{\Delta k}{2}L\right) + \frac{|g_{\parallel}(\omega_p,\Omega)|P_p L}{2\pi}\left(m_{th}(|\Omega|) + \nu(\Omega)\right). \quad (76)$$

The Raman contribution to the scattered light is exactly the one given by Eq. 5: this means that FPS has no impact on RS. The reverse is not true: RS as an influence on FPS since it modifies the total susceptibility $\chi(\Omega)$ appearing in the first term. In absence of RS ($f_R = 0$), $\chi(\Omega) = 1$ and one recovers the spectral density of power given by Eq. (14). Since $-1 < \chi_R'(\Omega) \leq 1$ and $-1.4 < \chi_R''(\Omega) < 1.4$, $|\chi(\Omega)| = \sqrt{(1 - f_R(1 - \chi_R'(\Omega)))^2 + (f_R\chi_R''(\Omega))^2}$ is always close to one. For this reason, spontaneous (not amplified ) FPS and RS can be considered has uncoupled phenomena.
If the power is high enough ($|g(\Omega)|P_pL > 1$), amplification of the spontaneous scattering takes place. The general formula (74) is well approximated by

$$\frac{S(L,\omega_p-\Omega)}{\hbar\,(\omega_p-\Omega)} = \frac{e^{2\Re[g(\Omega)]L}}{8\pi}\left[\left|\frac{\chi(\Omega)\gamma P_p}{g(\Omega)}\right|^2 + \frac{|g_{\parallel}(\omega_p,\Omega)|\gamma P_p}{2\Re[g(\Omega)]}\frac{|g(\omega)+\mathrm{i}\,\mathrm{sign}(\Omega)\Delta k/2|^2}{|g(\Omega)|^2}\right]. \tag{77}$$

In striking contrast with the analysis of Sec. 2.1, this result shows that the Raman anti-Stokes wave *grows at the same rate as the Raman Stokes wave* instead of saturating at the power density value $S(L,\omega_p+|\Omega|) = \frac{\hbar\omega}{2\pi}m_{\mathrm{th}}(|\Omega|)$. This effect is due to the coupling of the RS with the FWM. Its detailed explanation can be found in (Brainis et al., 2007; Coen et al., 2002). On the other hand, the exponential amplification of the Stokes wave is completely quenched at frequencies satisfying $\Delta k_L(\Omega) = 0$ because the gain $g(\Omega)$ vanishes in that case, see Eq. (72) (Golovchenko et al., 1990; Vanholsbeeck et al., 2003).

### 4.2 Dual pump configuration

In the dual pump configuration, the coupled-mode equations (66) do not form a closed set. For this reason, one cannot write an explicit solution as in Sec. 4.1. To study the spontaneous photon scattering, we apply the first-order perturbation technique introduced in (Brainis, 2009).
We first notice that the first term of the right-hand side of (66) represents the phase evolution of scattered field, including the cross-phase modulation due to the two pumps. This phase modulation has no impact on the population of the frequency modes and can be factored out by writing the total scattered E-field

$$E_{\mathrm{sc}}(z,t) = \sqrt{\frac{\hbar\omega_0}{4\pi\epsilon_0 n_0 c}}\int a(z,\Omega)\,e^{\mathrm{i}[k_L(\omega_0-\Omega)+B(\Omega-\Omega_p)\gamma P_{p1}+B(\Omega+\Omega_p)\gamma P_{p2}]z}e^{-\mathrm{i}(\omega_0-\Omega)t}d\Omega, \tag{78}$$

where $a(z,\Omega)$ is the annihilation operator of the frequency mode $\omega - \Omega$. Because the exact phase evolution of the $a(z,\Omega)$ has been factored out, the $z$ dependence of $a(z,\Omega)$ is only due to the FPS effect (Brainis, 2009). On the other hand, the scattered field can be written as

$$E_{\mathrm{sc}}(z,t) = \frac{1}{2\pi\sqrt{2\epsilon_0 n_0 c}}\,e^{\mathrm{i}k_L(\omega_0)z-\mathrm{i}\omega_0 t}\int \tilde{A}_{\mathrm{sc}}(z,\Omega)e^{\mathrm{i}\Omega t}d\Omega, \tag{79}$$

where we used the fact that $E(z,t) = \sqrt{1/(2\epsilon_0 n_0 c)}A(z,t)e^{\mathrm{i}k_L(\omega_0)z-\mathrm{i}\omega_0 t}$ and Eq. (46). Comparing Eqs. (78) and (79), one sees that

$$\tilde{A}_{\mathrm{sc}}(z,\Omega) = \sqrt{2\pi\hbar\omega_0}\,a(z,\Omega)\,e^{\mathrm{i}[k_L(\omega_0-\Omega)-k_L(\omega_0)+B(\Omega-\Omega_p)\gamma P_{p1}+B(\Omega+\Omega_p)\gamma P_{p2}]z}. \tag{80}$$

In the following, we make the approximation that $B(\Omega) \approx 2$, see Eq. (55) and $\chi(\Omega) \approx 1$, see Eq. (48). This approximation consists in neglecting the dispersion of the non linearity. Using Eq. (80), one obtains the evolution equation for the annihilation operators:

$$\begin{aligned}
\frac{\partial}{\partial z}a(z,\Omega) &= \mathrm{i}\,\gamma\ P_{p1}\,\mathrm{e}^{-\mathrm{i}\Delta k_{11}(\Omega)z}\,a^{\dagger}(z,2\Omega_p-\Omega) + \mathrm{i}\,\gamma\ P_{p2}\,\mathrm{e}^{-\mathrm{i}\Delta k_{22}(\Omega)z}\,a^{\dagger}(z,-2\Omega_p-\Omega) \\
&\quad + \mathrm{i}\,2\gamma\,\sqrt{P_{p1}P_{p2}}\,\mathrm{e}^{-\mathrm{i}\Delta k_{12}(\Omega)z}\,a^{\dagger}(z,-\Omega) \\
&\quad + \mathrm{i}\,\gamma\,\sqrt{P_{p1}P_{p2}}\,\mathrm{e}^{\mathrm{i}\Delta k_a z}\,a(z,-2\Omega_p) + \mathrm{i}\,\gamma\,\sqrt{P_{p1}P_{p2}}\,\mathrm{e}^{\mathrm{i}\Delta k_b z}\,a(z,2\Omega_p) \\
&\quad + \mathrm{i}\,\gamma\,\sqrt{P_{p1}P_{p2}}\,\mathrm{e}^{\mathrm{i}\Delta k_c z}\,a(z,\Omega+2\Omega_p) + \mathrm{i}\,\gamma\,\sqrt{P_{p1}P_{p2}}\,\mathrm{e}^{\mathrm{i}\Delta k_d z}\,a(z,\Omega-2\Omega_p) \\
&\quad + \mathrm{i}\sqrt{\frac{f_R\gamma P_{p1}}{2\pi\hbar\omega_0}}\,\mathrm{e}^{\mathrm{i}[k(\omega_0-\Omega_p)-k(\omega_0-\Omega)-\gamma P_{p1}]z}N_R(z,\Omega-\Omega_p) \\
&\quad + \mathrm{i}\sqrt{\frac{f_R\gamma P_{p2}}{2\pi\hbar\omega_0}}\,\mathrm{e}^{\mathrm{i}[k(\omega_0+\Omega_p)-k(\omega_0-\Omega)-\gamma P_{p2}]z}N_R(z,\Omega+\Omega_p)
\end{aligned} \tag{81}$$

where

$$\Delta k_{11}(\Omega) = k_L(\omega_0-\Omega) + k_L(\omega_0-2\Omega_p+\Omega) - 2k_L(\omega_0-\Omega_p) + 2\gamma P_{p1} \tag{82}$$

$$\Delta k_{22}(\Omega) = k_L(\omega_0-\Omega) + k_L(\omega_0+2\Omega_p+\Omega) - 2k_L(\omega_0+\Omega_p) + 2\gamma P_{p2} \tag{83}$$

$$\Delta k_{12}(\Omega) = k_L(\omega_0-\Omega) + k_L(\omega_0+\Omega) - k_L(\omega_0-\Omega_p) - k_L(\omega_0+\Omega_p) + \gamma P_{p1} + \gamma P_{p2} \tag{84}$$

$$\Delta k_a(\Omega) = k_L(\omega_0+2\Omega_p) + k_L(\omega_0+\Omega_p) - k_L(\omega_0-\Omega_p) - k_L(\omega_0-\Omega) + \gamma P_{p1} - \gamma P_{p2} \tag{85}$$

$$\Delta k_b(\Omega) = k_L(\omega_0-2\Omega_p) + k_L(\omega_0-\Omega_p) - k_L(\omega_0+\Omega_p) - k_L(\omega_0-\Omega) - \gamma P_{p1} + \gamma P_{p2} \tag{86}$$

$$\Delta k_c(\Omega) = k_L(\omega_0-\Omega-2\Omega_p) + k_L(\omega_0-\Omega_p) - k_L(\omega_0+\Omega_p) - k_L(\omega_0-\Omega) + \gamma P_{p1} - \gamma P_{p2} \tag{87}$$

$$\Delta k_d(\Omega) = k_L(\omega_0-\Omega+2\Omega_p) + k_L(\omega_0-\Omega_p) - k_L(\omega_0+\Omega_p) - k_L(\omega_0-\Omega) - \gamma P_{p1} + \gamma P_{p2} \tag{88}$$

Eq. (81) can be written

$$\mathrm{i}\hbar\frac{\partial}{\partial z}a(z,\Omega) = [G(z), a(z,\Omega)] + \mathrm{i}\hbar L(z,\Omega), \tag{89}$$

where

$$\begin{aligned}
L(z,\Omega) = \mathrm{i}\sqrt{\frac{f_R\gamma}{2\pi\hbar\omega_0}}\Big[&\sqrt{P_{p1}}\,\mathrm{e}^{\mathrm{i}[k(\omega_0-\Omega_p)-k(\omega_0-\Omega)-\gamma P_{p1}]z}N_R(z,\Omega-\Omega_p) \\
&+\sqrt{P_{p2}}\,\mathrm{e}^{\mathrm{i}[k(\omega_0+\Omega_p)-k(\omega_0-\Omega)-\gamma P_{p2}]z}N_R(z,\Omega+\Omega_p)\Big]
\end{aligned} \tag{90}$$

and

$$\begin{aligned}
G(z) &= \frac{\hbar}{2}\gamma P_{p1}\int_{-\infty}^{\infty} d\Omega' e^{-i\Delta k_{11}(\Omega')z}\, a^{\dagger}(z,\Omega')a^{\dagger}(z,2\Omega_p-\Omega') + h.c. \\
&+ \frac{\hbar}{2}\gamma P_{p2}\int_{-\infty}^{\infty} d\Omega' e^{-i\Delta k_{22}(\Omega')z}\, a^{\dagger}(z,\Omega')a^{\dagger}(z,-2\Omega_p-\Omega') + h.c. \\
&+ \hbar\gamma\sqrt{P_{p1}P_{p2}}\int_{-\infty}^{\infty} d\Omega' e^{-i\Delta k_{12}(\Omega')z}\, a^{\dagger}(z,\Omega')a^{\dagger}(z,-\Omega') + h.c. \\
&+ \hbar\gamma\sqrt{P_{p1}P_{p2}}\int_{-\infty}^{\infty} d\Omega' e^{i\Delta k_{a}(\Omega')z}\, a^{\dagger}(z,\Omega')a(z,-2\Omega_p) + h.c. \\
&+ \hbar\gamma\sqrt{P_{p1}P_{p2}}\int_{-\infty}^{\infty} d\Omega' e^{i\Delta k_{b}(\Omega')z}\, a^{\dagger}(z,\Omega')a(z,2\Omega_p) + h.c. \\
&+ \hbar\gamma\sqrt{P_{p1}P_{p2}}\int_{-\infty}^{\infty} d\Omega' e^{i\Delta k_{c}(\Omega')z}\, a^{\dagger}(z,\Omega')a(z,\Omega'-2\Omega_p) + h.c. \\
&+ \hbar\gamma\sqrt{P_{p1}P_{p2}}\int_{-\infty}^{\infty} d\Omega' e^{i\Delta k_{d}(\Omega')z}\, a^{\dagger}(z,\Omega')a(z,\Omega'+2\Omega_p) + h.c.
\end{aligned} \tag{91}$$

In Eq. (89), $L(z,\Omega)$ represents the Raman scattering from both pumps. Raman scattered photons contribute incoherently,

$$\begin{aligned}
\frac{S_R(L,\omega_p-\Omega)}{\hbar(\omega-\Omega)} &= \frac{|g_{\parallel}(\omega_{p1},\Omega-\Omega_p)|P_{p1}L}{2\pi}\left(m_{\text{th}}(|\Omega-\Omega_p|)+\nu(\Omega-\Omega_p)\right) \\
&+ \frac{|g_{\parallel}(\omega_{p2},\Omega+\Omega_p)|P_{p2}L}{2\pi}\left(m_{\text{th}}(|\Omega+\Omega_p|)+\nu(\Omega+\Omega_p)\right),
\end{aligned} \tag{92}$$

to the total scattered photon flux. The $[G(z),a(z,\Omega)]$ part of Eq. (89) represents several FPS processes taking place simultaneously: (i) $2\omega_{p1} \to (\omega_{p1}+\Omega_p-\Omega)+(\omega_{p1}-\Omega_p+\Omega)$, (ii) $2\omega_{p2} \to (\omega_{p2}-\Omega_p-\Omega)+(\omega_{p2}+\Omega_p+\Omega)$, (iii) $\omega_{p1}+\omega_{p2} \to (\omega_0-\Omega)+(\omega_0+\Omega)$. To see this explicitly, we writing down the evolution of the quantum state of light in the *interaction* picture. The interaction picture is chosen such that the phase evolution of the modes is part of the operator evolution, while energy transfer from mode to mode is part of the state evolution. In this interaction picture $a^{(I)}(z,\Omega) = a(0,\Omega)$ (Brainis, 2009). Therefore the first order perturbation Dyson expansion gives:

$$\begin{aligned}
|\psi(L)\rangle &= \left(1+\frac{i}{\hbar}\int_0^L G^{(I)}(z)dz\right)|0\rangle = |0\rangle + \int_{-\infty}^{\infty} d\Omega \\
&\left(\xi_{11}(L,\Omega)|1_{\Omega},1_{2\Omega_p-\Omega}\rangle + \xi_{12}(L,\Omega)|1_{\Omega},1_{-\Omega}\rangle + \xi_{22}(L,\Omega)|1_{\Omega},1_{-2\Omega_p-\Omega}\rangle\right)
\end{aligned} \tag{93}$$

where

$$\xi_{11}(L,\Omega) = i\frac{1}{2}(\gamma P_{p1}L)\, e^{-i\Delta k_{11}(\Omega')\frac{L}{2}}\, \text{sinc}\left(\Delta k_{11}(\Omega')\frac{L}{2}\right) \tag{94}$$

$$\xi_{12}(L,\Omega) = i(\gamma\sqrt{P_{p1}P_{p2}}L)\, e^{-i\Delta k_{12}(\Omega')\frac{L}{2}}\, \text{sinc}\left(\Delta k_{12}(\Omega')\frac{L}{2}\right) \tag{95}$$

$$\xi_{22}(L,\Omega) = i\frac{1}{2}(\gamma P_{p2}L)\, e^{-i\Delta k_{22}(\Omega')\frac{L}{2}}\, \text{sinc}\left(\Delta k_{22}(\Omega')\frac{L}{2}\right). \tag{96}$$

The threefold entanglement is a clear signature of the interference between three independent FPS processes. The spectral density of power due to these FPS processes can be deduced from the matrix element $\langle\psi(L)|a^{\dagger}(0,\Omega)a(0,\Omega)|\psi(L)\rangle$ (Brainis, 2009).

## 5. Conclusion

In this chapter, the physics of Raman, Brillouin, and four-photon scattering processes in silica fibers has been reviewed, as well as their theoretical modeling. It has been shown that a complete quantum field theory is needed to understand the coupling of theses processes in the stimulated and spontaneous regimes. Two examples of coupling have been discussed. The first one was the coupling of the Raman and four-photon scattering processes in a single pump configuration. In that case, it has been shown that the coupling may have spectacular consequences in the amplified spontaneous regime, where an unexpected exponential growth of the anti-Stokes wave is seen. In the second example, the interaction of three FPS processes in a dual pump configuration has been considered. It has been shown that this configuration leads to the generation of a threefold entangled bi-photon state of light.
Spontaneous scattering processes are of great importance the context of quantum light generation and quantum information processing. The methods presented in the chapter apply to the design of quantum source based on optical fibers: engineering the working principle (usually four-photon scattering processes) and estimating their figure of merit (usually limited by the Raman process).

## 6. Acknowledgement

This research was supported by the Interuniversity Attraction Poles program of the Belgian Science Policy Office, under grant IAP P6-10 "photonics@be" and by the Belgian Fonds de la Recherche Scientifique — FNRS under grant FRFC - 2.4.638.09F.

## 7. References

Agrawal, G. P. (2007). *Nonlinear Fiber Optics*, fourth edn, Academic Press, Amsterdam.

Alahbabi, M. N., Cho, Y. T. & Newson, T. P. (2005a). 150-km-range distributed temperature sensor based on coherent detection of spontaneous Brillouin backscatter and in-line raman amplification, *J. Opt. Soc. Am. B* 22(6): 1321–1324.

Alahbabi, M. N., Cho, Y. T. & Newson, T. P. (2005b). Simultaneous temperature and strain measurement with combined spontaneous Raman and Brillouin scattering, *Opt. Lett.* 30(11): 1276–1278.

Amans, D., Brainis, E., Haelterman, M., Emplit, P. & Massar, S. (2005). Vector modulation instability induced by vacuum fluctuations in highly birefringent fibers in the anomalous-dispersion regime, *Opt. Lett.* 30(9): 1051–1053.

Aoki, Y. (1988). Properties of fiber Raman amplifiers and their applicability to digital optical communication systems, *J. Ligthwave Technol.* 6(7): 1225–1239.

Benabid, F., Couny, F., Knight, J. C., Birks, T. A. & Russell, P. S. J. (2005). Compact, stable and efficient all-fibre gas cells using hollow-core photonic crystal fibres, *Nature (London)* 434: 488–491.

Benedek, G. B. & Fritsch, K. (1966). Brillouin scattering in cubic crystals, *Phys. Rev.* 149(2): 647–662.

Bloembergen, N. & Shen, Y. R. (1964). Coupling between vibrations and light waves in Raman laser media, *Phys. Rev. Lett.* 12: 504–507.

Boivin, L., Kärtner, F. X. & Haus, H. A. (1994). Analytical solution to the quantum field theory of self-phase modulation with a finite response time, *Phys. Rev. Lett.* 73(2): 240–243.

Bonfrate, G., Pruneri, V., Kazansky, P. G., Tapster, P. & Rarity, J. G. (1999). Parametric fluorescence in periodically poled silica fibers, *Appl. Phys. Lett.* 75(16): 2356–2358.

Brainis, E. (2009). Four-photon scattering in birefringent fibers, *Phys. Rev. A* 79(2): 023840.

Brainis, E., Amans, D. & Massar, S. (2005). Scalar and vector modulation instabilities induced by vacuum fluctuations in fibers: Numerical study, *Phys. Rev. A* 71(2): 23808.

Brainis, E., Clemmen, S. & Massar, S. (2007). Spontaneous growth of Raman Stokes and anti-Stokes waves in fibers, *Opt. Lett.* 32(19): 2819–2821.

Carter, S. J., Drummond, P. D., Reid, M. D. & Shelby, R. M. (1987). Squeezing of quantum solitons, *Phys. Rev. Lett.* 58(18): 1841.

Coen, S., Wardle, D. A. & Harvey, J. D. (2002). Observation of non-phase-matched parametric amplification in resonant nonlinear optics, *Phys. Rev. Lett.* 89(27): 273901.

Corwin, K. L., Newbury, N. R., Dudley, J. M., Coen, S., Diddams, S. A., Weber, K. & Windeler, R. S. (2003). Fundamental noise limitations to supercontinuum generation in microstructure fiber, *Phys. Rev. Lett.* 90(11): 113904.

Crosignani, B., Di Porto, P. & Solimeno, S. (1980). Influence of guiding structures on spontaneous and stimulated emission: Raman scattering in optical fibers, *Phys. Rev. A* 21(2): 594–598.

Dakin, J., Pratt, D., Bibby, G. & Ross, J. (1985). Distributed optical fibre Raman temperature sensor using a semiconductor light source and detector, *Electron. Lett.* 21(13): 569–570.

Digonnet, M. J. F. (ed.) (2001). *Rare-Earth-Doped Fiber Lasers and Amplifiers*, second edn, Marcel Dekker, New-York.

Dougherty, D. J., Kärtner, F. X., Haus, H. A. & Ippen, E. P. (1995). Measurement of the Raman gain spectrum of optical fibers, *Opt. Lett.* 20(1): 31–33.

Drummond, P. D. & Carter, S. J. (1987). Quantum-field theory of squeezing in solitons, *J. Opt. Soc. Am. B* 4(10): 1565.

Drummond, P. D. & Corney, J. F. (2001). Quantum noise in optical fibers. I. Stochastic equations, *J. Opt. Soc. Am. B* 18(2): 139–152.

Dudley, J. M., Genty, G. & Coen, S. (2006). Supercontinuum generation in photonic crystal fiber, *Rev. Mod. Phys.* 78(4): 1135–1184.

Dyer, S. D., Stevens, M. J., Baek, B. & Nam, S. W. (2008). High-efficiency, ultra low-noise all-fiber photon-pair source, *Opt. Express* 16: 9966.

Fan, J. & Migdall, A. (2007). A broadband high spectral brightness fiber-based two-photon source, *Opt. Express* 15: 2915.

Farahani, M. A. & Gogolla, T. (1999). Spontaneous Raman scattering in optical fibers with modulated probe light for distributed temperature Raman remote sensing, *J. Ligthwave Technol.* 17(8): 1379.

Golovchenko, E., Mamyshev, P. V., Pilipetskii, A. N. & Dianov, E. M. (1990). Mutual influence of the parametric effects and stimulated Raman scattering in optical fibers, *IEEE J. Quantum Electron.* 26: 1815–1820.

Hellwarth, R., Cherlow, J. & Yang, T.-T. (1975). Origin and frequency dependence of nonlinear optical susceptibilities of glasses, *Phys. Rev. B* 11(2): 964–967.

Huy, K. P., Nguyen, A. T., Brainis, E., Haelterman, M., Emplit, P., Corbari, C., Canagasabey, A., Ibsen, M., Kazansky, P. G., Deparis, O., Fotiadi, A., Mégret, P. & Massar, S. (2007). Photon pair source based on parametric fluorescence in periodically poled twin-hole silica fiber, *Opt. Express* 15(8): 4419–4426.

Kärtner, F. X., Dougherty, D. J., Haus, H. A. & Ippen, E. P. (1994). Raman noise and soliton squeezing, *J. Opt. Soc. Am. B* 11(7): 1267–1276.

Kazansky, P. G., Russell, P. S. J. & Takebe, H. (1997). Glass fiber poling and applications, *J. Ligthwave Technol.* 15: 1484–1493.

Kennedy, T. A. B. (1991). Quantum theory of cross-phase-modulational instability: Twin-beam correlations in a $\chi^{(3)}$ process, *Phys. Rev. A* 44: 2113–2123.
Kennedy, T. A. B. & Wabnitz, S. (1988). Quantum propagation: Squeezing via modulational polarization instabilities in a birefringent nonlinear medium, *Phys. Rev. A* 38(1): 563.
Kennedy, T. A. B. & Wright, E. M. (1988). Quantization and phase-space methods for slowly varying optical fields in a dispersive nonlinear medium, *Phys. Rev. A* 38(1): 212–221.
Lee, J. H., Tanemura, T., Kikuchi, K., Nagashima, T., Hasegawa, T., Ohara, S. & Sugimoto, N. (2005). Experimental comparison of a Kerr nonlinearity figure of merit including the stimulated Brillouin scattering threshold for state-of-the-art nonlinear optical fibers, *Opt. Lett.* 30(13): 1698–1700.
Lee, K. F., Chen, J., Liang, C., Li, X., Voss, P. L. & Kumar, P. (2006). Generation of high-purity telecom-band entangled photon pairs in dispersion-shifted fiber, *Opt. Lett.* 31: 1905.
Li, X., Chen, J., Voss, P., Sharping, J. & Kumar, P. (2004). All-fiber photon-pair source for quantum communications: Improved generation of correlated photons, *Opt. Express* 12(16): 3737–3744.
Lin, Q., Yaman, F. & Agrawal, G. P. (2006). Photon-pair generation by four-wave mixing in optical fibers, *Opt. Lett.* 31: 1286.
Lin, Q., Yaman, F. & Agrawal, G. P. (2007). Photon-pair generation in optical fibers through four-wave mixing: Role of Raman scattering and pump polarization, *Phys. Rev. A* 75: 023803.
Mahgerefteh, D., Butler, D. L., Goldhar, J., Rosenberg, B. & Burdge, G. L. (1996). Technique for measurement of the Raman gain coefficient in optical fibers, *Opt. Lett.* 21(24): 2026–2028.
McElhenny, J. E., Pattnaik, R. & Toulouse, J. (2008). Polarization dependence of stimulated Brillouin scattering in small-core photonic crystal fibers, *J. Opt. Soc. Am. B* 25(12): 2107–2115.
Mochizuki, K., Edagawa, N. & Iwamoto, Y. (1986). Amplified spontaneous Raman scattering in fiber Raman amplifiers, *J. Ligthwave Technol.* 4(9): 1328–1333.
Olsson, N. & Hegarty, J. (1986). Noise properties of a Raman amplifier, *J. Ligthwave Technol.* 4(4): 396–399.
Pi, Y., Zhang, W., Wang, Y., Huang, Y. & Peng, J. (2008). Temperature dependence of the spontaneous Brillouin scattering spectrum in microstructure fiber with small core, *Tsinghua Science & Technology* 13(1): 43 – 46.
Rarity, J. G., Fulconis, J., Duligall, J., Wadsworth, W. J. & Russel, P. S. J. (2005). Photonic crystal fiber source of correlated photon pairs, *Opt. Express* 13: 534.
Shelby, R. M., Levenson, M. D. & Bayer, P. W. (1985a). Guided acoustic-wave Brillouin scattering, *Phys. Rev. B* 31(8): 5244–5252.
Shelby, R. M., Levenson, M. D. & Bayer, P. W. (1985b). Resolved forward Brillouin scattering in optical fibers, *Phys. Rev. Lett.* 54(9): 939–942.
Smith, R. G. (1972). Optical power handling capacity of low loss optical fibers as determined by stimulated Raman and Brillouin scattering, *Appl. Opt.* 11(11): 2489–2494.
Stolen, R. H. (1979). Polarization effects in fiber Raman and Brillouin lasers, *IEEE J. Quantum Electron.* 15: 1157–1160.
Stolen, R. H., Lee, C. & Jain, R. K. (1984). Development of the stimulated Raman spectrum in single-mode silica fibers, *J. Opt. Soc. Am. B* 1(4): 652–657.
Stolen, R. & Ippen, E. (1973). Raman gain in glass optical waveguides, *Appl. Phys. Lett.* 22(6): 276–278.

Takesue, H. (2006). Long-distance distribution of time-bin entanglement generated in a cooled fiber, *Opt. Express* 14: 3453.

Vanholsbeeck, F., Emplit, P. & Coen, S. (2003). Complete experimental characterization of the influence of parametric four-wave mixing on stimulated Raman gain, *Opt. Lett.* 28: 1960–1962.

Wait, P. C., Souza, K. D. & Newson, T. P. (1997). A theoretical comparison of spontaneous Raman and Brillouin based fibre optic distributed temperature sensors, *Opt. Commun.* 144(1-3): 17 – 23.

Wardle, D. A. (1999). *Raman Scattering in Optical Fibres*, PhD thesis, University of Auckland.

Yeniay, A., Delavaux, J.-M. & Toulouse, J. (2002). Spontaneous and stimulated Brillouin scattering gain spectra in optical fibers, *J. Lightwave Technol.* 20(8): 1425.

# 3

# Optical Solitons in a Nonlinear Fiber Medium with Higher-Order Effects

Deng-Shan Wang
*School of Science, Beijing Information Science and Technology University, Beijing, 100192 China*

## 1. Introduction

Nowadays we can see many interesting applications of solitons in different areas of physical sciences such as plasma physics (1), nonlinear optics (2; 3), Bose-Einstein condensate (4; 5), fluid mechanics (6), and so on. Solitons are so robust particles that they are unlikely to breakdown under small perturbations. The most interesting factor about the soliton, however, is that their interactions with the medium through which it propagates is elastic. Recent researches on nonlinear optics have shown that dispersion-managed pulse can be more useful if the pulse is in the form of a power series of a stable localized pulse which is called soliton.
Optical solitons have been the objects of extensive theoretical and experimental studies during the last four decades, because of their potential applications in long distance communication. In 1973, the pioneering results of Hasegawa and Tappert (7) proved that the major constraint in the optical fiber, namely, the group velocity dispersion (GVD) could be exactly counterbalanced by the self-phase modulation (SPM). SPM is the dominant nonlinear effect in silica fibers due to the Kerr effect. The theoretical results of Hasegawa and Tappert were greatly supported by the experimental demonstration of optical solitons by Mollenauer et al. (8) in 1980. Since then many theoretical and experimental works have been done to achieve a communication system based on optical solitons.
The solitons, localized-in-time optical pulses, evolve from a nonlinear change in the refractive index of the material, known as Kerr effect, induced by the light intensity distribution. When the combined effects of the intensity-dependent refractive index nonlinearity and the frequency-dependent pulse dispersion exactly compensate for one another, the pulse propagates without any change in its shape, being self-trapped by the waveguide nonlinearity. The propagation of optical solitons in a nonlinear dispersive optical fiber is governed by the well-known completely integrable nonlinear Schrödinger (NLS) equation

$$i\frac{\partial q}{\partial z}+\epsilon\frac{\partial^2 q}{\partial \tau^2}+|q|^2q=0, \qquad \epsilon=\pm 1, \tag{1}$$

where $q$ is the complex amplitude of the pulse envelope, $\tau$ and $z$ represent the spatial and temporal coordinates, and the $+$ and $-$ sign of $\epsilon$ before the dispersive term denote the anomalous and normal dispersive regimes, respectively. In the anomalous dispersive regime, this equation possesses a bright soliton solution, and in the normal dispersive regime it possesses dark solitons. The bright soliton and dark soliton solutions can be derived by the inverse-scattering transform method with vanishing (9; 11) and nonvanishing boundary conditions (10).

However, if optical pulses are shorter, the standard NLS equation becomes inadequate. Therefore, some higher-order effects such as third-order dispersion, self-steepening, and stimulated Raman scattering, will play important roles in the propagation of optical pulses. In such a case, the governing equation is the one known widely as the higher-order NLS equation, first derived by Kodama and Hasegawa (12). The effect of these effects in uncoupled and coupled systems for bright solitons is well explained (13; 14). Inelastic Raman scattering is due to the delayed response of the medium, which forces the pulse to undergo a frequency shift which is known as a self-frequency shift. The effect of self-steepening is due to the intensity-dependent group velocity of the optical pulse, which gives the pulse a very narrow width in the course of propagation. Because of this, the peak of the pulse will travel more slowly than the wings.

In practice, the refractive index or the core diameter of the optical fiber are fucntions of the axial coordinate, which means that the fiber is actually axially inhomogeneous. In this case, the parameters which characterize the dispersive and nonlinear properties of the fiber exhibit variations and the corresponding nonlinear wave equations are NLS equations with variable coefficients. Moreover, the problem of ultrashort pulse propagation in nonlinear and axially inhomogeneous optical fibers near the zero dispersion point is more complicated because the high order effects have to be taken into account as well. In order to understand such phenomena, we consider the higher-order NLS (HNLS) equation with variable coefficients

$$\frac{\partial u}{\partial z} = i(d_1 \frac{\partial^2 u}{\partial \tau^2} + d_2 |u|^2 u) + d_3 \frac{\partial^3 u}{\partial \tau^3} + d_4 \frac{\partial (u|u|^2)}{\partial \tau} + d_5 u \frac{\partial |u|^2}{\partial \tau} + d_6 u, \tag{2}$$

where $u$ is the slowly varying envelope of the pulse, $d_1, d_2, d_3, d_4, d_5$ and $d_6$ are the $z$-dependent real parameters related to GVD, SPM, third-order dispersion (TOD), self-steepening, and stimulated Raman scattering (SRS), and the heat-insulating amplification or loss, respectively. Though Eq. (2) was first derived in the year 1980s, only for the past few years, it has attracted much attention among the researchers from both theoretical and experimental points of view. For example, Porsezian and Nakkeeran (13) derive all parametric conditions for soliton-type pulse propagation in HNLS equation using the Painlevé analysis, and generalize the Ablowitz-Kaup-Newell-Segur method to the $3 \times 3$ eigenvalue problem to construct the Lax pair for the integrable case. Papaioannou et al. (15) give an analytical treatment of the effect of axial inhomogeneity on femtosecond solitary waves near the zero dispersion point which governed by the variable-coefficient HNLS equation. The exact bright and dark soliton wave solutions of this variable-coefficient equation are derived and their behaviors in the presence of the inhomogeneity are analyzed. Mahalingam and Porsezian (16) analyze the propagation of dark solitons with higher-order effects in optical fibers by Painlevé analysis and Hirota bilinear method. Xu et al. (17) investigate the modulation instability and solitons on a cw background in an optical fiber with higher-order effects. In addition, there have recently been several papers giving W-shaped solitary wave solution in the HNLS equation.

However, in recent years the studies of Eq. (2) have not been widespread. In this chapter, we consider equation (2) again and derive some exact soliton solutions in explicit form for specified soliton management conditions. We first change the variable-coefficient HNLS equation into the well-known constant-coefficient HNLS equation through similarity transformation. Then the Lax pairs for two integrable cases of the constant-coefficient HNLS equation are constructed explicitly by prolongation technique, and the novel exact bright N-soliton solutions for the bright soliton version of HNLS equation are obtained by Riemann-Hilbert formulation. Finally, we examine the dynamics and present the features of the optical solitons. It is seen that the bright two-soliton solution of the HNLS equation behaves in an elastic manner characteristic of all soliton solutions. These results are useful in the design of transmission lines with spatial parameter variations and soliton management to future research.

## 2. Similarity transformation

A direct and efficient method for investigating the variable-coefficient nonlinear wave equation is to transform them into their constant-coefficient counterparts by similarity transformation. To do so, we firstly take the similarity transformation (18; 19)

$$u = \rho\, q\,(T, X)\, e^{i(a_1\tau + a_2)}, \tag{3}$$

to reduce Eq. (2) to the constant-coefficient HNLS equation

$$\frac{\partial q}{\partial T} = i(\alpha_1 \frac{\partial^2 q}{\partial X^2} + \alpha_2 |q|^2 q) + \epsilon(\alpha_3 \frac{\partial^3 q}{\partial X^3} + \alpha_4 \frac{\partial(q|q|^2)}{\partial X} + \alpha_5 q \frac{\partial |q|^2}{\partial X}), \tag{4}$$

where $q = q(T, X)$ is the complex amplitude of the pulse envelope, the parameter $\epsilon$ ($0 < \epsilon < 1$) denotes the relative width of the spectrum that arises due to the quasi-monochromocity, $\alpha_1, \alpha_2, \alpha_3, \alpha_4$ and $\alpha_5$ are the real constant parameters. In Eq. (3), $\rho, T, a_1$ and $a_2$ are functions of $z$, and $X$ is a function of $\tau$ and $z$.
Substituting Eq. (3) into Eq. (2) and asking $q\,(T, X)$ to satisfy the constant-coefficient HNLS equation (4), we have a set of partial differential equations (PDEs)

$$\begin{gathered}
d_1 X_{\tau\tau} + 3\, d_3 X_{\tau\tau} a_1 = 0,\;\; d_3 {X_\tau}^3 = \alpha_3 T_z,\;\; \rho_z = d_6 \rho, \\
2\, d_1 X_\tau a_1 + X_z + 3\, d_3 X_\tau {a_1}^2 = d_3 X_{\tau\tau\tau},\;\; \rho^2 d_4 a_1 + \rho^2 d_2 = \alpha_2 T_z, \\
2\, \rho^2 X_\tau d_4 + \rho^2 X_\tau d_5 = 2\, \alpha_4 T_z + \alpha_5 T_z,\;\; \rho^2 X_\tau d_4 + \rho^2 X_\tau d_5 = \alpha_4 T_z + \alpha_5 T_z, \\
d_3\, {a_1}^3 + \, a_{1z}\tau + \, a_{2z} + d_1\, {a_1}^2 = 0,\;\; 3\, d_3 {X_\tau}^2 a_1 + d_1 {X_\tau}^2 = \alpha_1 T_z,\;\; X_{\tau,\tau} = 0,
\end{gathered}$$

where the subscript denotes the derivative with respect to $z$ and $\tau$. Solving this set of PDEs, we have $X = k\tau + f$ and

$$a_1 = c,\;\; d_1 = \frac{T_z\,(k\alpha_1 - 3\,\alpha_3 c)}{k^3},\;\; d_2 = \frac{T_z\,(\alpha_2 k - \alpha_4 c)}{\rho^2 k},\;\; d_3 = \frac{\alpha_3 T_z}{k^3},$$

$$d_4 = \frac{\alpha_4 T_z}{\rho^2 k},\;\; d_5 = \frac{\alpha_5 T_z}{\rho^2 k},\;\; f = \frac{c\,(3\,\alpha_3 c - 2\, k\alpha_1)\, T}{k^2},\;\; a_2 = \frac{(2\,\alpha_3 c - k\alpha_1)\, c^2 T}{k^3},$$

where $\rho = \rho_0\, e^{\int d_6 z}$, $k, \rho_0$ and $c$ are constants, and $T$ and $d_6$ are arbitrary functions of $z$. So the similarity transformation (3) becomes

$$u = \rho_0\, e^{\int d_6 dz} q\left(T, \frac{k^3\tau - 2\, ck\alpha_1 T + 3\, c^2 \alpha_3 T}{k^2}\right) e^{ic(k^3\tau + 2\, c^2\alpha_3 T - ck\alpha_1 T)/k^3}. \tag{5}$$

Therefore, if we can get the exact soliton solutions of the constant-coefficient HNLS equation (4) we can obtain the exact soliton solutions for HNLS equation (2) through Eq. (5). In the next section, we will investigate the integrable condition of equation (4) by prolongation technique.

## 3. Prolongation structures of the constant-coefficient HNLS equation

In this section, we investigate the prolongation structures of the constant-coefficient HNLS equation (4) by means of the prolongation technique (20–22). Firstly, the complex conjugate of the dependent variable $q$ in Eq. (4) is denoted as $q^* = u$. Then, Eq. (4) and its conjugate become

$$i\alpha_1 q_{XX} + i\alpha_2 q^2 u + \epsilon[\alpha_3 q_{XXX} + (\alpha_4 + \alpha_5) q^2 u_X + (2\alpha_4 + \alpha_5) q u q_X] - q_T = 0, \tag{6a}$$

$$-i\alpha_1 u_{XX} - i\alpha_2 u^2 q + \epsilon[\alpha_3 u_{XXX} + (\alpha_4 + \alpha_5) u^2 q_X + (2\alpha_4 + \alpha_5) q u u_X] - u_T = 0. \tag{6b}$$

Next we introduce four new variables $p, r, v$ and $w$ by

$$q_X = p, \;\; p_X = r, \;\; u_X = v, \;\; v_X = w, \tag{7}$$

and define a set of differential 2-form $I = \{\theta_1, \theta_2, \theta_3, \theta_4, \theta_5, \theta_6\}$ on solution manifold $M = \{T, X, u, v, w, p, q, r\}$, where

$$\theta_1 = dq \wedge dT + p dT \wedge dX, \;\; \theta_2 = dp \wedge dT + r dT \wedge dX,$$

$$\theta_3 = du \wedge dT + v dT \wedge dX, \;\; \theta_4 = dv \wedge dT + w dT \wedge dX,$$

$$\theta_5 = dq \wedge dX + \alpha_3 dr \wedge dT + \rho_1 dX \wedge dT, \;\; \theta_6 = du \wedge dX + \alpha_3 dw \wedge dT + \rho_2 dX \wedge dT,$$

with

$$\rho_1 = i\alpha_1 r + i\alpha_2 q^2 u + \epsilon[(\alpha_4 + \alpha_5) q^2 v + (2\alpha_4 + \alpha_5) q u p],$$

$$\rho_2 = -i\alpha_1 w - i\alpha_2 u^2 q + \epsilon[(\alpha_4 + \alpha_5) u^2 p + (2\alpha_4 + \alpha_5) q u v].$$

When these differential 2-forms restricted on the solution manifold $M$ become zero, we recover the original constant-coefficient HNLS equation (4). It is easy to verify that $I$ is a differential closed idea, i.e. $dI \subset I$.
We further introduce $n$ differential 1-forms

$$\Omega^i = d\zeta^i - \tilde{F}^i dX - \tilde{G}^i dT, \tag{8}$$

where $i = 1, 2, \cdots, n$, $\tilde{F}^i$ and $\tilde{G}^i$ are functions of $u, v, w, p, q, r, \zeta^i$ and are assumed to be both linearly dependent on $\zeta^i$, namely $\tilde{F}^i = F^i \zeta^i, \tilde{G}^i = G^i \zeta^i$. For the sake of simplification, we drop the indices by rewriting $\zeta^i$ as $\zeta$, $F^i$ as $F$ and $G^i$ as $G$. When restricting on solution manifold, the differential 1-forms $\Omega^i$ are null, i.e. $\Omega^i = 0$ which is just the linear spectral problem $\zeta_X = F\zeta$ and $\zeta_T = G\zeta$.
Following the well-known prolongation technique, the extended set of differential form $\tilde{I} = I \bigcup \left\{\Omega^i\right\}$ must be a closed ideal under exterior differentiation, i.e. $d\tilde{I} \subset \tilde{I}$. Because $dI \subset I \subset \tilde{I}$, we only need to let $d\left\{\Omega^i\right\} \subset \tilde{I}$, which denotes that

$$d\Omega^i = \sum_{j=1}^{6} f_j^i \theta^j + \eta^i \wedge \Omega^i, \qquad i = 1, 2, \cdots, n, \tag{9}$$

where $f_j^i$ $(j = 1, 2, 3, 4, 5, 6)$ are functions of $(T, X)$, and $\eta^i = g^i(T, X) dX + h^i(T, X) dT$ are differential 1-forms.

When Eq. (9) is written out in detail, after dropping the indices we have the following PDEs about $F$ and $G$ as

$$G_r = \epsilon\alpha_3 F_q, \quad G_w = \epsilon\alpha_3 F_u,$$

$$G_q p + G_p r + G_u v + G_v w - F_q \left[ i\alpha_1 r + i\alpha_2 q^2 u + \epsilon\,(\alpha_4 + \alpha_5)\, q^2 v + \epsilon\,(2\,\alpha_4 + \alpha_5)\, qup \right] \tag{10}$$
$$-F_u \left[ -i\alpha_1 w - i\alpha_2 u^2 q + \epsilon\,(\alpha_4 + \alpha_5)\, u^2 p + \epsilon\,(2\,\alpha_4 + \alpha_5)\, quv \right] - [F,G] = 0,$$

with $[F,G] = FG - GF$.
Solving Eq. (10), we have the expressions of $F$ and $G$ as

$$F = x_0 + x_1 q + x_2 u, \tag{11}$$

$$G = \epsilon\,\alpha_3 x_1 r + \epsilon\,\alpha_3 x_2 w + vq\epsilon\,\alpha_3 x_5 + v\epsilon\,\alpha_3 x_4 - pu\epsilon\,\alpha_3 x_5 + p\epsilon\,\alpha_3 x_3 - iqu\alpha_1 x_5$$
$$+ipx_1\alpha_1 + qx_2\epsilon\, u^2\alpha_4 + \frac{2}{3}\, qx_2\epsilon\, u^2\alpha_5 + \frac{1}{2}\,\epsilon\,\alpha_3 u^2 x_{13} + q^2\epsilon\, x_1 u\alpha_4 + \frac{2}{3}\, q^2\epsilon\, x_1 u\alpha_5$$
$$+que\,\alpha_3 x_8 + qu\epsilon\,\alpha_3 x_{10} + \epsilon\,\alpha_3 x_7 u + \frac{1}{2}\, q^2\epsilon\,\alpha_3 x_9 + q\epsilon\,\alpha_3 x_6 + iq\alpha_1 x_3 - i\alpha_1 x_4 u - ivx_2\alpha_1 + x_{15},$$

where $L = \{x_0, x_1, x_2, \cdots, x_{15}\}$ is an incomplete Lie algebra which is called prolongation algebra and it satisfies the following commutation relations

$$[x_2, x_5] = x_{14}, \;\; x_2\alpha_5 = 3\,\alpha_3 x_{14}, 2\,x_8 + x_{10} = x_{12}, x_1\alpha_5 + 3\,\alpha_3 x_{11} = 0,$$

$$[x_0, x_1] = x_3, \;\; [x_0, x_2] = x_4, \;\; [x_0, x_3] = x_6, \;\; [x_0, x_4] = x_7, \;\; [x_0, x_5] = x_8,$$
$$[x_1, x_2] = x_5, \;\; [x_1, x_3] = x_9, \;\; [x_1, x_4] = x_{10}, \;\; [x_1, x_5] = x_{11}, \;\; [x_2, x_3] = x_{12},$$
$$[x_2, x_4] = x_{13}, \;\; \epsilon\,\alpha_3[x_0, x_9] + 2\,i\alpha_1 x_9 + 2\,\epsilon\,\alpha_3[x_1, x_6] = 0, \;\; [x_1, x_{15}] + i\alpha_1 x_6 + \epsilon\,\alpha_3[x_0, x_6],$$
$$\alpha_3[x_1, x_9] = 0, \;\; \alpha_3[x_2, x_{13}] = 0, \;\; [x_0, x_{15}] = 0, \;\; 2\,\epsilon\,\alpha_3[x_2, x_7] + \epsilon\,\alpha_3[x_0, x_{13}] = 2\,i\alpha_1 x_{13},$$
$$[x_2, x_{15}] + \epsilon\,\alpha_3[x_0, x_7] = i\alpha_1 x_7, \;\; 6\,\epsilon\,\alpha_3([x_0, x_{10}] + [x_1, x_7] + [x_2, x_6] + [x_0, x_8]) + 6\,i\alpha_1 x_8 = 0,$$
$$(6\,\epsilon\,[x_1, x_8] + 6\,\epsilon\,[x_1, x_{10}] + 3\,\epsilon\,[x_2, x_9])\,{\alpha_3}^2 + (4\,\epsilon\,\alpha_5 x_3 + 6\,\epsilon\,\alpha_4 x_3 + 6\,ix_1\alpha_2)\,\alpha_3 + 2\,i\alpha_1 x_1\alpha_5 = 0,$$
$$(6\,\epsilon\,[x_2, x_{10}] + 3\,\epsilon\,[x_1, x_{13}] + 6\,\epsilon\,[x_2, x_8])\,{\alpha_3}^2 + (4\,\epsilon\,\alpha_5 x_4 + 6\,\epsilon\,\alpha_4 x_4 - 6\,ix_2\alpha_2)\,\alpha_3 - 2\,i\alpha_1 x_2\alpha_5 = 0.$$

It is known that nontrivial matrix representations of prolongation algebra $L$ correspond to nontrivial prolongation structures. To find the matrix representation of $L$, following the procedure of Fordy (23), we try to embed it into Lie algebra $sl(n, C)$. Starting from the case of $n = 2$, we found that $sl(2, C)$ is the whole algebra for some special coefficients $\alpha_j (j = 1, 2, 3, 4, 5)$. For the case of $n = 3$, we can also find that $sl(3, C)$ will be the whole algebra for some other special coefficients $\alpha_j (j = 1, 2, 3, 4, 5)$. In this paper, we only examine the case of $sl(2, C)$ algebra.
From the above commutation relations, we have the special relations among elements $x_1, x_2$ and $x_5$ as

$$[x_2, x_5] = \frac{\alpha_5}{3\alpha_3} x_2, \;\; [x_1, x_5] = -\frac{\alpha_5}{3\alpha_3} x_1, \;\; [x_1, x_2] = x_5, \tag{12}$$

from which we know that $x_1$ and $x_2$ are nilpotent elements and $x_5$ is a neutral element. So we have $\alpha_5 = \pm 6\,\delta^2\alpha_3$ and

$$x_1 = \begin{pmatrix} 0 & \delta \\ 0 & 0 \end{pmatrix}, \;\; x_2 = \begin{pmatrix} 0 & 0 \\ \pm\delta & 0 \end{pmatrix}, \;\; x_5 = \begin{pmatrix} \pm\delta^2 & 0 \\ 0 & \mp\delta^2 \end{pmatrix}, \tag{13}$$

with $\delta$ a nonzero constant. Substituting (13) into the commutation relations of prolongation algebra $L$, we finally get the $2\times 2$ matrix representations of $F$ and $G$. Therefore, we obtain two integrable HNLS equations with $2\times 2$ spectral problems.
When $\alpha_2 = 2\delta^2\alpha_1, \alpha_4 = 6\delta^2\alpha_3$ and $\alpha_5 = -6\delta^2\alpha_3$, Eq. (4) becomes the bright soliton version of Hirota equation

$$q_T = i\alpha_1 q_{XX} + 2i\alpha_1\delta^2|q|^2 q + \epsilon\alpha_3 q_{XXX} + 6\epsilon\delta^2\alpha_3|q|^2 q_X, \tag{14}$$

with linear spectral problem

$$\zeta_X = F\zeta, \quad \zeta_T = G\zeta, \tag{15}$$

and

$$F = \begin{pmatrix} -i\lambda & \delta q \\ -\delta q^* & i\lambda \end{pmatrix}, \tag{16}$$

$$\begin{aligned} G = {} & 4i\alpha_3\epsilon\lambda^3 \begin{pmatrix} 1 & 0 \\ 0 & -1 \end{pmatrix} - 2\lambda^2 \begin{pmatrix} i\alpha_1 & 2\epsilon\alpha_3\delta q \\ -2\epsilon\alpha_3\delta q^* & -i\alpha_1 \end{pmatrix} \\ & + 2\lambda \begin{pmatrix} -i\epsilon\alpha_3\delta^2|q|^2 & \alpha_1\delta q - i\epsilon\alpha_3\delta q_X \\ -\alpha_1\delta q^* - i\epsilon\alpha_3\delta q^*_X & i\epsilon\alpha_3\delta^2|q|^2 \end{pmatrix} \\ & + \begin{pmatrix} \epsilon\alpha_3\delta^2 q_X q^* - \epsilon\alpha_3\delta^2 q^*_X q + i\alpha_1\delta^2|q|^2 & \epsilon\alpha_3\delta q_{XX} + i\alpha_1\delta q_X + 2\epsilon\delta^3\alpha_3|q|^2 q \\ i\alpha_1\delta q^*_X - \epsilon\alpha_3\delta q^*_{XX} - 2\epsilon\delta^3\alpha_3|q|^2 q^* & \epsilon\alpha_3\delta^2 q^*_X q - \epsilon\alpha_3\delta^2 q_X q^* - i\alpha_1\delta^2|q|^2 \end{pmatrix}, \end{aligned} \tag{17}$$

where $\lambda$ is a spectral parameter and $\zeta(T, X, \lambda)$ is a vector or matrix function.
When $\alpha_2 = -2\delta^2\alpha_1, \alpha_4 = -6\delta^2\alpha_3$ and $\alpha_5 = 6\delta^2\alpha_3$, Eq. (4) becomes the dark soliton version of Hirota equation

$$q_T = i\alpha_1 q_{XX} - 2i\delta^2\alpha_1|q|^2 q + \epsilon\alpha_3 q_{XXX} - 6\epsilon\delta^2\alpha_3|q|^2 q_X, \tag{18}$$

with linear spectral problem Eq. (15) and

$$F = \begin{pmatrix} -i\lambda & \delta q \\ \delta q^* & i\lambda \end{pmatrix}, \tag{19}$$

$$\begin{aligned} G = {} & 4i\epsilon\lambda^3\alpha_3 \begin{pmatrix} 1 & 0 \\ 0 & -1 \end{pmatrix} - 2\lambda^2 \begin{pmatrix} i\alpha_1 & 2\epsilon\alpha_3\delta q \\ 2\epsilon\alpha_3\delta q^* & -i\alpha_1 \end{pmatrix} \\ & + 2\lambda \begin{pmatrix} i\epsilon\alpha_3\delta^2|q|^2 & -i\epsilon\alpha_3\delta q_X + \alpha_1\delta q \\ \alpha_1\delta q^* + i\epsilon\alpha_3\delta q^*_X & -i\epsilon\alpha_3\delta^2|q|^2 \end{pmatrix} \\ & + \begin{pmatrix} \epsilon\alpha_3\delta^2 q^*_X q - \epsilon\alpha_3\delta^2 q_X q^* - i\alpha_1\delta^2|q|^2 & \epsilon\alpha_3\delta q_{XX} + i\alpha_1\delta q_X - 2\epsilon\delta^3\alpha_3|q|^2 q \\ \epsilon\alpha_3\delta q^*_{XX} - i\alpha_1\delta q^*_X - 2\epsilon\delta^3\alpha_3|q|^2 q^* & \epsilon\alpha_3\delta^2 q_X q^* - \epsilon\alpha_3\delta^2 q^*_X q + i\alpha_1\delta^2|q|^2 \end{pmatrix}. \end{aligned} \tag{20}$$

## 4. The bright soliton solutions for Eq. (14)

In this section, we propose the $N$-bright soliton solutions of Eq. (14) using the Riemann-Hilbert formulation (24–28). Let us consider Eq. (14) for localized solutions, i.e. assuming that potential function $q$ decay to zero sufficiently fast as $X, T \to \pm\infty$. In the Riemann-Hilbert formulation, we treat $\zeta$ as a fundamental matrix of the two linear equations in (15). From (15) we note that when $X, T \to \pm\infty$, one has $\zeta = e^{-i\lambda\Lambda X + (4i\alpha_3\epsilon\lambda^3 - 2i\lambda^2\alpha_1)\Lambda T}$ with $\Lambda = \mathrm{diag}(1, -1)$. This motivates us to introduce the variable transformation

$$\zeta = J e^{-i\lambda\Lambda X + (4i\alpha_3\epsilon\lambda^3 - 2i\lambda^2\alpha_1)\Lambda T}, \tag{21}$$

where $J$ is $(X, T)$-independent at infinity. Inserting (21) into (15) with (16)-(17), we get

$$J_X = -i\lambda[\Lambda, J] + \delta Q J, \tag{22a}$$

$$J_T = -(2i\alpha_1\lambda^2 - 4i\alpha_3\epsilon\lambda^3)[\Lambda, J] + VJ, \tag{22b}$$

with

$$Q = \begin{pmatrix} 0 & q \\ -q^* & 0 \end{pmatrix}, \quad V = (2\lambda\alpha_1\delta - 4\lambda^2\epsilon\alpha_3\delta)Q + 2\lambda\begin{pmatrix} -i\epsilon\alpha_3\delta^2|q|^2 & -i\epsilon\alpha_3\delta q_X \\ -i\epsilon\alpha_3\delta q_X^* & i\epsilon\alpha_3\delta^2|q|^2 \end{pmatrix}$$
$$+\begin{pmatrix} \epsilon\alpha_3\delta^2 q_X q^* - \epsilon\alpha_3\delta^2 q_X^* q + i\alpha_1\delta^2|q|^2 & \epsilon\alpha_3\delta q_{XX} + i\alpha_1\delta q_X + 2\epsilon\delta^3\alpha_3|q|^2 q \\ i\alpha_1\delta q_X^* - \epsilon\alpha_3\delta q_{XX}^* - 2\epsilon\delta^3\alpha_3|q|^2 q^* & \epsilon\alpha_3\delta^2 q_X^* q - \epsilon\alpha_3\delta^2 q_X q^* - i\alpha_1\delta^2|q|^2 \end{pmatrix}.$$

Here $[\Lambda, J] = \Lambda J - J\Lambda$ is the commutator, $\mathrm{tr}(Q) = \mathrm{tr}(V) = 0$ and

$$Q^\dagger = -Q, \quad V^\dagger = -V, \tag{23}$$

where $\dagger$ represents the Hermitian of a matrix.
In what folows, we consider the scattering problem of the Eq. (22a). By doing so, the variable $T$ is fixed and is a dummy variable. We first introduce the matrix Jost solutions $J_\pm(X, \lambda)$ of (22a) with the asymptotic condition

$$J_\pm \to I, \quad \text{when } X \to \pm\infty, \tag{24}$$

where $I$ is a $2 \times 2$ unit matrix. Here the subscripts in $J_\pm$ refer to which end of the $X$-axis the boundary conditions are set. Then due to $\mathrm{tr}(Q) = 0$ and Abel's formula we have $\det(J_\pm) = 1$ for all $X$. Next we denote $E = e^{-i\lambda\Lambda X}$. Since $\Psi \equiv J_+ E$ and $\Phi \equiv J_- E$ are both solutions of the first equation in (15), they must be linearly related, i.e.

$$J_- E = J_+ E S(\lambda), \qquad \lambda \in \mathbb{R} \tag{25}$$

where

$$S(\lambda) = \begin{pmatrix} s_{11} & s_{12} \\ s_{21} & s_{22} \end{pmatrix}, \qquad \lambda \in \mathbb{R}$$

is the scattering matrix, and $\mathbb{R}$ is the set of real numbers. Notice that $\det(S(\lambda)) = 1$ since $\det(J_\pm) = 1$. If we denote $(\Phi, \Psi)$ as a collection of columns,

$$\Phi = [\phi_1, \phi_2], \quad \Psi = [\psi_1, \psi_2], \tag{26}$$

by using the same formulation as (24; 25; 27), we have the Jost solution

$$P^{+} = [\phi_1, \psi_2]e^{i\lambda\Lambda X} = J_{-}H_1 + J_{+}H_2, \tag{27}$$

is analytic in $\lambda \in C_{+}$, and Jost solution

$$P^{-} = e^{-i\lambda\Lambda X}\begin{bmatrix}\hat{\phi}_1 \\ \hat{\psi}_2\end{bmatrix} = H_1 J_{-}^{-1} + H_2 J_{+}^{-1}, \tag{28}$$

is analytic in $\lambda \in C_{-}$, with

$$\Phi^{-1} = \begin{bmatrix}\hat{\phi}_1 \\ \hat{\phi}_2\end{bmatrix}, \quad \Psi^{-1} = \begin{bmatrix}\hat{\psi}_1 \\ \hat{\psi}_2\end{bmatrix},$$

and

$$H_1 = \mathrm{diag}(1,0), \qquad H_2 = \mathrm{diag}(0,1).$$

In addition, it is easy to see that

$$P^{+}(X,\lambda) \to I, \quad \text{as } \lambda \in \mathbb{C}_{+} \to \infty, \tag{29}$$

and

$$P^{-}(X,\lambda) \to I, \quad \text{as } \lambda \in \mathbb{C}_{-} \to \infty. \tag{30}$$

In addition, if we express $S^{-1}$ as

$$S^{-1} = \begin{pmatrix}\hat{s}_{11} & \hat{s}_{12} \\ \hat{s}_{21} & \hat{s}_{22}\end{pmatrix}, \qquad \lambda \in \mathbb{R},$$

from $\det(S(\lambda)) = 1$ we have

$$\hat{s}_{11} = s_{22}, \;\; \hat{s}_{22} = s_{11}, \;\; \hat{s}_{12} = -s_{12}, \;\; \hat{s}_{21} = -s_{21}. \tag{31}$$

Hence we have constructed two matrix functions $P^{+}$ and $P^{-}$ which are analytic in $C_{+}$ and $C_{-}$, respectively. On the real line, using Eqs. (25), (27) and (28), it is easily to see that

$$P^{-}(X,\lambda)P^{+}(X,\lambda) = G(X,\lambda), \qquad \lambda \in \mathbb{R}, \tag{32}$$

with

$$G = E(H_1 + H_2 S)(H_1 + S^{-1}H_2)E^{-1} = E\begin{pmatrix}1 & \hat{s}_{12} \\ s_{21} & 1\end{pmatrix}E^{-1}.$$

This determines a matrix Riemann-Hilbert problem with asymptotics

$$P^{\pm}(X,\lambda) \to I, \quad \text{as } \lambda \to \infty, \tag{33}$$

which provide the canonical normalization condition for this Riemann-Hilbert problem. If this problem can be solved, one can readily reconstruct the potential $q(X,T)$ as follows. Notice that $P^{+}$ is the solution of the spectral problem (22a). Thus if we expand $P^{+}$ at large $\lambda$ as

$$P^{+}(X,\lambda) = I + \frac{1}{\lambda}P_1^{+}(X) + O(\lambda^{-2}), \quad \lambda \to \infty, \tag{34}$$

and inserting this expansion into (22a), then comparing $O(1)$ terms in (34), we find that

$$\delta Q = i[\Lambda, P_1^{+}] = \begin{pmatrix}0 & 2iP_{12} \\ -2iP_{21} & 0\end{pmatrix}. \tag{35}$$

Thus, recalling the definition of $Q$ the potentials $q$ is reconstructed immediately as

$$q = 2iP_{12}/\delta, \tag{36}$$

where $P_1^+ = (P_{ij})$. In addition, from the definitions of $P^+, P^-$ and Eq. (25) we have

$$\det P^+ = \hat{s}_{22} = s_{11}, \quad \det P^- = s_{22} = \hat{s}_{11}. \tag{37}$$

The symmetry properties of the potential $Q$ and $V$ in (23) give rise to symmetry properties in the scattering matrix as well as in the Jost functions. In fact, after some computation we have $J_\pm$ satisfies the involution property

$$J_\pm^\dagger(X, \lambda^*) = J_\pm^{-1}(X, \lambda), \tag{38}$$

analytic solutions $P^\pm$ satisfy the involution property

$$(P^+)^\dagger(\lambda^*) = P^-(\lambda), \tag{39}$$

and $S$ satisfies the involution property

$$S^\dagger(\lambda^*) = S^{-1}(\lambda). \tag{40}$$

Let $\lambda_k$ and $\bar{\lambda}_k$ are zero points of $\det P^+$ and $\det P^-$, respectively. We see from (37) that $(\lambda_k, \bar{\lambda}_k)$ are zeros of the scattering coefficients $\hat{s}_{22}(\lambda)$ and $s_{22}(\lambda)$. Due to the above involution property, we have the symmetry relation

$$\bar{\lambda}_k = \lambda_k^*. \tag{41}$$

For simplicity, we assume that all zeros $\{(\lambda_k, \bar{\lambda}_k), k = 1, 2, \cdot, N\}$ are simple zeros of $\hat{s}_{22}(\lambda)$ and $s_{22}(\lambda)$, then each kernal of $P^+(\lambda_k)$ and $P^-(\bar{\lambda}_k)$ contains only a single column vector $v_k$ and row vector $\bar{v}_k$,

$$P^+(\lambda_k)v_k = 0, \quad \bar{v}_k P^-(\bar{\lambda}_k) = 0.$$

Taking the Hermitian of the above equations and using the involution properties, we have

$$\bar{v}_k = v_k^\dagger. \tag{42}$$

To obtain the soliton solutions, we set $G = I$ in (32). In this case, the solutions to this special Riemann-Hilbert problem have been derived in (25; 26) as

$$P_1^+(T, X, \lambda) = \sum_{j,k=1}^{N} v_j \left(M^{-1}\right)_{jk} \bar{v}_k, \tag{43}$$

where

$$M_{jk} = \frac{\bar{v}_j v_k}{\bar{\lambda}_j - \lambda_k}. \tag{44}$$

The zeros $\lambda_k$ and $\bar{\lambda}_k$ are $T$-independent. To find the spatial and temporal evolutions for vectors $v_k(T, X)$, we take the $X$-derivative to equation $P^+ v_k = 0$. By using (22a), one gets

$$P^+(X, \lambda_k)\left(\frac{\partial v_k}{\partial X} + i\lambda_k \Lambda v_k\right) = 0, \tag{45}$$

thus we have

$$\frac{dv_k}{dX} + i\lambda_k \Lambda v_k = 0. \tag{46}$$

Similarly, taking $T$-derivative to equation $P^+ v_k = 0$ and using (22b), one has

$$P^+(T, X, \lambda_k)(\frac{\partial v_k}{\partial T} + (2i\,\alpha_1 \lambda_k{}^2 - 4\,i\alpha_3\epsilon\lambda_k{}^3)v_k) = 0, \tag{47}$$

thus we have

$$\frac{\partial v_k}{\partial T} + (2i\,\alpha_1 \lambda_k{}^2 - 4\,i\alpha_3\epsilon\lambda_k{}^3)v_k = 0. \tag{48}$$

Solving (46) and (48) we get

$$v_k(T, X) = e^{-i\lambda_k \Lambda X + (4\,i\alpha_3\epsilon\lambda_k{}^3 - 2i\,\lambda_k{}^2\alpha_1)\Lambda T} v_{k0}, \tag{49a}$$

$$\bar{v}_k(T, X) = \bar{v}_{k0} e^{i\lambda_k \Lambda X + (-4\,i\alpha_3\epsilon\lambda_k{}^3 + 2i\,\lambda_k{}^2\alpha_1)}, \tag{49b}$$

where $(v_{k0}, \bar{v}_{k0})$ are constant vectors.

In summary, the $N$-bright soliton solutions to Eq. (14) are obtained from the analytical functions $P_1^+$ in (43) together with the potential reconstruction formula (36) as

$$q(T, X) = 2iP_{12}/\delta = 2i \left( \sum_{j,k=1}^{N} v_j \left(M^{-1}\right)_{jk} \bar{v}_k \right)_{12} /\delta, \tag{50}$$

where the vectors $v_j$ are given by (49). Without loss of generality, we take $v_{k0} = [b_k, 1]'$ with $b_k$ constants. And if we denote

$$\xi_k = -i\lambda_k X + (4\,i\alpha_3\epsilon\lambda_k{}^3 - 2i\,\lambda_k{}^2\alpha_1)T, \tag{51}$$

the general $N$-soliton solution to Eq. (14) can be written out explicitly as

$$q(T, X) = \frac{2i}{\delta} \sum_{j,k=1}^{N} b_j e^{\xi_j - \xi_k^*} (M^{-1})_{jk}, \tag{52}$$

with

$$M_{jk} = \frac{1}{\lambda_j^* - \lambda_k} \left( b_j^* c_k e^{\xi_k + \xi_j^*} - e^{-\xi_k - \xi_j^*} \right). \tag{53}$$

In what follows, we investigate the dynamics of the one-soliton and two-soliton solutions in Eqs. (14) in detail.

### 4.1 Examples of single and two bright solitons in Eq. (14)

To get the single bright soliton solution for Eq. (14), we set $N = 1$ in (52) to have

$$q(T, X) = \frac{2i(\lambda_1^* - \lambda_1)}{\delta} \frac{b_1 e^{\xi_1 - \xi_1^*}}{e^{-\xi_1 - \xi_1^*} + |b_1|^2 e^{\xi_1 + \xi_1^*}}. \tag{54}$$

If setting $\lambda_1 = \zeta_1 + i\eta_1$, $b_1 = e^{-2\eta_1 X_0 + i\omega_0}$, the single soliton solution (54) can be rewritten as

$$q(T, X) = \frac{2\eta_1}{\delta} \mathrm{sech}[2\,\eta_1(X + \left(4\,\alpha_3\epsilon\,\eta_1{}^2 + 4\,\alpha_1\zeta_1 - 12\,\alpha_3\epsilon\,\zeta_1{}^2\right) T - X_0)] \exp^{i\theta}, \tag{55}$$

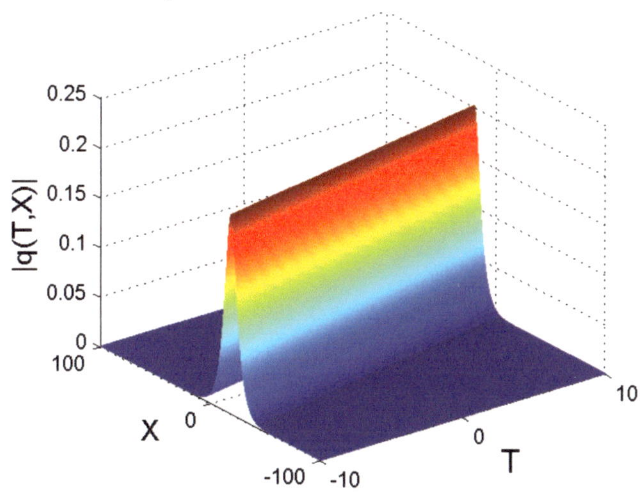


Fig. 1. (color online). Evolution of single soliton $|q(T,X)|$ in (55) with parameters (56). It is similar to single soliton in standard NLS equation.

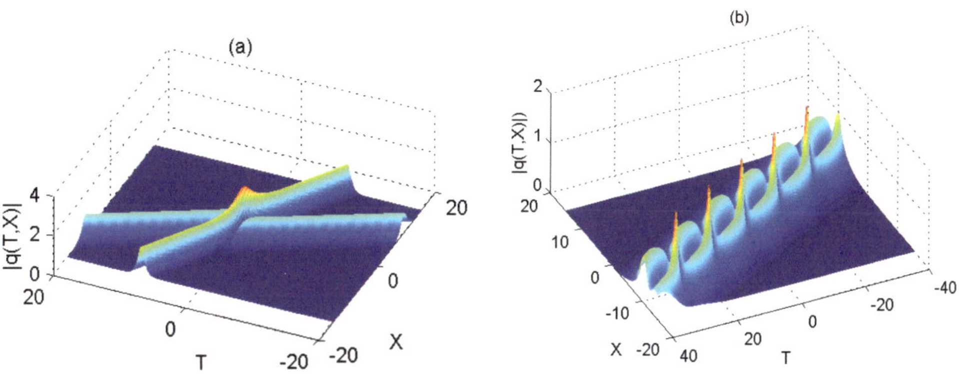


Fig. 2. (color online). The shapes of two-soliton solutions $|q(T,X)|$ in (52) with (53). (a) soliton collision with parameters (57); (b) bound state with parameters (58).

with $\theta = -2\zeta_1 X + \left(-4\,\alpha_1\zeta_1{}^2 + 4\,\alpha_1\eta_1{}^2 + 8\,\alpha_3\epsilon\,\zeta_1{}^3 - 24\,\alpha_3\epsilon\,\zeta_1\,\eta_1{}^2\right) T + \omega_0$, and $X_0, \omega_0$ are constants. This solution is similar to the solitary wave solution in the standard NLS equation (1). Its amplitude function has the shape of a hyperbolic secant with peak amplitude $2\eta_1/\delta$, and its velocity depends on several parameters, which is $12\,\alpha_3\epsilon\,\zeta_1{}^2 - 4\,\alpha_3\epsilon\,\eta_1{}^2 - 4\,\alpha_1\zeta_1$. The phase $\theta$ of this solution depends linearly both on space $X$ and time $T$. We show this single soliton solution in Fig. 1 with parameters

$$\zeta_1 = 0.5, \eta_1 = 0.1, X_0 = 1.5, \omega_0 = 2, \delta = 1, \alpha_1 = 0.5, \alpha_3 = 1, \epsilon = 1. \tag{56}$$

The two-soliton solution in Eq. (14) corresponds to $N = 2$ in the general $N$-soliton solution (52) with (53). This solution can also be written out explicitly, however, we prefer to showing

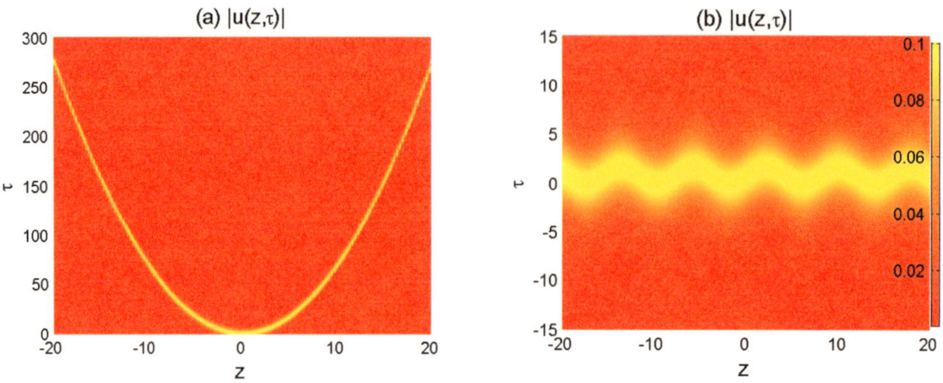


Fig. 3. (color online). Evolution of single soliton solutions $|u(z,\tau)|$ in HNLS equation (2) with controlable coefficients (59) and (62), respectively. (a) Soliton solution (61) with parameter (56) and $\rho_0 = 0.5, c = 1, k = 2$. (b) Soliton solution (64) with parameter (59) and $\rho_0 = 0.5, c = 1, k = 2$.

their behaviors by figures, see Fig. 2(a)-(b). Below we take $\lambda_1 = \zeta_1 + i\eta_1$ and $\lambda_2 = \zeta_2 + i\eta_2$ and examine this solution with various velocity parameters: one is $12\,\alpha_3\epsilon\,\zeta_1{}^2 - 4\,\alpha_3\epsilon\,\eta_1{}^2 - 4\,\alpha_1\zeta_1 = 12\,\alpha_3\epsilon\,\zeta_2{}^2 - 4\,\alpha_3\epsilon\,\eta_2{}^2 - 4\,\alpha_1\zeta_2$, i.e. the collision between two solitons, and the other is $12\,\alpha_3\epsilon\,\zeta_1{}^2 - 4\,\alpha_3\epsilon\,\eta_1{}^2 - 4\,\alpha_1\zeta_1 \neq 12\,\alpha_3\epsilon\,\zeta_2{}^2 - 4\,\alpha_3\epsilon\,\eta_2{}^2 - 4\,\alpha_1\zeta_2$, i.e. bound state. In Fig. 2(a), the two soliton parameters in Eq. (52) with (53) are

$$\alpha_1 = 0.5,\ \alpha_3 = 0.8,\ \epsilon = 1,\ \delta = 1,\ \lambda_1 = 0.2 + 0.7i,\ \lambda_2 = -0.1 + 0.5i,\ b_1 = 1,\ b_2 = 1. \tag{57}$$

Under these parameters, the velocity of the two solitons are different. It is observed that interactions between two soliton don't change the shape and velocity of the solitons, and there is no energy radiation emitted to the far field. Thus the interaction of these solitons is elastic, which is a remarkable property which signals that the HNLS equation (14) is integrable.
Fig. 2(b) displays a bound state in Eq. (14), and the soliton parameters here are

$$\alpha_1 = 0.5,\ \alpha_3 = 0.8,\ \epsilon = 1,\ \delta = 1,\ \lambda_1 = 0.3i,\ \lambda_2 = -0.1 + 0.4272i,\ b_1 = 1,\ b_2 = 1. \tag{58}$$

Under these parameters, the two constituent solitons have equal velocities, thus they will stay together to form a bound state which moves at the common speed. It can be seen that the width of this solution changes periodically with time, thus this solution is called breather soliton.

## 5. Dynamics of solitons in HNLS equation (2)

In what follows, we investigate the dynamic behavior of solitons in the variable-coefficients HNLS equation (2) with special soliton management parameters $d_j (j = 1, 2, 3, 4, 5, 6)$.

### 5.1 Single soliton solutions

We choose two cases of soliton management parameters $d_j (j = 1, 2, 3, 4, 5, 6)$ to study the dynamics of the single solitons in HNLS equation (2). Firstly, if we take the soliton management parameters to satisfy

$$d_1 = 1.6\ (k\alpha_1 - 3\,\alpha_3 c)\, z/k^3,\ d_2 = 1.6\ (\alpha_2 k - \alpha_4 c)\, z/\rho_0{}^2 k,$$

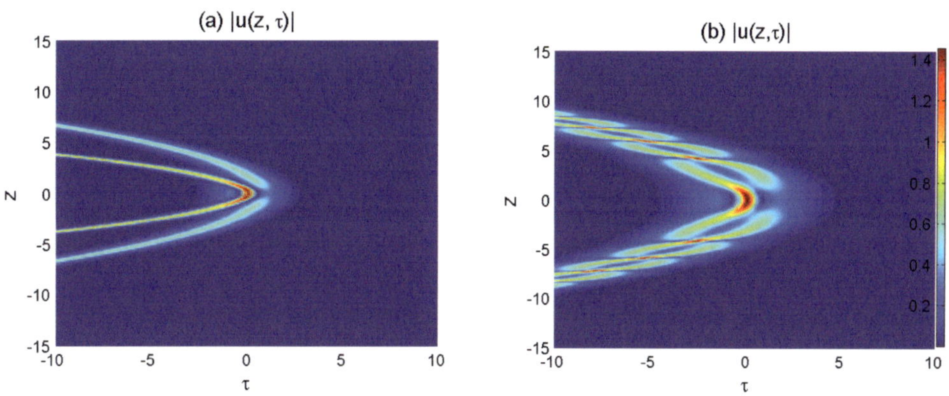


Fig. 4. (color online). The two-soliton solutions $|u(z,\tau)|$ in HNLS equation (2) with coefficients (59). (a) soliton collision with parameter (57) and $\rho_0 = 0.5, c = 1, k = 2$; (b) bound state with parameter (58) and $\rho_0 = 0.5, c = 1, k = 2$.

$$d_3 = 1.6\,\alpha_3 z/k^3,\; d_4 = 1.6\,\alpha_4 z/\rho_0{}^2 k,\; d_5 = 1.6\,\alpha_5 z/\rho_0{}^2 k,\; d_6 = 0, \tag{59}$$

the variables $\rho, T$ and $X$ in similarity transformation (3) are

$$\rho = \rho_0,\;\; T = 0.8\,z^2,\;\; X = k\tau + (2.4\,c^2\alpha_3 - 1.6\,ck\alpha_1)z^2/k^2. \tag{60}$$

So the single soliton solution in HNLS equation (2) with coefficients (59) is

$$u(z,\tau) = \rho_0 q\,(T,X)\,e^{ic(k^3\tau + 1.6\,c^2\alpha_3\,z^2 - 0.8\,ck\alpha_1 z^2)/k^3}, \tag{61}$$

where $q\,(T,X)$ satisfies Eq. (55) and $T, X$ satisfy Eq. (60).
Secondly, if we take the soliton management parameters to satisfy

$$d_1 = 0.8\,\cos{(0.8\,z)}\,(k\alpha_1 - 3\,\alpha_3 c)\,/k^3, d_2 = 0.8\,\cos{(0.8\,z)}\,(\alpha_2 k - \alpha_4 c)\,/\rho_0{}^2 k, d_6 = 0,$$

$$d_3 = 0.8\,\alpha_3\cos{(0.8\,z)}\,/k^3,\; d_4 = 0.8\,\alpha_4\cos{(0.8\,z)}\,/\rho_0{}^2 k,\; d_5 = 0.8\,\alpha_5\cos{(0.8\,z)}\,/\rho_0{}^2 k, \tag{62}$$

the variables $\rho, T$ and $X$ in similarity transformation (3) are

$$\rho = \rho_0,\;\; T = \sin{(0.8\,z)},\;\; X = k\tau + (3\,c^2\alpha_3 - 2\,ck\alpha_1)\sin{(0.8\,z)}\,/k^2. \tag{63}$$

In this case the single soliton solution in HNLS equation (2) with coefficients (62) is

$$u(z,\tau) = \rho_0 q\,(T,X)\,e^{ic(k^3\tau + 2\,c^2\alpha_3\sin(0.8\,z) - ck\alpha_1\sin(0.8\,z))/k^3}, \tag{64}$$

where $q\,(T,X)$ satisfies Eq. (55) and $T, X$ satisfy Eq. (63).
In Fig. 3, we show the single soliton solutions (61) and (64) in HNLS equation (2) with coefficients (59) and (62), respectively. Here the solution parameters are given in (56) and $\rho_0 = 0.5, c = 1, k = 2$. It is observed that when the soliton management parameters $d_j (j = 1,2,3,4,5)$ are linearly dependent on variable $z$ and $d_6 = 0$ (see Eq. (59)), the trajectory of the optical soliton is a localized parabolic curve, as shown in Fig. 3(a). When the soliton management parameters $d_j (j = 1,2,3,4,5)$ are periodically dependent on variable $z$ and $d_6 = 0$ (see Eq. (62)), the trajectory of the optical soliton is a periodical localized nonlinear wave, as shown in Fig. 3(b).

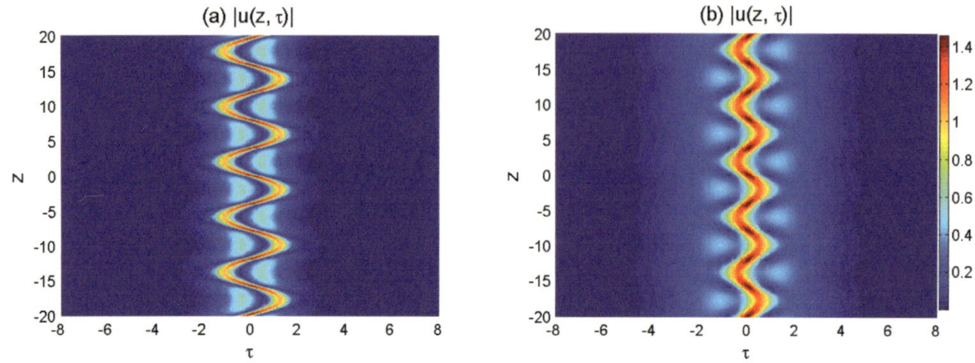


Fig. 5. (color online). The two-soliton solutions $|u(z,\tau)|$ in HNLS equation (2) with coefficients (62). (a) soliton collision with parameter (57) and $\rho_0 = 0.5, c = 1, k = 2$; (b) bound state with parameter (58) and $\rho_0 = 0.5, c = 1, k = 2$.

### 5.2 Collisions of the two-solitons

We now demonstrate various collision scenarios in HNLS equation (2) with coefficients (59) and (62), respectively. As in Section 4.1, we consider the two-soliton collisions and bound states in equation (2).

When the coefficients of equation (2) satisfies (59), its two-soliton solution is

$$u(z,\tau) = \rho_0 q\,(T,X)\, e^{ic(k^3\tau + 1.6\,c^2\alpha_3\, z^2 - 0.8\, ck\alpha_1 z^2)/k^3}, \tag{65}$$

where $T, X$ satisfy Eq. (60), and $q\,(T,X)$ satisfies Eq. (52) with (53) and $N = 2$.

When the coefficients of equation (2) satisfies (62), its two-soliton solution is

$$u(z,\tau) = \rho_0 q\,(T,X)\, e^{ic(k^3\tau + 2\,c^2\alpha_3 \sin(0.8\,z) - ck\alpha_1 \sin(0.8\,z))/k^3}, \tag{66}$$

where $T, X$ satisfy Eq. (63), and $q\,(T,X)$ satisfies Eq. (52) with (53) and $N = 2$.

In Fig. 4, we display the evolutions of the two-soliton solutions (65) in HNLS equation (2) with coefficients (59). Fig. 4(a) shows the soliton collision with parameter (57) and $\rho_0 = 0.5, c = 1, k = 2$, and Fig. 4(b) shows the bound state with parameter (58) and $\rho_0 = 0.5, c = 1, k = 2$. In Fig. 5, we display the evolutions of the two-soliton solutions (66) in HNLS equation (2) with coefficients (62). Fig. 5(a) shows the soliton collision with parameter (57) and $\rho_0 = 0.5, c = 1, k = 2$, and Fig. 5(b) shows the bound state with parameter (58) and $\rho_0 = 0.5, c = 1, k = 2$.

## 6. Conclusions

In summary, we have studied the variable-coefficient higher order nonlinear Schrödinger equation which describes the wave propagation in a nonlinear fiber medium with higher-order effects such as third order dispersion, self-steepening and stimulated Raman scattering. By means of similarity transformation, we first change this variable-coefficient equation into the constant-coefficient HNLS equation. Then we investigate the integrability of the constant-coefficient HNLS equation by prolongation technique and find two Lax integrable HNLS equations. The exact bright N-soliton solutions for the bright soliton version of HNLS equation are obtained using Riemann-Hilbert formulation. Finally, the dynamics of the optical solitons in both constant-coefficient and variable-coefficient HNLS equations is examined and the effects of higher-order effects on the velocity and shape of the optical soliton are observed. In addition, it is seen that the bright two-soliton solution of the HNLS equation behaves in an elastic manner characteristic of all soliton solutions.

## 7. Acknowledgments

This work was supported by NSFC under grant No. 11001263 and China Postdoctoral Science Foundation.

## 8. References

[1] K. E. Lonngren, Soliton experiments in plasmas, Plas. Phys. 25, 943 (1983).
[2] G.P. Agrawal, Nonlinear Fiber Optics, Academic Press, San Diego (1989).
[3] A. Hasegawa and Y. Kodama, Solitons in Optical Communications, Clarendon, Oxford, 1995.
[4] J. Denschlag, J. E. Simsarian, D. L. Feder, Charles W. Clark, L. A. Collins, J. Cubizolles1, L. Deng, E. W. Hagley, K. Helmerson, W. P. Reinhardt1, S. L. Rolston, B. I. Schneider and W. D. Phillips, Generating Solitons by Phase Engineering of a Bose-Einstein Condensate, Science, 287, 97-101 (2000).
[5] B. P. Anderson, P. C. Haljan, C. A. Regal, D. L. Feder, L. A. Collins, C. W. Clark, and E. A. Cornell, Watching Dark Solitons Decay into Vortex Rings in a Bose-Einstein Condensate, Phys. Rev. Lett. 86, 2926-2929 (2001).
[6] M. A. Helal, Soliton solution of some nonlinear partial differential equations and its applications in fluid mechanics, Chaos, Solitons and Fractals, 13, 1917-1929 (2002).
[7] A. Hasegawa and F. Tappert, Transmission of stationary nonlinear optical pulses in dispersive dielectric fibers. I. Anomalous dispersion, Appl. Phys. Lett. 23, 142-144 (1973); Transmission of stationary nonlinear optical pulses in dispersive dielectric fibers. II. Normal dispersion, Appl. Phys. Lett. 23, 171-172 (1973).
[8] L. F. Mollenauer, R. H. Stolen, and J. P. Gordon, Experimental observation of picosecond pulse narrowing and solitons in optical fibers, Phys. Rev. Lett. 45, 1095-1098 (1980).
[9] V. E. Zakharov and A. B. Shabat, Exact theory of two-dimensional self-focusing and one-dimensional self-modulation of manes in nonlinear media, Zh. Eksp. Teor. Fiz. 61, 118 (1971) (Sov. Phys. JETP 34, 62, 1972).
[10] V. E. Zakharov and A. B. Shabat, On interactions between solitons in a stable medium, Zh. Eksp. Teor. Fiz. 64, 1627 (1973) (Sov. Phys. JETP 37, 823, 1974).
[11] J. Yang, Nonlinear waves in integrable and non-integrable systems, Society for Industrial and Applied Mathematics, U.S. (2010).
[12] Y. Kodama and A. Hasegawa, Nonlinear pulse propagation in a monomode dielectric guide, IEEE J. Quantum Electron. 23, 510 (1987).
[13] K. Porsezian and K. Nakkeeran, Optical Solitons in Presence of Kerr Dispersion and Self-Frequency Shift, Phys. Rev. Lett. 76, 3955 (1996).
[14] K. Nakkeeran, K. Porsezian, P. Shanmugha Sundaram, and A. Mahalingam, Optical solitons in N-coupled higher order nonlinear Schrodinger equations, Phys. Rev. Lett. 80, 1425 (1998).
[15] E. Papaioannou, D. J. Frantzeskakis, K. Hizanidis, An analytical treatment of the effect of axial inhomogeneity on femtosecond solitary waves near the zero dispersion point, 32, 145-154 (1996).
[16] A. Mahalingam and K. Porsezian, Propagation of dark solitons with higher-order effects in optical fibers, Phys. Rev. E 64, 046608 (2001).

[17] Z. Y. Xu, L. Li, Z. H. Li, and G. S. Zhou, Modulation instability and solitons on a cw background in an optical fiber with higher-order effects, Phys. Rev. E 67, 026603 (2003).
[18] D. S. Wang, X. H. Hu, J. P. Hu, and W. M. Liu, Quantized quasi-two-dimensional Bose-Einstein condensates with spatially modulated nonlinearity, Phys. Rev. A 81, 025604 (2010).
[19] D. S. Wang, X. H. Hu, and W. M. Liu, Localized nonlinear matter waves in two-component Bose-Einstein condensates with time- and space-modulated nonlinearities, Phys. Rev. A 82, 023612 (2010).
[20] H.D. Wahlquist and F.B. Estabrook, Prolongation structures of nonlinear evolution equations. J. Math. Phys. 16, 1-7 (1975).
[21] F.B. Estabrook and H.D. Wahlquist, Prolongation structures of nonlinear evolution equations II, J.Math.Phys.17, 1293-1297 (1976 ).
[22] D. S. Wang, Integrability of the coupled KdV equations derived from two-layer fluids: Prolongation structures and Miura transformations, Nonlinear Analysis, 73, 270-281 (2010). Deng-Shan Wang
[23] R. Dodd and A. P. Fordy, The prolongation structure of quasi-polynomial flows Proc. R. Soc. London A 385, 389-429 (1983).
[24] V.S. Gerdjikov, Basic aspects of soliton theory, Sixth International Conference on Geometry, Integrability and Quantization June 3-10, 2004, Varna, Bulgaria Ivailo M. Mladenov and Allen C. Hirshfeld, Editors, Sofia 1-48 (2005).
[25] V. E. Zakharov, S. V. Manakov, S. P. Novikov, and L. P. Pitaevskii, *The Theory of Solitons: The Inverse Scattering Method*, Consultants Bureau, New York, 1984.
[26] V.S. Shchesnovich and J. Yang, General soliton matrices in the Riemann-Hilbert problem for integrable nonlinear equations. J. Math. Phys. 44, 4604 (2003).
[27] E.V. Doktorov, J. Wang, and J. Yang, Perturbation theory for bright spinor Bose-Einstein condensate solitons, Phys. Rev. A. 77, 043617 (2008).
[28] D. S. Wang, D. J. Zhang and J. Yang, Integrable properties of the general coupled nonlinear Schrodinger equations, J. Math. Phys. 51, 023510 (2010).

# 4

# Progress in Continuous-Wave Supercontinuum Generation

Arnaud Mussot and Alexandre Kudlinski
*Laboratoire PhLAM, IRCICA, Université Lille 1*
*France*

## 1. Introduction

Supercontinuum (SC) light sources are nowadays a very common way to access a large span of wavelengths, usually ranging from the near ultraviolet (around 400 nm) to the infrared (around 2.4 $\mu$m). It corresponds to a range of interest for many applications in optics for measuring transmission, dispersion, or in biophotonics for achieving fluorescence microscopy, optical coherence tomography... Indeed these sources are really promising because they should allow to replace the $N$ laser sources used in these experimental setups to access to all theses wavelengths by a single broad one and a spectral filtering apparatus. Most of these results have been obtained by using powerful pump lasers of several kilo-Watts peak power, operating from the femtosecond (Titane:Saphire) to the nanosecond regimes (Nd:YAG), launched in the low dispersion region of a microstructured optical fiber. Although these fibers are short enough (typically from 1 to 10 m) to neglect the linear absorption during the propagation of the pump, the spectral power density is relatively low (few hundreds of $\mu$w/nm) which could limit the implementation of SC sources in many application devices. This is related to a technological limitation of the pump source because it is not easy to combine strong peak power and high average power. One of the simplest solutions to increase the the spectral power density of SC sources is to replace pulsed sources with continuous-wave (CW) light sources whose available average powers are much more important. We will see that the dynamics of SC formation is considerably different in this case, requiring to perform intensive numerical studies to optimize the fiber parameters. Indeed, longer fibers are required (from tens to hundreds of meters) which heightens sensitivity to fiber attenuation, namely of the OH pic absorption, that strongly impacts the soliton evolution. However extremely powerful SCs have been reported with more than 10 mW/nm of spectral power density. Furthermore, these pump sources are usually all-fiber that leads to a second advantage against most of pulsed SC because CW pump can be directly spliced on the PCF. It is also important to point out that these SC sources have different temporal properties than the ones of pulsed SCs.

The first experimental demonstration of CW SC have been realized at the end of the nineties with a Raman laser launched in a standard telecommunication fiber. The spectral broadening was relatively restricted (around 200 nm) because it was mainly due to Raman effect Gonzalez-Herraez et al. (2003); Persephonis et al. (1996); Prabhu et al. (2000). A breakthrough was reached a few years later when stronger pump lasers (from more than one order of magnitude) based on Ytterbium doped fibers were combined with photonic crystal fibers

(PCFs) owing a low group-velocity dispersion (GVD) value around the pump wavelength Avdokhin et al. (2003). With these setups, SC generation was mainly due to solitonic effects like in pulsed SC. A renew of interest for these sources started from 2007 where first numerical demonstrations of very broad CW SC were reportedMussot et al. (2007), just followed by experimental demonstrationsCumberland et al. (2008a); Kudlinski & Mussot (2008).
As an example, a typical and simple experimental setup used for SC generation is schematized in Fig. 1. In our experimental configuration used for the experiments hereafter, the PCFs were

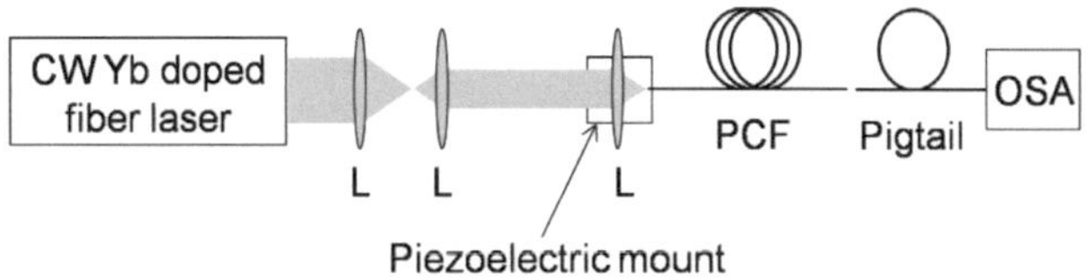


Fig. 1. Scheme of the experimental setup used for the SC generation experiments. L : lens.

pumped with ytterbium-doped fiber laser delivering either 20 W at 1064 nm with a full width at half maximum of 0.5 nm, or 50 W or 100 W at 1070 nm with a full width at half maximum of 1 nm. The output beam diameter of the laser was reduced with an afocal setup and the beam was then launched into the fiber with appropriate aspheric lenses (of a few mm focal length). All lenses were antireflection coated and a heat dissipater was carefully placed on top of the V-groove supporting the PCF in order to manage thermal issues caused by the high power laser. This allowed to greatly improve the temporal stability of the injection setup and no noticeable change in coupling efficiency was observed for several tens of minutes at full pump power. The coupling efficiency in these conditions was typically 70%–80%. The output of the fiber was butt-coupled to a pigtail to reduce the power launched inside the optical spectrum analyzer (OSA). All-fiber schemes are also used Cumberland et al. (2008a) but splicing issues are usually more -time-consuming for a lab experiment than free-space coupling.

## 2. Basic mechanisms of continuous-wave supercontinuum generation

Mechanisms at the origin of SC are now well knownCumberland et al. (2008b); Dudley et al. (2006); Kobtsev & Smirnov (2005); Mussot et al. (2007); Travers et al. (2008); Vanholsbeeck et al. (2005). The modulationnal instability (MI) process is at the origin of the formation of CW SC. It originates from the perfect balance between linear and nonlinear effects experienced by a strong field, the pump, and a small perturbation, the noise, when working in anomalous GVD region of an optical fiber. At the beginning of the fiber, this small perturbation is amplified. The typical signature of this process in the spectral domain is two symmetric side lobes located around the pump (Figs. 2-(a) and (b)). By further propagating into the fiber, this small periodic perturbation is amplified to become a train of solitonic pulses (Figs. 2-(c)and (d)). Note that these pulses have not identical characteristics as they originate from a process that is seeded by noise. On the other hand, we remind that in the case of a single soliton propagating in an optical fiber, it is well known that during its propagation, it is disturbed by higher order dispersion orders effect and as a consequence it shed energy to radiations called dispersive waves (DWs) which verify a phase matching condition. This leads to the generation of DWs on the short wavelength side of the soliton and to a red shift of the soliton, called *spectral recoil* for momentum conservation (Figs. 3-(a) and (b)). The consequence of the phase matching condition, is that solitons and DWs do not travel at the same velocity. If no additional effect is experienced by these waves, they will no longer interact again during their respective

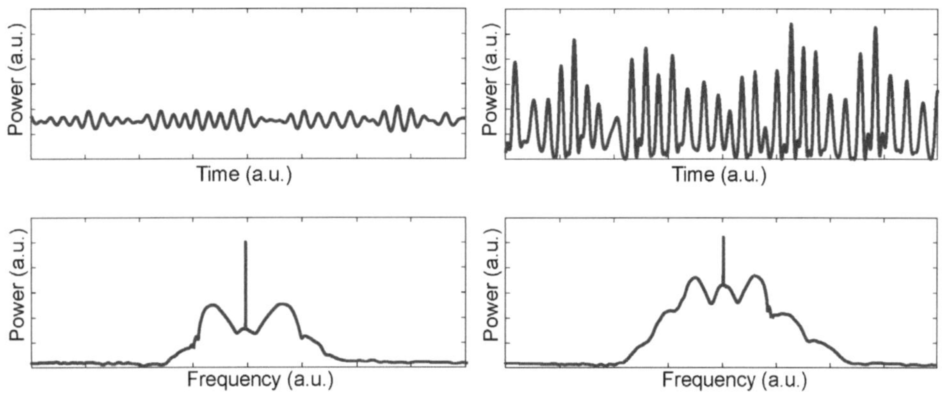


Fig. 2. Schemes illustrating the first steps of modulationnal instability, (a)-(b) at the beginning of the process when the small perturbation starts to grow and (c)-(d) when the solitonic train is created. (a) and (c) correspond to the time domain, and (b) and (d) correspond to the frequency domain.

propagation. However, during the propagation of solitons, because they are strong waves, the Raman effect induces an additional and continuous red shift of their central frequencies . This effect is called soliton self-frequency shift (SSFS)Gordon (1986) and decelerates the soliton because the group index increases with frequency in usual fibers (Fig. 3-(c)). As a consequence, as the velocity of DWs has not changed, the solitons can interact again with them via the cross-phase modulation effectGenty et al. (2004) which shifts DWs toward short wavelengths. Finally, it is important to understand that this group velocity matching is the rule that allows to connect lower and upper limits of SCs (Figs. 3-(d))Stone & Knight (2008). This is different in fibers with two zero dispersion wavelengths (ZDWs). When the soliton approaches the second ZDW, it still generates DWs but on the long wavelength side of the SCGenty et al. (2004); Mussot et al. (2007). The spectral recoil tends now to shift it on the opposite side that the one of the SSFS. An equilibrium is reached and the frequency shift of the soliton is cancelledSkryabin et al. (2003). In this case solitons and DWs will no interact together and no trapping mechanism will occur like it is the case in fiber with a single ZDW.

## 3. Bandwidth-limited near infrared continuous-wave supercontinuum

The interest of limiting the spectral extension of CW SC is to concentrate of the available power in the desired spectral span. This is achieved in fibers with two ZDWs in which the SSFS can be cancelled by the spectral recoil effect experienced by solitons located just below the second zero dispersion wavelength. By this way, increasing the power leads to an increase of the power spectral density.

### 3.1 Double zero-dispersion wavelength photonic crystal fibers

It is possible to design PCFs with two ZDWs from part to part of the pump wavelength at 1064 nm, and a low anomalous dispersion region at this wavelength. Such group-velocity dispersion (GVD) curves can be achieved with a relatively small hole-to-hole spacing $\Lambda$ in the order of 1.7 $\mu$m and a $d/\Lambda$ value in the range of 0.4–0.5 ($d$ is the hole diameter) Mussot et al. (2007); Tse et al. (2006). It is well known that a microstructured cladding with these geometrical properties would lead to relatively high confinement losses at wavelengths around 1.5 $\mu$m which is the reason why 10 periods of holes were necessary between the core

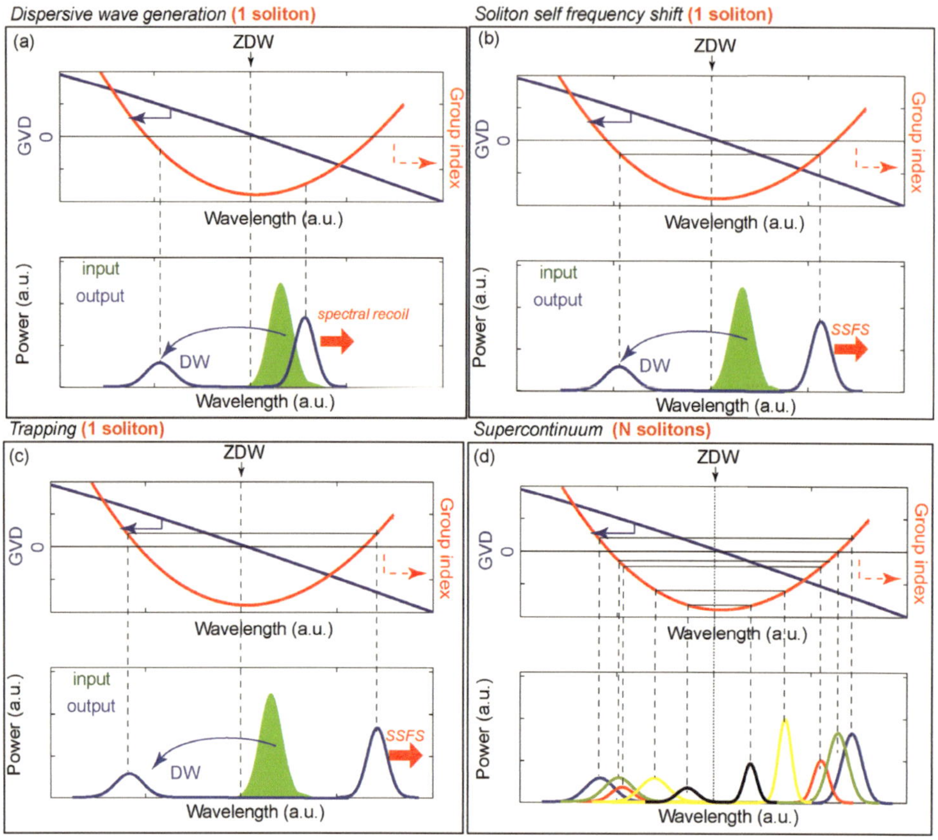


Fig. 3. Schemes illustrating (a) the DW generation from a single soliton, (b) the SSFS effect and (c) the trapping of teh DW by the soliton, *for a single soliton*. In each cases, the dispersion curve is represented in blue and the group index curve in red. (d) Schemes illustrating the formation of a CW SC, *involving N solitons*.

and the external jacket to decrease confinement losses to an acceptable values of a few dB/km at 1550 nm, at which the SC is expected to be generated. Figure 4(a) displays a scanning electron microscope (SEM) image of such a PCF (labelled fiber C in what follows, see Table 1). Another important issue in CW SC generation is the absorption of the water band centered at 1380 nm Cumberland et al. (2008a). We thus performed a chemical cleaning of the stacked preform under halogenic atmosphere to reduce surface contamination and to lower the water content. This allowed to decrease the peak attenuation at 1380 nm from typically 600 dB/km (without any special treatment) to about 120 dB/km. A typical attenuation spectrum is shown in Fig. 4(b) for fiber C, which SEM is represented in Fig. 4(a). The background loss being around 30 dB/km at 1380 nm, the contribution of the water contamination is about 90 dB/km at 1380 nm.

Three different PCF samples (labeled A, B and C) are investigated here. These fibers are characterized by slightly differing $\Lambda$ and $d/\Lambda$ values so that the GVD curve of each fiber is slightly different. The GVD curves are represented in Fig. 5(a), where the vertical line represents the pump wavelength. All GVD curves have been calculated with a finite-elements

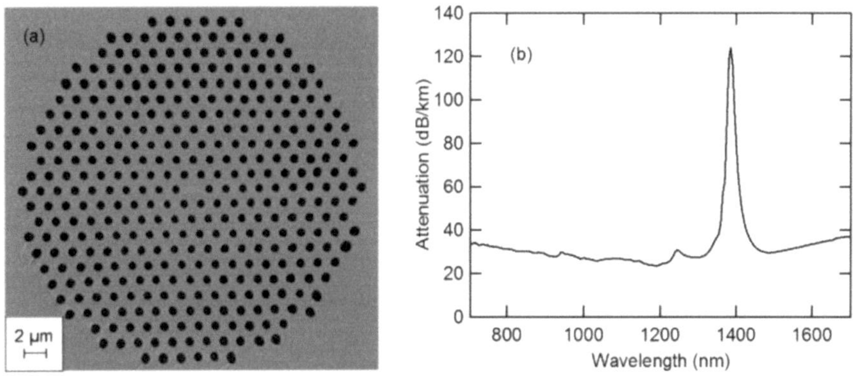


Fig. 4. (a) Typical SEM image of PCFs used for infrared CW supercontinuum. (b) Typical attenuation spectrum.

| Parameter | Fiber A | Fiber B | Fiber C |
|---|---|---|---|
| ZDW1 (nm) | 1012 | 958 | 903 |
| ZDW2 (nm) | 1236 | 1346 | 1570 |
| $\gamma$ at 1064 nm ($W^{-1}.km^{-1}$) | 22 | 24 | 30 |
| GVD at 1064 nm (ps/nm/km) | +4 | +12 | +30 |

Table 1. Parameters of the PCFs under investigation in this section.

method (FEM) from high resolution SEM images of the PCFs. Table 1 summarizes the properties of the three fibers under investigation.

### 3.2 Control of the supercontinuum long-wavelength edge

The experiments reported in this section were performed in 100 m-long samples of each PCF described in Table 1. In the launching conditions described in the previous paragraph, the output power were respectively 7.35 W, 7.25 W and 7.16 W for fibers A to C, with a 20 W CW fiber laser at 1064 nm. Figure 5(b) shows the spectra obtained in all fibers. The green curve corresponds to fiber A, with the closest ZDWs. As expected from Ref. Mussot et al. (2007), the spectral width is limited by the second ZDW. The dispersion value at the 1064 nm pump wavelength is very low (+4 ps/nm/km). The spectral broadening is thus initially dominated by MI, with the anti-Stokes MI sideband overlapping with the normal GVD region. The short wavelength extension just below the first ZDW results from a spectral overlap of MI sidebands and blue-shifted dispersive wave in the normal GVD region, as analyzed in Cumberland et al. (2008b). Since the spectral position of the Stokes MI sidebands is just below the second ZDW of the fiber, the SSFS is very short and the spectrum remains consequently quite symmetric. The power generated above the second ZDW (depicted by a vertical line) is attributed to the generation of red-shifted dispersive waves accompanying the cancelation of the SSFS. The red curve in Fig. 5(b) corresponds to the spectrum measured for fiber B, with both ZDWs separated by about 400 nm. In this fiber, the long-wavelength ZDW is very close to the center of the water absorption band (1380 nm). It is well known that the SSFS is canceled if a second ZDW is present at longer wavelengths Skryabin et al. (2003), which is the case here. This is seen in the red spectrum of Fig. 5(b) as a peak centered at 1310 nm, which corresponds

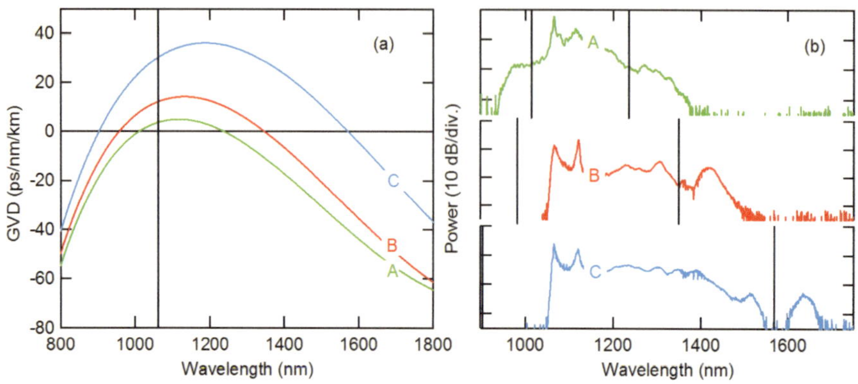


Fig. 5. (a) GVD curves of the three PCFs under investigation in this section. The vertical line depicts the pump wavelength. (b) Corresponding output spectra recorded for a fiber length of 100 m and a pump power of 12.5 W launched into the PCFs. Vertical lines represent the ZDWs.

to an accumulation of all solitons stopped by the second ZDW. Additionally, a large part of energy is transferred to a dispersive wave centered at 1420 nm, which is phase-matched with the solitons. In fiber C displayed in Fig. 5(b) in blue, the flatness of the spectrum is clearly affected by the water absorption at 1380 nm. This can be seen as a spectral decrease at wavelengths higher than 1380 nm. The higher energy solitons are able to tunnel through the water attenuation band and are then stopped by the second ZDW located at 1570 nm in fiber C. The less energetic ones stop just below 1380 nm due to the water absorption peak. As in fiber B, an accumulation of solitons is seen just before the second ZDW, and a dispersive wave which is phase-matched with the solitons is generated at 1630 nm. Note that the peak located around 1120 nm is due to Raman lasing because of Fresnel reflection at both fiber faces. It is also important to note that, unlike in fiber A, no short wavelength extension is observed in fibers B and C. Indeed, in these fibers, the amount of energy transferred from solitons to blue-shifted dispersive waves is negligible because there is no spectral overlap between solitons and dispersive waves Akhmediev & Karlsson (1995).

### 3.3 Dynamics of the supercontinuum formation

To go further into the detailed dynamics of SC formation, we performed a cut-back measurement on fiber C. The spectrum was measured every 5 m for fiber lengths between 0.5 and 100 m. The results are displayed in Fig. 6, where the output spectra are represented as a function of fiber length. The solitonic waves created by MI are progressively red-shifted by SSFS during the first 30 m of propagation. They are then stopped by the second ZDW located at 1570 nm (depicted by the white dotted vertical line). The soliton build-up due to spectral recoil before the second ZDW can be seen as an increase in spectral power. The red-shifted dispersive wave is also observed from this propagation length of 30 m. For more important fiber lengths, the spectrum extension remain almost constant. The experimental results displayed in Figs. 5 and 6 illustrate the possibility of tailoring the spectrum extent in the context of multi-watt and relatively flat SC generation. The long-wavelength edge of the spectrum is limited by red-shifted dispersive waves, whose spectral location is imposed by the second ZDW.

## 4. Extension towards visible wavelengths

Another long term issue of CW-pumped SC concerns the lack of short wavelengths generation when pumping at 1 $\mu$m. The generation of visible wavelengths would be of great interest for a substantial number of applications including high resolution imaging, metrology or spectroscopy. One possible approach to achieve this is to take advantage of the process of dispersive wave trapping by solitons Nishizawa & Goto (2002). This process leads to an extra blue shift of dispersive waves in the spectral domain Genty et al. (2004; 2005); Gorbach & Skryabin (2007a;b); Gorbach et al. (2006); Travers (2009); Travers & Taylor (2009). Experimentally, this phenomenon has been proved to be of primary importance to generate short wavelengths in pulsed pumping regime Stone & Knight (2008). It has also been combined with dispersion-engineered PCFs to further extend SC to the UV in nanosecond and picosecond pumping schemes Kudlinski et al. (2006). This idea consists in modifying the dispersion curve along the fiber so that group-velocity matching conditions for trapped dispersive waves continuously evolve along propagation. This leads to the generation of new wavelengths as the ZDW decreases along propagation. The present work is based on this idea which has been adapted to CW pumping conditions.

### 4.1 Zero-dispersion wavelength decreasing photonic crystal fibers

The dispersion-engineered PCF firstly used within this framework consists of a 100 m-long section with a constant dispersion followed by a 100 m-long section with decreasing ZDW, as illustrated in Fig.7(c). The total attenuation of the 200 m-long PCF is 1.5 dB at 1064 nm. A SEM image of the input and output faces of the PCF is represented in Figs. 7(a) and (b) respectively, with the same scale. The input outer diameter is 125 $\mu$m, the hole-to-hole spacing $\Lambda$ is 4.7 $\mu$m and the hole diameter $d$ is 2.6 $\mu$m. The dispersion curves at the PCF input and output have been computed with a finite elements method from high resolution SEMs and are represented

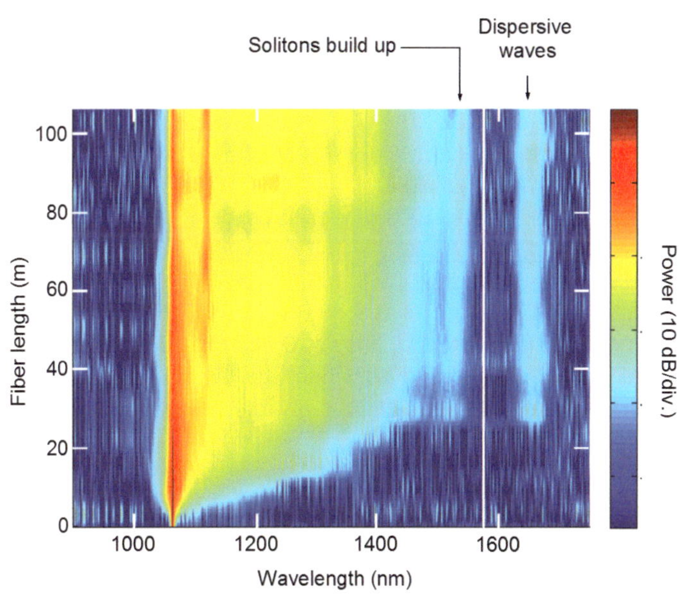


Fig. 6. Experimental measurement of the SC dynamics as a function of fiber length, in fiber C. The white vertical line represent the second ZDW.

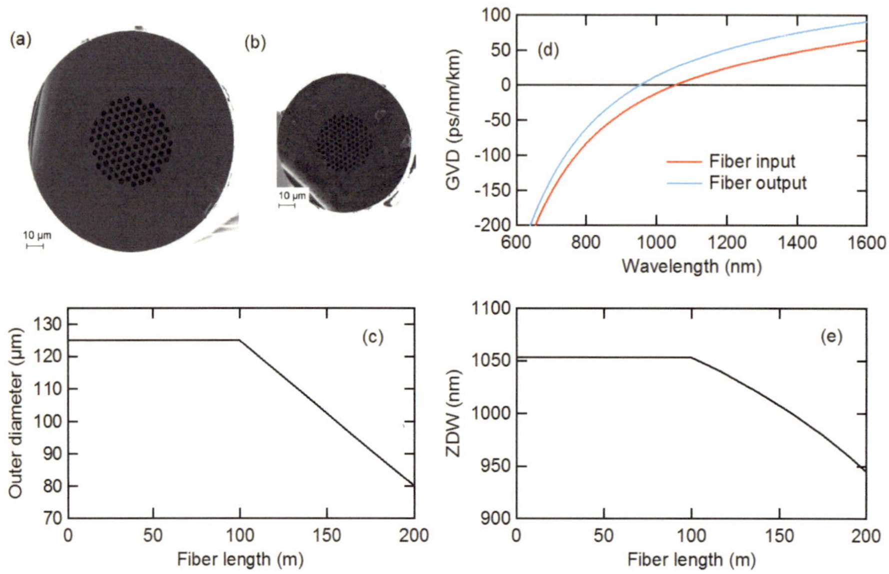


Fig. 7. (a),(b) SEMs of the input and output faces of the ZDW decreasing fiber. Respective outer diameters are 125 and 80 $\mu$m. (c) Outer diameter versus fiber length. (d) GVD curves at the input (red line) and output (blue line) of the ZDW decreasing fiber. (e) ZDW versus fiber length.

in Fig. 7(d) in red and blue lines, respectively. The input ZDW is located at 1053 nm, just below the pump wavelength of 1064 nm. To decrease the size of the microstructure along propagation and consequently shift the ZDW toward shorter wavelengths, the outer diameter of the fiber has been approximately linearly reduced to a final diameter of 80 $\mu$m (see Fig. 7(c)). This was done by gradually increasing the drawing speed during the fiber fabrication whilst keeping the preform feed rate constant. The pitch $\Lambda$ at the fiber output was reduced to 3.1 $\mu$m and the $d/\Lambda$ ratio was kept constant along the whole PCF, so that the output ZDW is shifted down to 950 nm. The longitudinal evolution of the ZDW of the PCF is represented in Fig. 7(e). In the first 100 m, the ZDW is fixed to 1053 nm, and it drops to 950 nm in a quasi-linear way along the last 100 m.

#### 4.2 Generation of visible light

The setup used to pump the fabricated PCF is shown in Fig. 1. The beam from a 20 W CW fiber laser at 1064 nm was collimated and launched into the fiber with a lens of 4.5 mm focal length. The coupling efficiency was 75%, corresponding to a power of 13.5 W launched into the PCFs.

The SC spectrum measured at full pump power in the ZDW-decreasing PCF described above displayed in Fig. 8 in red line. It is ranging from 670 nm to 1350 nm with an output power of 9.5 W. Additional spectral components are located around 550 nm and the visible part of the SC was easily observable with naked eye at the PCFs output. The inset of Fig 8 displays a far-field image of the whole visible spot observable at the PCF output. Since the modal distribution of the SC does not look single mode, the far-field profile was investigated as a function of wavelength by using 10 nm bandpass filters. Right insets of Fig. 8 show far-field

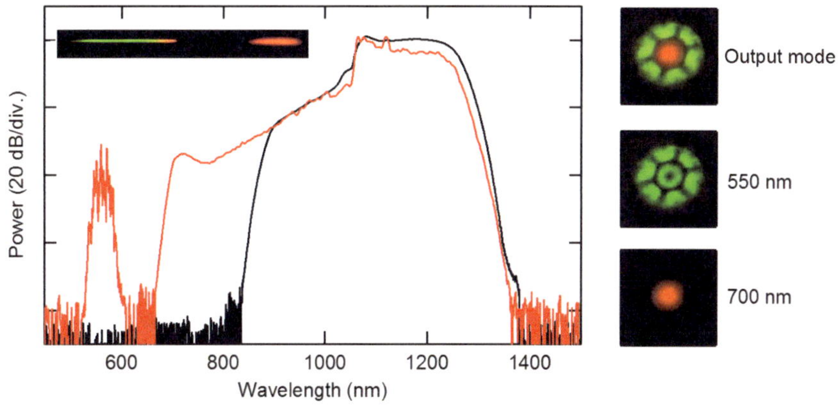


Fig. 8. Output spectra obtained in the uniform PCF (black line) and in the ZDW-decreasing one (red line). Inset: output beam dispersed by a prism. Right: output far-field without any filter (top), with a 550 nm filter (center) and with a 700 nm filter (bottom).

images centered at 550 nm and 700 nm . The green spectral components located around 550 nm are clearly generated in higher-order modes. This part of the spectrum was not expected from our design and is probably due to a phase-matching condition satisfied with higher-order modes Efimov et al. (2003); Omenetto et al. (2001). The far-field image recorded at 700 nm shows that the red spectral components are generated in a fundamental mode. We also checked that the mode was fundamental-like in the whole spectrum above 700 nm for all fibers.

A uniform PCF with dispersion comparable to the ZDW-decreasing fiber input was used for comparison. The output spectrum obtained in the same conditions than above is displayed in Fig. 8, in grey line. It extends up to 1355 nm into the infrared, which is very similar to the spectrum obtained in the ZDW-decreasing PCF. However, the short-wavelength edge is located at about 840 nm, which is much less spectacular than in in the ZDW-decreasing PCF where it reaches 670 nm. This clearly shows that the extra 170 nm bandwidth toward the visible is generated thanks to the decreasing ZDW.

In order to generate even shorter wavelengths, the ZDW-decreasing PCF has been pumped with a more powerful Yb fiber laser delivering 50 W at 1070 nm with a full width at half maximum of 1 nm. With the same setup as described above, we were able to launch a maximum power of 35 W in the fiber, corresponding to a coupling efficiency of 70 %. The resulting experimental spectrum is displayed in Fig. 9. For the highest pump power of 35 W, the SC is ranging from 650 nm to 1380 nm with a 19.5 W output power.

### 4.3 Discussion and numerical modelling

As claimed above, the basics of using a ZDW decreasing PCF to extend the spectrum towards short wavelengths was to use progressively red-shifted solitons to trap dispersive waves in the visible. The results displayed in Fig. 8 indeed suggest that the long- and short-wavelength edges of the spectra are correlated. In order to have a further insight into the mechanisms of the visible SC formation, the power dynamics of the spectral broadening has been investigated in the ZDW-decreasing PCF for launched powers of 8.2 W, 11.3 W and 13.5 W (see Ref. **?**). As expected, a broadening of the output spectrum occurs on both sides with increasing

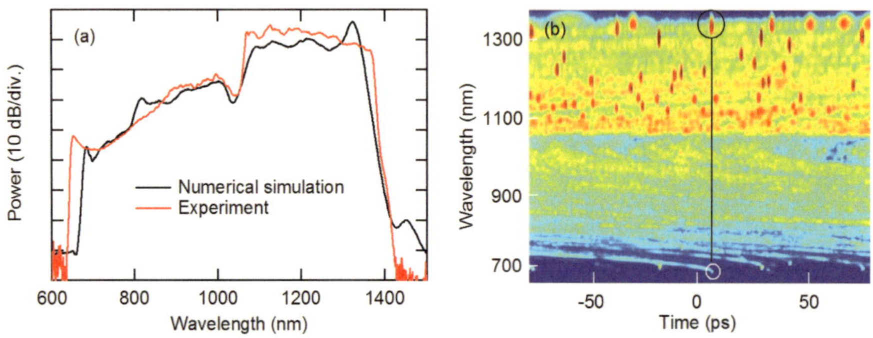


Fig. 9. Experimental (red line) and numerical (black line) output spectra obtained in the ZDW-decreasing PCF for a launched pump power of 35 W. The average power at the fiber output is 19.5 W. Spectrogram of a single-shot simulation at the output of the 200 m-long ZDW-decreasing PCF. The vertical line joins a soliton and its corresponding trapped dispersive wave. The color scale ranges over 30 dB.

launch power inside the fiber. The long and short wavelengths sides were identified by measuring the wavelength of typical spectral features on both edges of the spectrum. The spectral broadening on the long-wavelength side stops at 1140, 1160 and 1250 nm for respective increasing pump powers. The short-wavelength edge is progressively blue-shifted with increasing pump powers and extends to respectively 763, 751 and 720 nm. These experimental results have been compared with the computed group-index curve calculated for the end (small diameter) of the ZDW-decreasing PCF. This is illustrated in Fig. 10, where the group-index curve is plotted as a function of wavelength. Markers represents the long and short wavelength edges experimentally measured for launched powers of 8.2 W (green), 11.3 W (blue) and 13.5 W (red) respectively. The corresponding points for a fixed power are joined by nearly horizontal lines on the plot, which means that these radiations travel at almost the same group velocity when they go out of the fiber. This provides a strong support to the process of group-velocity matching between the most blue- and red-shifted spectral components of each spectrum and this evidences the benefit of the ZDW-decreasing fiber for the generation of shorter wavelengths.

In order to get a deeper understanding of the nonlinear mechanisms originating the visible extension, we performed numerical simulations. We integrated the generalized nonlinear Schrödinger equation including the experimental attenuation curve and all experimental parameters. The numerical method used to model the pump laser was fully described in Ref. Mussot et al. (2007). The simulations were very time consuming (1 week on a standard PC) so we did not perform the usual averaging procedure Mussot et al. (2007) required to account for the experimental measurements. The output spectra resulting from a single simulation is plotted in black line in Fig. 9(a). By performing several other simulations we checked that there was no significant modification from simulation to simulation. The agreement between numerical simulations and experimental results is excellent in terms of shape and extension of the spectrum. It is important to note that numerical simulations have been performed without any free parameters. In both cases, the stop of the spectral broadening at the long-wavelength is due to a relatively high OH absorption around 1380 nm (measured to be about 300 dB/km). The slight discrepancy with experiments observed at short wavelengths (680 nm for the simulations against 650 nm for the experiments) is

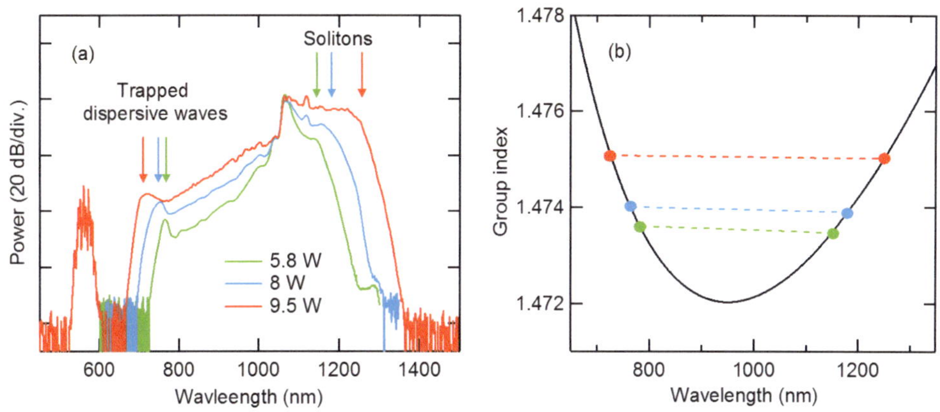


Fig. 10. (a) Output spectra recorded in the ZDW-decreasing PCF for pump powers of 8.2 W (green), 11.3 W (blue) and 13.5 W (red). Output powers are indicated on the graph. (b) Group-index curve calculated at the ZDW-decreasing PCF output (black line). Markers indicate extreme wavelengths of the corresponding SC spectrum experimentally recorded.

probably due to the uncertainty of the calculated GVD curve, leading to a slightly different group-velocity matching condition between red-shifted solitons and trapped dispersive waves. To further illustrate the trapping mechanism responsible for the short wavelength part of the spectrum, the numerical spectrogram of the optical field at the output of the fiber is represented in Fig. 9(b). It corresponds to the numerical spectrum displayed in black line in Fig. 9(a). In the spectrogram, one can see a whole spectral region full of solitons (represented as red dots) originating from the initial MI process. This region extends from the pump wavelength (1070 nm) to the upper limit of the spectrum. Some of the solitons are close to the pump wavelength and some other ones exhibit an important red-shift due to SSFS. The most shifted ones are stopped by the important OH absorption peak at 1380 nm. The region between 680 nm and the pump wavelength corresponds to dispersive waves generated from solitons Travers (2009); Travers et al. (2008). For the most red-shifted solitons, a blue-shifted trapped dispersive wave can be observed just below 700 nm, both travelling at the same group-velocity. An example of a soliton group-velocity matched with a dispersive wave is highlighted in Fig. **??**, where both are joined by a black line. We can see that the trapped dispersive wave is exactly at the vertical of the soliton which confirms that both waves travel at the same velocity inside the fiber. It thus confirms that the extension of the SC towards short wavelengths is mainly due to the trapping of dispersive waves by red-shifted solitons Nishizawa & Goto (2002) rather than by the basic dispersive wave generation process Akhmediev & Karlsson (1995). It should be noted then that the generation of even shorter wavelengths must be possible with the mechanism of dispersive waves trapping by reducing the OH absorption peak, which can be achieved by a careful cleaning treatment during the fiber fabrication process, and/or by enhancing the fiber nonlinearity.

## 5. White-light continuous-wave supercontinuum

### 5.1 Benefit of $GeO_2$ doping

As explained above, CW SC generation is intimately linked to the propagation of fundamental solitons generated from MI. In order to optimize the SC bandwidth, it is thus necessary to optimize the soliton self-frequency shift effect. One of the most natural solitons to do that is to use $GeO_2$-doped fibers, because this doping is well known to enhance both Kerr and Raman nonlinearities. However, in order to be usable in CW SC generation experiments, it is important that the ZDW remains slightly lower than the pump wavelength. By adjusting the microstructured cladding properties, it is possible to find some designs with greatly enhance nonlinearity, and still controlled dispersion Barviau et al. (2011). Figure 11 illustrates this. The blue line in (a) correspond to the GVD curve and nonlinear coefficient of a pure silica PCF with a 1060 nm ZDW, and the red line corresponds to a $GeO_2$-doped PCF with a parabolic profile and a maximum refractive index difference of 20 mol.%. The microstructured cladding parameters have been adjusted so that the doped PCF has the same ZDW of 1060 nm, but in this case, the nonlinear coefficient is enhanced by a factor of about 4 at 1064 nm. Indeed, it reached about 38 $W^{-1}.km^{-1}$ in the $GeO_2$-doped PCF, against 10 $W^{-1}.km^{-1}$ in the pure silica one. Moreover, Fig. 11(b) shows the enhancement of the material Raman gain due to the presence of $GeO_2$ as compared to pure silica. A $GeO_2$-doped PCF corresponding to this design has thus been fabricated, and the pure silica PCF has also been used for comparison.

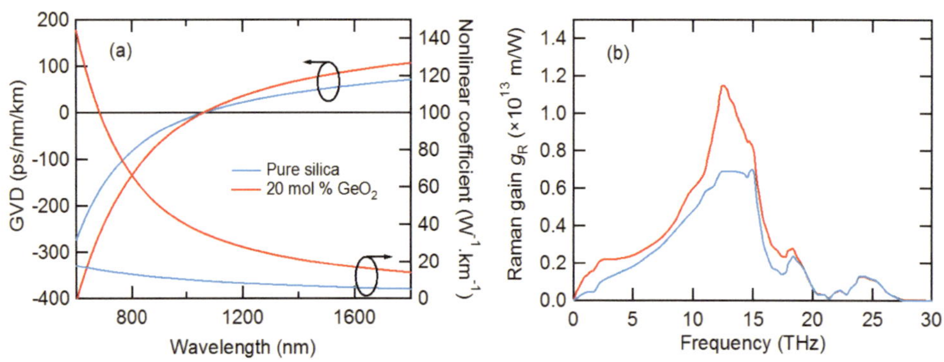


Fig. 11. (a) GVD curves (left axis) and nonlinear coefficient (right axis) calculated for a pure silica PCF (blue) and a PCF doped with a $GeO_2$ content of 20 mol.%. (b) Raman gain spectra $g_R$ for pure silica (blue) and 20 mol.% $GeO_2$-doped silica (for a 164 nm pump).

### 5.2 Spectral extension to the blue

In order to highlight the benefit of using $GeO_2$-doped PCFs in the context of CW SC generation, both fibers were pumped with a CW fiber laser at 1064 nm in similar conditions. Figure 12(a) shows output spectra obtained for a pump power of 13 W and a length of 300 m for the $GeO_2$-doped PCF and 400 m for the pure silica one. The SC spectrum looks much broader in the $GeO_2$-doped PCF than in the pure silica one, yet longer.

As mentioned above, long and short-wavelength SC edges are fixed by a group-velocity matching condition between solitons and trapped dispersive waves Genty et al. (2004; 2005); Gorbach & Skryabin (2007a;b); Gorbach et al. (2006); Travers (2009); Travers & Taylor (2009). Bottom curves in Fig. 12 show group index curves of both fibers, and blue lines illustrates the group-index matching between both SC edges in the pure silica fiber. In the $GeO_2$-doped one, a dip in the spectral power density appears just below 1380 nm because of soliton accumulation just below the OH absorption band. As a consequence, a dip in the spectral power density can be observed at the corresponding group-velocity matched wavelength (around 805 nm) because of trapped dispersive wave accumulation. From the measurement of

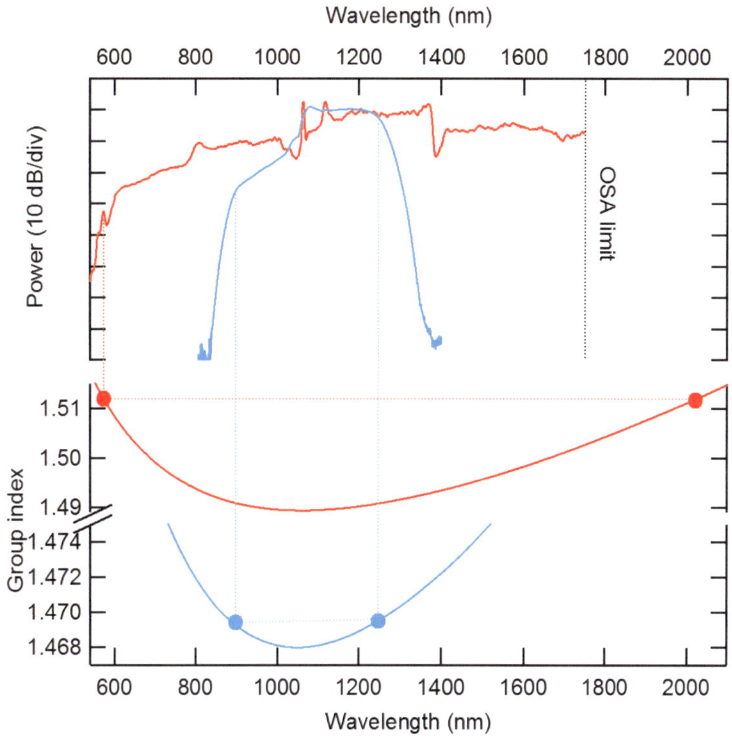


Fig. 12. Top: Supercontinuum generated in the pure silica PCF (blue line) and in the $GeO_2$-doped one (red line) for a pump power of 13 W. Bottom: corresponding calculated group index curves.

the short wavelength edge (570 nm for the $GeO_2$-doped PCF) together with the group-index matching, it is thus possible to estimate the long-wavelength one to 2040 nm (not reachable with our optical spectrum analyzer). In this case, there is thus a threefold enhancement of the SC bandwidth (in frequency) as compared to the pure silica PCF.

Note that comparable results in terms of spectral extent have been reported in pure silica PCFs Travers et al. (2008), but with a much higher pump power.

### 5.3 White-light generation

With the aim of still enhancing the SC bandwidth, it is possible to associate the benefits of $GeO_2$ doping and fiber tapering presented above. We have thus fabricated a $GeO_2$-doped ZDW decreasing PCF, characterized by a 50 m long uniform section followed by a 130-m long section over which the outer diameter linearly decreases from 135 to 85 $\mu$m. Figure 13(a) shows the spectrum obtained with this fiber for a 45 W pump power. It spans from 470 nm to more than 1750 nm, with an output power of 10 W. A picture of this experiment showing white-light generation is displayed in Fig. 13(b). This is the first demonstration of CW white-light supercontinuum generation.

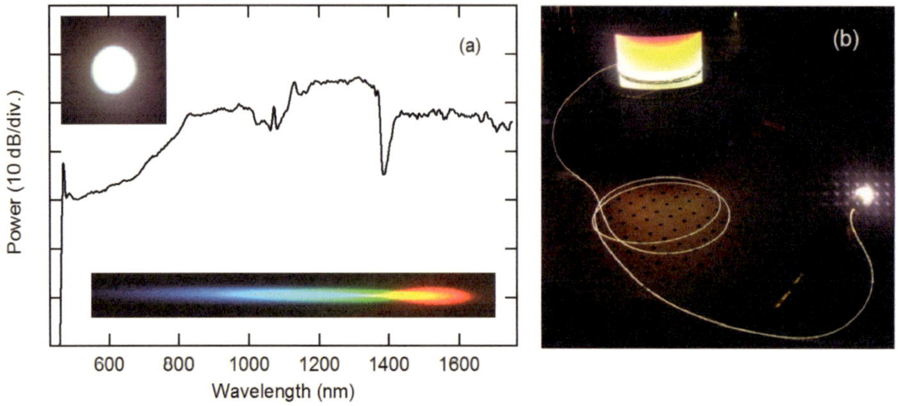


Fig. 13. (a) Supercontinuum generated in a $GeO_2$-doped ZDW decreasing PCF, with a pump power of 45 W. Top inset: photograph of the output beam far-field. Bottom inset: photograph of the output beam dispersed by a prism. (b) Photograph of the experiment.

## 6. Conclusion

It is now possible to generate continuous wave supercontinua ranging from the near ultraviolet to the near infrared with spectral power densities in the order of tens of mW/nm. Most of the manipulation of the spectra were carried out by a fine control of the fiber microstructure. These all-fiber sources are very promising for many applications requiring stable and extremely powerful sources.

## 7. References

Akhmediev, N. & Karlsson, M. (1995). Cherenkov radiation emitted by solitons in optical fibers, *Phys. Rev. A* 51(3): 2602–2607.

Avdokhin, A. V., Popov, S. V. & Taylor, J. R. (2003). Continuous-wave, high-power, Raman continuum generation in holey fibers, *Opt. Lett.* 28(15): 1353–1355.

Barviau, B., Vanvincq, O., Mussot, A., Quiquempois, Y., Melin, G. & Kudlinski, A. (2011). Enhanced soliton self-frequency shift and CW supercontinuum generation in GeO(2)-doped core photonic crystal fibers, *JOURNAL OF THE OPTICAL SOCIETY OF AMERICA B-OPTICAL PHYSICS* 28(5): 1152–1160.

Cumberland, B. A., Travers, J. C., Popov, S. V. & Taylor, J. R. (2008a). 29 W high power CW supercontinuum source, *Opt. Express* 16(8): 5954–5962.

Cumberland, B. A., Travers, J. C., Popov, S. V. & Taylor, J. R. (2008b). Toward visible cw-pumped supercontinua, *Opt. Lett.* 33(18): 2122–2124.

Dudley, J. M., Genty, G. & Coen, S. (2006). Supercontinuum generation in photonic crystal fiber, *Rev. Mod. Phys.* 78(4): 1135–1184.

Efimov, A., Taylor, A., Omenetto, F., Knight, J., Wadsworth, W. & Russell, P. (2003). Phase-matched third harmonic generation in microstructured fibers, *OPTICS EXPRESS* 11(20): 2567–2576.

Genty, G., Lehtonen, M. & Ludvigsen, H. (2004). Effect of cross-phase modulation on supercontinuum generated in microstructured fibers with sub-30 fs pulses, *Opt. Express* 12(19): 4614–4624.

Genty, G., Lehtonen, M. & Ludvigsen, H. (2005). Route to broadband blue-light generation in microstructured fibers, *Opt. Lett.* 30(7): 756–758.

Gonzalez-Herraez, M., Martin-Lopez, S., Corredera, P., Hernanz, M. & Horche, P. (2003). Supercontinuum generation using a continuous-wave Raman fiber laser, *Opt. Commun.* 226(1-6): 323–328.

Gorbach, A. V. & Skryabin, D. V. (2007a). Light trapping in gravity-like potentials and expansion of supercontinuum spectra in photonic-crystal fibres, *Nature Photon.* 1(11): 653–657.

Gorbach, A. V. & Skryabin, D. V. (2007b). Theory of radiation trapping by the accelerating solitons in optical fibers, *Phys. Rev. A* 76(5).

Gorbach, A. V., Skryabin, D. V., Stone, J. M. & Knight, J. C. (2006). Four-wave mixing of solitons with radiation and quasi-nondispersive wave packets at the short-wavelength edge of a supercontinuum, *Opt. Express* 14(21): 9854–9863.

Gordon, J. P. (1986). Theory of the soliton self-frequency shift, *Opt. Lett.* 11(10): 662–664.

Kobtsev, S. & Smirnov, S. (2005). Modelling of high-power supercontinuum generation in highly nonlinear, dispersion shifted fibers at CW pump, *Opt. Express* 13(18): 6912–6918.

Kudlinski, A., George, A., Knight, J., Travers, J., Rulkov, A., Popov, S. & Taylor, J. (2006). Zero-dispersion wavelength decreasing photonic crystal fibers for ultraviolet-extended supercontinuum generation, *Opt. Express* 14(12): 5715–5722.

Kudlinski, A. & Mussot, A. (2008). Visible cw-pumped supercontinuum, *Opt. Lett.* 33(20): 2407–2409.

Mussot, A., Beaugeois, M., Bouazaoui, M. & Sylvestre, T. (2007). Tailoring CW supercontinuum generation in microstructured fibers with two-zero dispersion wavelengths, *Opt. Express* 15(18): 11553–11563.

Nishizawa, N. & Goto, T. (2002). Pulse trapping by ultrashort soliton pulses in optical fibers across zero-dispersion wavelength, *Opt. Lett.* 27(3): 152–154.

Omenetto, F., Taylor, A., Moores, M., Arriaga, J., Knight, J., Wadsworth, W. & Russell, P. (2001). Simultaneous generation of spectrally distinct third harmonics in a photonic crystal fiber, *OPTICS LETTERS* 26(15): 1158–1160.

Persephonis, P., Chernikov, S. & Taylor, J. (1996). Cascaded CW fibre Raman laser source 1.6-1.9 $\mu$m, *Electron. Lett.* 32(16): 1486–1487.

Prabhu, M., Kim, N. & Ueda, K. (2000). Ultra-broadband CW supercontinuum generation centered at 1483.4 nm from Brillouin/Raman fiber laser, *Jpn. J. Appl. Phys. Part 2 - Letters* 39(4A): L291–L293.

Skryabin, D., Luan, F., Knight, J. & Russell, P. (2003). Soliton self-frequency shift cancellation in photonic crystal fibers, *Science* 301(5640): 1705–1708.

Stone, J. M. & Knight, J. C. (2008). Visibly "white" light generation in uniform photonic crystal fiber using a microchip laser, *Opt. Express* 16(4): 2670–2675.

Travers, J. C. (2009). Blue solitary waves from infrared continuous wave pumping of optical fibers, *Opt. Express* 17(3): 1502–1507.

Travers, J. C., Rulkov, A. B., Cumberland, B. A., Popov, S. V. & Taylor, J. R. (2008). Visible supercontinuum generation in photonic crystal fibers with a 400W continuous wave fiber laser, *Opt. Express* 16(19): 14435–14447.

Travers, J. C. & Taylor, J. R. (2009). Soliton trapping of dispersive waves in tapered optical fibers, *Opt. Lett.* 34(2): 115–117.

Tse, M., Horak, P., Poletti, F., Broderick, N., Price, J., Hayes, J. & Richardson, D. (2006). Supercontinuum generation at 1.0 mu m in holey fibers with dispersion flattened profiles, *Opt. Express* 14(10): 4445–4451.

Vanholsbeeck, F., Martin-Lopez, S., Gonzalez-Herraez, M. & Coen, S. (2005). The role of pump incoherence in continuous-wave supercontinuum generation, *Opt. Express* 13(17): 6615–6625.

5

# Slow Light in Optical Fibers

Shanglin Hou[1] and Wei Qiu[2]
[1]*School of Science, Lanzhou University of Technology, Lanzhou,*
[2]*Department of Physics, Liaoning University, Shenyang*
*China*

## 1. Introduction

The group velocity at which light pulses propagate through a dispersive material system is very different from the vacuum speed of light *c*, One refers to light as being "slow" for $v_g$ << *c* (Boyd & Gauthier, 2009) or "fast" for $v_g$ > c or $v_g$ <0 (Stenner et al, 2003 ). For $v_g$ <0, the pulse envelope appears to travel backward in the material (Gehring et al, 2006), and hence it is sometimes referred to as "backward light."
The subject of slow light has caused keen interest in the past decade or more, and it is possible to control the group velocity of light pulses in the dispersive materials. Interest in slow and fast light dates back to the early days of the 20th century. Sommerfeld and Brillouin (Sommerfeld & Brillouin, 1960) were intrigued by the fact that theory predicts that $v_g$ can exceed c, which leads to apparent inconsistencies with Einstein's special theory of relativity. Experimental investigations of extreme propagation velocities were performed soon after the invention of the laser (Faxvog and et al, 1970). In 1999, Harris's group research work greatly stimulated researchers' interests, which showed that light could be slowed down to 17m/s. The result was obtained in ultra cold atom clouds with the use of electromagnetically induced transparency (EIT), which induces transparency in a material while allowing it to retain strong linear and nonlinear optical properties (Hau et al, 1999). Slow light can also be obtained through the use of the optical response of hot atomic vapors (Philips et al, 2001). These early research works require hard conditions and the slow light cannot operate in room temperature.
Recently, researchers found ways to realize slow light operating in room temperature and solid-state materials, which are more suited for many practical applications, namely slow light via stimulated Brillouin scattering(SBS), slow light via coherent population oscillations (CPO), tunable time delays based on group velocity dispersion or conversion/ dispersion(C/D), slow light in fiber Bragg gratings and so on. In this chapter, we describe some of the physical mechanisms that can be used to induce slow and fast light effects in room-temperature solids (Bigelow et al, 2003) and some of the exotic propagation effects that can thereby be observed. We also survey some applications of slow and fast light within the fields of quantum electronics and photonics.

## 2. Fundamentals of slow and fast light

Slow and fast light refer to the group velocity of a light wave. The group velocity is the velocity most closely related to the velocity at which the peak of a light pulse moves through an optical dispersive material (Milonni, 2005), and is given by the standard result

$$v_g = \frac{c}{n_g}, \qquad n_g = n + \omega \frac{dn}{d\omega} \tag{1}$$

where $n$ is the refractive (phase) index and $\omega$ is the angular frequency of the carrier wave of the light field. One refers to light as being slow light for $v_g << c$, fast light for for $v_g >$ c, and backwards light for $v_g <0$ or $v_g$ is negative. Extreme values of the group velocity invariably rely on the dominance of the second contribution to the group index of Equation (1). This contribution of course results from the frequency dependence of the refractive index, and for this reason extreme values of the group velocity are usually associated with the resonant or near-resonant response of material systems.
According to this theory, slow light is expected in the wings of an absorption line and fast light is expected near line center, and the spatial dispersion, that is, the non-locality in space of the medium response, is another mechanism that can lead to slow light, as has been predicted (Kocharovskaya et al,2001) and observed (Strekalov et al, 2001). We catalogue the main methods to realize slow light in room temperature solid, namely slow light via stimulated Brillouin scattering(SBS), slow light via coherent population oscillations (CPO), tunable time delays based on group velocity dispersion or conversion/ dispersion(C/D), slow light in fiber Bragg gratings and so on. We will describe CPO and SBS slow light in more details.

## 3. Slow light via coherent population oscillations (CPO)

### 3.1 Introduction of CPO

The technique of coherent population oscillation(CPO) is also exploited to reduce the group velocity.The process of CPO allows the reduction of absorption and simultaneously provides a steep spectral variation of the refractive index which leads to a strong reduction of the optical group velocity,i.e.,slow light propagation.This process is easily achieved in a two-level system which interacts with a signal whose amplitude is periodically modulated.The population of the ground state of the medium will be induced to oscillate at the modulation frequency. This oscillation creates an arrow hole in the absorption spectrum, whose linewidth is proportional to the inverse of the relaxation lifetime of the excited level. CPO is highly insensitive to dephasing processes in contrast to what happens in other schema such as EIT, where the width of the spectral hole burned in the absorption profile is proportional to the inverse of the dephasing time of the ground state. That makes CPO an appropriate technique to easily achieve slow light propagation in solid-state materials at room temperature.

### 3.2 Theoretical mode

Making use of this technology, we observe optical pulse delay and advancement propagation in an erbium-doped optical fiber. Compared to other solid material, erbium-doped optical fiber allows for long interaction lengths, which can be desirable in producing strong influence(Schwartz& Tan, 1967). We obtain the controllable pulse delay continuously from positive to negative by using a separate pump laser.
When pumped at 980nm, the erbium-doped fiber acts as a three-level molecular system. The relaxation time from the metastable state to the ground state is much greater than the one from the excited state to the metastable state. The energy levels and pumping scheme that we employed to observe slow and fast light in erbium-doped fiber is shown in Fig 1.

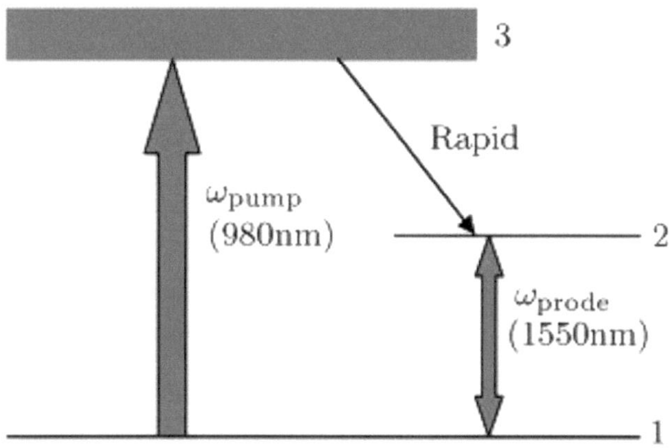


Fig. 1. The energy levels and pumping scheme we employed to observe slow and fast light in erbium-doped fiber.

The relaxation time from the excited state associated with the signal frequency to the ground state is much greater than the one from the excited state associated with the pump frequency to the excited state associated with the signal frequency. The population of upper excited level is approximately equal to zero, which indicates $n_1 + n_2 = \rho$. The population density of the ground state will accord with the rate equation(Novak & Gieske, 2002)

$$\frac{\partial n_1}{\partial t} = -R_{13}n_1 - W_{12}n_1 + W_{21}(\rho - n_1) + \frac{\rho - n_1}{T_1}, \tag{2}$$

where $W$ denotes transition rates associated with signal and $R$ presents transition rates associated with pump. $T_1$ is the lifetime of the excited state and $\rho$ is the erbium ion density. The transition rates are also functions of $t$ and $z$, as well as being proportional to the pump and signal powers. According to the equations for the transition rates and neglecting the losses through the fiber, we have

$$\frac{\partial}{\partial t}N_1 = I_p(L,t) - I_p(0,t) + I_s(L,t) - I_s(0,t) + \frac{N_0 - N_1}{T_1}. \tag{3}$$

If we modulate the signal(~1550nm) intensity as

$$I(0,t) = I_s^0(0)(1 + \mu_s \cos \delta t). \tag{4}$$

here, $I_s^0(0)$ is the average input power at the input (Z=0) and $I_s^0(0)\mu_s = I_m(0)$ is the modulation amplitude. A single intensity-modulated beam contains only a carrier wave (to act as the pump) and two sidebands (to act as probes) on the output spectrum, which induce population oscillation. The population of the ground state is given by

$$N_1(t) = N_1^0\left[1 + \xi \cdot \cos(\delta t + \phi)\right], \tag{5}$$

where $N_1^0$ is the mean (un-modulated) steady-state population. We next find the steady-state solution to Eq. (3). Finally, We can determine the expression of the delay of the optical signal

$$\Delta t = \frac{1}{\delta \times 2\pi} \arctan\left( \frac{\delta}{\frac{\delta^2_{eff} + \delta^2}{\eta} - \delta_{eff}} \right). \tag{6}$$

### 3.3 Experimental results

The signal optical field from a distributed feedback laser diode operating at 1550nm through the attenuator is divided into two parts: one part of laser (98%) goes through an erbium-doped optical fiber and then to an InGaAs photodetector with 10MHz bandwidth. The other part of laser output signal (2%) is sent directly to an identical photodetector to be used as reference.Transmitted signals are received by photodetectors, and sent into a digital oscillograph for recording. Then the comparison between the reference signal and the EDOF signal is made in a computer (Sargent,1978, Boyd & Gauthier, 2005). The group velocity in fibers can be inferred. In the experiments, the injection current of the laser is sinusoidally modulated by a function generator. We use single mode, $Al_2SiO_5$-glass-based erbium-doped optical fibers at several ions density. The experimental setup is shown in Fig.2.

The absorption coefficient $\alpha$ and the emission coefficient $\beta$ are related to their cross sections respectively and shown by the following

$$\alpha_s = \Gamma_s \sigma_{12} \rho \quad \alpha_p = \Gamma_p \sigma_{13} \rho \quad \beta_s = \Gamma_s \sigma_{21} \rho \ . \tag{7}$$

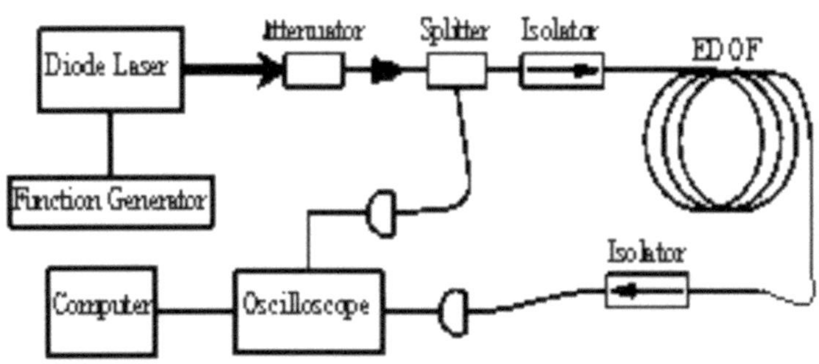


Fig. 2. The experimental setup used to observe slow light in an erbium-doped optical fiber.

Parameters used for the calculation are $\alpha_s$ =31.71 dB/m, $\alpha_p$ =42.3dB/m, $\beta_s$ =47.665 dB/m, $T_1 = 10.5$ ms, $L$=2m, and $\rho$ =6.3×10$^{25}$. Our experimental results are obtained through use of modulation techniques such that the optical field contains only a carrier wave (to act as the pump)and two sidebands (to act as probes). Because the decay time is so long (about 10.5ms), this oscillation will only occur if the beat frequency ($\delta$) between the pump and probe beams is small so that $\delta T_1 \sim 1$.When this condition is fulfilled, the pump wave can efficiently scatter off the temporally modulated ground state population into the probe wave, resulting in reduced absorption of the probe wave. We consider the Kramers-Kronig relations, which show that a narrow hole in an absorption spectrum will produce strong normal dispersion. Fig.3 illustrates the results of our experiments.

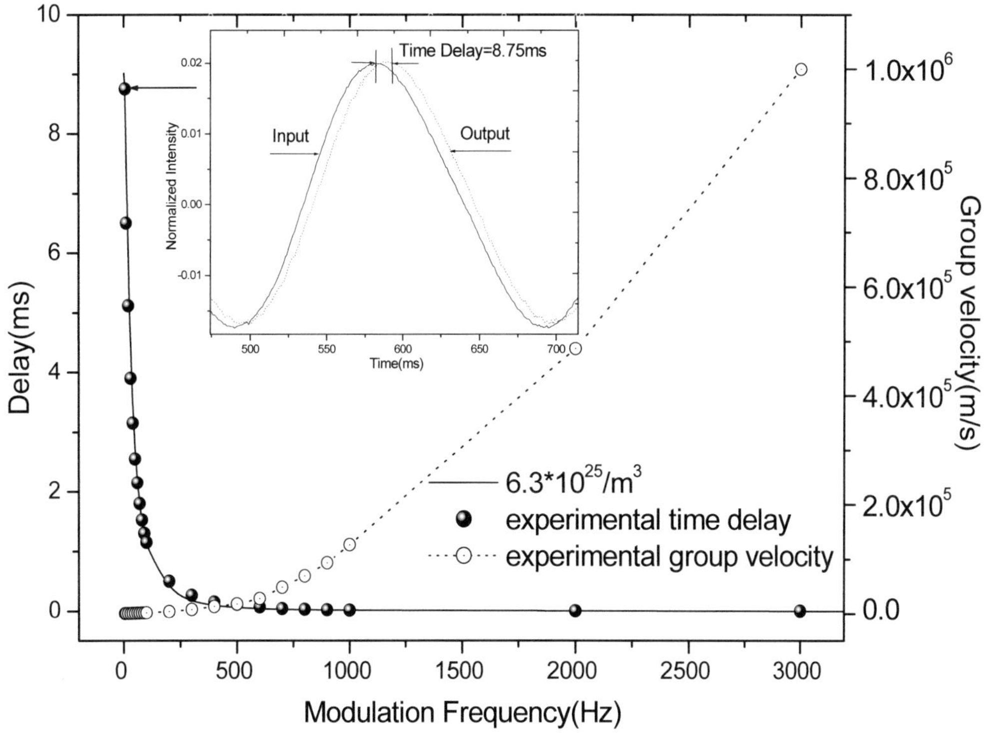


Fig. 3. Observed time delay as a function of the modulation frequency for input power of 1.85mW. The solid line is the theoretical fit to the experimental data. The open circles represent the measured group velocity. The inset shows the normalized 5Hz input (solid line) and output (dashed line) signal. The signal is delayed 8.75ms corresponding to a group velocity as low as 228.57m/s.

In Fig.3, we show the measured delay in an optical fiber with erbium ion density of $6.3\times10^{25}m^{-3}$ and compare it with the numerical solution of Eq.(6) for input power of 1.85mW. We observe the largest delay, 8.75ms, which corresponds in the inset of Fig.3. The inferred group velocity is as low as 228.57m/s. A maximum fractional delay of 0.129 is observed at the modulation frequency of 60Hz. Fig.3 shows the delay as a function of modulation frequency in the low frequency region.

## 4. Slow light via stimulated Brillouin scattering (SBS)

### 4.1 Introduction

Slow light based on stimulated Brillouin scattering (SBS) in optical fibers has attracted much more interests for its potential application in optical buffering, data synchronization, optical memories and optical signal processing. Compared with previously demonstrated slow-light techniques (Gehring et al, 2008, Zhu et al, 2007), such as electromagnetically induced transparency (EIT) (Hau et al, 1999) and coherent population oscillations (CPO) (Bigelow et al, 2003), it has a lot of advantages, for instance, the simple, flexible and easy-to-handle SBS can be realized in room temperature; the optical fiber components based on it can easily integrated with the existing telecommunications infrastructure; the slow-light resonance can be tunable within the optical communications wavelength windows; the use of optical fiber allows for a relaxed pump-power requirement owning to long interaction length, small effective mode area and so on.

However, the SBS-induced group index change is always so small in standard single mode fiber and dispersion shift fibers (DSFs) (Song et al, 2005) to delay the time very little. In order to explore suitable optical fibers served as slow light generation with much efficiency, some special optical fibers, such as chalcogenide fiber (Abedin,2005, Song et al, 2006), tellurite fiber (Abedin,2006), bismuth fiber (Jauregui et al, 2006) and so on, have been extensively studied, these kinds of optical fiber are usually with large gain coefficient and low loss coefficient. Though long pulse delay can be obtained using cascaded fiber segments joined by unidirectional optical attenuators to overcome pump depletion (gain saturation) and amplified spontaneous Brillouin emission (ASBE), it's always accompanied with serious pulse distortion (Song et al, 2005). So gain tailoring is used in pulse distortion management to keep a balance between time delay and pulse distortion (Stenner et al, 2007). To overcome the narrow band spectral resonance of SBS which limits the maximum data rate of the optical system, a simple and inexpensive pump spectral broadening technique is used in broadening the SBS slow light bandwidth (Herraez et al, 2006), which paves the way towards real applications based on SBS slow light.

Numerical studies of SBS slow light focusing on different pulse parameters were also studied (Kalosha et al, 2006), which provide an insight into the SBS slow light process, but we can't learn a lot about how the optical fiber structures and Brillouin gain parameters influence on the SBS process, the time delay and the pulse shape. In this section, the SBS model in optical fiber is described and the three coupled SBS equations are solved by the method of finite difference with prediction-correction, the effects of gain coefficient, gain bandwidth and effective mode area on time delay and pulse broadening are demonstrated. Considering the injected stokes pulse shape, the influence of its sharpness, magnitude and duration on delay time and pulse broadening factor was observed mainly, and its reason was analyzed. These results provide base for designing optical buffer, time delay line or other optical components based on the SBS slow light technologies.

### 4.2 Theory foundation and numerical model

The process of SBS is the interaction of two counter-propagating waves, a strong pump wave and a weak Stokes wave. If a particular frequency relation is satisfied

$$v_{pump} = v_{Stokes} + v_B , \tag{8}$$

Where $v_{pump}$ and $v_{Stokes}$ are the frequency of pump wave and stocks wave respectively, $v_B$ is the Brillouin frequency. Then an acoustic wave is generated which scatters photons from the pump to the Stokes wave and the interference of these two optical waves in turn stimulates the process. From a practical point of view, the process of SBS can be viewed as a narrowband amplification process, in which a continuous-wave pump produces a narrowband gain in a spectral region around $v_{pump} - v_B$. In this paper, the Stokes pulse is set on the SBS gain line center to achieve the maximum delay.

For simply describing the SBS process, assume: (1) Transverse field variations are neglected, Stokes and pump fields are assumed to vary with time t and space z only. (2) The slowly varying envelope approximation (SVEA) SBS model is used, i.e., the field amplitudes are assumed to vary slowly in time and space as compared with their temporal and spatial frequencies. (3) The initial (t=0) phonon field is zero and the Stokes output grows from an injected Stokes field at z=0. (4) The frequency difference between the pump and Stokes wave is set to the Brillouin shift of the fiber, i.e., the Stokes pulse is on the SBS line center.

Considering a Brillouin amplifier where the pump wave counter-propagates through the fiber with respect to the Stokes pulse, the SBS process can be described by one-dimensional coupled wave equations involving a backward pump wave (−z direction), a forward Stokes wave (+z direction), and a backward acoustic wave. Under the slowly varying envelope approximation (SVEA) and neglecting the transverse field variations, the equations are written as follows (Damzen et al,2003)

$$-\frac{\partial A_p}{\partial z}+\frac{n}{c}\frac{\partial A_p}{\partial t}=-\frac{\alpha}{2}A_p+ig_2A_sQ\,, \tag{9}$$

$$\frac{\partial A_s}{\partial z}+\frac{n}{c}\frac{\partial A_s}{\partial t}=-\frac{\alpha}{2}A_s+ig_2A_pQ^*\,, \tag{10}$$

$$\frac{\partial Q}{\partial t}+\frac{\Gamma_B}{2}Q=ig_1A_pA_s^*\,, \tag{11}$$

where $A_p$, $A_s$ ,and $Q$ are the amplitudes of the pump wave, the Stokes wave, and the acoustic wave, respectively; $n$ is the group refractive index when SBS is absent; $\alpha$ is the loss coefficient of the fiber; $\Gamma_B/2\pi$ is the bandwidth (FWHM) of the Brillouin gain; $g_1$ is the coupled coefficient between the pump wave and the Stokes wave, $g_2$ is the coupled coefficient between the pump (Stokes) wave and the acoustic wave, $g_0=4g_1g_2/\Gamma_B$ is the peak value of the Brillouin gain coefficient.

According to the small signal steady state theory of stimulated Brillouin scattering, the pump power $P_{critical}$ required to reach Brillouin threshold in a single pass scheme is related to the Brillouin gain coefficient $g_0$ by the following equation:

$$g_0(P_{critical}/A_{eff})L_{eff}\cong 21\,, \tag{12}$$

where $P_{critical}$ is the power corresponding to the Brillouin threshold, $L_{eff}$ is the effective length defined as $L_{eff}=\alpha^{-1}[1-\exp(-\alpha L)]$, from Eq.(12) we can obtain the threshold pump intensity

$$I_{critical}=P_{critical}/A_{eff}\cong 21/(g_0L_{eff})\,. \tag{13}$$

Once reaching the threshold pump intensity, a large part of the pump power is transferred to the Stokes wave, resulting in the generation of Stokes wave at the output depletes the pump seriously and leads to serious Stokes pulse distortion. In our simulations, we consider the pump intensity is near the Brillouin threshold and obtain the Stokes gain around 16 using the previous parameters, here the Stokes gain is defined as:

$$Gain=\log\left(\frac{P_{out}}{P_{in}}\right), \tag{14}$$

where $P_{out}$ and $P_{in}$ are the output and input of the Stokes power, respectively.
Let us assume that pump wave is continuous wave and stokes field is sufficiently weak. The group index is the function of frequency described as follow(Zhu et al, 2005)

$$n(\omega_s) = n_g + \frac{cg_B I_p}{\Gamma_B} \frac{1 - 4\delta\omega^2 / \Gamma_B{}^2}{(1 + 4\delta\omega^2 / \Gamma_B{}^2)^2} \tag{15}$$

where $I_p$ is the optical intensity of pump wave; $\delta\omega$ is the margin between the angular frequency of stokes pulse and the center angular frequency of the gain bandwidth; $g_B = \dfrac{\gamma_e{}^2 \omega_p{}^2}{n_g v_B c^3 \rho_0 \Gamma_B}$ is the line-center gain factor which is associated with the material physical properties, $v_B$ is the velocity of acoustic wave.
Delay time $T_d$ is defined to describe the difference of the arrival time when the output stokes pulse reach its maximum between when SBS occurs and doesn't occurs, $T_{rd} = T_d / T$ to describe the relative delay time with T which is the FWHM of the injected stokes pulse. According to the weak signal theory, delay time and B are given by (Velchev et al, 1999)

$$T_d = G / \Gamma_B \tag{16}$$

$$B = \sqrt{1 + \frac{16 \ln 2}{T^2 \Gamma_B{}^2} G} \tag{17}$$

where $G = g_B I_p L$ is the weak signal gain parameter; L is the fiber length. For the purpose of indicating how much the pump wave energy contributes to the stokes wave energy, we define real gain as

$$G_r = \log(P_{\text{out}} / P_{\text{in}}) \tag{18}$$

where $P_{out}$ and $P_{in}$ are power of the output and input stokes wave, respectively.
We assume that injected stokes pulse is super-Gaussion shaped

$$U(t) = \exp\left( -\frac{1}{2} \left( \frac{t}{T_0} \right)^{2m} \right) \tag{19}$$

where t is the time of pulse transmission, U(t) is normalized amplitude, $T_0$ is the half width of pulse (at 1/e-intensity point). The parameter m controls the degree of edge sharpness (for m=1, it is Gaussion-shaped; for m>1, it is super-gaussion-shaped, and the degree of edge sharpness is increased with m). For different m, pulse with same FWHM can be written as

$$U(t) = \exp\left( -2^{2m} (\ln 2) \left( \frac{t}{T} \right)^{2m} \right) \tag{20}$$

Furthermore, we define average intensity of normalized super-Gaussion pulse with FWHM of T as

$$\overline{I}_s = \frac{1}{T}\int_{-T/2}^{T/2} (U(t))^2 dt = \frac{1}{T}\int_{-T/2}^{T/2}\left(\exp\left(-2^{2m}(\ln 2)\frac{t^{2m}}{T^{2m}}\right)\right)^2 dt \quad (21)$$

In our numerical processing, applying the slowly-varying envelope approximation (SVEA) for both pump and stokes fields, firstly, we obtained the values of $\rho$ at some time by solving the Eq.(11) with Fourier transformation and inverse Fourier transformation. Secondly, we transform the Eqs (9-10) into single variable partial differential equations by using characteristics. Finally, applying the value of $\rho$ we obtained to the single variable partial differential equations, we can calculate the values of $E_p$ and $E_s$ at next time by using the fourth-order Runge-Kutta formula. And we set these results as the initialization value to achieve the value of $\rho$ at next time. Repeating above steps, we can achieve the output stokes pulse at anytime.

### 4.3 Numerical simulation results

In order to study the situation where the pump is depleted, we solve the Eqs. (9)-(11) numerically using the method of implicit finite difference with prediction-correction to determine how gain coefficient, gain bandwidth and effective mode area influence SBS slow light.

In our simulation, the parameters are considered from the common single-mode fiber, and select: fiber length L=25m, pump wavelength $\lambda$ =1550nm, group refractive index n=1.45, effect mode area $A_{eff}$ =50 $\mu m^2$, loss coefficient $\alpha$ =0.2dB/km, gain bandwidth (FWHM) $\Gamma_B / 2\pi$ =40MHz, gain coefficient $g_0 = 5\times 10^{-11}$ m/W. We assume the pump wave is CW and the Stokes wave is Gaussian shaped with the peak power of 0.1 $\mu W$ and the FWHM pulse width of 120ns (its FWHM bandwidth in frequency domain is around 3.7MHz which is much smaller than that of SBS gain bandwidth we use).

#### 4.3.1 Influence of gain coefficient on time delay and pulse broadening

The curves of the pulse delay and pulse broadening factor as a function of the gain with different gain coefficient $g_0$ were shown in Fig.4. It can be seen from Fig.4(a) that the time delay increases linearly with Stokes gain when the gain is small (≤10), that's because the pump isn't completely affected when the gain is small. For larger gain, pump depletion becomes more and more seriously, the time delay increases slowly with gain and reaches its maximum before decreasing with gain. At the same time, for larger gain coefficient, the time delay decreases with increasing gain more quickly and even leads to pulse advancement which can be explained by gain saturation. It can also be seen that the smaller gain coefficient reaches the gain saturation at a larger gain and the maximum time delay is accordingly larger, the gain saturation limits the maximum time delay for a Stokes pulse at a given input power.

The pulse broadening factor for different gain coefficients as a function of gain was shown in Fig.4(b). It shows that the pulse broadening factor is also increasing linearly with gain when the small signal regime holds. As the gain further increases, the pulse broadening factor increases slowly with gain and then gradually decreases to less than 1, it means that the pulse become more and more narrower, the pulse with larger gain coefficient narrows more seriously than the smaller one. The time delay is always accompanied with pulse distortion, the Stokes pulse broadens a little in the small signal regime but can narrow

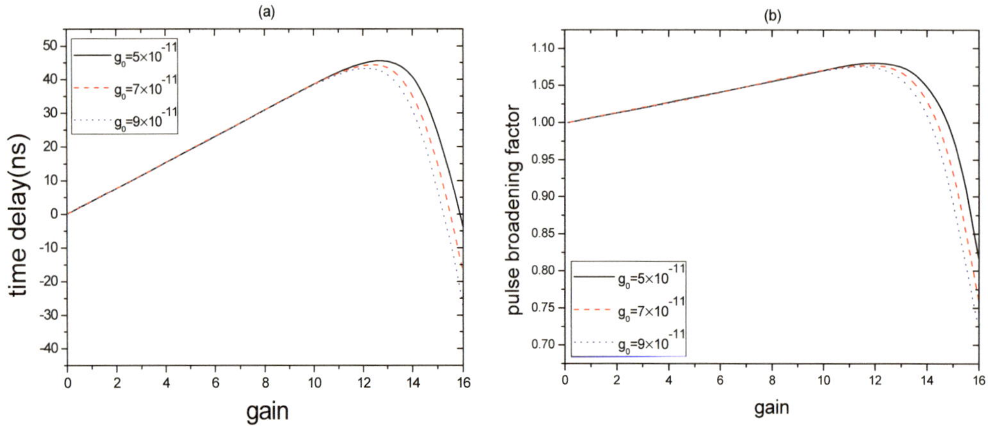


Fig. 4. (a) Time delay and (b) pulse broadening as a function of gain with different gain coefficients.

largely in the gain saturation regime. Fig.5 shows the normalized output pulse shapes with the gain coefficient $g_0 = 5\times10^{-11}$ m/W at gain=0, 12, and 17, respectively. The output Stokes pulse with a maximum time delay ~45 ns at gain=12 with a little distortion while the output Stokes pulse is advanced by 42.9 ns at gain=17 but is distorted substantially.

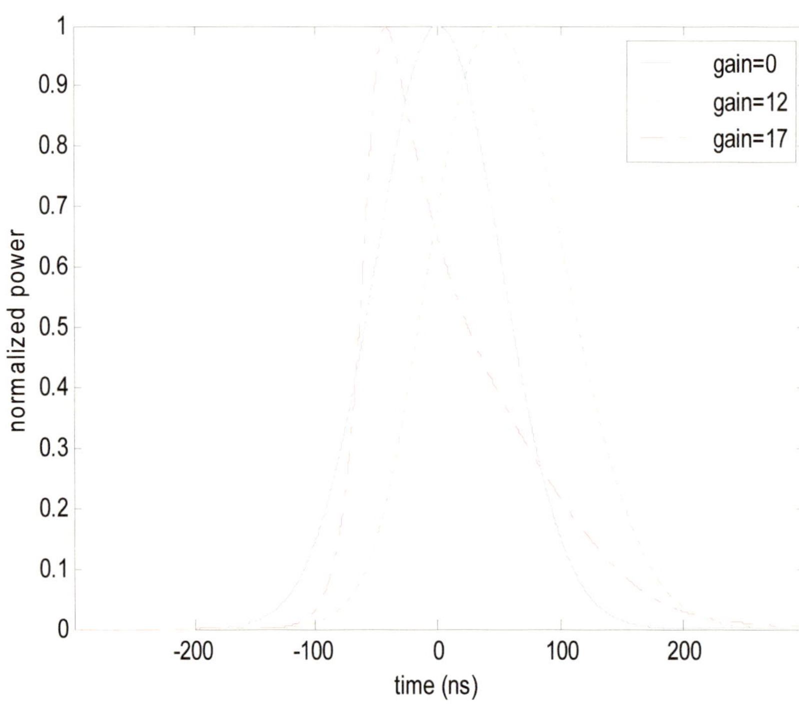


Fig. 5. Normalized output Stokes pulse at gain=0, 12, 17 with the gain coefficient $g_0 = 5\times10^{-11}$ m/W.

Next, we consider the time delay and pulse broadening factor varying with the gain coefficient at a given pump peak power 0.125W shown in Fig.6. From Fig.6(a) we can see that the time delay increases with the increasing gain coefficient in a linear fashion. Fig.6(b) shows that the pulse broadening factor also increases with the increasing gain coefficient. Note that the maximum gain parameter is 6.25 at the gain coefficient $g_0 = 1 \times 10^{-10}$ m/W, which satisfies the small signal condition.

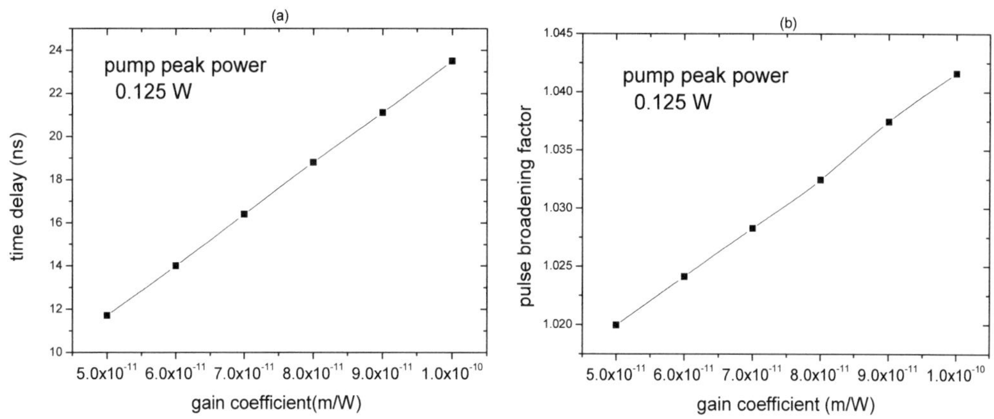


Fig. 6. (a) Time delay and (b) pulse broadening as a function of gain coefficient at a given pump power.

### 4.3.2 Influence of gain bandwidth on time delay and pulse broadening

Fig.7(a) shows the time delay as a function of gain with different gain bandwidths. The time delay increases with the gain linearly when the gain is small, but for smaller gain bandwidth, the time delay increases with the gain more quickly and reaches the saturation at a larger gain, the maximum time delay is accordingly larger. Once reaching the gain saturation, the time delay also decreases more quickly for smaller gain bandwidth.

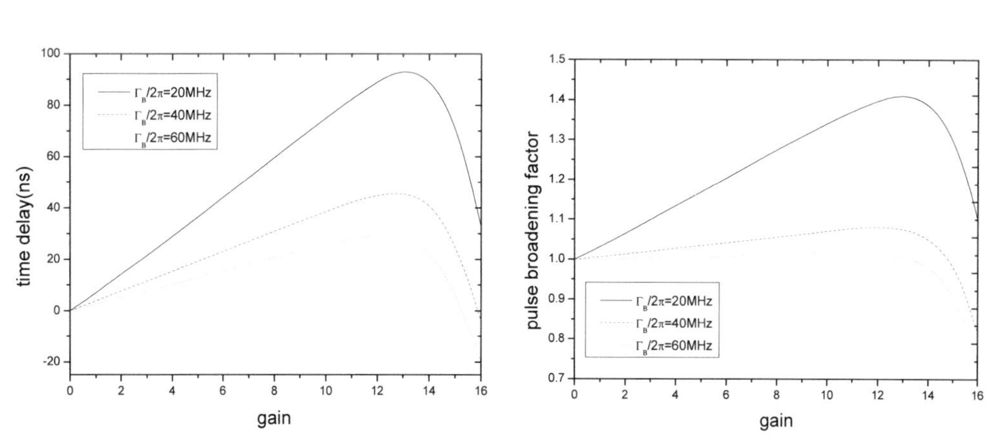


Fig. 7. (a) Time delay and (b) pulse broadening as a function of gain with different gain bandwidths.

The pulse broadening factor as a function of gain with different gain bandwidths was shown in Fig.7(b). It can be seen that the pulse broadening factor increases with the gain before gain saturation and then it decreases with the increasing gain which is similar with time delay versus gain in Fig.7(a). The smaller the gain bandwidth is, the more quickly the broadening factor increases with the gain in the small signal regime and decreases with the gain in the gain saturation. It indicates that the pulse with smaller gain bandwidth always obtains the longer time delay but at the cost of much larger pulse broadening.

We also calculated the time delay and pulse broadening factor as a function of gain bandwidth at a given pump peak power 0.125W. As we can see in Fig.8, both the time delay and pulse broadening factor decrease with the increasing gain bandwidth and keep an inverse proportion to it. The maximum gain parameter is 6.25, which is also in the small signal regime for these different gain bandwidths.

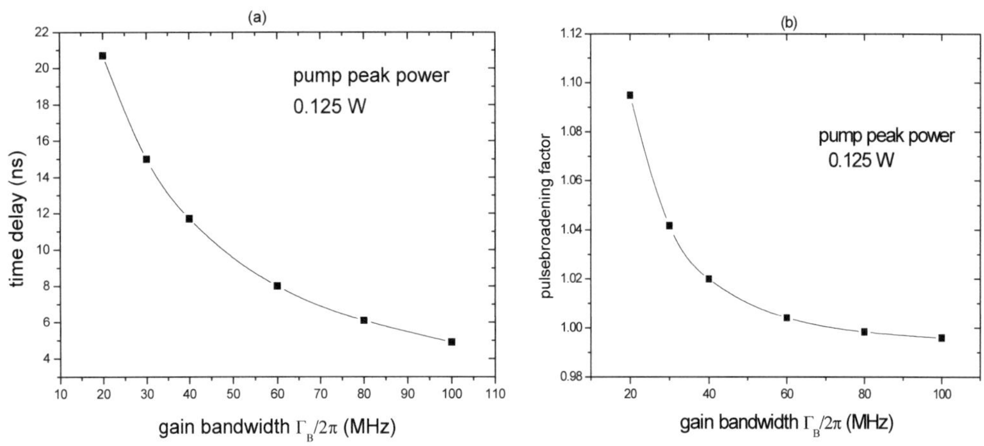


Fig. 8. (a) Time delay and (b) pulse broadening as a function of gain bandwidth at a given pump power.

### 4.3.3 Influence of effective mode area on time delay and pulse broadening

For the pulse with same peak power, it has a larger intensity for the smaller effective mode area, which increases its intensity in the other way, so it can also influence the gain saturation obviously. As can be seen from Fig.9(a), in the small signal regime, the time delay still increases with the gain linearly for different effective mode areas, the pulse with smaller effective mode area reaches the gain saturation at a smaller gain, and the maximum time delay is accordingly smaller. Once reaching the gain saturation, the pulse with smaller effective mode area also decrease more quickly than the others.

Fig.9(b) shows the pulse broadening factor versus gain, the pulse broadening factor increases linearly with the increasing gain in the small signal regime, which is the same as the time delay versus gain. As has been said before, the pulse with larger effective mode area reaches the gain saturation at a larger gain and its maximum pulse broadening factor is accordingly larger. In the gain saturation regime, the pulse with smaller effective mode area narrows more seriously than the others at a fixed gain. We can even see that the pulse broadening factor begins to increase for the gain around 16.

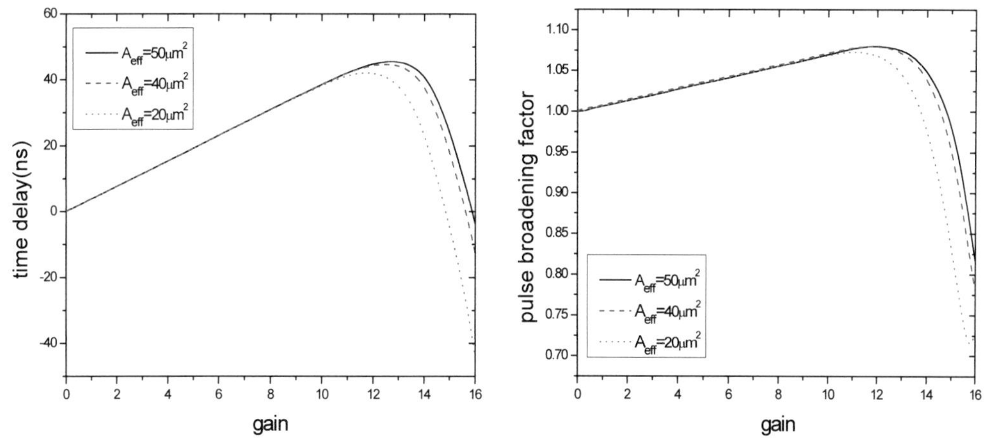


Fig. 9. (a) Time delay and (b) pulse broadening as a function of gain with different effective mode areas.

We also investigate the time delay and pulse broadening factor as a function of effective mode area at a given pump power 0.125W. It can be seen from Fig.10, both the time delay and pulse broadening factor decrease with the increasing effective mode area and keep an inverse proportion to it. As previously mentioned, we also make sure that the corresponding gain parameter is within the small signal regime for these different effective mode areas.

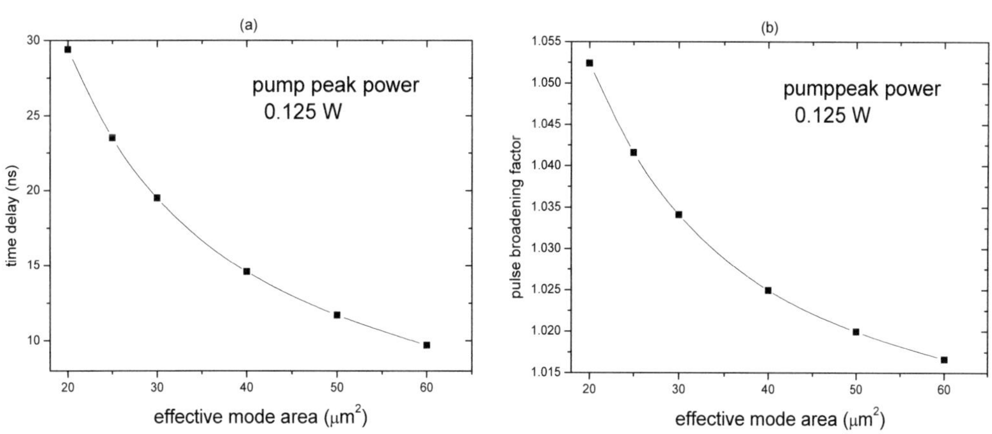


Fig. 10. (a) Time delay and (b) pulse broadening as a function of effective mode area at a given pump power.

### 4.3.4 The influence of stokes pulse with different m on SBS slow light

We first consider the pulse time delay and pulse broadening factor as a function of parameter real gain for super-Gaussion-shaped pulse with different m, which is indicated in Fig.11. In this case, T=120ns, $P_{in} = 0.1\mu W$ . Fig.11 (a) shows that with the increase of $G_r$, $T_d$ increases accordingly and reaches its maximum. Then it decreases with further increasing $G_r$, even becomes negative. Comparing with different m, we can see that maximum $G_r$ and the time when maximum $G_r$ obtains decrease with m because when m changes from 0.5 to 3, $\overline{Is}$ is equal to 0.5410, 0.6805, 0.7559 and 0.8705, respectively, i.e., for the same duration and apex, the super-Gaussion-shaped pulse with higher m is easier to reach gain saturation. And $T_d$ when $T_d$ reaches its maximum will decrease, which leads to the reducing of maximum $T_d$.

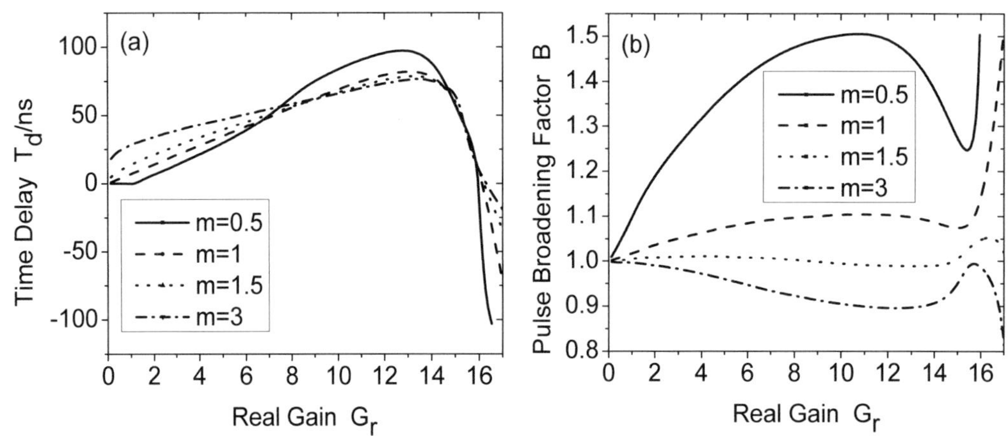


Fig. 11. Parameters of output stokes pulse versus real gain for different parameter m. a) Time delay $T_d$ versus real gain; b) Pulse broadening factor B versus real gain.

We observe an advantageous phenomenon for practical applications. When m=1.5, pulse broadening factor B is close to 1. While m=3, B decreases with increasing $G_r$ and reaches its peak value at gain=13. Then it increase with $G_r$ and reaches its maximum at $G_r$ =15.5, it decrease with further increasing $G_r$. The reason why B has the rule can be explained well by Fig.12. Considering SBS process and no SBS process, respectively, for three different m, the normalized output stokes pulses at $G_r$ =5 and $G_r$ =13 are shown in Fig.12, where t is the time axis. It is indicated from Fig.12 (a) that with increasing m the leading edge and the trailing edge of super-Gaussion-shaped pulse become steeper and steeper in time domain. So they will become broader and broader in the frequency domain. Considering equation (2), we can conclude that the difference among the speeds of points at the leading edge will increase and the leading edge will be compressed more. Moreover, the increasing m leads to the increasing energy of pulse with same peak value. And most energy of pump wave is depleted in the leading edge of stokes pulse, the trailing edge only can get less energy from pump wave, the broadening of the trailing edge of stokes pulse is limited. All of this can contribute to decreasing B and result in B is almost close to 1 before stokes pulse is near saturation. When $G_r$ increases to saturation gain step by step, the trailing edge gets more and more energy, resulting in broadening of the trailing, i.e., B will increase, like shown in Fig.12(b). When $G_r$ go on increasing, it is out of the range of weak signal, the output stokes pulse is distortion seriously.

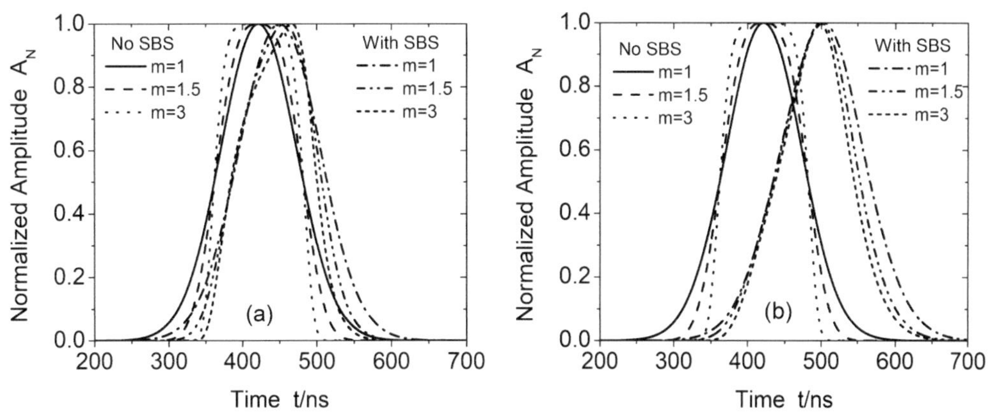


Fig. 12. Output stokes pulse for different real gain. a) $G_r$ =5; b) $G_r$ =13.

#### 4.3.5 The influence of stokes pulse with different power and FWHM on SBS slow light

Based on the result of above that when m=1.5 B is very close to 1, the next numerical simulation will select different power and duration of super-Gaussion pulse with m=1.5 as the injected stokes pulse. Fig.13 shows that delay and B as a function of parameter $G_r$ for super-Gaussion-shaped pulse with different power. It can be seen form Fig.13 (a) that maximum $T_d$ and $G_r$ needed for obtaining maximum $T_d$ increase with injected power. And before entering gain-saturation regime $T_d$ is equal to each other. The reason is in the condition of weak signal delay is in direct proportion to gain approximately. As we can see from Fig.13(b) that when $G_r$ is smaller than saturation gain, B is close to 1 and B of the pulse which has the largest power will reach the peak shown in Fig.11(b) firstly with increasing $G_r$. The peak value becomes lager and larger, which correspond to high power pulse is easy to enter saturation regime.

Fig.14 shows that $T_{rd}$ and B as a function of parameter $G_r$ for super-Gaussion-shaped pulse with different T. The power of injected stokes pulse is 0.1 $\mu W$ and other parameter is the same like above. It is indicated that smaller duration pulse can obtain lager relative delay. The $G_r$ needed for obtaining maximum $T_{rd}$ becomes lager and larger with the decreasing T. The reason is for same peak value the smaller duration pulse contains less energy. Then when it enters gain-saturation regime it need more energy and higher saturation gain results in higher $T_{rd}$. The changing rule of pulse broadening factor of the pulse which duration is less than 120ns is different from the one which duration is 120ns, however, it is the same with the one when injected stokes pulse is Gaussion-shaped. The main reason is the energy

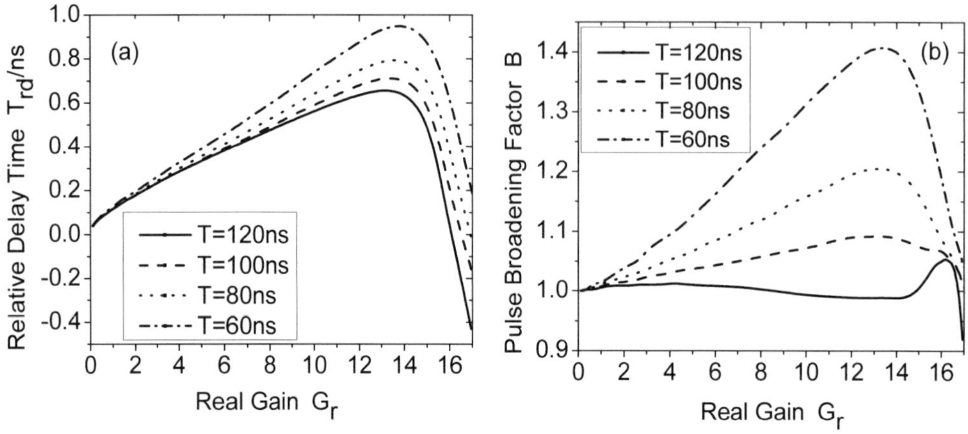


Fig. 13. Parameters of output stokes pulse versus real gain for different power. a)Time delay $T_d$ versus real gain; b) Pulse broadening factor B versus real gain.

which pulse contains decrease with the decreasing duration for same peak value. The energy getting from pump wave decreases, too. So the tailing edge of stokes pulse can obtain more energy than the one when T=120ns, resulting in the tailing edge broaden widely. This counteracts the compression corresponding to the steep leading edge. It can be predicted that increasing m can make the leading edge steeper and can decrease B. Our numerical result proved it. For the pulse with T=60ns, B at maximum $T_{rd}$ decreases from 1.406 to 1.295 when m is changed from 1.5 to 5.

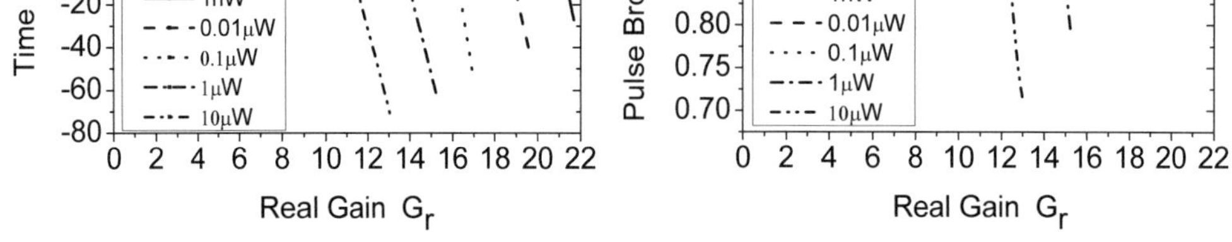


Fig. 14. Parameters of output stokes pulse versus real gain for different duration. a) Relative time delay $T_{rd}$ versus real gain; b) Pulse broadening factor B versus real gain.

## 5. Conclusion

We make a numerical study of the SBS slow light in optical fibers, and consider the influences of gain coefficient, gain bandwidth and effective mode area on time delay and pulse broadening. In the small signal regime, we find that the time delay and the pulse broadening factor increase with the increasing gain, but for pulse with the smaller gain bandwidth has a larger slope than the others. In the gain saturation regime, the pulse with larger gain coefficient, smaller gain bandwidth, smaller effective mode area begins to decrease more quickly in the gain range of 0~16. For the gain larger than 16, the pulse advancement becomes more obviously and the distortion also becomes more seriously, which may render the delay useless. We also investigate the time delay and pulse broadening factor vary with the increasing gain coefficient, gain bandwidth and effective mode area at a given pump power whose gain parameter is in the small signal regime, and find that the time delay and pulse broadening factor are proportional to the gain coefficient, whereas inversely proportional to the gain bandwidth and the effective mode area.

According to the above numerical calculation and theory analysis, we find that decreasing the power and duration of injected stokes pulse induces increasing delay time and pulse broadening factor; using super-Gaussion-shaped pulse as the injected stokes pulse can contribute evidently to decreasing pulse broadening factor in low frequency. Selecting pretty m can get perfect delay time and pulse broadening factor. Though this adjusting effect will become weaker for a shorter pulse, this reform still takes advantage of decreasing the error rate of all-optical-buffer.

## 6. Acknowledgment

This work was supported by the National Natural Science Foundation of China (Grant No. 61167005), the Provincial Natural Science Foundation of Gansu Province (Grant No. 1010RJZA036), the Provincial Natural Science Foundation of Guangdong Province (Grant No. 10451170003004948) and Dongguan Science and Technology Program (Grant No.2008108101002).

## 7. References

Boyd R.W. and Gauthier D. J.(2009), Controlling the Velocity of Light Pulses, *Science*, Vol.326,No.5956,pp.1074-1077, ISSN 0036-8075.

Stenner M. D, Gauthier D. J, Neifeld M. A(2003). The speed of information in a 'fast-light' optical medium, *Nature*, Vol. 425, pp. 695-698.ISSN0028-0836.

Gehring G. M, Schweinsberg A., BarsC. i, Kostinski N., Boyd R. W.(2006) Observation of Backward Pulse Propagation Through a Medium with a Negative Group Velocity, *Science*, Vol.312,No.5775,pp.895-897, ISSN 0036-8075.

Brillouin L(1960), *Wave Propagation and Group Velocity* ,Academic Press, New York.

Faxvog F. R, Chow C. N. Y, Bieber T., and Carruthers J. A(1970). Measured Pulse Velocity Greater Than *c* in a Neon Absorption Cell, *Appl. Phys. Lett.*, Vol.17, No.5, pp.192-193,ISSN0003-6951.

Hau L. V., Harris S. E., Dutton Z. et al(1999). Light speed reduction to 17 metres per second in an ultracold atomic gas. *Nature*, Vol.397, pp.594–598, ISSN0028-0836.

Philips D. F., Fleischhauer A., Mair A. et al(2001). Storage of light in atomic vapor. *Phys. Rev. Lett.*, Vol. 86,No.5,pp.783–786. ISSN1079-7114.

Bigelow M. S., Lepeshkin N. N., Boyd R. W(2003). Superluminal and slow light propagation in a room-temperature solid. *Science*, Vol.301,No.5630,pp.200–202, ISSN 0036-8075.

Milonni P. W(2005), *Fast Light, Slow Light, and Left-Handed Light*, Institute of Physics Publishing, Bristol, UK.

Kocharovskaya, O.; Rostovtsev, Y.; Scully, M.O(2001). Stopping light via hot atoms, *Phys. Rev. Lett.* Vol. 86,No.4,pp. 628-631, ISSN1079-7114.

Strekalov, D.; Matsko, A.B.; Yu, N.; Maleki, L.J. J(2004). Influence of inhomogeneous broadening on group velocity in coherently pumped atomic vapour, *Mod. Opt.* Vol.51, No.16,pp.2571-2578, ISSN0950-0340.

Schwartz S.E, Tan T.Y(1967). Wave Interactions in Saturable Absorbers. *Appl. Phys. Lett.* Vol. 10,No.1,pp.4~7, ISSN0003-6951.

Hillman L.W., Boyd R.W., Krasinski J., and Stroud C.R(1983). Observation of a Spectral Hole Due to Population Oscillations in a Homogeneously Broadened Optical Absorption Line. *Opt. Commun.* Vol. 45,No.6,pp. 416~419, ISSN0030-4018.

Bigelow M. S., Lepeshkin N. N., and Boyd R. W(2003). Observation of Ultraslow Light Propagation in a Ruby Crystal at Room Temperature. *Phys. Rev. Lett.* Vol.90, No.11,pp. 113903, ISSN1079-7114.

Boyd R.W(1992). *Nonlinear Optics* , Academic Press, San Diego.

A. Schweinsberg, N. N. Lepeshkin, M. S. Bigelow, R.W.Boyd, and S.Jarabo. Observation of Superluminal and Slow light Propagation in Erbium-Doped Optical Fiber. Europhys. Lett. 2006,73(2):218-224

Shin H., Schweinsberg A., Gehring G., Schwertz K., Chang H. J., and Boyd R. W(2007). Reducing Pulse Distortion in Fast-Light Pulse Propagation through an Erbium-Doped Fiber Amplifier. *Opt. Lett.*, Vol.32,No.8,pp.906-908,ISSN0146-9592.

Song K. Y., Herráez M. G., and Thévenaz L.(2005). Long Optically Controlled Delays in Optical Fibers. *Opt. Lett.*, Vol.30,No.14,pp.1782-1784, ISSN0146-9592.

Novak S., Gieske R(2002). Analytic Model for Gain Modulation in EDFAs. *J Lightwave Technol.*, Vol.20,No.6,pp.975-985,ISSN0733-8724.

Sargent III M(1978). Spectroscopic techniques based on Lamb's laser theory. *Phys. Rep.*,Vol. 43,No.5,pp. 223-265,ISSN0370-1573.

Boyd,R. W, Gauthier D. J, Gacta A.L, and Willner A.E(2005). Maximum Time Delay Achievable on Propagation through a Slow-light Medium. *Phys. Rev. A.*, Vol. 71,No.2,pp.023801-1-023801-4, ISSN1094-1622.

Gehring G. M., Boyd R. W.,. Gaeta A.L., Gauthier D. J. and Willner A. E(2008). Fiber-Based Slow-Light Technologies, *Journal of Lightwave Technology*, Nol.26,No.23,pp. 3752-3762 , ISSN0733-8724 .

Zhu Z., Gauthier D. J., Boyd R.W(2007). Stored Light in an Optical Fiber via Stimulated Brillouin Scattering, *Science*, Vol.318,No.5857,pp.1748-1750, ISSN 0036-8075.

Song K. Y., Herraez M. G., and Thevenaz L(2005). Observation of pulse delaying and advancement in optical fibers using stimulated Brillouin scattering, *Opt. Express,* Vol.13,No.1,pp.82-88 , eISSN1094-4087.

Abedin K. S.(2005), Observation of strong stimulated Brillouin scattering in single-mode As2Se3 chalcogenide fiber, *Opt. Express,* Vol.13,No.25,pp.10266-10271,eISSN1094-4087.

Song K. Y, Abedin K. S, Hotate K., Herraez M. G and Thevenaz L(2006), Highly efficient Brillouin slow and fast light using As2Se3 chalcogenide fiber, *Opt. Express,* Vol.14,No.13, pp.5860–5865, eISSN1094-4087.

Abedin K. S(2006), Stimulated Brillouin scattering in single-mode tellurite glass fiber, *Opt. Express*, Vol.14,No.24, pp.11766–11772, eISSN1094-4087.

Jauregui C., Ono H., Petropoulos P. and Richardson D. J(2006), Four-fold reduction in the speed of light at practical power levels using Brillouin scattering in a 2-m bismuth-oxide fiber, in *Proc. of Conference on Optical Fiber Communication (OFC 2006),* Paper PDP2 .

Stenner M. D., Neifeld M. A., Zhu Z., Dawes A. M. C. and Gauthier D. J(2005). Distortion management in slow-light pulse delay, *Opt. Express,* Vol.13, No.25, pp. 9995-10002 , eISSN1094-4087.

Schneider T., Henker R., Lauterbach K. U., and Junker M(2007). Comparison of delay enhancement mechanisms for SBS-based slow light systems, *Opt. Express,* Vol.15,No.15),pp.9606-9613, eISSN1094-4087.

Schneider T., Henker R., Lauterbach K. U., and Junker M(2008). Distortion reduction in Slow Light systems based on stimulated Brillouin scattering, *Opt. Express,* Vol.16,No.11,pp. 8280-8285, eISSN1094-4087.

Herraez M. G, Song K. Y., and Thévenaz L. (2006), Arbitrary-bandwidth Brillouin slow light in optical fibers, *Opt. Express* , Vol.14, No.4, pp.1395-1400, eISSN1094-4087.

Zhu Z., Dawes A. Gauthier M. C., D. J., Zhang L., and Willner A. E. (2006), 12-GHz-Bandwidth SBS Slow Light in Optical Fibers, in *Proc. of OFC 2006,* paper PD1.

Song K. Y., Hotate K. (2007), 25 GHz bandwidth Brillouin slow light in optical fibers, *Opt. Lett.* , Vol.32 , No.3, pp.217-219, ISSN0146-9592.

Zhu Z., Gauthier D. J., and et al(2005), Numerical study of all-optical slow-light delays via stimulated Brillouin scattering in an optical fiber, *J. Opt. Soc. Am. B,* Vol.22 , No.11, pp.2378-2384, ISSN0740-3224.

Kalosha V. P., Chen L, and Bao X. Y(2006), Slow and fast light via SBS in optical fibers for short pulses and broadband pump, *Opt. Express,* Vol.14, No.26, pp. 12693-12703, eISSN1094-4087

Damzen M. J., Vlad V. I., Babin,V. and Mocofanescu A(2003), *Stimulated Brillouin Scattering: Fundamentals and Applications,* IOP Publishing, ISBN0 7503 0870 2.

Dane C. B., Neuman W. A. and Hackel L. A. (1994), High-energy SBS pulse compression, *IEEE J. Quantum Electro,* Vol.30, No.8, pp.1907-1915, ISSN0018-9197.

Zhu Z., Gauthier D. J., Okawachi Y, *et al*(2005). Numerical study of all-optical slow-light delays via stimulated Brillouin scattering in an optical fiber, *J. Opt. Soc. Am. B,* Vol.22, No.11, pp. 2378- 2384, ISSN0740-3224.

Velchev I., Neshev D., Hogervorst W., and Ubachs W(1999). Pulse compression to the subphonon lifetime region by half-cycle gain in transient stimulated Brillouin scattering. *IEEE J. Quantum Electron*, Vol.35, No.12, pp.1812-1816, ISSN0018-9197.

# Part 2

## Other Impairments in Optical Fibers

# 6

# Polarization Losses in Optical Fibers

Hassan Abid Yasser
*Thi-Qar University,*
*Iraq Republic*

## 1. Introduction

Very long span optical communications are mainly limited by the chromatic dispersion (CD) or group velocity dispersion (GVD), fiber nonlinearities, and optical amplifier noise (Agrawal 2005). Different frequencies of a pulse travel with their own velocities, which involves a pulse spreading. In a fiber-optic communication system, information is transmitted within a fiber by using a coded sequence of optical pulses whose width is determined by the bit rate of the system. The CD induced broadening of pulses is undesirable phenomenon since it interferes with the detection process leading to errors in the received bit pattern (Kogelnik & Jopson 2002; Mechels et al. 1997). Clearly GVD will limit the bit rate and the transmission distance of a fiber-optic communication system. GVD is basically constant over time, and compensation can be set once and forgotten (Karlsson 1994).

When the signal channel bit rates reached beyond 10 Gb/s, polarization mode dispersion (PMD) becomes interesting to a larger technical community. PMD is now regarded as a major limitation in optical transmission systems in general, and an ultimate limitation for ultra-high speed signal channel systems based on standard single mode fibers (Mahgerftech & Menyuk 1999). PMD arises in optical fibers when the cylindrical symmetry is broken due to noncircular symmetric stress. The loss of such symmetry destroys the degeneracy of the two eigen-polarization modes in fiber, which will cause different GVD parameters for these modes. In standard single mode fibers, PMD is random, i.e. it varies from fiber to fiber. Moreover, at the same fiber PMD will vary randomly with respect to wavelength and ambient temperature (Lin & Agrawal 2003b; Sunnerud et al. 2002). The differential group delay (DGD) between two orthogonal states of polarization called the principal states of polarization (PSP's) causes the PMD (Tan et al. 2002; Wang et al. 2001). As a pulse propagates through a light-wave transmission system with a PMD, the pulse is spilt into a fast and slow one, and therefore becomes broadened. This kind of PMD is commonly known as first-order PMD. Under first-order PMD, a pulse at the input of a fiber can be decomposed into two pulses with orthogonal states of polarization (SOP). Both pulses will arrive at the output of the fiber undistorted and polarized along different SOP's, the output SOP's being orthogonal (Chertkov et al. 2004; Foshchini & Poole 1991). Both the PSP's and the DGD are assumed to be frequency independent when only first-order PMD is being considered (Lin & Agrawal 2003c; Gordon & Kogelnik 2000).

Second-order PMD effects account for the frequency dependence of the DGD and the PSP's. The frequency dependence of the DGD introduces an effective chromatic dispersion of opposite sign on the signals polarized along the output PSP's (Elbers et al. 1997; Ibragimv &

Shtenge 2002). Fiber PMD causes a variety of impairments in optical fiber transmission systems. First of all there is the inter-symbol interference (ISI) impairment of a single digital transmission channel. The ISI impairment is caused by the DGD between the two pulses propagating in the fiber when the input polarization of the signal does not match one of the PSP's of the fiber PMD impairments due to inter-channel effects that occur in polarization-multiplexed transmission systems (Agrawal 2005; Yang et al. 2001).
There are two polarization effects that lead to impairments in the long-haul optical fiber transmission systems: PMD and polarization dependent loss (PDL) (Chen et al. 2003; Chen et al. 2007). The WDM systems whose channels are spread over a large bandwidth rapidly change their state of polarizations (SOP's) due to PMD so that the overall DOP of the system is nearly zero (Agrawal 2005; Kogelnik & Jopson 2002). At the same time different channels expiries different amounts of PDL, and since the amplifiers maintain the total signal power nearly constant, individual channels undergo a kind of random walk so that it is possible for some channels to fade (Shtaif & Rosenberg 2005; Menyuk et al. 1997). Calculating the impairments due to the combination of PMD and PDL in WDM systems is a formidable theoretical challenge (Phua & Ippen 2005). Physically, light pulses polarized along these PSP's propagate without polarization-induced distortion. When there is no PDL, the two PSP's are orthogonal and correspond to the fastest and slowest pulses, which can propagate in the fiber (Yasser 2010; Yaman et al. 2006). They thus constitute a convenient basis for polarization modes. When the system includes PDL, the Jones formalism is still applicable, but several of the above facts are not valid anymore. The notion of PSP's is still correct, but the two PSP's are not orthogonal nor do they represent the fastest and slowest pulses (Yoon & Lee 2004).
In this chapter, the analysis of Jones and Stokes vectors and the relation between them were discussed in section 2. The statistics of PMD are presented in section 3. The pulse broadening in presence of PMD and CD were illustrated in section 4. In section 5, the principal comparison between PMD and birefringence vector will be obtained. The combined effects of PMD and PDL are presented in section 6. Finally, section 7 will summarize the effects of nonlinearity on the effective birefringence vector.

## 2. Polarization dynamics

The representation of polarization in Jones and Stokes spaces and the connection between the two spaces will be presented in this section. Throughout this chapter, it is assumed that the usual loss term of the fiber has been factored out so that one can deal with unitary transmission matrices. Light in optical fibers can be treated as transverse electromagnetic waves. Considering the two perpendicular and linearly polarized light waves propagating through the same region of space in the z-direction, the two fields can be represented in complex notation as (Azzam & Bashara 1989)

$$\vec{E}_x(z,t) = \hat{x}\ E_{x0}\ e^{i(kz-wt+\varphi_x)} \tag{1a}$$

$$\vec{E}_y(z,t) = \hat{y}\ E_{y0}\ e^{i(kz-wt+\varphi_y)} \tag{1b}$$

where $\varphi_x$ and $\varphi_x$ are the phases of the two field components, and $k$ is the propagation constant. The resultant optical field is the vector sum of these two perpendicular waves, i.e.

$$\vec{E}(z,t) = \vec{E}_x(z,t) + \vec{E}_y(z,t) \tag{2}$$

The polarization state can be represented in terms of Jones vectors as

$$\hat{A} = \begin{bmatrix} a_x e^{i\varphi_x} \\ a_y e^{i\varphi_y} \end{bmatrix} \tag{3}$$

where $a_x = E_{x0} / \sqrt{E_{x0}^2 + E_{y0}^2}$ , $a_y = E_{y0} / \sqrt{E_{x0}^2 + E_{y0}^2}$ , and $\sqrt{a_x^2 + a_y^2} = 1$. Here $E_{x0}$ and $E_{y0}$ are the initial amplitude components of the light. The familiar form of Jones vector is denoted as ket vector as (Gordon & Kogelnik 2000)

$$|s> = \begin{bmatrix} s_x \\ s_y \end{bmatrix} = \begin{bmatrix} a_x e^{i\varphi_x} \\ a_y e^{i\varphi_y} \end{bmatrix} \tag{4}$$

whereas the bra $<s|$ indicates the corresponding complex conjugate row vector, i.e. $<s| = [s_x^* \quad s_y^*]$, where * indicates complex conjugation. The bra-ket notation is used to distinguish Jones vectors from another type of vectors that will be used in this chapter which is called the Stokes vectors. Partial correlation yields partial polarization and total correlation gives total polarization (Karlsson 1994; Sunnerud et al. 2002). When the light is coherent, Jones vectors are all of unit magnitude, i.e. $<s|s> = s_x s_x^* + s_y s_y^* = 1$. Given the Jones vector, the values of the azimuth angle, $\psi$, and the ellipticity angle, $\eta$, can be found by solving the equations (Rogers 2008)

$$\tan\, 2\psi = \frac{2\,\mathrm{Re}(s_y / s_x)}{1 - |s_y / s_x|^2} \tag{5a}$$

$$\sin\, 2\eta = \frac{2\,\mathrm{Im}(s_y / s_x)}{1 + |s_y / s_x|^2} \tag{5b}$$

where Re and Im denote the real and imaginary parts, respectively. Fig.(1 a) illustrates Jones representation of polarization vector.

The Poincare sphere is a graphical tool in real three dimensional space that allows convenient description of polarized signals and polarization transformations caused by propagation through devices. Any SOP can be represented uniquely by a point on or within a unit sphere centered on a rectangular coordinates system. The coordinates of a point are the three normalized Stokes parameters describing the state of polarization (Azzam & Bashara 1989; Rogers 2008). Partially polarized light can be considered as a combination of purely polarized light and un-polarized light. Orthogonal polarizations are located diametrically opposite to the sphere. As shown in Fig.(1 b), linear polarizations are located on the equator. Circular states are located at the poles, with intermediate elliptical states continuously distributed between the equator and the poles (Karlsson 1994; Kogelnik & Jopson 2002). There are two angles (or degrees of freedom, i.e. $\psi$ and $\eta$) describing an arbitrary Jones vector. These angles can be interpreted as coordinates in a spherical coordinates system, and each polarization state can then correspond to a point, represented by a Stokes vector, $\hat{s} = (s_1, s_2, s_3)^t$ on the Poincare sphere, where $t$ represents the transpose. The three Cartesian components can be defined as (Gordon & Kogelnik 2000)

$$\begin{aligned} s_1 &= |E_x|^2 - |E_y|^2 = \cos 2\psi \cos 2\eta \\ s_2 &= E_x E_y^* + E_y E_x^* = \sin 2\psi \cos 2\eta \\ s_3 &= i(E_y E_x^* - E_x E_y^*) = \sin 2\eta \end{aligned} \tag{6}$$

Therefore, the angle $2\psi$ is the angle from the direction of $s_1$ to the projection of $\hat{s}$ on the $s_1 - s_2$ plane, and $2\eta$ is the angle from $s_1 - s_2$ plane to the vector $\hat{s}$, see Fig. (1 b). Given Stokes vector, the values of $\psi$ and $\eta$ are obtained by solving the equations $s_2 / s_1 = \tan 2\psi$, and $s_3 = \sin 2\eta$.

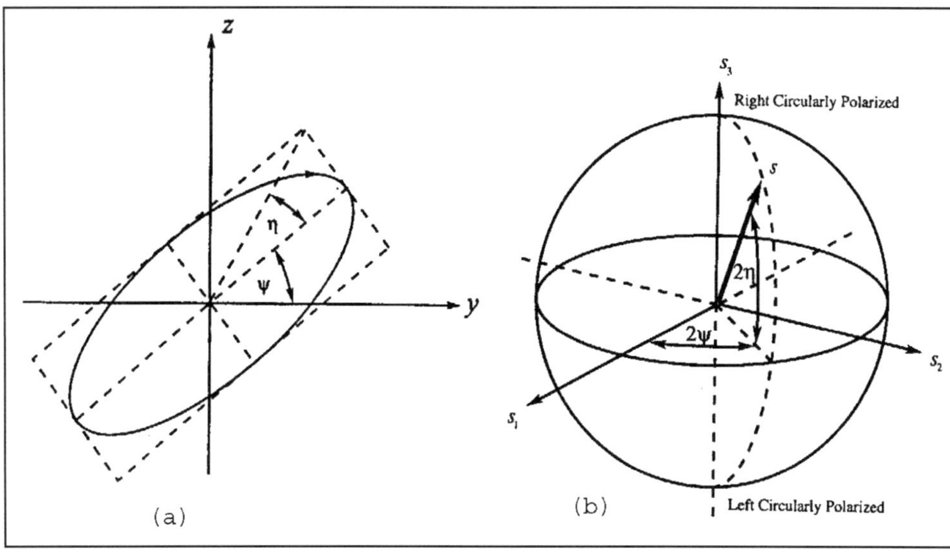


Fig. 1. Illustration of: a) Jones representation , b) Stokes representation.

Any Stokes vector $\hat{s}$ is related to another one $|s>$ in Jones space as $\hat{s} = <s|\vec{\sigma}|s>$, where $\vec{\sigma} = (\sigma_1, \sigma_2, \sigma_3)$ is the Pauli spin vector whose components are defined as (Levent et al. 2003)

$$\sigma_1 = \begin{bmatrix} 1 & 0 \\ 0 & -1 \end{bmatrix}, \quad \sigma_2 = \begin{bmatrix} 0 & 1 \\ 1 & 0 \end{bmatrix}, \quad \sigma_3 = \begin{bmatrix} 0 & -i \\ i & 0 \end{bmatrix} \tag{7}$$

It is important to note that if the angle between $\hat{p}$ and $\hat{s}$ in Stokes vector is $\theta$, then the angle between $|p>$ and $|s>$ in Jones space is $\theta / 2$. That is; if two vectors are perpendicular in Jones space then the corresponding two vectors in Stokes space are antiparallel. Each of these two spaces gives certain illustrations according to the case of study. For totally polarization, the value of polarization vector is unity, elsewhere, the value differs from unity. In general, the three components of $\hat{s}$ are not zero for elliptical polarization. The third component of $\hat{s}$ equals zero for linear polarization, whereas the first two components of $\hat{s}$ are zero for circular polarization. There is a unitary matrix, $T$, in Jones space which relates output to input via $|s>=T|t>$, where $|s>$ and $|t>$ are the output and input Jones vectors, respectively. On the other hand, a transformation matrix (Muller), $R$, in Stokes space relates output to input via $\hat{s}=R\,\hat{t}$, where $\hat{s}$ and $\hat{t}$ are the output and input Stokes vectors, respectively. The transmission matrices are related as $R\vec{\sigma}=T^{\dagger}\vec{\sigma}T$, where $\dagger$ denotes the transpose of the complex conjugate (Agrawal 2007; Chen et al. 2007).

## 3. Statistical managements

The effects of PMD are usually treated by means of the three-dimensional PMD vector that is defined as $\vec{\tau}=\tau_{pmd}\hat{p}$, where $\hat{p}$ is a unit vector pointing in the direction of slow PSP and $\tau_{pmd}$ is the DGD between the fast and slow components which is defined as (Mahgerftech & Menyuk 1999)

$$\tau_{pmd}=|\vec{\tau}|=\sqrt{\tau_1^2+\tau_2^2+\tau_3^2} \tag{8}$$

The PMD vector $\vec{\tau}$ in Stokes space gives the relation between the output SOP, $\hat{s}$, and the frequency derivative of the output SOP: $d\,\hat{s}(w)/dw=\vec{\tau}(w)\times\hat{s}(w)$. The PSP's are defined as the states that $\vec{\tau}(w)\times\hat{s}(w)=0$, so that no changes in output polarization can be observed close to these states at first order in $w$. To the first order, the impulse response of an optical fiber with PMD is defined as (Karlsson 1994)

$$h_{pmd}(T)=\gamma_+\delta(T-\tau_{pmd}/2)|p_+>+\gamma_-\delta(T+\tau_{pmd}/2)\ |p_->\tag{9}$$

where $\gamma_\pm$ are the splitting ratios and $|p_\pm>$ are the PSP's vectors. The factors $\gamma_\pm$ and $\tau_{pmd}$ vary depending on the particular fiber and its associated stresses, where the splitting ratios can range from zero to one. Note that, the function $h_{pmd}(T)$ is normalized in the range ($-\infty$ to $\infty$).

### 3.1 Splitting ratios

Consider that the PSP's occur with a uniform distribution over the Poincare sphere, and that $\hat{s}$ is aligned with the north pole of the sphere as shown in Fig.(2). The probability density of PSP's which is found in the range $d\theta$ about the angle $\theta$ relative to $\hat{s}$ is proportional to the differential area $2\pi\sin\theta\,d\theta$ sketched in the figure. As there is north/south symmetry in the differential area, the ranges ($0$ to $\pi/2$) and ($\pi/2$ to $\pi$) of $\theta$ are combined to obtain the combined probability density $p_\theta(\theta)=\sin\theta$. For the effective range ($0$ to $\pi/2$) describing the occurrence of PSP's with angle $\theta$ (and $\pi-\theta$) relative to $\pm\hat{s}$, the distribution $p_\theta(\theta)$ is properly normalized through the range ($0$ to $\pi/2$). The analyses of splitting ratios have led to a number of important fundamental advances as well as the technical point of view (Rogers 2008; Kogelnik & Jopson 2002). The splitting ratios $\gamma_\pm$ can be determined from the polarization vectors. In other words $\gamma_\pm$ represent the projection of $|p_+>$ and $|p_->$ onto $|s>$. Formally, $\gamma_\pm^2=|<s|p_\pm>|^2$, where $|s>$ and $|p_\pm>$ are the input SOP and the two PSP's vectors.
If the PSP's are defined as $|p_\pm>=[p_{\pm x}\ p_{\pm y}]^t$, then

$$|p_\pm><p_\pm|=\begin{bmatrix}p_{\pm x}\\ p_{\pm y}\end{bmatrix}\begin{bmatrix}p_{\pm x}^* & p_{\pm y}^*\end{bmatrix}=\begin{bmatrix}|p_{\pm x}|^2 & p_{\pm x}p_{\pm y}^*\\ p_{\pm y}p_{\pm x}^* & |p_{\pm y}|^2\end{bmatrix}\tag{10}$$

where $< p_{\pm} |$ are the transpose conjugation of $| p_{\pm} >$. Now, it is straightforward to show that

$$\pm \hat{p} =< p_{\pm} | \vec{\sigma} | p_{\pm} >= \begin{bmatrix} \pm p_1 \\ \pm p_2 \\ \pm p_3 \end{bmatrix} = \begin{bmatrix} < p_{\pm} | \sigma_1 | p_{\pm} > \\ < p_{\pm} | \sigma_2 | p_{\pm} > \\ < p_{\pm} | \sigma_3 | p_{\pm} > \end{bmatrix} = \begin{bmatrix} | p_{\pm x} |^2 - | p_{\pm y} |^2 \\ p_{\pm x} p_{\pm y}^* + p_{\pm y} p_{\pm x}^* \\ i(p_{\pm y} p_{\pm x}^* - p_{\pm x} p_{\pm y}^*) \end{bmatrix} \tag{11}$$

Comparing Eqs.(10) and (11), $| p_{\pm} >< p_{\pm} |= (I_2 \pm \hat{p} \cdot \vec{\sigma}) / 2$ can be extracted. In turn, the splitting ratios can be calculated by using Eq.(11) and the fact that $< a | \hat{p} \cdot \vec{\sigma} | a >= \hat{p} \cdot \hat{a}$ as follows

$$\gamma_+^2 =< s | p_+ >< p_+ | s >=< s | (I_2 + \hat{p} \cdot \vec{\sigma}) | s > / 2 = (1 + \hat{p} \cdot \hat{s}) / 2 = \cos^2(\theta / 2) \tag{12a}$$

$$\gamma_-^2 =< s | p_- >< p_- | s >=< s | (I_2 - \hat{p} \cdot \vec{\sigma}) | s > / 2 = (1 - \hat{p} \cdot \hat{s}) / 2 = \sin^2(\theta / 2) \tag{12b}$$

Until now, the relationship between the splitting ratios and elevation angle was calculated, where the ratios $\gamma_{\pm}$ are identical only for $\theta = \pi / 2$.

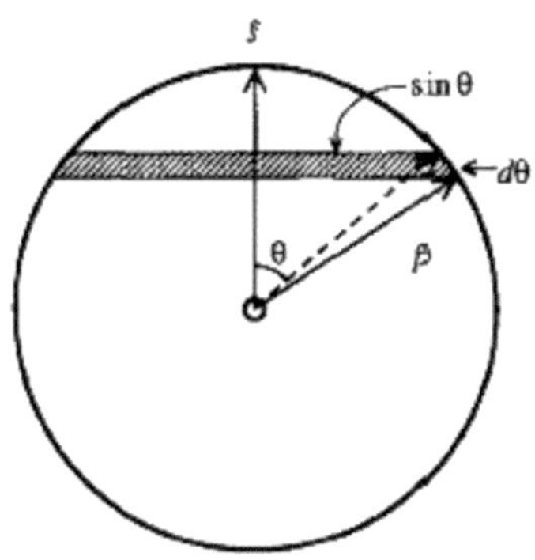


Fig. 2. Sketch of differential area on Poincare sphere as a function of elevation angle $\theta$.

### 3.2 Statistics of DGD

Throughout this subsection, the PMD statistics have been carefully analyzed since it causes a variation in the pulse properties. A proper measure of pulse width for pulses of arbitrary shapes is the root-mean square (rms) width of the pulse defined as $\tau_{rms} = \sqrt{< T^2 > - < T >^2}$. The PMD induced pulse broadening is characterized by the rms value of $\tau_{pmd}$. The $\tau_{rms}$ is obtained after averaging over random birefringence changes. The second moment of $\tau_{pmd}$ is given by (Fushchini & Poole 1991)

$$< \tau_{pmd}^2 >= \tau_{rms}^2 = 2(\Delta\beta_1)^2 \ \ell_c^2 \ [ L / \ell_c + e^{-L/\ell_c} - 1 ] \tag{13}$$

where $\ell_c$ is the correlation length that is defined as the length over which two polarization components remain correlated, $\Delta\beta_1 = v_{gx}^{-1} - v_{gy}^{-1}$ is related to the difference in group velocities along the two PSP's. For distances $L >> 1\ km$, a reasonable estimate of pulse broadening was obtained by taking the limit $L >> \ell_c$ in Eq.(13). The result is given by $\tau_{rms} \approx \Delta\beta_1 \sqrt{2L\ell_c} = D_p \sqrt{L}$, where $D_p$ is known as the PMD parameter that takes the values (0.01-10)ps/$\sqrt{\text{km}}$. The variable $\tau_{pmd}$ has been determined to obey a Maxwellian distribution of the form (Agrawal 2005)

$$p(\tau_{pmd}) = \sqrt{\frac{54}{\pi}} \frac{\tau_{pmd}^2}{\tau_{rms}^3} e^{-3\tau_{pmd}^2/2\tau_{rms}^2} \tag{14}$$

The mean of $\tau_{pmd}$ is done simply as $\bar{\tau}_{pmd} = \sqrt{8/3\pi}\,\tau_{rms}$. Using this result, the Maxwellian distribution will take the form

$$p(\tau_{pmd}) = \frac{32}{\pi^2} \frac{\tau_{pmd}^2}{\bar{\tau}_{pmd}^3} e^{-4\tau_{pmd}^2/\pi\bar{\tau}_{pmd}^2} \tag{15}$$

A cursory inspection of Eq.(15) reveals that the $p(\tau_{pmd})$ can be found if $\bar{\tau}_{pmd}$ is known. Here, a relationship for $\tau_{pmd}$ that will maximize $p(\tau_{pmd})$ can be found. The distribution $p(\tau_{pmd})$ has a maximum value at $\tau_{pmd} = \tau_{pmd}^{\max} = \sqrt{\pi}\bar{\tau}_{pmd}/2$ . This conclusion provides a method for calculating the maximum likelihood value of $\tau_{pmd}$ if $\bar{\tau}_{pmd}$ is known.

### 3.3 Statistics of impulse response

The rms width of the impulse response, $\tau_{eff}$, can be readily calculated by substituting Eq.(9) into $\tau_{rms} = \sqrt{<T^2> - <T>^2}$ to yield

$$\tau_{eff} = \sqrt{\int_{-\infty}^{\infty} T^2 h_{pmd}(T)dT - \left[\int_{-\infty}^{\infty} T h_{pmd}(T)dT\right]^2} = \sin\theta\ \tau_{pmd}/2 \tag{16}$$

Using the result $p_\theta(\theta) = \sin\theta$ and Eq.(16), the density distribution for $\theta$ can be transformed to the density for $\tau_{eff}$ as follows

$$p_{\tau_{eff}}(\tau_{eff}) = p_\theta(\theta(\tau_{eff}))\frac{d\theta}{d\tau_{eff}} = \frac{4\tau_{eff}}{\tau_{pmd}\sqrt{\tau_{pmd}^2 - 4\tau_{eff}^2}} \tag{17}$$

It is important to note that the probability density is a function of $\tau_{eff}$ and $\tau_{pmd}$. As a consequence of this dependence, Eq.(17) can not be integrated to determine $\tau_{eff}$ due to the presence of the other variable $\tau_{pmd}$. So, the next step is to seek about $p_{\tau_{eff}}(\tau_{eff})$ in order to determine the statistical properties of output pulses. The joint probability distribution $p(\tau_{eff}, \tau_{pmd})$ can be illustrated using Eqs.(14) and (17) as follows

$$p(\tau_{eff}, \tau_{pmd}) = \frac{64\tau_{eff}}{\pi\bar{\tau}_{pmd}^3} \frac{\tau_{pmd}}{\sqrt{\tau_{pmd}^2 - 4\tau_{eff}^2}} e^{-4\tau_{pmd}^2/\pi\bar{\tau}_{pmd}^2} \tag{18}$$

Recalling Eq.(16), it may be written as $\tau_{pmd} = 2\tau_{eff}/\sin\theta$. Since $0 \le \sin\theta \le 1$, such that $2\tau_{eff} \le \tau_{pmd} < \infty$. The probability distribution $p(\tau_{eff})$ can be found by integrating Eq.(18) about $\tau_{pmd}$ through the range $2\tau_{eff} \le \tau_{pmd} < \infty$ to obtain

$$p(\tau_{eff}) = \frac{32\tau_{eff}}{\pi\bar{\tau}_{pmd}^2} e^{-16\tau_{eff}^2/\pi\bar{\tau}_{pmd}^2} \tag{19}$$

At a basic level, Eq.(17) is the same as Eq.(19) but the latter is a function of $\tau_{eff}$ only, which can be integrated to obtain $\tau_{eff}$. However, both equations are normalized properly. The mean value of $\tau_{eff}$ is determined as $\theta = \pi / 2$. So, Eq.(19) may be written as

$$p(\tau_{eff}) = \frac{\pi\, \tau_{eff}}{2\, \bar{\tau}_{eff}^{2}}\, e^{-\pi\, \tau_{eff}^{2} / 4\, \bar{\tau}_{eff}^{2}} \tag{20}$$

The distribution $p(\tau_{eff})$ has a maximum value at $\tau_{eff} = \tau_{eff}^{\max} = \sqrt{\pi / 32}\bar{\tau}_{pmd}$. This is equivalent to find the maximum likelihood value of $\tau_{eff}$ if $\bar{\tau}_{pmd}$ is known.

### 3.4 Pulse characteristics

Using the PSP's as an orthogonal basis set, any input or output polarization can be expressed as the vector sum of two components, each aligned with a PSP. Within the realm of the first-order PMD, the output electric field from a fiber with PMD has the form (Rogers 2008)

$$| A_{out}(T) >= \gamma_{+}\, A_{in}(T - \tau_{pmd} / 2)\;\; | p_{+} > + \gamma_{-}\, A_{in}(T + \tau_{pmd} / 2)\;\; | p_{-} > \tag{21}$$

where $A_{in}(T)$ is the input electric field. To determine the output power $P_{out}(T) =< A_{out}(T) | A_{out}(T) >$, it is important to point out the orthogonality properties of Jones vectors, that is; $< p_{\pm} | p_{\mp} >= 0$ and $< p_{\pm} | p_{\pm} >= 1$. Note that, we perform the derivation using a normalized Gaussian pulse that takes the form $A_{in}(T) = D\exp(-T^{2} / 2T_{0}^{2})$, where $D = \sqrt{E_{in} / T_{0}\sqrt{\pi}}$, $T_{0}$ is the initial pulse width, and $E_{in}$ is the input pulse energy. For normalized power, we make $D^{2} = 1$. Therefore, according to Eq.(21), the shifted pulses will reshape as

$$A_{in}(T \pm \tau_{pmd} / 2) = D\exp\left[ -\frac{(T \pm \tau_{pmd} / 2)^{2}}{2T_{o}^{2}} \right] \tag{22}$$

Substituting Eqs.(12) and (22) into (21), using the output power definition, using the orthogonality properties of Jones vectors, and simplified the result, we obtain the following expression

$$P_{out}(T) = \left[ \cos^{2}(\theta\ / 2)\, e^{-T\tau_{pmd} / T_{0}^{2}} + \sin^{2}(\theta / 2)\, e^{T\tau_{pmd} / T_{0}^{2}} \right] e^{-(4T^{2} + \tau_{pmd}^{2}) / 4T_{0}^{2}} \tag{23}$$

The width of the output pulse $T_{1}$ can be determined as follows

$$T_{1} = \sqrt{\int_{-\infty}^{\infty} T^{2} P_{out}(T) dT - \left[ \int_{-\infty}^{\infty} T P_{out}(T) dT \right]^{2}} = \sqrt{T_{0}^{2} + (\tau_{pmd} / 2)^{2} \sin^{2}\theta} \tag{24}$$

The time jittering of the pulse can be found by determining the maximum value of $P_{out}(T)$. This maximum value will happen at $T = T_{peak} = \tau_{pmd} \cos(\theta) / 2$. The peak power, as a function of DGD and an angle $\theta$, at the pulse center can be determined by substituting the latter result into Eq.(23) to get

$$P_{peak}(\tau_{pmd}, \theta) = \cos^{2}(\theta/2)\, e^{-\sin^{2}(\theta/2)\tau_{pmd}^{2} / T_{0}^{2}} + \sin^{2}(\theta/2)\, e^{-\cos^{2}(\theta/2)\tau_{pmd}^{2} / T_{0}^{2}} \tag{25}$$

At this point, we drive formulas for the output power form, final width, time jittering (shifting), and peak power as functions of the random physical variables $\theta$ and $\tau_{pmd}$.

Fig.(3) illustrates the simulation with the parameters: $L = 50\ km$, $D_p = 0.5\ ps/\sqrt{km}$ , and $T_0 = 5\ ps$. The solid line represents the original pulse while the discrete lines represent the resulted pulses with different values of $\tau_{pmd}$ ranging from $0$ to $8ps$, where the closest to $T = 0$ is the pulse that has least value of $\vec{\tau}$. At the angle $\theta = 0$, one note that the pulse is faced only by a displacement to the right at $T_{peak} = \tau_{pmd} / 2$. Increasing $\theta$, the pulse width and distortion will be increased, while the power and shifting will be decreased. These variations are the greatest at $\theta = \pi / 2$. After $\theta = \pi / 2$, the effects are reversed. At $\theta = \pi$, again the pulse is faced only by a displacement but to the left at $T_{peak} = -\tau_{pmd} / 2$. It is clear that the penalty could be greater if $\theta = \pi / 2$ and will be zero at $\theta = 0\ or\ \pi$.

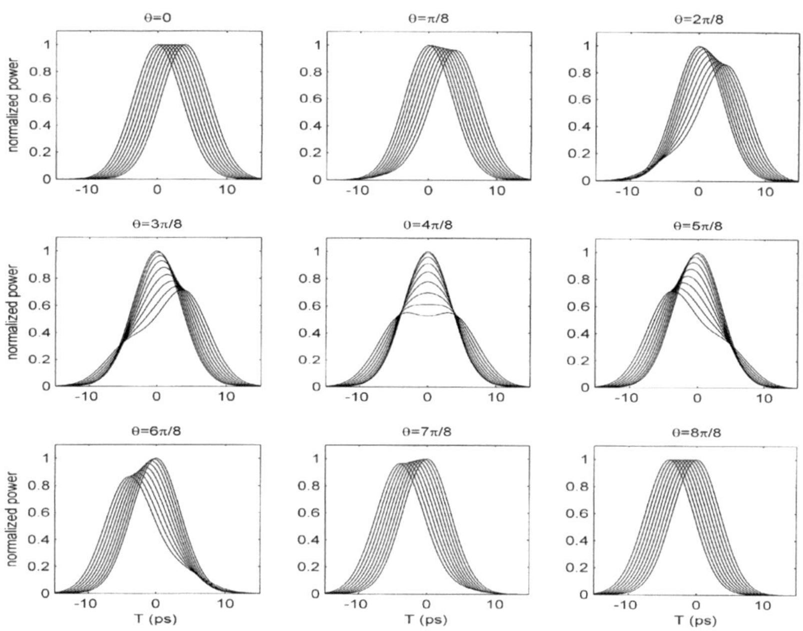


Fig. 3. Pulse shape with different values of $\tau_{pmd}$ and $\theta$ for different values of $\tau_{pmd}$; the lower value of $\tau_{pmd}$ is the closest to the pulse center.

## 4. Polarization mode dispersion and chromatic dispersion

The pulses that propagate through single mode fiber (SMF) are affected by two types of dispersion which are CD and PMD. Notice that the effects of the two types of dispersion happen at the same time, so to give a distinct sense of the two types of dispersion we decided to obtain the effects in the frequency domain. The initial pulse, $\tilde{A}(0,w) = \Im\{A(0,T)\}$, first faces the affect of CD (the transfer function $H_1(w)$) to obtain $H_1(w)\tilde{A}(0,w)$. The CD does not depend on SOP therefore the input SOP (the Jones vector $|a>$) will not change. Next, the pulse divides into two orthogonal components towards PSP's ( $|a^+>$ and $|a^->$) under the effects of PMD. The component in the direction $|a^+>$ will face the effects of the function $H_{2s}(w)$ to obtain the pulse $H_{2s}(w)H_1(w)\tilde{A}(0,w)$ and at the same time the SOP will change from $|a^{+}>$ to $|b^{+}>$. On the other hand, the pulse in the direction $|a^{-}>$ faces the effects of the function $H_{2f}(w)$ to yield $H_{2f}(w)H_{1}(w)\tilde{A}(0,w)$ and also the SOP will change from $|a^{-}>$ to $|b^{-}>$. The input or output PSP's does remain orthogonal when the PDL is absent. Finally, the vector sum of the two components will produce the final pulse $H_{2}(w)H_{1}(w)\tilde{A}(0,w)$. The transfer function of the CD of lossless fiber in frequency domain is $H_{1}(w)=\exp(i\,w^{2}\,\beta_{2}\,L/2)$, where $\beta_{2}=-\lambda^{2}d(\lambda)/2\pi\,c$, $d(\lambda)$ is fiber chromatic dispersion parameter, L is the fiber length, and $\lambda$ is light wavelength. Now, assume that there is negligible PDL, so that we can use the principal states model (Lin & Agrawal 2003b; Ibragimv & Shtenge 2002; Foshchini & Poole 1991) to characterize first-order PMD. Under this model, there exist a pair of orthogonal input PSP's, $|a^{+}>$ and $|a^{-}>$, and a pair of orthogonal output PSP's, $|b^{+}>$ and $|b^{-}>$, where all of PSP's are expressed as Jones vectors. If an arbitrary polarized field $\vec{A}_{a}(t)=A_{a}(t)\ |a>$ is input to the fiber, this input field can be projected onto the two PSP's as

$$\vec{A}_{a}(T)=\gamma_{+}\,A_{a}(T)\ \ |a^{+}>+\gamma_{-}\,A_{a}(T)\ \ |a^{-}> \tag{26}$$

In terms of first-order PMD, the output field of the fiber takes the form

$$\vec{A}_{b}(T)=\gamma_{+}A_{b}(T-\tau_{pmd}/2)|b^{+}>+\gamma_{-}\,A_{b}(T+\tau_{pmd}/2)\ \ |b^{-}> \tag{27}$$

According to Eq.(9), the fiber transfer functions for first-order in the time and frequency domains are given by

$$h_{2}(t)=\gamma_{+}\ \ \delta(T-\tau_{pmd}/2)\ \ |b^{+}>+\gamma_{-}\ \delta(T+\tau_{pmd}/2)|b^{-}> \tag{28a}$$

$$H_{2}(w)=\ \gamma_{+}e^{iw\tau_{pmd}/2}\ \ |b^{+}>\ +\gamma_{-}\ e^{-iw\tau_{pmd}/2}|b^{-}> \tag{28b}$$

The root mean square width of this impulse response which can be calculated as

$$\begin{aligned} <T>&=\int_{-\infty}^{\infty}T\,h_{2}(T)dT=(\gamma_{+}\ \tau_{pmd}|\,b^{+}>-\gamma_{-}\ \tau_{pmd}\ \ |b^{-}>)/2 \\ <T^{2}>&=\int_{-\infty}^{\infty}T^{2}h_{2}(T)dT=(\gamma_{+}\ \tau_{\mathrm{pmd}}^{2}\,|b^{+}>+\gamma_{-}\tau_{\mathrm{pmd}}^{2}\ \ |b^{-}>)/4 \\ \tau_{rms}^{\pm}&=[\sqrt{<T^{2}>-<T>^{2}}\,]_{\pm}=0 \end{aligned} \tag{29}$$

where the signs (+ , - ) mean that the impulse response in directions of $|b^{+}>$ or $|b^{-}>$, respectively. That is; the width of an impulse response in the direction of PSP's will be zero, while the width in the direction of $|b>$ will be $\tau_{rms}=\sin\theta\ \tau_{pmd}/2$. This represents the extra width that results due to the effects of PMD on the propagated signal. It is clear that, if the input SOP is in direction of PSP's, then the pulse will not suffer any broadening.

The Fourier transformation of the initial pulse takes the form $\tilde{A}(0,w)=D\sqrt{2\pi}T_{o}\exp(-w^{2}T_{o}^{2}/2)$. The total effects on the pulse shape can be obtained by using the convolution of the transfer functions of the combined PMD and CD with the input Gaussian signal in the time domain, or equivalently by using the inverse Fourier transform as follows

$$A(z,T)=\Im^{-1}\{\tilde{A}(0,w)\cdot H_{1}(w)\cdot H_{2}(w)\}=\cos(\theta/2)\,A_{+}(z,T)\ \ |b^{+}>+\sin(\theta/2)A_{-}(z,T)\ \ |b^{-}> \tag{30}$$

where

$$A_{\pm}(z,T) = D\frac{T_o}{\chi}\ \exp(-\frac{T_{\pm}^2}{2\ T_1^2})\ \ \exp(\ i\ \phi_{\pm}(z,T)\ )$$

$$\chi = \sqrt[4]{\mathrm{T_o^4} + (\beta_2\ z)^2}$$

$$\mathrm{T}_{\pm} = T \pm \tau_{pmd} / 2$$

$$T_1 = \sqrt{T_o^2 + (\beta_2 z / T_o^2)^2}$$

$$\phi_{\pm}(z,T) = -\frac{\beta_2 z}{2T_1^2}\frac{T_{\pm}^2}{T_o^2} + \frac{1}{2}\tan^{-1}(\beta_2 z / T_o^2)$$

The parameter $T_1$ represents the pulse width including CD effects where it is the same for the two orthogonal components. The width of each component will not increase under the effects of PMD, but the pulse which results from the vector sum of the two orthogonal components will face a broadening that can be determined by $\tau_{rms}$. The parameters $\phi_{\pm}(z,T)$ represent the nonlinear phases that generate through the propagation in optical fiber. The nonlinear phase as a function of time differs from one component to another by the amount $\tau_{pmd}$, but in the frequency domain they remain the same and add the same value of noise to both components. The frequency chirp can be written as

$$\delta\ w_{\pm}(T) = -\frac{\partial\phi_{\pm}(z,T)}{\partial T} = \frac{\beta_2 z}{T_1^2}\frac{T_{\pm}}{T_o^2} = \frac{\beta_2 z}{2T_1^2}\frac{T \pm \Delta\tau / 2}{T_o^2} \tag{31}$$

This means that the new frequencies generated are similar for the two components and the difference lies in $T \pm \tau_{pmd} / 2$ only, which means that one of the components advances the other by time $\tau_{pmd}$. Eq.(30) explains that the pulse amplitude will decrease by increasing the propagation distance, which will be converted to the same equations as in reference (Agrawal 2007) by ignoring the effects of PMD. The Jones vectors $|b^+>$ and $|b^->$ are orthogonal, i.e. $<b^+|b^->=0$. That is enough to assume a random form to one of them to find the other. For example, if $|b^+>=[x\quad iy]^t$ then $|b^->=[iy\quad x]^t$ keeping in mind that all the polarization vectors have unit values.
Now, the reconstructed width after including the effects of CD is $T_1$. Next, the input pulse has a width $T_1$ which will be increased by the amount $\tau_{rms}$ due to the PMD. Such that, the final width will be

$$T_f = \sqrt{T_1^2 + \tau_{pmd}^2 \sin^2\theta / 4} \tag{32}$$

Fig.(4 a) illustrates the shape of pulse for various values of $\beta_2$, assuming $\tau_{pmd} = 2\ ps$, $\theta = \pi / 2$, $T_o = 10$ ps, and $L = 60$ km. Since $\tau_{pmd}$ is constant for all cases, this implies that the time separation between the orthogonal components remains the same. The width of both components increases (under the effects of CD) by increasing $\beta_2$. Consequently, the width of the final pulse increases by increasing $\beta_2$, but the amplitude is decreased. The existence of CD causes a broadening factor (BR) of value $T_1 / T_o$, and the existence of PMD adds a BR of value $\tau_{rms} / T_1$. That is; the width of pulse will increase due to the existence of

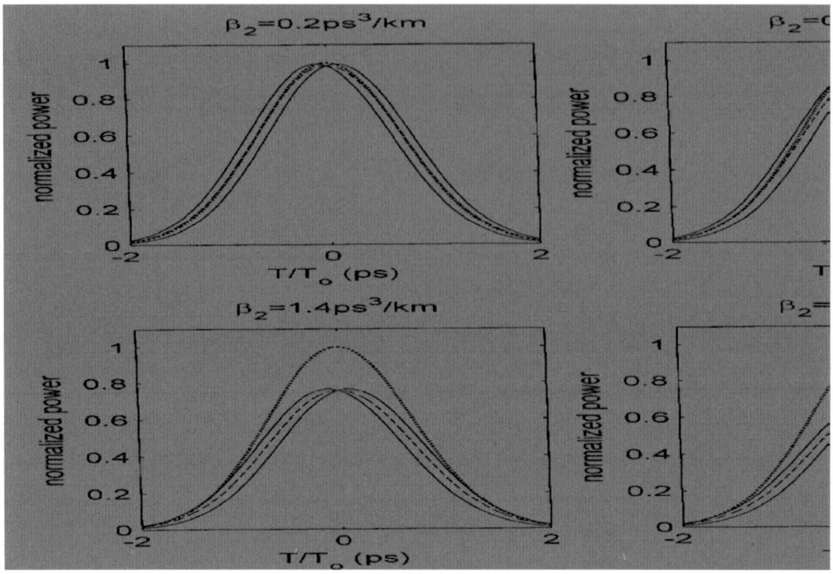


(a)

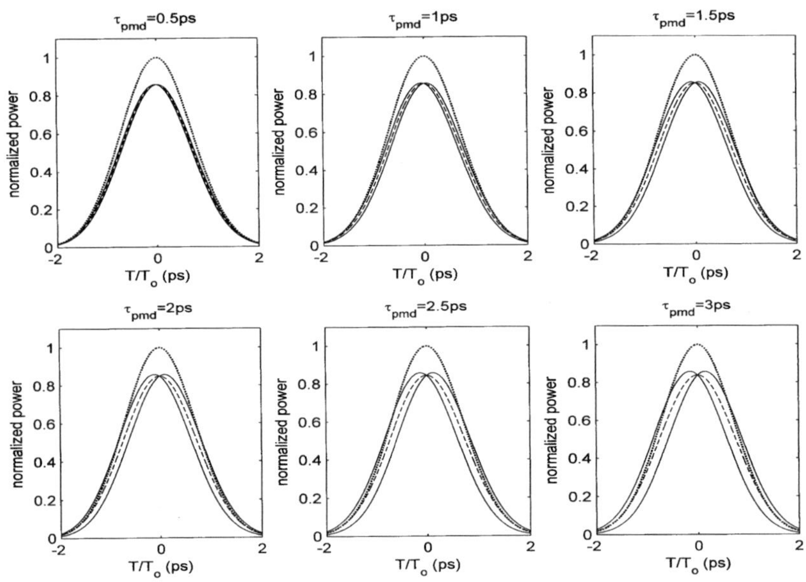


(b)

Fig. 4. Evolution of the pulse shape at $\theta = \pi / 2$, $T_o = 10$ ps, and $L = 60$ km: a) for various values of $\beta_2$ and $\tau_{pmd} = 2$ $ps$, b) for various values of $\tau_{pmd}$ and $\beta_2 = 1$ ps$^3$ / $km$. The dotted, continuous, and discrete lines refer to the initial pulse, two orthogonal components, and final pulse, respectively.

the two types of dispersion. In other words, the time separation between the two orthogonal components will be fixed, both amplitude and width of the pulse will change under the effects of CD. Fig.(4 b) illustrates the shape of pulse for various values of $\tau_{pmd}$, assuming $\beta_2 = 1$ ps$^3$ / $km$, $\theta = \pi / 2$, $T_o = 10$ ps, and $L = 60$ km. Since $\beta_2$ and $L$ are constants this implies that $T_1$ is constant also. That is; the width of both orthogonal components are similar for all $\tau_{pmd}$ values, but the difference appears as a time increase separation between the two components. This leads to adding a BR of value $\tau_{rms} / T_1$ to the reconstructed pulse.

## 5. Polarization mode dispersion and birefringence

In the optical fibers, the birefringence vector $\vec{\beta}$ may be defined in two forms as (Schuh et al. 1995)

$$\vec{\beta}_L = \begin{bmatrix} \Delta\beta\cos 2\alpha \\ \Delta\beta\sin 2\alpha \\ 0 \end{bmatrix} \quad or \quad \vec{\beta}_{NL} = \begin{bmatrix} \Delta\beta\cos 2\alpha \\ \Delta\beta\sin 2\alpha \\ \zeta T \end{bmatrix} \tag{33}$$

where $\alpha$ is the angle of birefringence in Jones space, $\Delta\beta$ is the magnitude of linear birefringence, i.e. $\Delta\beta = |\vec{B}_L|$ , $\zeta$ is the photo-elastic coefficient of glass, and T is the twist rate in (rad/m). The angle $\alpha$ is not constant along the fiber; also, $\Delta\beta$ and $T$ . This means that each segment of fiber has a birefringence vector differs from another position randomly, depending on the values of $\alpha$, $\Delta\beta$, and $T$. If $\Delta\beta = |\vec{B}_L|$, then $\vec{\beta}_L = \Delta\beta\,\hat{r}$, where $\hat{r}$ represents a unit vector in Stokes space. The vector $\hat{r}$ represents a rotation axis of the polarization vector, which differs from one section to another randomly.
Consequently, the PMD vector can be defined as a function of $\hat{r}$ and $\phi$ (Gordon & Kogelnik 2000)

$$\vec{\tau} = \phi_w \hat{r} + \hat{r}_w \sin\phi + \hat{r}_w \times \hat{r}(\cos\phi - 1) \tag{34}$$

where $\phi = \Delta\beta\Delta z$ represents the rotation angle of the polarization state vector $\hat{s}$ around the birefringence vector $\vec{\beta}$, and $\phi_w$ and $\hat{r}_w$ represent their first derivatives of frequency. Eq.(34) obtains that the angle and direction of rotation control the resultant vector $\vec{\tau}$ . Substituting the first definition in Eq.(33) into (34), yields

$$\vec{\tau} = \begin{bmatrix} \tau_1 \\ \tau_2 \\ \tau_3 \end{bmatrix} = \begin{bmatrix} \varepsilon\ \Delta z \cos(2\alpha) \\ \varepsilon\ \Delta z \sin(2\alpha) \\ 0 \end{bmatrix} + 2\frac{d\alpha}{dw} \begin{bmatrix} -\sin(\phi)\sin(2\alpha) \\ \sin(\phi)\cos(2\alpha) \\ 1-\cos(\phi) \end{bmatrix} \tag{35}$$

where $\varepsilon = d\Delta\beta / dw$ represents PMD parameter, and $\Delta z$ is the fiber segment length. On the other hand, $\vec{\tau}$ is a function of $w$, which may be written as a Taylor series around the central frequency $w_o$ as follows (Agrawal 2005)

$$\vec{\tau}(w) = \vec{\tau}(w_o) + \Delta w \frac{d\vec{\tau}}{dw}|_{w=w_o} + \frac{\Delta w^2}{2}\frac{d^2\vec{\tau}}{dw^2}|_{w=w_0} + \ldots\ldots\ldots\ldots\ldots\ldots \tag{36}$$

Comparing Eqs.(35) and (36), the first term on the right hand side of Eq.(35) will represent the first order of PMD vector, while the second term indicates all higher orders of PMD vector. Accounting that the higher orders depend on the value of $d\alpha / dw$ . For a very small variations of $\alpha$ with frequency, the second term on the right hand side of Eq.(35) may be neglected. Elsewhere, the higher order effects must be included through the determination of PMD vector.

### 5.1 Linear birefringence

Neglecting the higher order effects makes the PMD vector as follows

$$\vec{\tau} = \begin{bmatrix} \tau_1 \\ \tau_2 \\ \tau_3 \end{bmatrix} = \begin{bmatrix} \varepsilon\ \Delta z \cos(2\alpha) \\ \varepsilon\ \Delta z \sin(2\alpha) \\ 0 \end{bmatrix} = \varepsilon\ \Delta z\ \hat{r} = \frac{\varepsilon \Delta z}{\Delta\beta}\ \vec{\beta} = \text{const.}\ \vec{\beta} \tag{37}$$

This means, $\vec{\tau}$ coincides with the birefringence vector $\vec{\beta}$ if the intrinsic birefringence is linear and the higher order PMD effects are neglected. Elsewhere, the two vectors are never coincided. Using Eq.(37), we can obtained DGD of the fiber segment as

$$DGD_1 = \tau_{pmd}^{(1)} = |\vec{\tau}| = \varepsilon\ \Delta z \tag{38}$$

The value of $DGD_1$ represents the delay time between the two components of polarization in a single segment of the optical fiber. Since the DGD's of the fiber segments are random, so that $DGD_1$ can be calculated as $<\tau_{pmd}> = \frac{1}{N}\sum_{i=1}^{N}\tau_{pmd}^{i}$. For the case of wide frequency band, the higher order effects of the PMD must be included. The $DGD_2$ of this case can be obtained using Eq.(35) as follows

$$DGD_2 = \tau_{pmd}^{(2)} = \sqrt{(\varepsilon\ \Delta z)^2 + 8(1-\cos\phi)^2\alpha_w^2} \tag{39}$$

Clearly, the $DGD_2$ is related to the change of $\alpha$ with respect to frequency, and $\tau_{pmd}^{(1)} < \tau_{pmd}^{(2)}$. This means that the higher order effects increase the DGD. The angle between the two vectors $\vec{\tau}$ and $\vec{\beta}$ is determined as: $\psi = \cos^{-1}(\tau_{pmd}^{(1)} / \tau_{pmd}^{(2)})$. This means that the two vectors in the same direction if the higher order PMD is neglected, i.e. $\tau_{pmd}^{(1)} =< \tau_{pmd}^{(2)}$.

### 5.2 Nonlinear birefringence

For the nonlinear intrinsic birefringence, $\vec{\tau}$ can be calculated using the second definition in Eq.(33) and (34) as follows

$$\vec{\tau} = \begin{bmatrix} (a_1 + a_3\sin\phi)\cos(2\alpha) + a_6(\cos\varphi - 1)\sin(2\alpha) \\ (a_1 + a_3\sin\phi)\sin(2\alpha) - a_6(\cos\varphi - 1)\cos(2\alpha) \\ a_2 + a_5\sin\phi \end{bmatrix} + \frac{d\alpha}{dw}\begin{bmatrix} -a_4\sin\phi\sin(2\alpha) + a_7(\cos\phi - 1)\cos(2\alpha) \\ a_4\sin\phi\cos(2\alpha) + a_7(\cos\phi - 1)\sin(2\alpha) \\ a_8(\cos\phi - 1) \end{bmatrix} \tag{40}$$

where the parameters $a_1$ into $a_8$ are defined as

$$a_1 = \frac{\Delta\beta\varepsilon\Delta z}{\Delta\beta_{NL}} \qquad a_2 = -\frac{\Delta\beta\varepsilon\ \zeta\ T}{\Delta\beta_{NL}^2}$$

$$a_3 = a_5 = -\frac{\zeta\ T\Delta\beta\varepsilon}{K} \qquad a_4 = \frac{2\Delta\beta}{\Delta\beta_{NL}}$$

$$a_7 = \frac{\zeta\ T\ \Delta\beta^2\varepsilon}{\Delta\beta_{NL}^2} \qquad a_8 = -\frac{2b_L}{\Delta\beta_{NL}^2}$$

$$\Delta\beta_{NL} = \sqrt{\Delta\beta^2 + (\zeta\ T)^2} \qquad K = (\Delta\beta^2 + \zeta^2 T^2)^{3/2}$$

$$a_6 = \frac{2\zeta\ T\Delta\beta}{\Delta\beta_{NL}} - \frac{(\zeta\ T)^3\varepsilon}{K}$$

Eq.(40) represents a new formula of the PMD vector demonstrating the difficulties to compensate the noise that arises due to PMD when the pulse propagates through optical fibers. Many approaches have been proposed (McCurdy et al. 2004;Lima et al. 2001; Vanwiggeren & Ray 1999; Ibragimv & Shtenge 2002; Schuh et al. 1995), which deal only with the first order of PMD. This means that the compensation depends on the first term presented in the right hand side of Eq.(40) and assuming that the birefringence vector is linear.

The vector $\vec{\tau}$ can be found from $\vec{\beta}$. Ignoring the higher orders of the vector $\vec{\tau}$, the vector $\vec{\tau}$ is linear only if $\vec{\beta}$ is linear, otherwise they are different. When the distance is changed this implies to rotate $\hat{s}$ around $\vec{\beta}$ by an angle $\varphi$. On the other hand, the change of frequency causes to rotate $\hat{s}$ around $\vec{\tau}$ by an angle $\theta$. Fig.(5 a) illustrates the relation among the three vectors $\hat{s}$, $\vec{\beta}$, and $\vec{\tau}$ where the polarization vector $\hat{s}$ is rotating around $\vec{\beta}$ and $\vec{\tau}$. Adding the higher orders of $\vec{\tau}$, the vector $\vec{\tau}$ is now nonlinear which does not coincided with the vector $\vec{\beta}$ as illustrated in Fig.(5 b). The general case considers the birefringence vector is nonlinear and assuming all orders of $\vec{\tau}$ as illustrates in Fig.(5 c), which shows that each vector rotates in Stokes space.

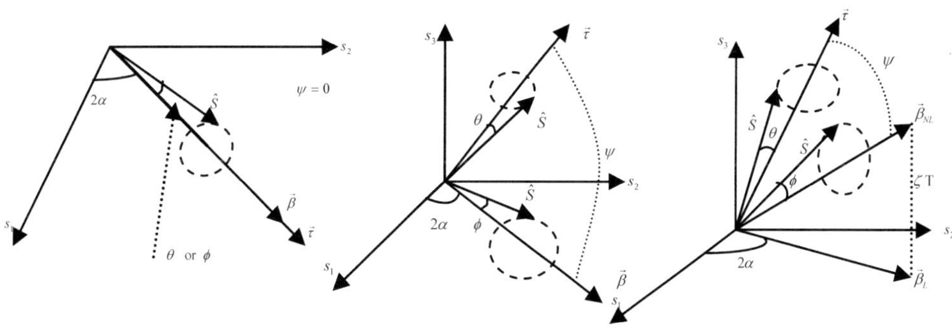


Fig. 5. Rotation of SOP around $\vec{\beta}$ and $\vec{\tau}$ : a) $\vec{\beta}$ and $\vec{\tau}$ are linear, b) $\vec{\beta}$ is linear and $\vec{\tau}$ is nonlinear, c) $\vec{\beta}$ and $\vec{\tau}$ are nonlinear.

## 6. Combined PMD and PDL effects

As far as the continuum limit at the end is set, the following simple arrangement are considered: each PMD element (having $\vec{\tau}_i$ vector) is followed by a PDL element (having $\vec{\alpha}_i$) leading to the following transmission Jones matrix (Yasser 2010)

$$T = T_{PDL}T_{PMD} = \exp(\frac{1}{2}\vec{\alpha}_j.\vec{\sigma})\exp(-\frac{iw}{2}\vec{\tau}_j.\vec{\sigma}) \tag{41}$$

where

$$\exp(\frac{1}{2}\ \vec{\alpha}_j.\vec{\sigma}) = \ [\cosh(\alpha_j / 2) + (\hat{\alpha}_j.\vec{\sigma})\sinh(\alpha_j / 2)]$$

$$\exp(-\frac{iw}{2}\vec{\tau}_j.\vec{\sigma}) = \ [\cos(w\tau_{pmd}^{(j)} / 2) - i(\hat{p}_j.\vec{\sigma})\sin(w\tau_{pmd}^{(j)} / 2)]$$

Here $\vec{\tau}_j = \tau_{pmd}^{(j)}\hat{p}_j$ represents the j-th PMD segment having DGD $\tau_{pmd}^{(j)}$ and the fast polarization axis is expressed by the unit vector $\hat{p}_j$ in the Stokes space. The PMD vector $\vec{\tau}$ is, generally, frequency dependent; the first term in the Taylor expansion of $\vec{\tau}(w)$ is conventionally referred to as the first-order PMD (Agrawal 2005). To clarify the notation used in this section, we attempt to keep the notation simple and transparent while linking to the notation already established as much as possible. The following is an abbreviated group of present notation: The letters C, c, S, and s represent $\cos(w\tau_{pmd}^{(j)} / 2)$, $\cosh(\alpha_j / 2)$, $\sin(w\tau_{pmd}^{(j)} / 2)$, and $\sinh(\alpha_j / 2)$, respectively.

Notice that, in this representation PDL matrix, the polarization component of the field that is parallel to $\vec{\alpha}_j$ experiences a gain $e^{\alpha}$, but the anti-parallel component is attenuated by $e^{-\alpha}$. The expressions $e^{\pm\alpha_j/2}$ represent the eigenvalues $\lambda_1, \lambda_2$ of PDL matrix. The vector $\vec{\alpha}_j = \alpha_j\hat{\alpha}_j$ stands for the j-th PDL segment with value expressed in dB by

$$PDL(dB) = 10\ell og_{10}(\frac{\lambda_1}{\lambda_2})^2 = 20\,|\,\alpha_j\,|\,\ell og_{10}(e) \tag{42}$$

The action of an optical component exhibiting PDL and PMD on a field can be described by (Chen et al. 2007)

$$|\,s> = T\,|\,t> = T_{PDL}T_{PMD}\,|\,t> \tag{43}$$

where $|\,s>$ and $|\,t>$ are output and input SOP, respectively. The eigenvalues of the matrix $T = T_{PDL}T_{PMD}$ are (Yasser 2010)

$$\lambda = [cC - i(\hat{\alpha}\cdot\hat{p})sS] \pm \sqrt{[cC - i(\hat{\alpha}\cdot\hat{p})sS]^2 - 1} \tag{44}$$

It was evident from Eq.(44) that the eigenvalues are complex, where the real part will control the new rotation angle of $\hat{s}$ around the PSP vector, and imaginary part can be used into Eq.(42) to obtain the PDL value in presence of PMD. Obviously, the new eigenvalues in presence of the combined PMD and PDL effects are different from that obtained for each effect separately.

### 6.1 Special cases

**1.** In presenting PDL only, the eigenstates of the PDL matrix are orthogonal, the output Stokes vector can be obtained as follows: combining the relations $|\,s> = T_{PDL}\,|\,t>$, and $<s\,| = <t\,|\,T_{PDL}^{\dagger}$ into $\hat{s} = <s\,|\,\vec{\sigma}\,|\,s>$, and using the facts (Yasser 2010; Gordon & Kogelnik 2000)

$$(\vec{\alpha}\,.\,\vec{\sigma})(\vec{\beta}\,.\,\vec{\sigma}) = \vec{\alpha}\,.\,\vec{\beta} + i\vec{\alpha}\times\vec{\beta}\,.\,\vec{\sigma} \tag{45a}$$

$$(\vec{\beta}\,.\,\vec{\sigma})(\vec{\alpha}\,.\,\vec{\sigma}) = \vec{\alpha}\,.\,\vec{\beta} - i\vec{\alpha}\times\vec{\beta}\,.\,\vec{\sigma} \tag{45b}$$

$$(\vec{\beta}\cdot\vec{\sigma})\vec{\sigma} = \vec{\beta}\text{-i}\vec{\beta}\times\vec{\sigma} \tag{45c}$$

$$\vec{\sigma}(\vec{\beta}\cdot\vec{\sigma}) = \vec{\beta} + \mathrm{i}\vec{\beta}\times\vec{\sigma} \tag{45d}$$

$$(\vec{\beta}.\vec{\sigma})\vec{\sigma}(\vec{\beta}.\vec{\sigma}) = 2\vec{\beta}(\vec{\beta}\cdot\vec{\sigma}) - \beta^2\vec{\sigma} \tag{45e}$$

$$T_{PDL}^{\dagger} = T_{PDL} = c + (\vec{\alpha}.\vec{\sigma})s \tag{45f}$$

a useful relation can be deduced

$$\hat{s} = c^2\hat{t} - s^2\hat{t} + 2sc\hat{\alpha} + 2s^2\hat{\alpha}(\hat{\alpha}\cdot\hat{t}) \tag{46}$$

The output SOP which is a combination of the vectors $\hat{t}$ and $\hat{\alpha}$, i.e. $\hat{t}$ does not rotate around $\hat{\alpha}$. If the input SOP is parallel or anti-parallel to PDL then the output SOP takes the form $e^{-\alpha}\hat{t}$ or $e^{\alpha}\hat{t}$. The first component, that is parallel to PDL vector, experience a gain $e^{\alpha}$ and the other, that is anti-parallel to PDL vector, is attenuated by $e^{-\alpha}$.

**2.** Similarly, in presence of PMD only, the eigenstates of the PMD matrix are also orthogonal and the output SOP can be determined as follows: combining the relations $<s| = <t|T_{PMD}^{\dagger}$ and $|s> = T_{PMD}|s>$ into $\hat{s} = <s|\vec{\sigma}|s>$, using Eqs.(45) with the facts that $T_{PMD} = C - i(\hat{p}\cdot\vec{\sigma})S$ and $T_{PMD}^{\dagger} = T_{PMD}^{-1} = C + i(\hat{p}\cdot\vec{\sigma})S$ to yield

$$\hat{s} = C^2\hat{t} - S^2\hat{t} + 2SC(\hat{p}\times\hat{t}) + 2S^2\hat{p}(\hat{p}\cdot\hat{t}) \tag{47}$$

This equation refers to the input SOP that are parallel or anti-parallel to PMD vector which experiences no change, i.e. $\hat{s} = \hat{t}$ along the optical fiber. Notice that, the PMD causes a rotation of the SOP around $\vec{\tau}$, which is presented through the third term.

**3.** Finally, in presenting the combined PDL-PMD effects, determining $\hat{s}$ as a function of $\hat{t}$, $\hat{\alpha}$, and $\hat{p}$ which is very complicated, is beyond the scope of this chapter.

### 6.2 The output power

The normalized Gaussian pulse before entering the PMD and PDL components has the form

$$|A_{in}(T)> = D\, e^{-T^2/2T_o^2}\,|a> \tag{48}$$

where $T_o$ is the initial pulse width, and $|a>$ is the Jones vector of the signal. Clearly, the normalized input power is found to be $\vec{P}_{in}(T) = <A_{in}(T)|\vec{\sigma}|A_{in}(T)> = e^{-T^2/T_o^2}\hat{s}$, where $\hat{s}$ is the input Stokes vector. The Fourier transform of Eq.(48) is

$$|A_{in}(w)> = \Im\{|A_{in}(T)>\} = D\frac{T_o}{\sqrt{2\pi}}e^{-w^2T_o^2/2}\,|a> \tag{49}$$

As far $|A_{out}(w)> = T_{PDL}T_{PMD}(w)|A_{in}(w)>$, the output field which can be illustrated by the inverse Fourier transformation as follows

$$|A_{out}(T)> = D\frac{T_o}{\sqrt{2\pi}}T_{PDL}\Im^{-}\{e^{-w^2T_o^2/2}e^{-i(w/2)(\vec{\tau}\cdot\vec{\sigma})}\}\,|a> = e^{-\frac{T^2+\tau_{pmd}^2}{2T_o^2}}e^{\vec{\alpha}\cdot\vec{\sigma}}e^{-T\vec{\tau}\cdot\vec{\sigma}/2T_o^2}\,|a> \tag{50}$$

In order to compute the output power from this equation. The vector $\vec{n}$ was set to equal $\vec{n} = (\vec{\alpha} - \vec{\tau}\;T/T_o^2)/2$, such that

$$|A_{out}(T)>= e^{-\frac{T^2+\tau_{pmd}^2}{2T_o^2}} e^{\vec{n}\cdot\vec{\sigma}}|a> \tag{51}$$

The new vector $\vec{n}$ is a random. Its value is $n=\sqrt{\alpha^2+T^2\tau_{pmd}^2/T_o^4-2T\alpha\tau_{pmd}^2\cos\theta/T_o^2}/2$, where $\theta$ is the angle between $\vec{\alpha}$ and $\vec{\tau}$, while the direction is $\hat{n}=(\vec{\alpha}-T\vec{\tau}/T_o^2)/n$. Substituting Eq.(51) into the definition $\vec{P}_{out}(T)=<A_{out}|\vec{\sigma}|A_{out}>$ and introducing the fact $(\vec{n}\cdot\vec{\sigma})^{\dagger}=\vec{n}\cdot\vec{\sigma}$, yields

$$\vec{P}_{out}(T)= e^{-\frac{T^2+\tau_{pmd}^2}{T_o^2}} <a|(\cosh n+\hat{n}\cdot\vec{\sigma}\sinh n)\vec{\sigma}(\cosh n+\hat{n}\cdot\vec{\sigma}\sinh n)|a> \tag{52}$$

Considering Eqs.(45), the last equation may be written as

$$\vec{P}_{out}(t)=e^{-\frac{T^2+\tau_{pmd}^2}{T_o^2}}[\hat{s}+2\hat{n}(\sinh n\cosh n+\sinh^2 n\cos\phi_1)] \tag{53}$$

where $\phi_1$ is the angle between the random vector $\hat{n}$ and the input SOP, $\hat{s}$. To visualize the situation more easily, Eq.( 53) was written as

$$\vec{P}_{out}(T)= e^{-\frac{T^2+\tau_{pmd}^2}{T_o^2}} \ .\ f(\mathrm{T},\tau_{\mathrm{pmd}},\alpha)\ .\ \hat{\mathrm{s}}_{\mathrm{out}} \tag{54}$$

where $f(\mathrm{T},\tau_{\mathrm{pmd}},\alpha)$ and $\hat{\mathrm{s}}_{out}$ are the value and direction of the expression inside the square brackets. Eq.(54) represents the output power in presenting of PMD and PDL, which may be written in certain cases as in the following subsections.

### 6.2.1 PMD only

In this case, $\vec{n}=-T\vec{\tau}/2T_o^2$ and $\hat{n}=-\hat{p}$, hence, Eq.(53) can be simplified as

$$\vec{P}_{out}(T)=e^{-\frac{T^2+\tau_{pmd}^2}{T_o^2}}[\hat{s}-2\hat{p}(\sinh n\cosh n+\sinh^2 n\cos\phi_2)]=f(T,\tau_{pmd})e^{-\frac{T^2+\tau_{pmd}^2}{T_o^2}}\hat{\mathrm{s}}_{out} \tag{55}$$

Here $\phi_1$ is replaced by $\phi_2$ which represents the angle between $\hat{\tau}$ and $\hat{s}$. If $\vec{\tau}=0$, then $\vec{P}_{in}=\vec{P}_{out}$. That is; the power and SOP are not affected in absence of PMD. The PSP's are the states that are parallel or antiparallel to $\hat{p}$, so the powers in the PSP's direction are $\vec{P}_{out}(T)_{PSP}=\exp(-(T^2\pm\tau_{pmd}^2)/T_o^2)\hat{s}$. The parallel or antiparallel SOP to $\hat{p}$ will not be changed through the propagation, but the position of the pulse components will be shifted by $\pm\tau_{pmd}/2$.

### 6.2.2 PDL only

Here, $\vec{n}=\vec{\alpha}/2$ and $\hat{n}=\hat{\alpha}$, hence, Eq.(53) will be

$$\vec{P}_{out}(T)=e^{-\frac{T^2}{T_o^2}}[\hat{s}+2\hat{\alpha}(\sinh n\cosh n+\sinh^2 n\cos\phi_3)]=f(\alpha)e^{-\frac{T^2}{T_o^2}}\hat{\mathrm{s}}_{out} \tag{56}$$

Here $\phi_1$ is replaced by $\phi_3$ which represents the angle between $\hat{\alpha}$ and $\hat{s}$. If $\vec{\alpha}=0$, then $\vec{P}_{in}=\vec{P}_{out}$. That is; the power and SOP are not affected in absence of PDL. There are two important SOP's that are parallel or antiparallel to $\hat{\alpha}$. For these SOP's, Eq.(56) will be reduced to $\vec{P}_{out}(T) = e^{\pm\alpha} e^{-T^2/T_o^2} \hat{s}$ . This means that, the power will be affected by the factor $e^{\pm\alpha}$ but the pulse shape and SOP will not change.

### 6.3 The complex PSP vector

Before discussing the impact of PMD and PDL on the dynamical equation of SOP, we notice: First, without including PDL, the transmission matrix of the fiber is always unitary. However, when the fiber PMD is intertwined with PDL elements, the transmission matrix losses its unitary property. Nevertheless, by the polar decomposition theorem (Kogelnik & Jopson 2002), a complex $2 \times 2$ matrix can be decomposed into $T = T_{PDL} T_{PMD}$, where $T_{PDL}$ is a positive definite Hermitian matrix, i.e. $T_{PDL}^{\dagger} = T_{PDL}$, and $T_{PMD}$ is a unitary matrix, i.e. $T_{PMD}^{\dagger} T_{PMD} = I$. Second, the PDL vector may by frequency dependent. This will influence the PDL induced waveform distortion effect in an optic link. Considering that such frequency dependent waveform distortion is not so important in a system with realistic parameters (Shtaif & Rosenberg 2005; Phua & Ippen 2005), the PDL vector was approximated as a frequency independent.

As pulses are described by wave packets with a finite frequency band, the frequency dependence of $|s>$ should be considered now. A fixed input polarization was assumed, i.e. $|t>_w = 0$ hence $\hat{t}_w = 0$, as is appropriate for a pulse entering the fiber at zero time. Now, by differentiating Eq.(43) with respect to frequency and eliminating $|t>$, the change of the output Jones vector was obtained

$$\frac{d|s>}{dw} = T_{PDL} T'_{PMD} T_{PMD}^{-1} T_{PDL}^{-1} |s> \tag{57}$$

where $T'_{PMD}$ represents the derivative of $T_{PMD}$ with respect to frequency. Eq.(57) tell us that for most input polarizations, the output polarization will change with frequency in the first order. Notice that, if $|s>$ either of the two eigenstates of the operator $T_{PDL} T'_{PMD} T_{PMD}^{-1} T_{PDL}^{-1}$ then $|s>_w = 0$. The dynamical equation of SOP in Stokes space can be obtained by using Eq.(57) as, see (Yasser 2010)

$$\hat{s}_w = [(c^2 + s^2)\vec{\tau} - 2s^2(\vec{\tau} \cdot \hat{\alpha})\hat{\alpha} + 2isc(\vec{\tau} \times \hat{\alpha})] \times \hat{s} \tag{58}$$

Many published studies (Chen et al. 2007; Wang & Menyuk 2001; Shtaif & Rosenberg 2005) related to the theoretical treatment of the combined effects of PMD and PDL, which are introduced in many forms of the frequency derivative of Stokes vector, but all these forms may be considered as a partial form of Eq.(58) above.

The expression between brackets in the right hand side of the last equation represents the complex PSP vector which can be decomposed as real and imaginary parts as follows

$$\vec{W} = \vec{\Omega} + i\vec{\Lambda} \tag{59}$$

where $\vec{\Omega}$ and $\vec{\Lambda}$ represent the new vectors in presenting of PMD and PDL. The two new vectors take the forms

$$\vec{\Omega} = (c^2 + s^2)\vec{\tau} - 2s^2(\vec{\tau} \cdot \hat{\alpha})\hat{\alpha} \tag{60a}$$

$$\vec{\Lambda} = 2sc(\vec{\tau} \times \hat{\alpha}) \tag{60b}$$

There are many features that can be deduced from Eq.(59): if $\vec{\tau}$ is parallel or anti-parallel to $\hat{\alpha}$ then $\vec{\Omega} = \vec{\tau}$, i.e. the old and new PMD vectors are identical, and $\vec{\Lambda} = 0$, i.e. the PDL effects will disappear. If $\vec{\tau}$ is perpendicular on $\hat{\alpha}$ then $\vec{\Omega} = (c^2 + s^2)\vec{\tau}$, i.e. the old and new PMD vector have the same direction but distinct values, and $\vec{\Lambda} = 2sc\vec{\xi}$ (where $\vec{\xi} = \vec{\tau} \times \hat{\alpha}$) that means the new PDL vector is perpendicular to the plane that contains $\vec{\tau}$ and $\hat{\alpha}$. If $\vec{\tau} = 0$ then both vectors $\vec{\Omega}$ and $\vec{\Lambda}$ are zero. Remembering that, the absence of PMD will not permit the emergence of two components, as a result there is no PDL but the reverse is not correct. Since the PSP vector is complex, then the fast and slow PSP's are not orthogonal. If $\hat{\alpha} = 0$, i.e. no PDL, then $\vec{\Omega} = \vec{\tau}$. The new DGD takes the form $\tau_{pmd}^{new} = \mathrm{Re}\sqrt{\vec{W} \cdot \vec{W}} = \tau_{pmd}^{old}$, where the meaning of DGD over infinite frequency is called the scalar PMD. Thereafter, the SOP rotates around the PSP vector by an angle $\tau_{pmd}^{new} w$. The new DAS takes the form $\alpha_{new} = \mathrm{Im}\sqrt{\vec{W} \cdot \vec{W}}$. Accordingly, the new PDL value is $20 \,|\, \alpha_{new} \,|\, \ell og_{10}(e)$.

## 7. Birefringence and nonlinearity

To formulate the birefringence effects more precisely, considering the nonlinear Helmholtz equation (Agrawal 2007)

$$\nabla^2 \vec{\tilde{E}} + \frac{w^2 \ddot{\varepsilon}_s}{c^2} \vec{\tilde{E}} = -\frac{w^2}{c^2 \varepsilon_o} \vec{\tilde{P}}_{NL} \tag{61}$$

where the tilde denotes the Fourier transformation, $\varepsilon_o$ is the vacuum permittivity, and $\ddot{\varepsilon}_s$ is the linear part of the dielectric constant. Notice that the tensorial nature is important to account for the PMD effects that have their origin in the birefringence of silica fibers, while its frequency dependence leads to chromatic dispersion. Assuming that the instantaneous electronic response dominates and neglecting Raman contribution (Lin & Agrawal 2003 a), the third order nonlinear polarization in a medium as silica glass is found to be

$$\vec{\tilde{P}}_{NL}(w) = \frac{\varepsilon_o \chi_{xxxx}^{(3)}}{4} \left[ (\vec{\tilde{E}} \cdot \vec{\tilde{E}}) \vec{\tilde{E}}^* + 2(\vec{\tilde{E}}^* \cdot \vec{\tilde{E}}) \vec{\tilde{E}} \right] \tag{62}$$

The electric field vector evolves along the fiber length and its SOP changes because of the birefringence. It is assumed here that the z-axis is directed along the fiber length and The electric field vector lies in the x-y plane. This assumption amounts to neglect the longitudinal component of the vector and is justified in practice as long as the spatial size of the fiber mode is longer than the optical wavelength. In Jones-matrix notation, the field at any point $r$ inside the fiber can be written as (Kogelnik & Jopson 2002)

$$\vec{\tilde{E}}(r,w) = F(x,y) \,|\, A(z,w) > e^{ikz} \tag{63}$$

where $F(x,y)$ represents the fiber mode profile, $k$ is the propagation constant, and Jones vector $|\, A>$ is a two-dimensional column vector representing the two components of the electric field in the x-y plane. Since $F(x,y)$ does not change with z, one needs to consider only the evolution of $|\, A>$ along the fiber.

Substituting Eq.(63) into Eq.(62), inserting the result into Eq.(61), and integrate over the transverse mode distribution in the x-y plane, assuming $|A>$ to be slowly varying function of z so that neglecting their second-order derivative with respect to z. With these simplifications, the equation governing the evolution of $|A>$ takes the form

$$\frac{d|A>}{dz}+\left(\frac{w^2\ddot{\varepsilon}_s}{2ikc^2}+i\frac{k\sigma_o}{2}\right)|A>=\frac{i\gamma}{3}\left[2<A|A>+|A^*><A^*|\right]|A> \tag{64}$$

where $\sigma_o$ is a unit matrix. To proceed Eq.(64) further, the dielectric constant tensor $\ddot{\varepsilon}_s$ may be represented in the basis of Pauli matrices as (Lin & Agrawal 2003 a)

$$\frac{w^2\ddot{\varepsilon}_s}{c^2}=\left[k+i\frac{\alpha}{2}\right]^2\sigma_o-k\vec{\beta}.\vec{\sigma} \tag{65}$$

The vector $\vec{\beta}$ accounts for the fiber birefringence and its frequency dependence produces PMD. The vector $\vec{\sigma}$ is formed as $\vec{\sigma}=\hat{e}_1\sigma_1+\hat{e}_2\sigma_2+\hat{e}_3\sigma_3$, where $\hat{e}_1$, $\hat{e}_2$, and $\hat{e}_3$ are a three unit vectors in the Stokes space. Substituting Eq.(65) into (64) leads to the following vector equation

$$\frac{d|A>}{dz}+\frac{\alpha}{2}\sigma_o|A>=-\frac{i}{2}\vec{\beta}\bullet\vec{\sigma}|A>+\frac{i\gamma}{3}\left[2<A|A>+|A^*><A^*|\right]|A> \tag{66}$$

Eq.(66) can be put in simplified form by neglecting the second term on the left hand side, by proposing that the medium is lossless; then, using the following identity

$$|A^*><A^*| \;=\; [<A|A>+<A|\vec{\sigma}|A>\cdot\vec{\sigma}]/2-<A|\sigma_3|A>\sigma_3 \tag{67}$$

into Eq.(66) yields the following elegant equation that describes the evolution of Jones vector through the optical fiber

$$\frac{d|A>}{dz}=(-\frac{i}{2}\vec{\beta}\bullet\vec{\sigma}+\frac{i\gamma}{6}\;[\hat{s}\bullet\vec{\sigma}\;]^t\;)|A> \tag{68}$$

where the proportionality term $|A>$ affects only the global phase and can be neglected, $\hat{s}=<A|\vec{\sigma}|A>$ is the normalized power (Stokes vector). Using Eq.(68), it is not difficult to obtain

$$\frac{d\hat{s}}{dz}=(\vec{\beta}+2\gamma(0,0,s_3)^t\,/\,3)\times\hat{s} \tag{69}$$

Eq.(69) presents the effect of nonlinearity. Introducing $\gamma$ effect is considered as the main contribution of this section, because it is a phenomenon that can not be neglected in the study of the evolution of polarization through the optical fibers. However, the rotation axis in presence of nonlinearity is $\vec{\beta}+2\gamma(0,0,s_3)^t/3$ instead of $\vec{\beta}$. The simplest case, without nonlinearity effect, has been studied by many researches using different approaches, see for example (Gordon & Kogelnik 2000; Agrawal 2005; Vanwiggeren and Roy 1999).

## 8. Conclusions

In conclusion, we have achieved the following: an important mathematical relationship between PMD and birefringence are presented and all possible assumptions are discussed. The statistics of PMD are simply analyzed. The combined effect of PMD and chromatic dispersion causes an additional amount of pulse broadening. Interaction of PMD and PDL makes the two PSP's are not orthogonal nor do they represent the fastest and slowest pulses, which causes a change in DGD and PDL compared with the impact of each individual. Nonlinearity causes a change in the rotation axis and therefore it changes the properties of polarization state during the propagation. Finally, all results are generally subject to random changes as long as most of the causes random.

## 9. References

Agrawal G. P. (2007). *Nonlinear Fiber Optics*, 4th Edition, Academic Press, USA.

Agrawal G. P. (2005). *Lightwave Technology: Telecommunication Systems*, 1st Edition, Wiley Interscience, USA.

Azzam R. M. and Bashara N. M. (1989). *Ellipsometry and Polarized Ligh*t, Elsevier, Amsterdam.

Chen L., Hadjifaradji S., Waddy D., and Baw X. (2003). Effect of Local PMD and PDL Directional on the SOP Vector Autocorrelation, *Optics Express*, Vol.11, No.23, pp.3141-3146.

Chen L., Zhang Z., and Bao X. (2007). Combined PMD-PDL Effects on BERs in Simplified Optical System: an Analytical Approach, *Optics Express*, Vol.15, No.5, pp.2106-2119.

Chertkov M., Gabitov I., Kolokolov I., and Schafer T. (2004). Periodic Compensation of Polarization Mode Dispersion, *J. Opt. Soc. Am. B*, Vol.21, No.3, pp.486-497.

Elbers J., Glingener C., Duser M., and Voges E. (1997). Modeling of Polarization Mode Dispersion in Single Mode Fibers, *Elect. Let.*, Vol.33, No.22, pp. 662-664.

Foshchini G. and Poole C. (1991). Statistical Theory of Polarization Dispersion in Single Mode Fibers", *J. Lightwave Tech.*, Vol.9, pp.1439-1456.

Gordon J. and Kogelnik H. (2000). PMD Fundamentals: Polarization Mode Dispersion in Optical Fibers", *Proc. Natl. Acad. Sci.*, Vol.97, No.9, pp.4541-4550.

Ibragimv E. and Shtenge G. (2002). Statistical Correlation Between First and Second Order PMD, *J. Lightwave Tech.*, 20(4): 586-590.

Karlsson M. (1994). Polarization Mode Dispersion Induced Pulse Broadening in Optical Fibers", *Optics Let.*, 23, pp.688-690.

Kogelnik H. and Jopson R. M. (2002). Polarization Mode Dispersion, *in Optical Fiber Telecommunications volume: IV B, I. P. Kaminov and T. Li, Eds. San Diego: Academic*, pp.725-861, USA.

Levent A., Rajeev S., Yaman F., and Agrawal G. P. (2003). Nonlinear Theory of Polarization Mode Dispersion for Fiber Solitons, *Phys. Rev. Let.*, Vol.90, No.1, pp.730-737.

Lima T., Khosravani R., and Menyuk C. R. (2001). Comparison of PMD Emulators", *J. Lightwave Tech.*, Vol.19, No. 12, pp.1872-1881, 2001.

Lin Q. and Agrawal G. P. (2003), Vector Theory of Stimulated Raman Scattering and Its Application to Fiber-based Raman Amplifier, *J. Opt. Soc. Am. B.,* Vol.20, No.8, pp.492-501.

Lin Q. and Agrawal G. P. (2003). Correlation Theory of Polarization Mode Dispersion", *J. Opt. Soc. Am. B,* Vol.20, No.2, pp-292-301.

Lin Q. and Agrawal G. P. (2003). Statistics of Polarization Dependent Gain in Fiber Based Raman Amplifiers, *Optics Let.,* Vol.28, No.4, pp.227-229,.

Mahgerftech D. and Menyuk C.R. (1999). Effects of First-Order PMD Compensation on the Statistics of Pulse Broadening in Fiber with Random Varying Birefringence, *IEEE Photo. Tech. Lett.* 13(3): 340-342.

McCurdy A., Sengupta A., and Glodis P. (2004). Compact Measurement of Low PMD Optical Telecommunication Fibers, *Optics Express,* Vol.12, No.6, pp.1109-1118.

Mechels S., Schlger J., and Franzen D. (1997). Accurate Measurements of the Zero Dispersion Wavelength in Optical Fibers, *J. Res. Natl. Inst. Stand. Tech.,* Vol.102, No.3, pp.333-347.

Menyuk C.R., Wang D., and Pilipetskii A. (1997). Re-polarization of Polarization Scrambled Optical Signals Due to PDL, *IEEE Pho. Tech. Let.,* Vol.9, No.9, pp.1247-1249.

Phua P. and Ippen E. (2005). A Deterministic Broad Band Polarization Dependent Loss Compensator", *J. Lightwave Tech.,* Vol.23, No. 2, pp.771-780.

Rogers A. (2008). *Polarization in Optical fibers,* Artech House, INC, USA.

Schuh R., Sikora E., Walker N., Siddiqui A., Gleeson L., and Bebbington D. (1995). Theoretical Analysis and Measurements of Fiber Twist on the Differential Group Delay of Optical Fibers, Electron Let., Vol.31, No.20, pp.1772-1773.

Shtaif M. and Rosenberg O. (2005). Polarization Dependent Loss as a Waveform Distortion Mechanism and Its Effect on Fiber Optic Systems, *J. Lightwave Tech.,* Vol.23, No. 2, pp.923-930.

Sunnerud H., Karlsson M., Xie C., and Andrekson P. (2002). Polarization Mode Dispersion in High Speed Fiber Optic Transmission Systems", J. Lightwave Tech., Vol.20, No.12, pp.2204-2219.

Tan Y., Yang J., Kath W., and Menyuk C. (2002). Transient Evolution of Polarization Dispersion Vector's Probability Distribution, *J. Opt. Soc. Am. B,* Vol.19, No.5, pp.992-1000.

Vanwiggeren G. and Roy R. (1999). Transmission of Linearly Polarized Light through a Single Mode Fiber with Random Fluctuations of Birefringence, *Applied Optics,* Vol.38, No.18, pp.3888-3892.

Wang D. and Menyuk C. (2001). Calculation of Penalties Due to Polarization Effects in Long-Haul WDM System Using a Stokes Parameter Mode, *J. Lightwave Tech.,* Vol.19, No. 4, pp.487-494.

Yaman F., Lin Q., Radic S., and Agrawal G. (2006). Fiber Optic Parametric Amplifiers in the Presence of Polarization Mode Dispersion and Polarization Dependent Loss, *J. Lightwave Tech.,* Vol.24, No. 8, pp.3088-3096.

Yang J., Kath W., and Menyuk C. (2001). Polarization Mode Dispersion Probability Distribution for Arbitrary Distances, *Optics Let.,* Vol.26, No.19, pp.1472-1474.

Yasser H. A. (2010). The Dynamics of State of Polarization in the Presence of Conventional Polarization Effects, *Optics & Laser Tech.*, 42, 1266-1268.

Yoon I. and Lee B. (2004). Change in PMD Due to the Combined Effects of PMD and PDL for a Chirped Gaussian Pulse, *Optics Express*, Vol.12, No.3, pp.492-501.

7

# Optical Fiber Birefringence Effects – Sources, Utilization and Methods of Suppression

Petr Drexler and Pavel Fiala
*Department of Theoretical and Experimental Engineering,*
*Brno University of Technology,*
*Czech Republic*

## 1. Introduction

The application area of optical fibers is quite extensive. Telecommunication applications were the primary field of fibers employment. The related area is the utilization of optical fibers for control purposes, which benefits from principal galvanic isolation between the transmitting and receiving part of the system. A minimal sensitivity of light propagation inside the fiber to electromagnetic field of common magnitudes allows use the fiber in systems with high level of electromagnetic disturbance. Regarding the physical aspects of light propagation in fibers, they find utilization possibility in physical quantities sensors. It is possible to modulate the phase and state of polarization of the wave inside fiber optical medium by means of external physical quantity. The interaction is described by electro-optical, magnet-optical and elasto-optical effects.

In order to achieve high data transmission rates in field of telecommunication applications the single-mode fibers are used exclusively. Similarly, single-mode fibers are used in the case of intrinsic fiber optic sensors. Intrinsic fiber optic sensors exploit the fiber itself to external quantity sensing and the fiber serves to signal transmission also. The reason of single-mode fiber utilization is the presence of basic waveguide mode - single wave with single phase and single polarization characteristic.

In spite of the fiber utilization advantages we have to take into account undesirable effects, which are present in real non-ideal optical fiber. In telecommunication and sensor application field the presence of inherent and induced birefringence is crucial. The presence of birefringence may cause an undesirable state of polarization change. In the case of high-speed data transmission on long distances the polarization mode dispersion may occur. Due to this effect the light pulses are broadened. This may result in inter-symbol interference. In the case of sensor application, when the state of polarization is a carrier quantity, the possibility of output characteristic distortion and sensors sensitivity decreasing may occur.

It's advantageous to consider fiber sensor application for purposes of birefringence origin and influence description, since the presence of linear and circular birefringence together is watched often. While the inherent circular birefringence is negligible in common single-mode fibers, the inherent and induced linear birefringence may be present in considerable rate. The inherent linear birefringence is mostly undesirable effect, when we exclude utilization in polarization maintaining fibers. Whereas, the induced linear

birefringence may be utilized for sensing purposes, e.g. for mechanical stress or pressure sensing. Similarly, induced circular birefringence is a principle effect for group of polarimetric sensors, e.g. polarimetric current sensor.
For the suppression of unwanted linear birefringence influence, inherent and induced also, several approaches and methods have been developed and published. They are often based on different principles. However, they differ in view of their properties and suitability for various applications.
The goal of this chapter is to present basic effects, which lead to occurrence of linear and circular birefringence in single-mode fibers. The methods, which may be used in order to suppress unwanted birefringence, will be presented also. Since the main manifestation of birefringence effects is the transformation of the polarization state of transmitted light, a brief recapitulation of basic polarization states and their illustrative visualizations are given. In following subchapters various mechanisms, which induce birefringence will be introduced together with corresponding relations and comprehensible illustrations. In the last descriptive subchapter the most significant methods for unwanted linear birefringence are presented with their properties and references to related literature.

## 2. Light polarization

The electromagnetic wave polarization represents how varies orientation, pertinently projection magnitude, of electric field component in a plane which is perpendicular to propagation direction. The polarization character of the wave may be described by means of the magnetic field component also. However, the interaction of matter with light wave is done mainly via the electric field component. Then the electric field intensity vector $\boldsymbol{E}$ is used for polarization states description usually. The general polarization state of the wave is the elliptical one. The special cases are the circular birefringence and linear birefringence.
Consider an electromagnetic wave which is described by electric field intensity vector $\boldsymbol{E}$ and which propagates in direction of $z$ axis. The wave may be represented as a superposition of two partial waves with mutually orthogonal linear polarizations and with the same frequency

$$\boldsymbol{E} = \boldsymbol{E}_x + \boldsymbol{E}_y, \tag{1}$$

where $xyz$ is orthogonal coordinate system and $\boldsymbol{E}_x$, $\boldsymbol{E}_y$ are vectors of electrical field intensity, which are aligned in $x$ axis direction and $y$ axis direction. It should be noted, that we consider the same frequency of both waves in all of the following analysis. In case of linear polarizations the electric field intensity vectors $\boldsymbol{E}_x$, $\boldsymbol{E}_y$ swing along a straight line. These two vectors may be assigned to two degenerated modes of the single-mode fiber, which is a dielectric circular waveguide. In a lossless medium hold for field components magnitude relations

$$\begin{aligned} E_x(z,t) &= E_{0,x}\cos\left(kz - \omega t + \phi_x\right), \\ E_y(z,t) &= E_{0,y}\cos\left(kz - \omega t + \phi_y\right), \end{aligned} \tag{2}$$

where $E_{0,x}$, $E_{0,y}$ are wave amplitudes, $\omega$ is angular frequency of the waves, $t$ is time and $\phi_x$, $\phi_y$ are phases of the wave, $k$ is magnitude of the wave vector. Amplitudes ratio of $E_{0,x}$ and $E_{0,y}$ and phase difference $\Delta\phi = \phi_x - \phi_y$ determine the state of the polarization of the resulting wave.

In case when $E_{0,x} = E_{0,y}$ and $\Delta\phi = 0$ the orthogonal waves are in phase with the same amplitude. We obtain a linearly polarized wave by their superposition. Its plane of polarization is in 45° to $y$ axis (or -45° to $x$ axis) as shown in Fig. 1. In Fig. 1 and following figures $\boldsymbol{k}$ represents the wave vector.

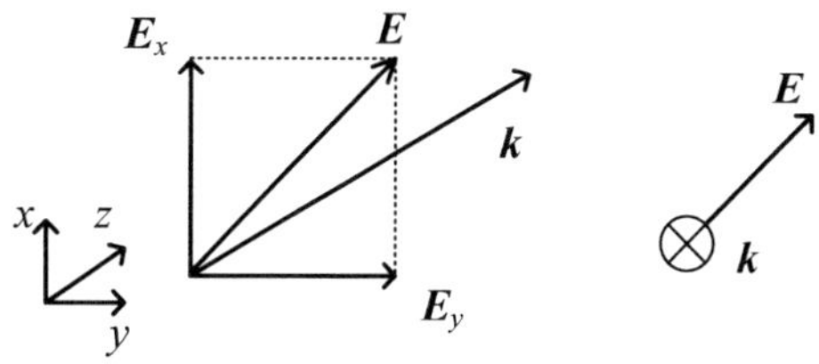


Fig. 1. Superposition of in-phase wave equal in amplitude results in linear polarized wave.

In case when $E_{0,x} = E_{0,y}$ and $\Delta\phi = \pm\pi/2$ the resultant wave has a circular polarization. The end point of $\boldsymbol{E}$ vector of circular polarized wave traces a circle. We differ between right-handed and left-handed circular polarized wave depending on the phase difference $\Delta\phi$ polarity, plus or minus. An illustration of right-handed circular polarized wave is shown in Fig. 2.

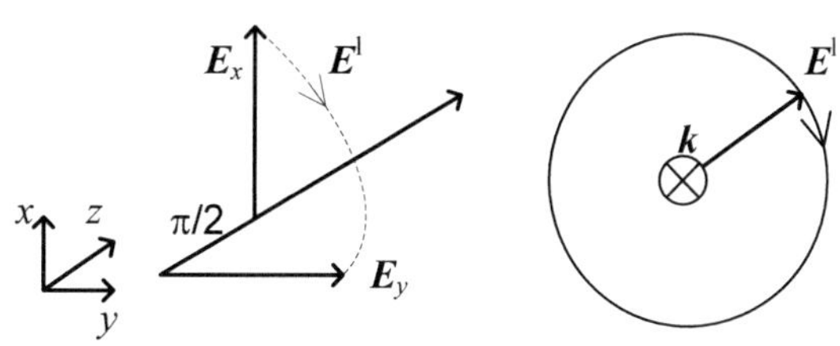


Fig. 2. Right-handed circular polarized wave.

When $\Delta\phi \neq 0, \pm\pi/2$ or $E_{0,x} \neq E_{0,y}$, we obtain an elliptically polarized wave, right-handed or left-handed, in dependence on phase difference $\Delta\phi$ polarity. The end point of $\boldsymbol{E}$ vector of elliptically polarized wave traces an ellipse. The case of left-handed elliptically polarized wave is shown in Fig. 3

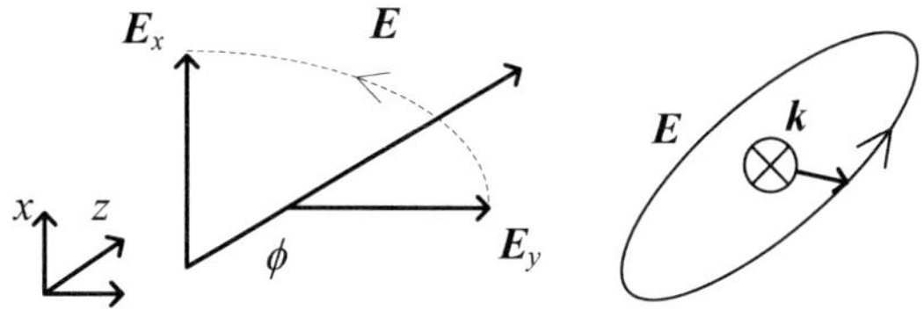


Fig. 3. Left-handed elliptically polarized wave.

As the light wave propagates in homogenous isotropic medium, its velocity remains constant independently on the propagation direction. The propagation velocity is given by the refractive index of the medium $n$. Refractive index is a ratio of wave velocity in vacuum c and wave phase velocity $v_p$ in the medium, $n = c/v_p$. However, medium may be of anisotropic character. This means, that the propagation velocity depends on the propagation direction, pertinently polarization. This effect is observed in birefringent materials. In common birefringent materials the optical properties are described by means of index ellipsoid, which is shown in Fig. 4. When linearly polarized wave travels in $z$ axis direction and it is polarized in $y$ axis direction, the wave phase velocity is given by refractive index $n_y$. When the same wave will be polarized in $x$ axis direction, the phase velocity will be given by refractive index $n_x$. Since $n_y > n_x$, the first wave will travel slower.

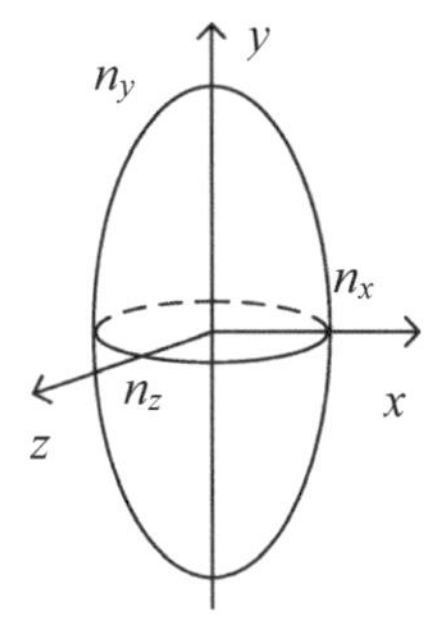


Fig. 4. Index ellipsoid example of birefringent material.

The velocity of wave polarized between $y$ and $x$ axis directions will be given by refractive index, which magnitude lies on ellipse in $xy$ plane. When a light wave travels in birefringent medium of such type described above, it may occur a phase shift between its orthogonal components, which are described by relation (2). This occurs due to different propagation velocities of the components. The resulting state of polarization depends on total phase difference $\Delta\phi$, which is a function of propagation length in birefringent medium also. An example of state of polarization change from linear to elliptical by birefringent medium crossing is shown in Fig. 5.

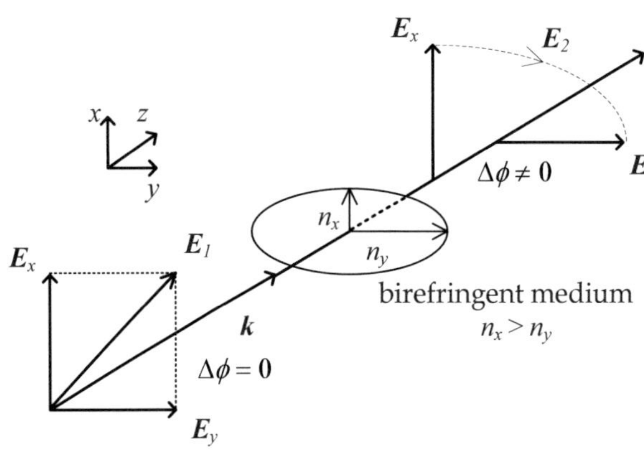


Fig. 5. State of polarization change in birefringent medium.

## 3. Linear birefringence in optical fiber

In previous section a polarization state of wave has been explained as a superposition result of two partial waves with certain phase shift and certain amplitude ratio. The similar concept may be used for description of polarization state transformation in single-mode optical fiber. As has been mentioned above, two degenerate modes $HE_{11}^{x}$ and $HE_{11}^{y}$ may exist in the fiber with circular core cross-section. The superposition of the two modes, which are orthogonal, results in the wave propagating in the fiber. And, the phase shift of these two modes determines the polarization state of the wave in the fiber. A deeper analysis of mode theory of fibers is out of the scope of this chapter and may be found in relevant literature (Iizuka, 2002).

The phase velocities of two orthogonal modes in the fiber $v_{f,x}$ and $v_{f,y}$ are given by magnitudes of wave numbers $\beta_x$ and $\beta_y$ of the modes

$$v_{f,x} = \frac{2\pi f}{\beta_x}, \tag{3}$$

$$v_{f,y} = \frac{2\pi f}{\beta_y}, \tag{4}$$

where $f$ is the frequency of the wave. An ideal single-mode fiber with circular core cross-section along its length, made from homogenous isotropic material, will exhibit the same refractive index $n$ for both of the modes. The wave numbers $\beta_x$ and $\beta_y$ will be equal also. The modes will propagate with the same phase speed $v_f$. At this condition the modes remain degenerate and the resulting polarization state will be preserved. A non-ideal fiber has not a constant circular core cross-section along its length or it exhibits anisotropy due to bending or other mechanical stress. As a consequence, the loss of modes degeneracy occurs. The fiber will behave as a birefringent medium with different refractive indices $n_x$ and $n_y$ and different phase velocities $v_{f,x}$ and $v_{f,y}$. In case of constant core cross-section and constant anisotropy, we can designate $\beta_y$ as a wave number for a fast mode and $\beta_x$ for a slow mode. Corresponding axes $x$ and $y$ may be designated as a fast axis and a slow axis of the fiber.

If a linear polarized wave is coupled into the birefringent fiber with gradually varying core cross-section or varying anisotropy, it is not possible to designate one mode as a fast one and second as a slow one. The mode phase shift $\Delta\phi$, which determines the output polarization state, is dependent on average wave number magnitudes and on the fiber length

$$\Delta\phi = \left(\overline{\beta_x} - \overline{\beta_y}\right) l_v. \tag{5}$$

The output polarization state will not be stable, when one would manipulate with the fiber or when the ambient temperature fluctuates. Since the wave number will be changing. This fact complicates the utilization of single-mode fibers in application with defined polarization state, as the fiber lasers or fiber sensors. Further, photodetectors, which are used in the field of fiber optic telecommunication, are not sensitive to polarization state. However, owing to the fiber birefringence, the phase shift of partial modes, pertaining to individual pulses, occurs. This effect causes a broadening of the impulses resulting in inter-symbol interferences.

The fiber birefringence rate is characterized by beat length $l_b$. It is possible to deduce from (5), that the state of polarization will transform periodically, as shows Fig. 6. Linear polarization of the wave with the polarization plane at angle 45° to $x$ axis gradually transforms across right-handed elliptical polarization to right-handed circular polarization. It transforms further across right-handed elliptical and linear polarization perpendicular to the original one. Then it transforms across left-handed elliptical and left-handed circular to perpendicular left-handed elliptical polarization and finally to original linear polarization. At this point, the total phase shift of the modes is $\Delta\phi = 2\pi$ and the corresponding fiber length is the fiber beat length $l_b$.

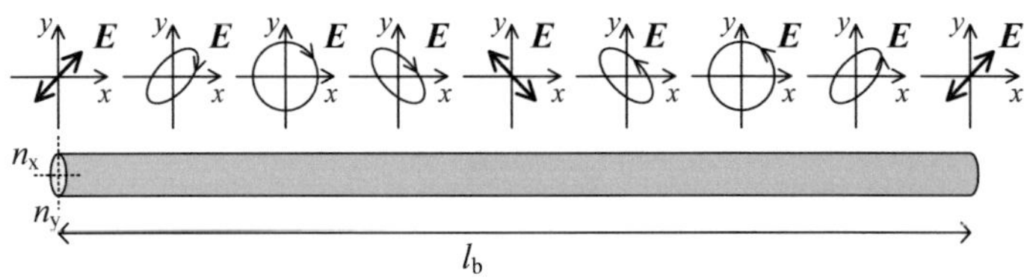


Fig. 6. Periodical transformation of the state of polarization in fiber with beat length $l_b$.

$$l_{\mathrm{b}} = \frac{2\pi}{\beta_x - \beta_y} = \frac{\lambda}{n_{x,\mathrm{eff}} - n_{y,\mathrm{eff}}} = \frac{\lambda}{\Delta n_{\mathrm{eff}}}, \tag{6}$$

where $n_{x,\mathrm{eff}}$ and $n_{y,\mathrm{eff}}$ are effective refractive indices in $x$ axis and in $y$ axis, $\Delta n_{\mathrm{eff}}$ is the difference of effective refractive indices and $\lambda$ is light wavelength.

### 3.1 Linear birefringence owing to elliptical fiber core cross-section

As mentioned above, the linear birefringence may be of latent or induced nature. The main cause of latent linear birefringence in real fiber is the manufacture imperfection. The cross-section of the fiber core is not ideally circular but slightly elliptical, as shown in Fig. 7.

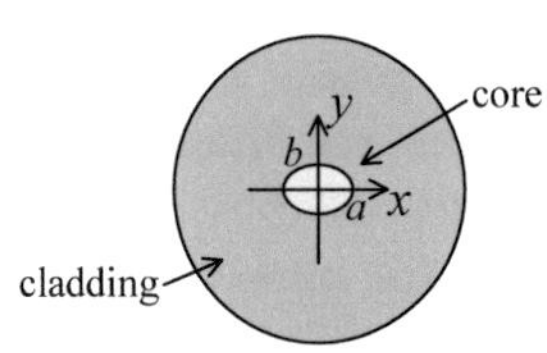


Fig. 7. Elliptical cross-section of the non-ideal fiber core.

Let the major axis of the ellipse representing core cross-section lies in the $x$ axis direction and the minor axis lies in the $y$ axis direction. The wave number of the mode, which propagates in $x$ axis direction, will be of a larger magnitude than the wave number of the $y$ axis mode. The difference of effective refractive indices $\Delta n_{\mathrm{eff}}$ is determined by the ellipticity ratio $a/b$. In case of small ellipticity rate, when $a \approx b$, holds the relation

$$\Delta n_{\mathrm{eff}} = 0.2\left(\frac{a}{b} - 1\right)(\Delta n)^2, \tag{7}$$

where $\Delta n = n_{\mathrm{co}} - n_{\mathrm{cl}}$ is the difference of core refractive index $n_{\mathrm{co}}$ and cladding refractive index $n_{\mathrm{cl}}$. For specific phase shift of the modes it may be derived from (7) a relation

$$\Delta\phi = \frac{0.4\pi}{\lambda}\left(\frac{a}{b} - 1\right)(\Delta n)^2. \tag{8}$$

In order to attain a large beat length the fiber core should approach the circular cross-section as much as possible. It may be derived by means of (6) and (8) a demand for relative deviation from ideal circularity

$$\left(\frac{a}{b}-1\right)\cdot 100\% = \frac{\lambda}{0.2 l_b \left(n_{co}-n_{cl}\right)^2}\cdot 100. \tag{9}$$

For typical single-mode fiber with $n_c = 1.48$, $n_{cl} = 1.46$ and operating wavelength $\lambda = 633$ nm the required deviation from ideal circularity achieves 0.016%. This demand is very hard to accomplish in fiber manufacture. The common fibers maintain the polarization state close to the initial only a few meters along.

### 3.2 Inner mechanical stress induced linear birefringence

The fiber core ellipticity is not a single source of fiber birefringence imposed by the manufacture. A second important source, which may take effect, is the presence of inner mechanical stress on the core. This may be caused by non-homogeneity of cladding density in area close to the core. In order to simplify the analysis we can consider an elliptical density distribution owing to imperfect technology process of fiber drawing from hot preform. The far area of the cladding influences the inner area by centripetal pressure after the fiber cooled down. Since the core-close area has a non-homogenous density, the pressures on core, $p_x$ and $p_y$, will act non-uniformly as illustrates Fig. 8.

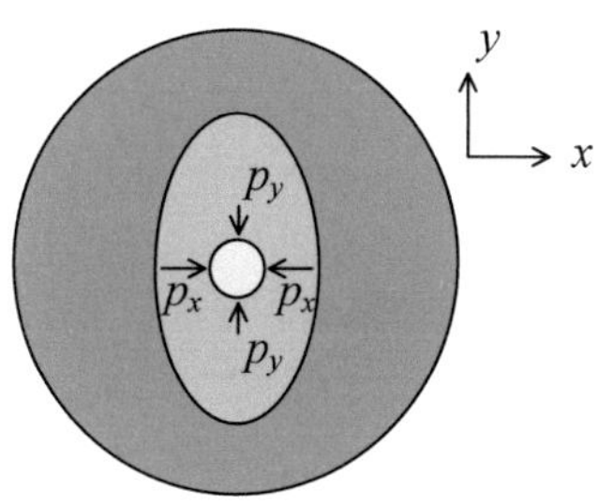


Fig. 8. Non-uniform stress on fiber core owing to imperfect inner structure.

Due to the photo-elastic effect, which causes pressure dependent anisotropy, the fiber core becomes a birefringent medium. Then, the difference of effective refractive indices in $x$ axis and $y$ axis is

$$\Delta n_{eff} = \frac{C_f}{1-\mathbf{v}_c}\Delta \upsilon \Delta T \frac{{p_x}/{p_y}-1}{{p_x}/{p_y}-1}, \tag{10}$$

where $\mathrm{v}_c$ is Poisson constant of the core, $\Delta\upsilon$ is difference of expansion coefficients of outer and inner cladding areas, $\Delta T$ is difference between softening temperature of the cladding and the ambient temperature. Coefficient $C_f$ is characteristic for given fiber, given by

$$C_f = \frac{1}{2}\left(\frac{n_{co}-n_{cl}}{2}\right)^3 \left(r_{11}-r_{12}\right)\left(1-\mathbf{v}_{co}\right), \tag{11}$$

where $r_{11}$, $r_{12}$ are components of photo-elastic tensor matrix of the fiber material. Photo-elastic matrix description is above the scope of this chapter and may be found in (Huard, 1997). For specific phase shift of the modes it may be derived from (10) a relation

$$\Delta\phi = \frac{2\pi}{\lambda}\frac{C_\mathrm{f}}{1-\mathbf{v}_\mathrm{c}}\Delta\upsilon\Delta T\frac{p_x/p_y - 1}{p_x/p_y - 1}. \tag{12}$$

The influence of inner stress induced linear birefringence is weak in compare to birefringence owing to elliptical core cross-section in common single-mode fibers. However, the inner stress induced linear birefringence may be imposed intentionally in case of polarization maintaining (PM) fibers manufacture.

### 3.3 Outer mechanical stress induced linear birefringence

Linear birefringence in single mode fiber may be induced by outer influence also. It is caused by outer mechanical stress (pressure or tensile force) on fiber cladding. Cladding transfers the mechanical stress on the core and similar effect described above uprises. In practice, an action of force in one dominant direction appears usually. It induces origin of two axis of symmetry, $x$ and $y$, with two refractive indices $n_x$ and $n_y$ again.

One of the possible effects causing linear birefringence is fiber bending, which is illustrated in Fig. 9. A fiber with cladding diameter $d_\mathrm{cl}$ is bended with diameter $R$. The fiber axis is equal to $y$ axis direction. The pressure imposed on core in $x$ axis increase refractive index $n_x$ in compare to $n_y$ due to the photo-elastic effect. In this configuration, the slow mode propagates in the bending plane $xy$ and the fast mode propagates in plane $yz$.

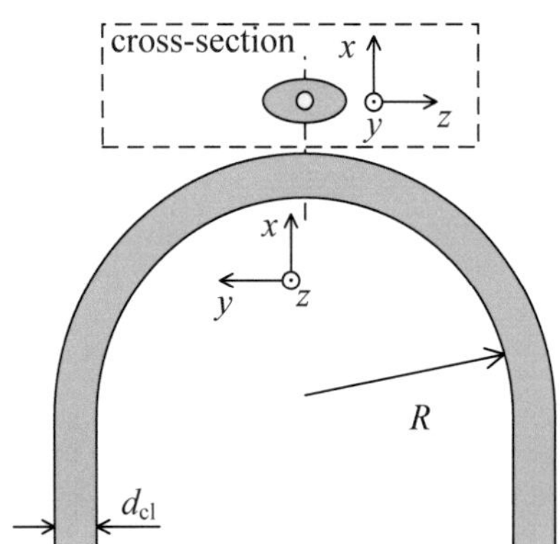


Fig. 9. Geometric relation of fiber bending causing induced linear birefringence.

The difference of effective refractive indices in $x$ axis and $y$ axis will be (Huard, 1997)

$$\Delta n_\mathrm{eff} = \frac{1}{2}C_\mathrm{f}\frac{d_\mathrm{cl}^2}{R^2} + 2\zeta C_\mathrm{f}\frac{d_\mathrm{cl}}{R} \tag{13}$$

and related specific phase shift of the modes is expressed as

$$\Delta\phi = \frac{2\pi}{\lambda}\left[\frac{1}{2}C_f\frac{d_{cl}^2}{R^2} + 2\zeta C_f\frac{d_{cl}}{R}\right], \tag{14}$$

where $C_f$ is fiber coefficient given by (11), $\zeta$ is the rate of axis deformation caused by longitudinal tensile force. The terms on the right side of relations (13) and (14) represent a situation when an additive tensile force acts on the fiber. This may occur when the fiber is bended over a solid, as a coil core. If the fiber is bended without the additive tensile force relations (13) and (14) are simplified. Then, the resultant relation for specific phase shift is

$$\Delta\phi = \frac{\pi}{\lambda}C_f\frac{d_{cl}^2}{R^2}. \tag{15}$$

Relation (15) may be substituted by relation, where the fiber coefficient $C_f$ is replaced by the product of Young module of the fiber material $E_c$ and photo-elastic coefficient of the fiber core $\Re$ (Ulrich et al., 1980)

$$\Delta\phi = \frac{\pi}{\lambda}E_c\Re\frac{d_{cl}^2}{R^2}. \tag{16}$$

As in previous cases, it may be derived from (16) a relation for specific phase shift of the modes

$$\Delta\phi_{1z} = \frac{2\pi^2}{\lambda}E_c\Re\frac{d_{cl}^2}{R}. \tag{17}$$

When a one fiber turn with radius $R = 8$ cm would be formed from a typical single-mode fiber with $Ec = 7.45\cdot10^9$ Pa, $\Re = -3.34\cdot10^{-11}$ Pa$^{-1}$ (Namihira, 1983) and $d_{cl} = 125$ μm, the phase shift at wavelength $\lambda = 633$ nm achieves $\Delta\phi \approx -\pi/2$. In this case, for example linear polarization will be transformed into the circular. The original polarization state will be lost. A second significant effect, which induces linear birefringence in the fiber is a lateral pressure, which is illustrated in Fig. 10.

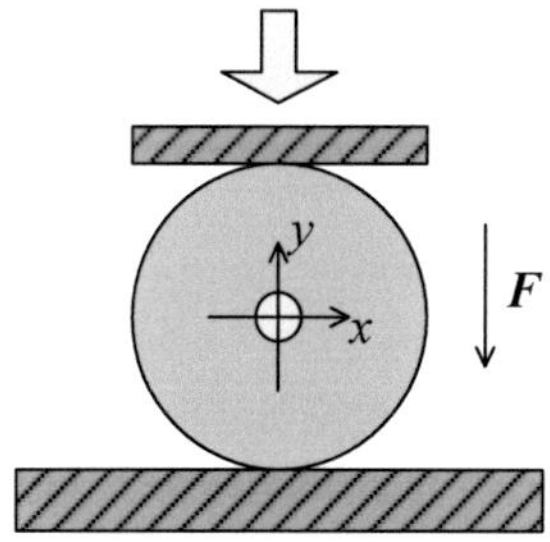


Fig. 10. Imposing a lateral pressure force on the fiber.

Induced anisotropy in the fiber is a result of photo-elastic effect, which is induced by compressing fiber between two planar solid slabs. If we consider $F_m$ as a force acting on unit length, the phase difference of the fiber modes will be (Huard, 1997)

$$\Delta\phi = \frac{2\pi}{\lambda}C_f\frac{4F_m}{\pi d_{cl}E_c}. \tag{18}$$

The fast mode will propagate along the $x$ axis and the slow one along the $y$ axis. Lateral pressure induced birefringence may occur by fiber assembly in to optical components, such as connectors.

At the close of this chapter, there should be mentioned another way to induce the fiber linear birefringence. It may be imposed by electro-optical effect in the fiber core. However, the fiber core is made from amorphous material and the electro-optic effect is of a very weak character (Wagner et al., 1992).

## 4. Circular birefringence in optical fiber

In the case of circular birefringence analysis we introduce a concept of chiral birefringent medium. It exhibits two refractive indices $n^{\mathrm{r}}$ and $n^{\mathrm{l}}$ for right-handed and left-handed circular polarized waves. Counter rotating waves, which propagate in this medium, travel with different phase velocities and they gain a phase shift. Both of the circular polarized waves we may decompose on two linear polarized waves with equal amplitudes and with $\pi/2$ or $-\pi/2$ phase shift. Thus, the right-handed circular polarized wave propagating in $z$ axis direction is a superposition of two orthogonal linear polarized waves described by components $\boldsymbol{E}_x^{\mathrm{r}}$ and $\boldsymbol{E}_y^{\mathrm{r}}$. For their magnitudes holds

$$\begin{aligned} E_x^{\mathrm{r}}(z,t) &= \frac{E_0}{2}\cos\left(\beta^{\mathrm{r}} z - \omega t - \frac{\pi}{2}\right), \\ E_y^{\mathrm{r}}(z,t) &= \frac{E_0}{2}\cos\left(\beta^{\mathrm{r}} z - \omega t\right), \end{aligned} \tag{19}$$

where $\beta^{\mathrm{r}}$ is wave number of right-handed circular polarized wave. Likewise, for left-handed circular polarized wave components holds

$$\begin{aligned} E_x^{\mathrm{l}}(z,t) &= \frac{E_0}{2}\cos\left(\beta^{\mathrm{l}} z - \omega t + \frac{\pi}{2}\right), \\ E_y^{\mathrm{l}}(z,t) &= \frac{E_0}{2}\cos\left(\beta^{\mathrm{l}} z - \omega t\right), \end{aligned} \tag{20}$$

where $\beta^{\mathrm{l}}$ is wave number of left-handed circular polarized wave. Wave numbers of waves propagating in fiber core are given as

$$\beta^{\mathrm{r}} = \frac{2\pi}{\lambda} n_{\mathrm{j}}^{\mathrm{r}}, \; \beta^{\mathrm{l}} = \frac{2\pi}{\lambda} n_{\mathrm{j}}^{\mathrm{l}}. \tag{21}$$

Due to the magnitude difference of refractive indices $n_{\mathrm{j}}^{\mathrm{r}}$ and $n_{\mathrm{j}}^{\mathrm{l}}$ in circular birefringent fiber core, the counter rotating waves travel with different phase velocities and they gain a phase shift

$$\Delta\phi = \left(\beta^{\mathrm{r}} - \beta^{\mathrm{l}}\right) l_{\mathrm{f}} = \frac{2\pi}{\lambda}\left(\beta_{\mathrm{j}}^{\mathrm{r}} - \beta_{\mathrm{j}}^{\mathrm{l}}\right) l_{\mathrm{f}} = \sigma_{\mathrm{c}} l_{\mathrm{f}}, \tag{22}$$

where $\sigma_{\mathrm{c}}$ is specific rotation of fiber core and $l_{\mathrm{f}}$ is fiber length.

When we superpose two circular polarized waves, which were described above, we obtain a linear polarized wave with a certain orientation of polarization plane. The change of polarization plane rotation angle $\Delta\alpha$ is equal to the phase shift $\Delta\phi$ from (22), $\Delta\alpha = \Delta\phi$.

It may be concluded, that the presence of circular birefringence in the fiber results in polarization plane rotation. When the fiber is free from linear birefringence and we couple a linear polarized wave into the fiber, we obtain a linear polarized wave with rotated polarization plane at the output. The angle of plane rotation is due to the circular birefringence rate and the fiber length.
In contrast to linear birefringence, circular birefringence of latent origin is negligible in common single-mode fiber. Nevertheless, it is possible to impose it in manufacturing process or induced it by outer influence. This can be attained by suitable applied mechanical stress or by magnetic field applying in the direction of fiber axis.

### 4.1 Outer mechanical stress induced circular birefringence

If a fiber section with length $l_f$ is exposed to torsion with specific torsion rate $\tau$

$$\tau = \frac{\delta}{l_{\mathrm{f}}}, \tag{23}$$

where $\delta$ is torsion angle as shown in Fig. 11, a sheer stress is imposed in plane perpendicular to fiber axis.

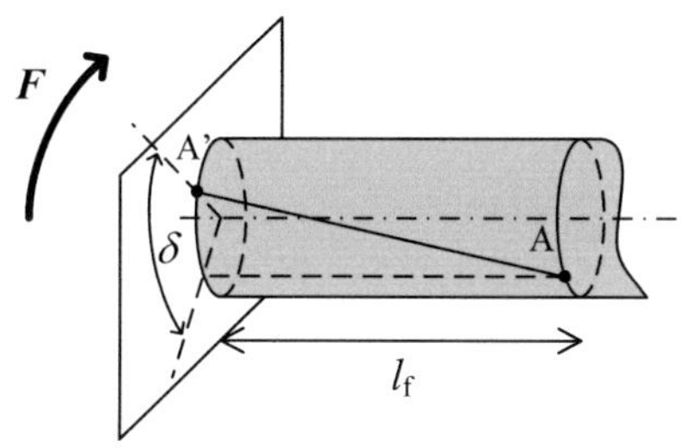


Fig. 11. A fiber section with length $l_f$ exposed to torsion with angle $\delta$.

Imposed sheer stress results in fiber core anisotropy owing to photo-elastic effect. In order to describe optical properties of anisotropic fiber core, it is useful to exploit tensor matrix of dielectric constant $\boldsymbol{\varepsilon}$ (Saleh & Teich, 1991)

$$\boldsymbol{\varepsilon} = \boldsymbol{\varepsilon}_{\mathrm{i}} + \Delta\boldsymbol{\varepsilon}_{\mathrm{t}} = \begin{bmatrix} \varepsilon & 0 & 0 \\ 0 & \varepsilon & 0 \\ 0 & 0 & \varepsilon \end{bmatrix} + \begin{bmatrix} 0 & -g\tau y & 0 \\ g\tau y & 0 & -g\tau x \\ 0 & g\tau x & 0 \end{bmatrix} = \begin{bmatrix} \varepsilon & -g\tau y & 0 \\ g\tau y & \varepsilon & -g\tau x \\ 0 & g\tau x & \varepsilon \end{bmatrix}, \tag{24}$$

where $\boldsymbol{\varepsilon}_{\mathrm{i}}$ is dielectric constant tensor of original medium and $\Delta\boldsymbol{\varepsilon}_{\mathrm{t}}$ is tensor of torsion contribution. Coordinates $x,y$ in matrix $\Delta\boldsymbol{\varepsilon}_{\mathrm{t}}$ belongs to point A′ in sheer stress plane, where is the dielectric constant expressed and for coefficient $g$ holds

$$g = r_{44} n_{\mathrm{c}}^{4} = \left(r_{11} - r_{12}\right) n_{\mathrm{c}}^{4}, \tag{25}$$

where $r_{11}$, $r_{12}$ and $r_{44}$ are components of photo-elastic matrix of the fiber core material. For further analysis, it is advantageous to exploit a Jones calculus (Jones, 1941) to characterize the influence of torsion modified medium on the polarization state of the wave. The relations of photo-elastic coefficients of the medium and Jones matrix of the medium are beyond the scope of this chapter and may be found for example in (Iizuka, 2002). The Jones matrix of the torsion modified medium is in the form

$$\boldsymbol{T}_{\mathrm{c}} = \begin{bmatrix} 0 & -\mathrm{j}g\tau \\ \mathrm{j}g\tau & 0 \end{bmatrix}, \tag{26}$$

where j is imaginary unit. Jones matrix $\boldsymbol{T}_{\mathrm{c}}$ describes the polarizing properties of circular birefringent medium. If we multiply matrix $\boldsymbol{T}_{\mathrm{c}}$ with Jones vector $\boldsymbol{J}_1$ of linear polarized wave, we obtain vector $\boldsymbol{J}_2$ with imaginary components. Both of the components represent left-handed and right-handed circular polarized waves.

$$\boldsymbol{J}_2 = \boldsymbol{T}_{\mathrm{c}} \cdot \boldsymbol{J}_1 = \frac{1}{\sqrt{2}} \begin{bmatrix} 0 & -\mathrm{j}g\tau \\ \mathrm{j}g\tau & 0 \end{bmatrix} \cdot \begin{bmatrix} 1 \\ 1 \end{bmatrix} = \frac{1}{\sqrt{2}} \begin{bmatrix} -\mathrm{j}g\tau \\ \mathrm{j}g\tau \end{bmatrix}. \tag{27}$$

The phase shift of circular polarized waves $\Delta\phi$ and corresponding polarization rotation angle $\Delta\alpha$ is proportional to the torsion rate $\tau$. Then, a twisted single mode fiber with length $l_{\mathrm{f}}$ acts as a polarization rotator with rotation angle

$$\Delta\alpha = g\tau l_{\mathrm{f}}. \tag{28}$$

### 4.2 Magnetic field induced circular birefringence

The second source of fiber circular birefringence is magneto-optical effect. Between three types of magneto-optical effect (Cotton-Mouton, Kerr, Faraday) (Craig & Chang, 2003), the Faraday effect is significant for silica fiber. It induces circular birefringence owing to magnetic field action in direction along the fiber axis. Analogous to fiber torsion, the Faraday magneto-optical effect modifies the dielectric constant tensor

$$\varepsilon = \varepsilon_{\mathrm{i}} + \Delta\varepsilon_{\mathrm{mo}} = \begin{bmatrix} \varepsilon & 0 & 0 \\ 0 & \varepsilon & 0 \\ 0 & 0 & \varepsilon \end{bmatrix} + \begin{bmatrix} 0 & -\mathrm{j}\eta B & 0 \\ \mathrm{j}\eta B & 0 & 0 \\ 0 & 0 & 0 \end{bmatrix} = \begin{bmatrix} \varepsilon & -\mathrm{j}\eta B & 0 \\ \mathrm{j}\eta B & \varepsilon & 0 \\ 0 & 0 & \varepsilon \end{bmatrix}, \tag{29}$$

where $\Delta\boldsymbol{\varepsilon}_{\mathrm{mo}}$ is tensor of magneto-optical effect contribution, $B$ is the magnitude of flux density of the external magnetic field, $\eta$ is coefficient, which is proportional to magneto-optic specific rotation coefficient (Huard, 1997). Again, dielectric constant tensor (29) describes a birefringent medium, where right-handed and left-handed circular polarized waves travel with different velocities. Here, the resulting phase shift of the waves is proportional to magnitude of magnetic flux density and the length of birefringent medium.

In order to explain the origin of Faraday magneto-optical effect, it is possible to model the effect as an electron oscillator movement in magnetic field (Waynant & Ediger, 2000). The effect itself results from interaction of outer magnetic field with oscillating electron, which is excited by the electric field of the light wave. Electrons represent harmonic oscillators. For them equations of forced oscillations hold. In the presence of external magnetic field with flux density $\boldsymbol{B}$, parallel to wave propagation direction, for the electron oscillator holds

$$m_{\mathrm{e}} \frac{\mathrm{d}^2\boldsymbol{u}}{\mathrm{d}t^2} + \kappa\boldsymbol{u} = -e\boldsymbol{E} - e\left[\frac{\mathrm{d}\boldsymbol{u}}{\mathrm{d}t} \times \boldsymbol{B}\right], \tag{30}$$

where $m_e$ is electron mass, $e$ is electron charge, $\boldsymbol{u}$ is vector, which determines the electron displacement, $\kappa\boldsymbol{u}$ is quasi-elastic force preserving electron in equilibrium position, $\boldsymbol{E}$ is electric field vector of propagating wave. Electric field of the wave polarizes the medium

$$\boldsymbol{P} = -N_e e \boldsymbol{u}, \tag{31}$$

where $N_e$ is the count of electrons in volume unit, which are deflected by the electric field of the wave. Substituting equation (31) into (30) we get

$$\frac{\mathrm{d}^2\boldsymbol{P}}{\mathrm{d}t^2} + \frac{e}{m_e}\left[\frac{\mathrm{d}\boldsymbol{P}}{\mathrm{d}t} \times \boldsymbol{B}\right] + \omega_0^2 \boldsymbol{P} = \frac{N_e e^2}{m_e}\boldsymbol{E}, \tag{32}$$

where $\omega_0$ is frequency of the electron oscillator. Equation (32) represents the system of two simultaneous differential equations. We obtain two terms by their solution. One for the right-handed, second for the left-handed circular polarized wave in the medium (Born & Wolf, 1999)

$$\begin{aligned} E^{\mathrm{r}} &= E_0^{\mathrm{r}} e^{\mathrm{j}\omega t}, \\ E^{\mathrm{l}} &= E_0^{\mathrm{l}} e^{\mathrm{j}\omega t}, \end{aligned} \tag{33}$$

where $\omega$ is frequency of circular polarized waves. The macroscopic relation for the medium polarization due to the electric field of circular polarized waves is in the form

$$\begin{aligned} P^{\mathrm{r}} &= \varepsilon_0 \chi^{\mathrm{r}} E^{\mathrm{r}}, \\ P^{\mathrm{l}} &= \varepsilon_0 \chi^{\mathrm{l}} E^{\mathrm{l}}, \end{aligned} \tag{34}$$

where $\chi^{\mathrm{r}}$ and $\chi^{\mathrm{l}}$ are dielectric susceptibilities for right-handed and left-handed circular polarized waves and $\varepsilon_0$ is dielectric constant of vacuum. Refractive index of the medium is related to dielectric susceptibility

$$n^2 = \varepsilon_{\mathrm{r}} = 1 + \chi, \tag{35}$$

where $\varepsilon_{\mathrm{r}}$ is relative dielectric constant of the medium. Substituting equations (34) into system (32) and by utilization of relation (35), we obtain relations for refractive indices of right-handed and left-handed circular polarized waves

$$\begin{aligned} \left(n_{\mathrm{c}}^{\mathrm{r}}\right)^2 &= 1 + \frac{N_e e^2}{\varepsilon_0 m_e} \cdot \frac{1}{\omega_0^2 - \omega^2 + \dfrac{e}{m_e} B\omega}, \\ \left(n_{\mathrm{c}}^{\mathrm{l}}\right)^2 &= 1 + \frac{N_e e^2}{\varepsilon_0 m_e} \cdot \frac{1}{\omega_0^2 - \omega^2 - \dfrac{e}{m_e} B\omega}. \end{aligned} \tag{36}$$

When we take into account certain simplifications, we can differentiate equations (36) and we can derive relation for polarization plane rotation in dependence on the outer magnetic field flux density $B$ and on the interaction length $l_f$ (fiber length in magnetic field)

$$\Delta\alpha = \Delta\phi = \frac{\pi}{\varepsilon_0 \lambda} \frac{N_e}{\overline{n_c}} \frac{e^3}{m_e^2} \frac{\omega}{\left(\omega_0^2 - \omega^2\right)^2} B l_f = V B l_f , \tag{37}$$

where $\lambda$ is wavelength of the wave, $\overline{n_c} = (n_c^r + n_c^l)/2$ is the mean refractive index, $\omega$ is angular frequency of the wave, $V$ is Verdet constant, which characterizes magneto-optic properties of medium. It is obvious, that Verdet constant depends on the wavelength.
The right part of equation (37) is the basic relation for Faraday magneto-optic effect. The effect is non-reciprocal. The polarization rotation direction depends on the mutual orientation of magnetic flux density $\boldsymbol{B}$ and the wave propagation direction. The polarization of wave propagating in the direction of $\boldsymbol{B}$ experiences a rotation $\Delta\alpha$. The polarization of wave propagating in the opposite direction to $\boldsymbol{B}$ experiences a rotation $-(\Delta\alpha)$. This non-reciprocal character is important for example for polarization mode conjugation as will be shown later. The illustration of polarization plane rotation in fiber section due to Faraday magneto-optic effect is shown in Fig. 12.

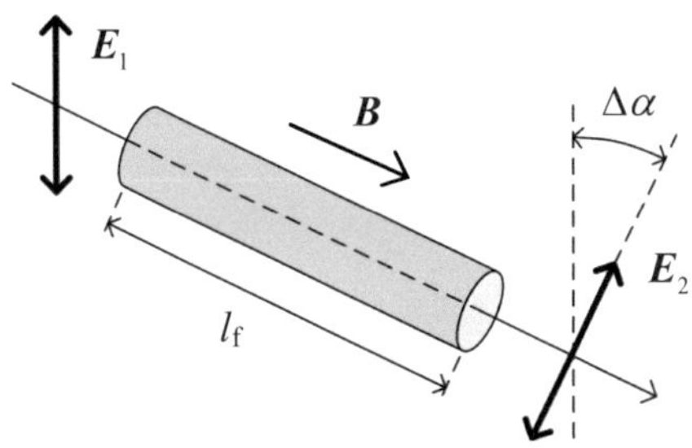


Fig. 12. Polarization plane rotation in fiber section due to Faraday magneto-optic effect.

## 5. Superposition of linear and circular birefringence in fiber

Both types of birefringence, linear and circular, may appear in single mode fiber. Both of them may be of latent or induced origin. The total phase shift of modes in fiber, determining the output polarization, is given by their geometrical average

$$\Delta\phi = \sqrt{\phi_c^2 + \left(\frac{\phi_l}{2}\right)^2} , \tag{38}$$

where $\phi_c$ is mode phase shift caused by circular birefringence and $\phi_l$ is mode phase shift caused by linear birefringence (Ripka, 2001). Generally, the polarization state of the output wave will be elliptical due to the linear birefringence and the orientation of axes of the polarization ellipse will rotate due to the circular birefringence. The analysis of the optical system with fiber, which exhibits both types of birefringence, may be performed by means of Jones calculus. Jones matrix of the fiber will be in the form (Tabor & Chen, 1968)

$$\boldsymbol{T}_f = \begin{bmatrix} \cos\Delta\phi + j\dfrac{\phi_l}{2}\dfrac{\sin\Delta\phi}{\Delta\phi} & -\phi_c\dfrac{\sin\Delta\phi}{\Delta\phi} \\ \phi_c\dfrac{\sin\Delta\phi}{\Delta\phi} & \cos\Delta\phi - j\dfrac{\phi_l}{2}\dfrac{\sin\Delta\phi}{\Delta\phi} \end{bmatrix} , \tag{39}$$

where $\Delta\phi$ results from (38). By means of (39), it is possible to study transformation of polarization state of the wave, which passed trough the fiber. Generally, in presence of both types of birefringence, the fiber behaves as phase retarder and polarization rotator simultaneously.

## 6. Techniques for unwanted fiber birefringence suppression

In previous chapters, it has been explained how the birefringence affects the polarization state of the wave in fiber. The transformation of polarization state is often unwanted, if we intend to use it as a carrier quantity.

It is important to suppress the polarization mode dispersion in telecommunication applications, in order to avoid pulses broadening. It is caused by linear birefringence. Important is also to avoid the unwanted birefringence in polarimetric sensors applications. It has to be ensured, that the polarization state will be modified by sensing quantity only. Polarimetric fiber optic sensor may be divided into two groups. Sensors of mechanical quantities (strain, pressure, vibrations) utilize induced linear birefringence. Sensors of magnetic field utilize induced circular birefringence. Since the inherent circular birefringence of common fibers is insignificant, the key parameter is the rate of linear birefringence.

The facts mentioned above place demands for methods for unwanted linear birefringence suppression. Following subchapters present a brief overview of the most significant selected methods, which are used to meet this requirement. The methods differ in view of its principle, efficiency or usability in various applications.

### 6.1 Polarization maintaining fibers

Polarization maintaining (PM) fibers have a specific inner structure, which allows maintaining polarization of the wave on long distances. In general view, polarization maintaining fibers may be divided in two groups. The first represents polarization maintaining fibers with low birefringence (PM LB). PM LB fiber approaches the concept of ideal fiber with constant circular cross-section and with very low linear birefringence. As has been mentioned above, these fibers are difficult to manufacture. Moreover, the manipulation (as bending or compressing) with fiber induces linear birefringence due to photo-elastic effect. In the second group belong polarization maintaining fibers with high birefringence (PM HB). A strong linear birefringence is imposed in the fiber by means of internal mechanical strain, which results in the loss of degeneracy of hybrid fiber modes $HE_{11}$. Therefore, the beat length of PM HB fibers is only a few millimeters. Hybrid modes propagate in fiber along the major and minor axis of ellipse, whose ellipticity is given by the ratio of mode wave numbers $\beta_x$ and $\beta_y$. If the light wave, for example with linear polarization, is coupled into the fiber with polarization plane in direction of one of the axes, the both orthogonal wave modes will experience equal wave numbers and the equal refractive indices. The wave will propagate along the fiber without the polarization state transformation (Kaminow & Ramaswamy, 1979). The sensitivity on bending and temperature fluctuations is greatly reduced. Nevertheless, the insensitive polarization state preserving is ensured only for one certain polarization plane orientation, when both wave modes experience same refractive indices.

The principle of PM HB fibers manufacturing consists in implementing of stress components in the fiber cladding. Stress components impose symmetrical defined pressure force on

circular fiber core. Stress components are implemented by doping of designated cladding areas with atoms of certain elements, typically boron atoms. In this way, areas with different thermal expansion coefficient are formed. After the drawn fiber cools down, the doped areas cause inner strain, which acts on the fiber core. Doped areas may have variety of shapes as shown in Fig. 13. The influence of fiber latent linear birefringence is strongly exceeded by the imposed birefringence, which temperature and bending dependence is very weak.
In order to characterize the properties of PM HB fibers the polarization crosstalk *CT* is defined (Noda et al., 1986). Polarization crosstalk, defined by relation (40), is given by logarithmic ratio of optical power of excited mode $P_x$ and optical power of coupled mode $P_y$. The polarization crosstalk is typically lower than -40 dB for fiber length of 100 meters (Senior, 2009).

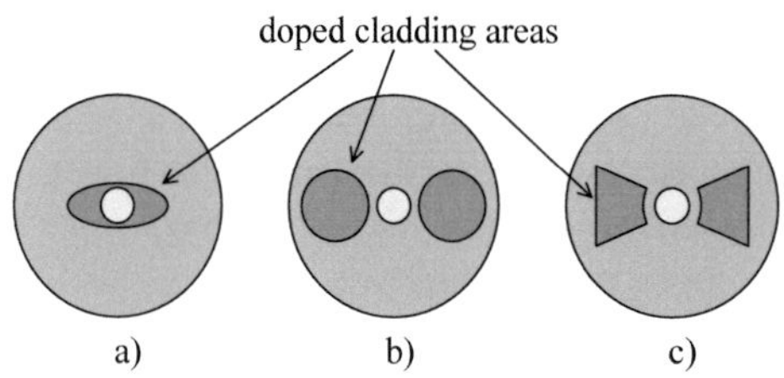


Fig. 13. Various profiles of PM HB fibers: a) elliptical, b) PANDA type, c) Bow-Tie type.

$$CT = 10\log\frac{P_y}{P_x}. \tag{40}$$

In the following chapters, fibers with intended polarization preserving properties are discussed. Although they exploit different principles, they may be considered as a special type of PM LB or PM HB fibers, as will be mentioned.

### 6.2 Fibers with high circular birefringence

The demands of telecommunication applications for polarization state preserving in fiber may be satisfactory covered by PM fibers. Since, PM fiber allows preservation of state polarization only for one certain polarization plane orientation, they are not well suited for applications in polarimetric fiber sensors. The polarization plane rotates due to the sensing quantity magnitude in case of these sensors. Therefore, the different fiber modifications were studied.

#### 6.2.1 Twisted fibers

One of the approaches is fiber twisting, which may impose a strong circular birefringence in the core. If the rate of induced circular birefringence will be much greater than the rate of linear birefringence, relation (38) may be modified

$$\Delta\phi = \sqrt{\phi_c^2 + \left(\frac{\phi_l}{2}\right)^2} \overset{\phi_c \gg \phi_l}{\cong} \sqrt{\phi_c^2} = \phi_c. \tag{41}$$

The influence of circular birefringence will dominate and the linear birefringence may be neglected. Since the effect of linear birefringence is canceled, twisted fibers belong to group of PM LB fibers. Twisted fiber will behave as polarization plane rotator, which preserves the polarization state of the wave. The polarization plane rotation is proportional to specific rotation of the fiber core and $\sigma_c$ and $l_f$ is fiber length

$$\Delta\alpha = \sigma_c l_f. \tag{42}$$

If this principle would be utilized in polarimetric fiber sensor application, the polarization rotation due to sensed physical quantity will be additive to inherent polarization rotation of the fiber.

The circular birefringence in fiber core is possible to impose by fiber twisting in a plane, which is perpendicular to fiber longitudinal axis. The rate of circular birefringence corresponds to photo-elastic properties of the core, core refractive index, fiber length and specific torsion rate (relation (23)). In order to minimize the influence of linear birefringence, the specific torsion rate should be maximized. However, this is limited by torsion limit of the fiber. When exceeds, the fiber may be broken.

The disadvantage of the twisted fiber utilization in fiber sensing applications is the temperature dependence of twisting imposed circular birefringence, due to temperature dependence of core anisotropy. The next issue is the fabrication difficulty of small fiber coils for magnetic field sensing around conductors. The bending induced linear birefringence achieves a higher magnitude for small fiber coils. Therefore, the large torsion rate of the fiber has to be used and it may exceed the torsion limit. The approximate torsion limit of common single mode fiber is 100 turns per 1 meter (Payne et al., 1982). Taking into account this limit, it is possible to fabricate fiber coils with minimal diameter of 15 cm (Laming & Payne, 1989).

### 6.2.2 Spun low- and high-birefringent fibers

The more sophisticated approach to linear birefringence suppression was development of spun fibers. Spun fiber fabrication consists in twisting of melted preform during fiber drawing. During the fiber drawing, all fiber imperfections, as deviation from circularity and other non-uniformities, are spread out in all directions. Therefore, phase retardations, which experience propagating modes, cancel each other out. Since the twisted preform is melted, no stress induced anisotropy is present in the core after the fiber cools down. Simultaneously, the fiber is free from temperature dependency effects. Hence, the spun fiber behaves as an ideal fiber with circular cross-section core, which retain any polarization state of coupled wave, from the input to the output. In principle, spun fibers belong to group of PM LB fibers. We designate them as low-birefringent spun fibers (spun LB). Spun LB fiber exhibits only a negligible latent circular birefringence, due to limited viscosity of the preform during the drawing. The principle of spun LB fiber implies their main disadvantage, which is the sensitivity on fiber bending. This results in stress induced linear birefringence and the limitation for small radius fiber coil fabrication remains (Payne et al., 1982).

A similar concept of spun LB fiber represents highly birefringent spun fibers (Spun HB). Spun HB fibers are manufactured by rotating of melted preform also. However, the preform is prepared as for classical PM HB fiber, e.g. Bow-Tie (Laming & Payne, 1989). Spun HB fiber transforms the input linear polarization on to the elliptical. By carefully chosen rotation rate of the preform in relation to the fiber linear birefringence rate, it is possible to attain quasi-circular birefringence with negligible residual linear birefringence (Payne et al., 1982). Generally, we may consider spun HB fibers as a type of PM HB fiber group. The advantage of spun HB fibers is considerable immunity to rising of linear birefringence by fiber bending or compressing. Since the quasi-circular birefringence in the fiber originates from twisting of stress components, the temperature dependence of anisotropy is indispensable. Therefore temperature compensation has to be used in applications utilizing spun HB fibers. On the present, spun HB fibers for telecommunication and sensing application are available for wavelengths from 600 nm to 1600 nm, with attenuation in order of ones of $dB{\cdot}km^{-1}$. They may be wind on fiber coils with radius above 20 mm.

In connection with recent advances in microstructured fibers research, new possibilities of spun fiber fabrication emerge. The concept of microstructured fiber allows designing and producing of fibers with specific parameters on selected wavelength, single mode or multi mode character, polarization transformation properties and others. A development of microstructured spun fiber with six air chambers around the core and attenuation below 5 $dB{\cdot}km^{-1}$ is reported in (Nikitov et al., 2009). The possibility of circular polarization transmission has been achieved by rotating of the microstructured preform, together with magneto-optic properties preservation. The fiber coils with diameter above 2.5 mm can be fabricated for current sensing application.

### 6.3 Annealed fibers

Drawback of twisted fibers and spun HB fibers, which is anisotropy temperature dependence, limits their applicability mainly in polarimetric current sensor applications. A method for suppression of temperature dependence of anisotropy together with the suppression of bending induced linear birefringence has been proposed and experimentally studied (Stone, 1988; Rose et al., 1996). The method utilizes annealing of fabricated fiber coil. The procedure consists in temperature treatment of the fiber coil, which is installed in ceramic labyrinth. The coil is then heated up with approximate temperature-time gradient $\Delta T/\Delta t = 8{\cdot}10^{-2}$ °C·s$^{-1}$. When 850 °C is reached, the temperature is maintained for roughly 24 hours. Then, the slow cooling follows with approximate gradient $\Delta T/\Delta t = -3{\cdot}10^{-3}$ °C·s$^{-1}$. Annealed fiber coil is then transferred into protective case, which is filled with low-viscosity gel in order to damp the vibrations. The annealing procedure leads to removing of bending induced stress on the fiber and the linear birefringence is greatly suppressed. Prior to the annealing procedure, the fiber jacket and buffer has to be removed, because its oxidation at the temperatures 500 - 600 °C would damage the fiber. Since the fiber jacket and buffer act as fiber strength element, their removal is difficult and fiber rupture impends. The outer layers removal is facilitated by etching in organic solvent. The oxidation proceeds then without the negative influence on the fiber cladding (Rose et al., 1996).

Due to considerable temperature stability, the annealing method is used for fabrication of fiber current sensors, which are installed in outdoor environment on high voltage systems. The need for reliable galvanic isolation and accuracy predominates the technological difficulties in fiber coil fabrication. The annealing procedure has to be carefully performed and it has to be handled a technology of fiber coil isolation from vibrations. Since the fiber strength outer layers have been removed, the fiber coils have increased sensitivity to vibrations. For outdoor installations on high voltage system an annealed fiber sensors with sensitivity variation smaller than 0.2 %, dynamic range of 80 dB and temperature range 20-80 °C were developed (Higuera-Lopez, 2002).

### 6.4 Reciprocal compensation of linear birefringence

In combination with fibers, which were described above and with common single mode fibers also, another perspective approach for linear birefringence suppression may be exploited. The approach is based on reciprocity of linear birefringence. Beyond this, the circular birefringence is of non-reciprocal character, which gives usability especially in polarimetric current fiber sensors. Utilizing this fact, a compensation of linear birefringence may be performed on sensor output signal level in case of counter-propagating of two light waves in fiber. Second possibility consists in compensation of modes phase shift on fiber level in case of back-propagating of light wave with ortho-conjugated polarization.

### 6.4.1 Compensation on sensor output signal level

Since the linear birefringence is of reciprocal character, its influence on polarization state of the wave in fiber is not dependent on the propagation direction. The wave will experience the same polarization state transformation with the same orientation, no matter the propagation direction. Conversely, the magnetic field induced circular birefringence is non-reciprocal. When the wave will propagate in one direction, it will gain a polarization rotation. When the wave will propagate in opposite direction, it will gain rotation in opposite direction. The total rotation will be the double of the rotation in one direction.
Setups, which exploit the reciprocity of linear birefringence, have been demonstrated as polarimetric current sensors. An example of the sensor setup is shown in Fig. 14 (Claus & Fang, 1996).

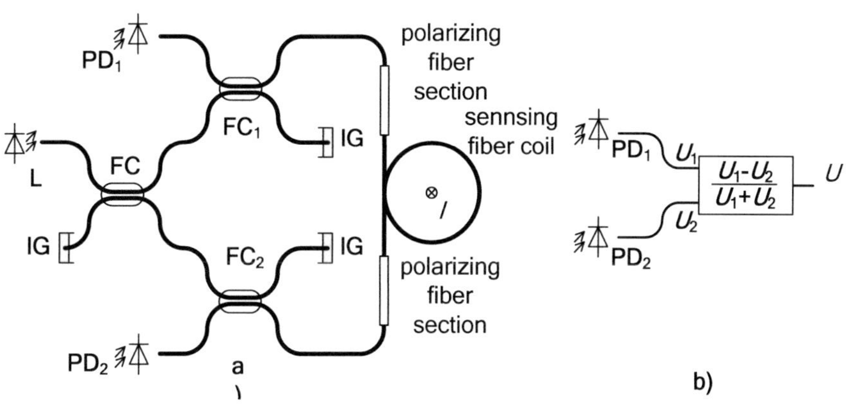


Fig. 14. Sensor setup with compensation on output signal level a), signal processing b).

In Fig. 14a), the signal from laser source L is divided into two channels by means of fiber coupler FC. After passing $FC_1$ and $FC_2$, the two optical signals propagate in opposite direction through the polarizing parts and the sensing part of the fiber. The unused coupler outputs are led into immersion gel to avoid reflections. The optical signals are sensed by photodetectors $PD_1$, $PD_2$. For output voltage signals of the detectors we can deduce

$$\begin{aligned} U_1 &= R_{\mathrm{U},1} \cdot FC_1 \cdot T_{\mathrm{ov}}^{-} \cdot FC_2 \cdot FC \cdot P_{\mathrm{o}}, \\ U_2 &= R_{\mathrm{U},2} \cdot FC_2 \cdot T_{\mathrm{ov}}^{+} \cdot FC_1 \cdot FC \cdot P_{\mathrm{o}}, \end{aligned} \tag{43}$$

where $P_o$ is the input optical power. $FC$, $FC_1$, $FC_2$ are split ratio of coupler FC, $FC_1$ $FC_2$, $T_{ov}^+$ and $T_{ov}^-$ are polarizing transfer functions of the fiber path in one and in second direction, $R_{U,1}$, $R_{U,2}$ are responsivities of photodetectors. For correct operation, it must hold $FC = FC_1 = FC_2 = 0{,}5$ and $R_{U,1} = R_{U,2} = R_U$. Relation (43) transforms into

$$U_1 = R_U \cdot FC^3 \cdot T_{ov}^-,$$
$$U_2 = R_U \cdot FC^3 \cdot T_{ov}^+. \tag{44}$$

In signal processing block, as shown in Fig. 14b), the normalized difference is computed

$$U = \frac{U_1 - U_2}{U_1 + U_2} = \frac{R_U \cdot FC^3 \cdot \left(T_{ov}^- - T_{ov}^+\right)}{R_U \cdot FC^3 \cdot \left(T_{ov}^- + T_{ov}^+\right)} = \frac{T_{ov}^- - T_{ov}^+}{T_{ov}^- + T_{ov}^+}. \tag{45}$$

Considering the presence of reciprocal linear birefringence only, the equality $T_{ov}^+ = T_{ov}^-$ holds. The output signal according to (45) will be zero. Since the magnetic field induced circular birefringence is non-reciprocal, the equality of polarizing transfer functions will not hold. Hence, the system will be responsive on varying rotation due to induced circular birefringence. Sensor utilizing above described principle dispose of considerable temperature stability and vibration insensitivity. More detailed description and sensor properties may be found in (Fang et al., 1994; Wilsch et al., 1996).

### 6.4.2 Orthogonal conjugation compensation

The reciprocity of undesirable birefringences in optical fiber may be used for their compensation exploiting the polarization ortho-conjugation of the wave modes. The method involves the back-propagation of light wave with conjugated modes through the same section of birefringent fiber.

As it has been stated above, imposing stress on fiber (pressure, bending) leads to origin of linear birefringent fiber core with two refractive indices, one lying in $x$ axis direction – $n_x$ and second lying in $y$ axis direction – $n_y$. We may designate the axis as the fast fiber axis, with lower refractive index, and the slow fiber axis, with higher refractive index. However, the orientation of the fast and slow axes system towards the geometrical coordinate system of the fiber changes along, in dependence on the bending or pressure force orientation. The modes propagating with different refractive indices gain a phase shift to each other, which results in wave polarization transformation. Since the instantaneous magnitudes of refractive indices in fast and slow axis may vary due to rate of bending or compressing, the resulting phase difference relies on average magnitudes of the refractive indices in both axes.

In order to restore the original polarization state, the wave at the output of the birefringent fiber has to be reflected and coupled back in the fiber together with modes conjugation. To accomplish this, Faraday polarization rotator and flat mirror is exploited, as shown in Fig. 15. The whole device is called ortho-conjugation reflector (OCR) or often Faraday

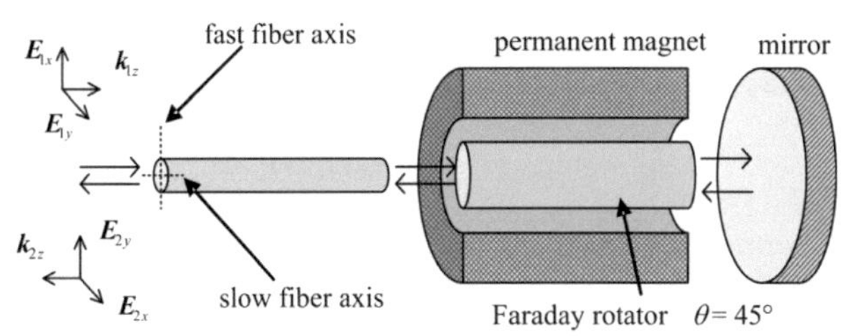


Fig. 15. Principle of Faraday rotation mirror.

rotation mirror (FRM). It consists of Faraday rotator, mounted inside a permanent magnet, flat mirror and collimator (for fiber optic application). The magnetic field magnitude, rotator dimension and its assembly is properly adjusted, that a light wave polarization is rotated with angle 45° while passing the rotator. Mirror allows reflection in perpendicular direction and collimator serves for fiber coupling.

Consider a birefringent fiber with a fast axis and a slow axis, Fig. 15. The modes of light wave, which travels in fiber, gain a phase shift. The mode in slow axis is retarded, while the mode in fast axis travels faster. At the fiber output, the wave is collimated to rotator. It rotates the wave polarization with angle $\theta/2 = 45°$ during a single pass. Then the wave is reflected back. After the second rotator pass, the rotation angle is $\theta = 90°$ due to the rotation non-reciprocity. The wave is coupled back into the fiber. Now, the wave propagates in fiber in backward direction. The wave mode, which traveled previously along the fast axis, travels now along the slow axis. Conversely, the mode, which traveled previously along the slow axis, travels now along the fast axis. The total phase shift of the modes is equalized and the original polarization state is restored. It should be mentioned, although the difference of average refractive indices will no be constant in time (for example by fiber manipulation), the final phase shift will remain zero thanks to reciprocal compensation. The temperature stability of the method is considerable also. However, this is true only when the temperature of FRM is stable, since the Verdet constant of rotator in FRM is temperature dependent.

Though the principle is not applicable for telecommunication purposes, it may be utilized in applications, where the polarization state preservation is desirable. This is often required in erbium-doped fiber amplifiers or tunable fiber lasers. Fiber optic sensors are another field of application. In case of fiber interferometers a polarization state of waves incoming from reference arm and sensing arm has to be preserved in order to interfere. Therefore, FRMs are used in both arms of interferometer. Fiber optic current sensors are another example of usage of FRM (Drexler & Fiala, 2008). Fiber current sensors exploit polarization rotation of the guided wave due to the magnet field actuation. Since the circular birefringence owing to Faraday magneto-optic effect is of non-reciprocal character, it will not be compensated during the backward propagation. Moreover, the polarization rotation will be double, which improves the sensitivity of the sensor. The example of fiber optic polarimetric current sensor utilizing FRM is shown in Fig. 16 (Drexler & Fiala, 2009). Laser beam from laser source L is collimated by means of collimator C and linear polarized by means of polarizer P. Beam passes a non-polarizing beam splitter NBS and it its collimated

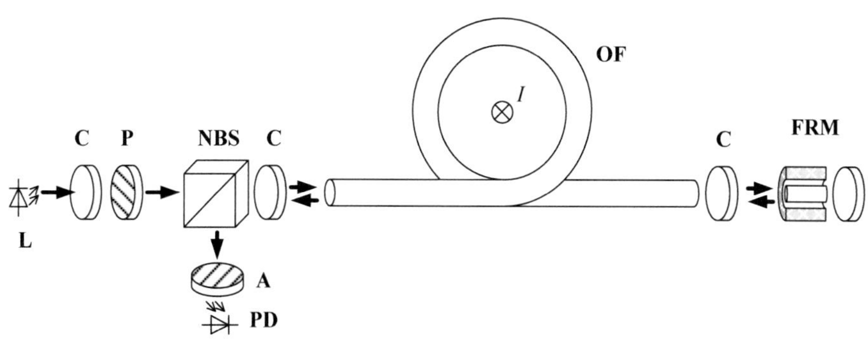


Fig. 16. Fiber optic polarimetric current sensor utilizing Faraday rotation mirror.

into the fiber by collimator C. The beam propagates through the magnetic field sensing fiber OF and exits the fiber. Collimator C collimates the beam into the FRM and collimates backward beam back into the fiber. The beam travels the fiber in opposite direction. After the collimation, part of the beam is deflected by beam splitter NBS. It passes the analyzer A and hits the photodetector PD. The output linear polarization state is perpendicular to the input linear polarization state. When magnetic field acts on the sensing part of the fiber, the polarization plane rotates. The polarization modulation is converted on intensity modulation by means of analyzer A.

In spite of the advantages of the FRM application, several drawbacks limits its usage. One of the drawbacks is temperature dependence of rotator Verdet constant (Santoyo-Mendoza & Barmenkov, 2003). Therefore the FRM unit has to be temperature stabilized. Simultaneously, it has to be shielded from outer magnetic field. Indispensable is the cost of this solution also owing to precise fabrication and adjusting of the FRM. Commercially available are FRMs in compact fiber pigtailed housing for longer wavelengths (1310 nm, 1550 nm). There are also available FRMs for shorter wavelengths (633 nm). However, they are bulky, because of the need of more powerful magnet.

## 7. Conclusion

Due to their unique properties, single mode fibers have found a huge application potential in various fields of industry and science. They are massively exploited in telecommunication technology, control and sensor systems, industrial laser systems and they represent an unsubstitutable tool for advanced science. The requirements for specific fiber properties differ for various applications. In lot of them, a transmission of light wave with preserving of state of polarization is demanded, which is often a weak point of common fibers. However, this drawback is possible to overcome with a suitable approach, depending on demands of particular application.

In order to evaluate the possibility of polarization state distortion, various influences have to be conceived. It is also very advantageous to be able to quantify them. According to this demand, the intention of the first part of the contribution is to specify the fundamental effects, which lead to fiber birefringence. The basic relations, which allow estimating the birefringence rate are presented also.

Once the fiber birefringence occurs, in lot of cases arises a need to suppress its undesirable consequences. A various approaches, which may be utilized, are the point of the interest of the second part of the contribution. The principles of the most significant methods are described. Their advantages and disadvantages are presented also. The suitability of a selected method is given by the application requirements. They differ in cost, complexity, temperature and mechanical stability and others. Because of the limited extent of this contribution, all of the methods properties and details could not be presented. Nonetheless, the chapter may be a convenient starting point for orientation in this field and details may be found in cited reference sources.

## 8. Acknowledgement

The contribution has been prepared with support of research project GA102/09/0314 of the Czech Science Foundation and project FEKT-S-10-13 of the Grant Agency of Brno University of Technology.

## 9. References

Born, M. & Wolf, E. (1999). *Principles of Optics*, Cambridge University Press, ISBN 978-0521642224, Cambridge

Claus, R. O. & Fang, X. (1996). Optimal Design of IRIS-Based Polarimetric Intrinsic Optic Current Sensor. *Journal of Lightwave Technology*, Vol. 14, No. 7, pp. (1664-1673), ISSN 0733-8724

Craig, A. E. & Chang, K. (Ed(s).). (2003). *Handbook of Optical Components and Engineering*, John Wiley & Sons, ISBN 0-471-39055-0, New Jersey

Drexler, P. & Fiala, P. (2008). Utilization of Faraday mirror in fiber optic current sensors. *Radioengineering*, Vol. 17, No. 4, pp. (101-107), ISSN 1210-2512.

Drexler, P. & Fiala, P. (2009). Suppression of polarimetric birefringence effect in optical fiber and its application for pulsed current sensing, *Proceedings of 2009 Waveform Diversity & Design Conference*, ISBN 978-1-4244-2971- 4, USA: Orlando, 2009

Fang, X., Wang, A., May, R. G. & Claus, R. O. (1994). A Reciprocal-Compensated Fiber Optic Current Sensor. *Journal of Lightwave Technology*, Vol. 12, No. 10, pp. (1882-1890), ISSN 0733-8724

Higuera-Lopez, J. M. (2002). *Handbook of Optical Fiber Sensing Technology*, John Wiley & Sons, ISBN 978-0-471-82053-6, New York

Huard, S. (1997). *Polarization of Light*, John Wiley & Sons, ISBN 2-225-85327-X, Paris

Iizuka, K. (2002). *Elements of Photonics, Volume II*, John Wiley & Sons, ISBN 0-471-83938-8, Toronto

Jones, R. C. (1941). New calculus for the treatment of optical systems. *Journal of the Optical Society of America*, Vol. 31, No. 7, ISSN 1862-6254

Kaminow, I. P. and Ramaswamy, V. (1979). Single-polarization optical fibers: slab model. *Applied Physics Letters*, No. 34, pp. (268-270), ISSN 0003-6951

Laming, R. I. & Payne, D. N. (1989). Electric current sensors employing spun highly birefringent optical fibers. *Journal of Lightwave Technology*, Vol. 7, No. 12, pp. (2084-2094), ISSN 0733-8724

Namihira, Y. (1983). Opto-elastic constants in single-mode optical fibers. *Journal of Lightwave Technology*, Vol. 3, No. 5, pp. (1078-1083), ISSN 0733-8724

Nikitov, S., Chamorovskiy, Y., Starostin, N., Ryabko, M., Morshnev, S., Morshnev, V. & Vorob'ev, I. (2009). Microstructured optical fibers for the fiber optics sensors, *Proceedings of International Conference on Materials for Advanced Technologies*, Singapore, June 2009

Noda, J., Okamoto, K. & Sasaki, J. (1986). Polarization-maintaining fibers and their applications. *Journal of Lightwave Technology*, Vol. 4, No. 8. pp. (1071-1089), ISSN 0733-8724

Payne, D. N., Barlow, A. J. & Hansen, J. J. R. (1982). Development of Low- and High-Birefringence Optical Fibers. *IEEE Transaction on Microwave Theory and Techniques*, Vol. 30, No. 4, pp. (323-334), ISSN 0018-9480

Ripka, P. (Ed(s).) (2001). *Magnetic sensors and magnetometers*, Artech House, ISBN 1-58053-057-5, London

Rose, A., Ren, Z. F. & Day, G. W. (1996). Twisting and annealing optical fiber for current sensors. *Journal of Lightwave Technology*, Vol. 14, No. 11, pp. (2492-2498), ISSN 0733-8724

Saleh, B. A. & Teich, M. C. (1991). *Fundamentals of Photonics*, John Wiley & Sons, ISBN 978-0471839651, New York

Santoyo-Mendoza, F. & Barmenkov, Y. O. (2003). Faraday plasma current sensor with compensation for reciprocal birefringence induced mechanical perturbations. *Journal of Applied Research and Technology*, Vol. 1, No. 2, pp. (157–163), ISSN 1665-6423

Senior, J. M., (2009). *Optical Fiber Communication: Principles and Practice*, Pearson Education, ISBN 978-0-13-032681

Stone, J. (1988). Stress-optic effect, birefringence, and reduction of birefringence by annealing in fiber Fabry-Perot interferometers. *Journal of Lightwave Technology*, Vol. 6, No. 7, pp. (1245-1248), ISSN 0733-8724

Tabor, W. J. & Chen, F. S. (1968). Electromagnetic Propagation Through Materials Processing both Faraday Rotation and Birefringence: Experiments with Ytterbium Orthoferrite. *Journal of Applied Physics*, Vol. 10, pp. (2760-2765), ISSN 0021-8979

Ulrich, R., Rashleigh, S. C. & Eickhoff, W. (1980). Bending-induced birefringence in single-mode fibers. *Optical Letters*, Vol. 5, No. 5, pp. (273 - 275), ISSN 1539-4794

Wagner, E., Dändliker, R. & Spenner, K. (Ed(s).). (1992). *Sensors, a comprehensive survey, Volume 6 – Optical Sensors*, VCH, ISBN 3-527-26772-7, Weinheim

Waynant, R. & Ediger, M. (2000). *Electro-Optics Handbook*, McGraw-Hill Professional, ISBN 978-0-07-068716-5, New York

Wilsch, M., Menke, P. & Bosselman, T. (1996). Magneto-optic current transformers for applications in power industry, *Proceedings of 2nd Congress of Optical Sensor Technology OPTO 1996*, Germany: Leipzig, 1996

8

# Spun Fibres for Compensation of PMD: Theory and Characterization

Lynda Cherbi and Abderrahmane Bellil
*Laboratory of Instrumentation LINS,*
*University of Sciences and Technology Houary Boumedienne, USTHB,*
*Algiers,*
*Algeria*

## 1. Introduction

The polarization is a relative property to the vibratory nature of light. In an optical fibre, light is a combination of two vibrations of perpendicular directions. Each direction represents one mode of polarization. Indeed, the optical fibres and the components of the optical fibres present a small difference in the refractive index in the pair of the polarization states, a property called the birefringence. This last one induces a difference of propagation speed between the two modes. So, light at the output, cannot be restored more faithfully. The birefringence can change the state of polarization (SOP) of light when it crosses the fibre. In a single mode fibre, the birefringence is combined with a random coupling of polarization modes. The delay measured at the output of the fibre between the two polarization modes is called the difference of group delay DGD (measured in picoseconds). The polarization modes dispersion (PMD) results from the variation of the DGD according to the wavelength and the environment conditions.
The typical tolerance of a system to the PMD is roughly 10% of the bit period, which gives 40 Ps for a system of 2.5 Gb/s, 10 Ps for a system of 10 Gb/s and only 2.5 Ps for a system of 40 Gbs/s (Noé et al.,1999). The PMD is a random phenomenon and constitutes an enormous obstacle ahead of the increase of the debits from 10 Gbit/s for a part of the networks of most telecommunication companies. Several solutions have been proposed to compensate the PMD as: The electronic compensation after a direct photo-detection that can only eliminate a part of the PMD effects since the information about the polarization and the phase get lost at the detection; the second solution is the electronic compensation in a coherent receptor with diversity of polarizations, and the third one is the optical compensation in at least a differential delay section. Other solutions are proposed by the Corning society and which rely on the use of spun fibres allowing the control of the coupling of the modes, therefore reducing the PMD; thus, giving differential group delays of order of Femtoseconds. In the past decade, some considerable efforts have been made to understand the origins of the PMD and to attenuate its effects in the systems. The PMD can be reduced in a fibre with two different manners. The first one consists in minimizing the asymmetries in the refractive index profile and the constraints, which implies improvements of the industrial process in the manufacture of the fibre in order to assure a better geometry and to reduce the rate of constraints in the fiber. The second method allows the control of the coupling of the modes

of the polarization in the fibre while spinning it during its manufacture. Indeed, the spinning has been used in the manufacture of fibres since the beginning of the 1990s, and it showed that it is an efficient technique to reduce the PMD in the fibre. First, we start by presenting in this work the spun fibres explaining their technology, their principle and their different types. Next, the description of the reduction of the PMD by using the spinning is developed by a mathematical formalism based on the theory of coupling and Jones's matrix. Moreover, the reduction of the PMD is verified in the spun fibres while applying the method of JME and the COTDR method (photon counting -Optical temporal Domain of Reflectometry) that allowed us to measure the DGD of the order of femtosecond (Cherbi et al., 2009). The comparison of the DGD found in this type of new generation of fibres with those of the standard ones, led us to confirm that the spun fibres offer effectively a smaller DGD than those of the standard fibres, emphasizing the importance of this type of fibres in the reduction of the PMD.

We present the different results already published (Cherbi et al., 2009) while using the reflectometers COTDR and POFDR (polarization- Optical Frequency Domain of Reflectometry) which are used to get the polarization characteristics of the spun fibres as the beat length and the PMD and to observe the spatial frequencies linked directly to the period of spinning.

## 2. Principle of spun fibres

### 2.1 Technologies of the spun fibres

There are more than two decades when the concept of the spun fibres has been proposed originally in an article published by (Barlow et al.,1981).

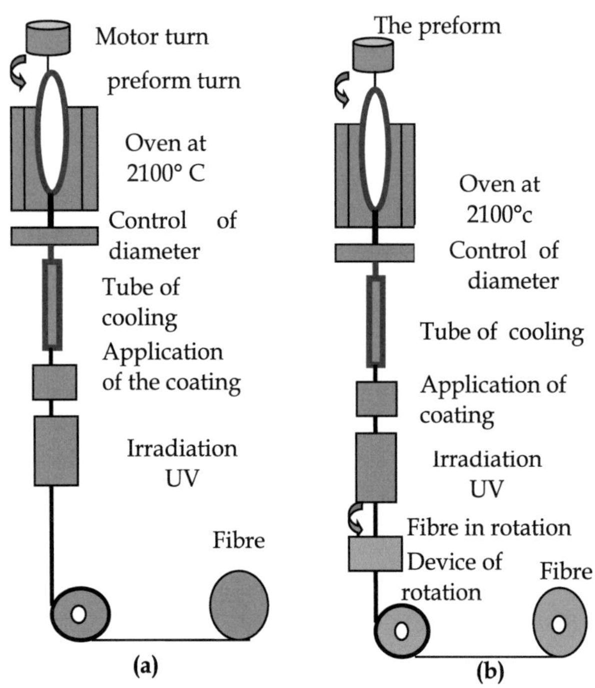


Fig. 1. Two approaches used to present the rotating fibres (a) turn preforms (b) turn the fibre.

The fibre spun is achieved by the rotation of the preform during the drawing of the fibre (figure 1.a). In this approach, the system of drawing is the same as that of the conventional standard fibre systems (OVD) except that a rotating motor is placed on the top of the preform. When the motor is set in motion with a predetermined speed, the preform starts turning dragging the rotations of the axes of the birefringence. The rotation will end up with the end of the drawing operation. This approach is quite simple and appropriate for the pulling of the fibre at low speed. However, this is not convenient for the production of the fibre with a high speed pulling because the rotation of the motor must be at very high speed as well. To illustrate this, we consider a rotation rate of the fibre of 3 turns/min, for a drawing speed of 1 m/s, the rotating speed of the preform is thus only 180 turns/min.

On the other hand, for a modern drawing device having a speed higher than 20m/s, the perform must turn at a speed greater than 3600 turns/mn, which is far from practical. For this reason the concept of the spun fibres has not been used in the production of fibres until the half of the nineties when methods of more adapted spinning have been proposed (Ming-Jun & Nolan, 1998). Moreover, the transmission systems as they appeared at low rate (<= 2.5 Gb/s), the PMD was not a major problem to seek fibres that perform this reduction.

Several convenient techniques have been suggested during the year 1990, for example, by (Hart et al., 1994] in order to make the fibre turn rather than the preform. Later on, this technique became the most adapted one for the manufacturing of the fibres performing the reduction of the PMD.

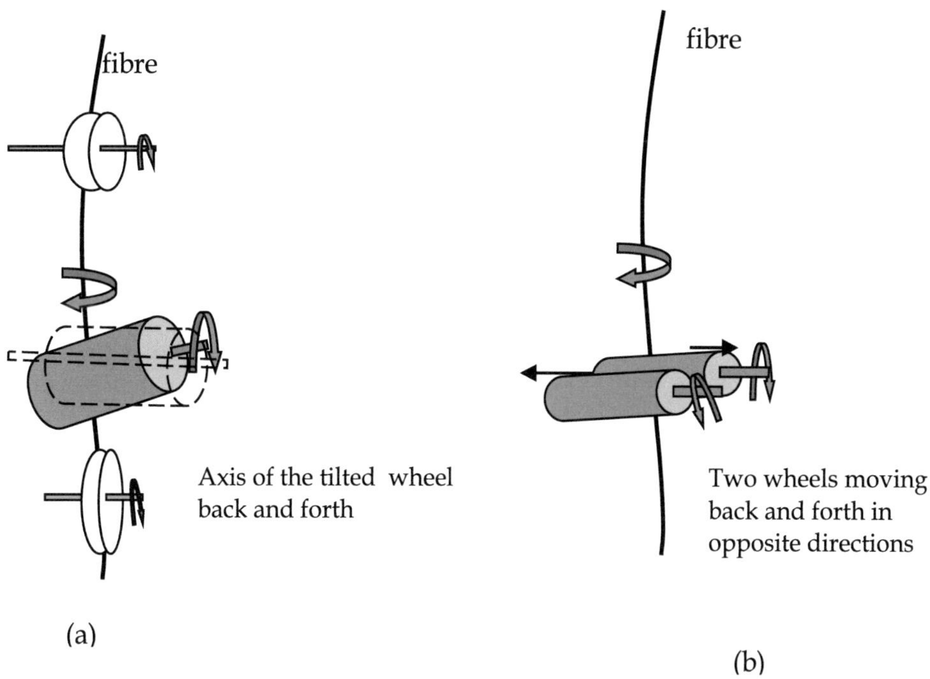


Fig. 2. Examples of fibre rotation systems: (a) tilted wheel, (b) two wheels moving in opposite directions.

In this case (fig 1.b), a rotating device of the fibre, is put along the way of fibrage to rotate the fibre directly. Two examples of this device are illustrated in figue 2. In the first example (fig 2.a), a wheel is in contact with the fibre and tilted with respect to its initial position, thus applying a moment of rotation to turn the fibre. In the second example (fig 2.b), two wheels are placed horizontally and are both in contact with the fibre (Blaszyk & Christoffand,2001). The two wheels move back and forth in opposite directions driving the rotational movement of the fibre. Imposing a direct motion to the fibre eliminates the problem of the preform when turning at high speed. Besides, this technique provides flexibility to control and implement different profiles of rotation for a better reduction of the PMD.

### 2.2 Theory of spun fibres

Two approaches have been suggested to model the reduction of the PMD in the spun fibres, one of which is based on the evolution of the polarization state (Galtarossa et al., 2001; Ming-Jun &Nolan, 1998). The evolution of the vector representing the polarization dispersion is ruled by the dynamic equation which is linked to the vector of the local birefringence. While solving the dynamic equation, the vector representing the polarization's dispersion is gotten and its module gives the delay of the differential group (DGD). Another approach is based on the theory of the coupled modes of Jones' matrix (Ming-Jun et al., 2002) where the complex amplitudes of the two modes of polarization are described by the equations of the modes coupling. While solving these equations, the complex amplitudes are derived and Jones' matrix is determined and the DGD can be computed from this matrix. Basically, the two approaches give equivalent results. Our survey of the spun fibres is founded on Jones' matrix formalism, where we notice that the analytical solutions obtained are simple.

#### 2.2.1 Equations of the mode coupling

As the birefringence in the fibres used in telecommunications is generally small, the formalism based on the theory of disruption (Ming-Jun et al., 2002), can be used to describe the different mechanisms of birefringence in the single mode fibres, including the birefringence due to the distortion of the core, constraints, curvature, rotation of the fibre and torsion. In what follows, we will present the theory of the coupled modes and we will show how to implement it in the different problems of birefringence. Indeed, the small birefringence of telecommunication fibres can be treated as an anisotropic disruption to a material originally isotrope. In the condition of weak guidance, the electric field E is described by the following wave equation (Dandliker,1992):

$$\Delta E - \mu_0 \, \varepsilon_0 \, \varepsilon \, E = \mu_0 p \tag{1}$$

Where $\varepsilon_0$ et $\mu_0$ are the dielectric and magnetic constants of vacuum respectively, ε is the relative dielectric constant of the non disrupted fibre, and p is the disruption term given by the following relation:

$$p = \varepsilon_0 \, \Delta\varepsilon \, E \tag{2}$$

Where $\Delta\varepsilon$ is the electric tensor describing the anisotropy of the medium. Without the term of disruption, the equation (1) has modal solutions of the following shape:

$$E_n(x,y,z) = e_n(x,y)\exp(-i\beta_0 z) \qquad n = 1,2 \tag{3}$$

where $e_n(x,y)$ is the distribution of the electric field. For a monomode fibre, n =1, 2 represent the two modes of polarization.
In absence of disruption, the two modes are degenerated and propagate with the same constant $\beta_0$. In presence of the disruption term, it is supposed that the electric field E(x,y,z) is given by the linear superposition of the two non disrupted modes (Ming-Jun et al., 2002):

$$E(x,y,z) = \sum_n A_n(z) e_n(x,y)\exp(-i\beta_0 z) \tag{4}$$

Where $A_n(z)$ are the complex coefficients describing the amplitudes and the phases of the two modes. Let's put equation (4) into equations (1) and (2) and use the relation of orthogonality between the two modes (Ming-Jun et al., 2002):

$$\int e_m(x,y)\,.\,e_n(x,y)dxdy = \begin{cases} N_m & m = n \\ 0 & m \neq n \end{cases} \tag{5}$$

Knowing that $N_m$ is a constant of normalization which can be calculated as follows:

$$N_m = \tfrac{1}{2}\int \vec{e}_m \times \vec{h}_m^{*}\, z\, ds = \frac{n_{coeur}}{2}\left(\frac{\varepsilon_0}{\mu_0}\right)^{1/2} \int e_m^2 ds \tag{6}$$

and using the condition of the weak coupling:

$$\frac{1}{\beta_0}\left|\frac{d^2 A_n}{dz^2}\right| << \left|\frac{dA_n}{dz}\right| \tag{7}$$

We get the equations of the coupled modes that describe the evolution of the complex amplitudes $A_n(z)$:

$$\frac{dA}{dz} = ik.A \tag{8}$$

where A is the complex amplitude vector taking the following form:

$$A = (A_1 \quad A_2)^T \tag{9}$$

and $k$ is the matrix of the coupling coefficients.

$$k = \begin{pmatrix} k_{11} & k_{12} \\ k_{21} & k_{22} \end{pmatrix} \tag{10}$$

The coupling coefficients are associated to the different types of disruptions:

$$k_{mn} = \frac{k_0}{2n_0N_0}\int e_n^*(x,y).\Delta\varepsilon(x,y,z).\,e_m(x,y)dxdy \tag{11}$$

Where $n_0$ is the effective refractive index of both non disrupted modes.

### 2.2.2 Jones matrix and the PMD of spun fibre

The evolution of the local polarization along the birefringent fibre is described by the equations of the modes coupling. The total change of polarization of an input signal, after having traveled a given distance in the fibre is better described by Jones' matrix. Let's assume that the losses in the fibres are negligible, the already predefined Jones matrix, can be put under another form which is:

$$T = \begin{bmatrix} A_1(z) & -A_2^*(z) \\ A_2(z) & A_1^*(z) \end{bmatrix} \text{ with } |A_1|^2 + |A_2|^2 = 1 \tag{12}$$

The four complex elements of Jones' matrix can be gotten while integrating the equations of the coupled modes with suitable initial conditions. Once Jones' matrix is known, the PMD can be calculated easily from the elements of the matrix (Chen, 2002; Ming-Jun et al., 2002):

$$\tau = 2\sqrt{\left|\frac{dA_1}{d\omega}\right|^2 + \left|\frac{dA_2}{d\omega}\right|^2} \tag{13}$$

In order to describe the reduction of the PMD, we define a parameter, named reduction factor of PMD (PMDRF) $\zeta$ as the ratio of the DGD of the spun fibres over the DGD of the standard fibre.

$$\zeta = \frac{\tau}{\tau_0} \tag{14}$$

Where the used lengths for the spun fibres and standard fibres are the same. For example, if $\zeta$ is equal to 1, the reduction of the PMD is not achieved and if $\zeta$ is equal to 0.5, a factor of two is obtained in the reduction of the PMD.

### 2.3 Different types of spun fibres

The coupling coefficients matrix depends upon the dielectric tensor of the disruption. The values of these elements are determined by the type of disruption, which means that they depend on the configuration of the fibre. In this section, we describe some configurations of the fibres and we give their coupling coefficients matrix. It is important to note that the coupling matrixes in this work are expressed on the basis of the circular polarization because it is more appropriate to process the rotating fibres (Ming-Jun et al., 2002).

### 2.3.1 The linearly birefringent fibre

The linear birefringence is a consequence of disruptions as the distortions of the core, the asymmetry of the lateral constraints, the curvature. In the case of the linear birefringence, the coupling coefficients matrix is given by (Ming-Jun et al., 2002):

$$k = \frac{1}{2}\begin{pmatrix} 0 & \Delta\beta e^{i2\varphi} \\ \Delta\beta\, e^{-i2\varphi} & 0 \end{pmatrix} \tag{15}$$

where Δβ is the linear birefringence, and Φ is the orientation of the birefringence with respect to a given axis

### 2.3.2 Spun fibres

In a spun fiber, the orientation of the birefringence takes place depending on the x axis. The rotation angle Φ accumulated is therefore a function of the fibre length 'z', which in turn is determined by the rate of rotation $\alpha(z)$:

$$\Phi = \int_0^z \alpha(z)\, dz \tag{16}$$

Replacing the equation (16) in equation (15), we get the coupling coefficients matrix of the rotating fibres, describing the disruption of the birefringence,:

$$k = \frac{1}{2}\begin{bmatrix} 0 & \Delta\beta\, e^{i2\int_0^z \alpha(z)dz} \\ \Delta\beta\, e^{-i2\int_0^z \alpha(z)dz} & 0 \end{bmatrix} \tag{17}$$

### 2.3.3 Twisted fibre

There are two effects in this type of fibres: The rotation of the birefringence and the mechanical torsion. The rotation of the birefringence is similar to that of the rotating fibre. If the rate of torsion is T, the angle Φ is calculated by

$$\Phi = T\, z \tag{18}$$

The rate of torsion is determined by the coefficients of photo - elasticity of the fibre. The torsion constraint induces the circular birefringence proportionally to the rate of torsion.

$$\delta = g.T \tag{19}$$

Where g is the coefficient determined by the coefficients of photo elasticity of the glass. The typical value of g for fibres in silica is 0.16. Combining both effects of rotation and torsion, the coupling matrix comes up with the following form:

$$k = \frac{1}{2}\begin{bmatrix} \delta & \Delta\beta\, e^{i2Tz} \\ \Delta\beta\, e^{-iTz} & -\delta \end{bmatrix} \tag{20}$$

### 2.4 Solutions of the coupled equations for different types of the spun fibres

Generally, the matrix of the coupling coefficients depends upon the variable z, and the analytic solutions of equation (8) have no existence in the majority of the cases. The numerical integration is always used to get numerical solutions. Different methods, as the method of the finite differences, the Runge-Kutta method (Chen, 2002), can be applied to solve the equation of the coupled modes. However, in the two following special cases, we can derive the analytic solutions which will be discussed in this section.

#### 2.4.1 Constant spinning rate

For a constant spinning rate, the function 'spin' (rotation) can be written as follows:

$$\alpha = \alpha_0 \tag{21}$$

Where $\alpha_0$ is a constant. In this case, the birefringence of a fibre is estimated in only one direction with a rate $\alpha_0$. For this reason, the constant spinning rate is often assigned to an unidirectional spinning. For a spun fibre, with a constant spinning rate, the integral of the coupling matrix can be calculated easily, and the coupled equations become (Hart, 1994):

$$\frac{dA_1}{dz} = \frac{1}{2} i\,\Delta\beta\, e^{i2\alpha_0 z} A_2 \tag{22}$$

$$\frac{dA_2}{dz} = \frac{1}{2} i\,\Delta\beta\, e^{i2\alpha_0 z} A_1 \tag{23}$$

With initial conditions $A_1(0) = 1,\ A_2(0) = 0$.

The solutions of equations (22) and (23) are:

$$A_1 = -\frac{\alpha_0 - \upsilon}{2\upsilon} e^{i(\alpha_0 + \upsilon)z} + \frac{\alpha_0 + \upsilon}{2\upsilon} e^{i(\alpha_0 - \upsilon)z} \tag{24}$$

$$A_2 = \frac{\Delta\beta}{4\upsilon} e^{i(-\alpha_0 + \upsilon)z} - \frac{\Delta\beta}{4\upsilon} e^{-i(\alpha_0 + \upsilon)z} \tag{25}$$

Where $\upsilon = \sqrt{\alpha_0^2 + \frac{1}{4}\Delta\beta^2}$

Using equations (12) and (13), we find that the DGD can be expressed by a simple equation for the spinning constant:

$$\tau(z) = \frac{\gamma_\omega}{2\upsilon} \sqrt{(\Delta\beta)^2 z^2 + \left( \frac{4\alpha_0}{\Delta\beta} \sin\left( \frac{\Delta\beta\, z}{2} \right) \right)^2} \tag{26}$$

Where $\gamma_\omega = \frac{d\Delta B}{d\omega} = \frac{\tau_0}{L}$ is the PMD of a uniform birefringent fibre without modes coupling at the z position. The sinusoidal term of equation (26) doesn't play an important role when the fibre is sufficiently long. On the other hand, for long fibres, the DGD is given by:

$$\tau(z) = \frac{\gamma_{\omega}.\Delta\beta.z}{2\upsilon} \quad (27)$$

Equation (27) indicates that the DGD progresses linearly with the length of the fibre, and the PMDRF takes the following form:

$$\zeta = \frac{\Delta\beta}{2\upsilon} \quad (28)$$

We notice that for spun fibres of constant rate, the PMDRF depends upon the length of beating or the birefringence.

### 2.4.2 The periodic spin function

For the functions of periodic spin, under some conditions, we can describe analytic solutions by using the theory of disruption (Chen et al., 2002) in which fibres are submitted to uniform disruptions only, or in the case of small lengths regime (typically smaller than 100 m) in order to fine down their analysis. Indeed, in this approach, the random characteristic of the variation of the disruption in case of important lengths regime is ignored. Using the initial conditions issued from the previous paragraph, the first order solutions of disruption for A1 (z) and A2(z) are as follows:

$$A_1(z) = 1 \quad (29)$$

$$A_2(z) = (i/2)\Delta\beta \int_0^z \exp[-2i\Theta(z')]dz' \quad (30)$$

Where $\Theta(z) = \int_0^z \alpha(z')dz'$

It becomes easier to obtain the DGD by using equation (13):

$$\tau(z) = \gamma_{\omega} \left| \int_0^z \exp[-2i\Theta(z')']dz' \right| \quad (31)$$

Based on the theory of disruption, the first order of the disruption's expansion is valid only when $\Delta\beta << 1$. This condition puts some limits on the application of equation (31) on fibres that have a low PMD.

The validity of this solution has been tested by (chen et al., 2002). When the length of beating is important (some meters), i.e. $\Delta\beta \approx 1$, and the period of spin is smaller than the length of beating, the theory of disruption of the first order can always be applied. For sinusoidal profiles of spin, the expression for the factor of the PMD reduction can be gotten from the solutions of the disruption equation. Let's notice that the profile of sinusoidal spin takes the following form:

$$\alpha(z) = \alpha_0 \cos(\eta z) \quad (32)$$

Where α0 is the spin amplitude , and η is the angular frequency of the spatial modulations, which is linked to the spin period Λ through the following relationship η=2π/Λ.

With the analytical solution of equation (31), we are able to assert which spin parameters give right to the optimization of the PMD performances. With the first observations, we remark that, when the length of beating of a fibre is bigger than some meters, the PMDRF is independent of the beating length, and therefore of the intrinsic birefringence of the fibre. In equation (31), the only contribution to the birefringence of the fibre, comes from $\gamma_{\omega}$, and the DGD is proportional to this size. Let's note that $\gamma_{\omega}$ is the PMD of the unspun fibers (non rotating). On the other hand, the PMDFR will be independent of $\gamma_{\omega}$.
This conclusion is also verified by the direct numerical integration of equation (8) with k given by equation (17). Some old fibres had beating lengths inferior to some meters; with the improvement of fibre manufacturing, the majority of these lengths were improved lately beyond some meters. The PMDRF independence from the intrinsic birefringence of the fibre, offers the advantage of simplicity in its conception because it is worthless to optimize the spin profiles for the different birefringences of the fibre. Moreover, we noticed that the DGD increases linearly when the length of the fibre increases (figure 3) despite the fact that we got some overlapping oscillations on the graph representing the variation of the DGD with respect to the distance.

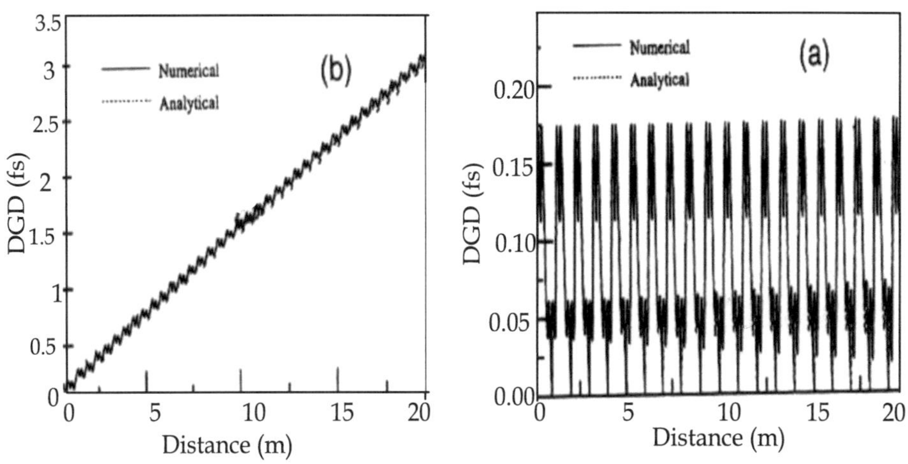


Fig. 3. Evolution of the DGD along a spun fibre.

We can also separate the real and imaginary contributions of the integral (31) in order to better analyze the variation of the DGD along the fibre. We express the equation (31) in an alternative way of the DGD for one spatial period T:

$$DGD(T) = \gamma_{\omega} \left| \int_0^T \cos[2\Theta(z')]dz' - i\int_0^T \sin[2\Theta(z')]dz' \right| \tag{33}$$

$$DGD(T) = \gamma_{\omega} \int_0^T \cos[2\Theta(z')]dz', \quad \Theta(z) \text{ is an even function}$$

$$DGD(T) = \gamma_{\omega} \int_0^T \sin[2\Theta(z')]dz', \quad \Theta(z) \text{ is an odd function}$$

We suppose that $\Theta(z)$ is a periodic function. When $\Phi(z)$ is an even function, $\int_0^T \sin[2\Theta(z')]dz'$ is equal to zero. When $\Theta(z)$ is an odd function, $\int_0^T \cos[2\Theta(z')]dz'$ is equal to zero. For multiple values of the period T, the DGD becomes $n[DGD(T)]$. For values in between, some oscillations are encrusted in the linear variation of the DGD. On the other hand, this survey based on [40] leads us to the conclusion that the dependence of the DGD of the standard fibres on the square root of their lengths, comes from the statistical nature of the random coupling of the two modes of polarization. The linear evolution of the DGD with respect to the length of the spun fibre is caused by the periodicity of the coupling induced by the spinning, thus we have a coupling mode better-controlled than in the case of standard fibres. However, it is possible that the DGD of the spun fibres follows a different evolution law in a region where the first order theory of disruption is not valid any more; for example, when the intrinsic birefringence of the fibre is high and / or the spin rate is high.

With the aforementioned results, it is rather simple to find the phase matching conditions for which the maximum reduction of the PMD can be obtained. In this case, the condition is fixed such that the PMDFR is equal to zero (chen et al., 2002):

$$\int_0^T \exp[-2i\Theta(z')]dz' = 0 \tag{34}$$

Equation (34) can be expressed in another way if we use the properties discussed previously for even and odd functions. We notice that when the phase matching conditions are satisfied, the evolution of the DGD along the spun fibre is periodic. The DGD doesn't increase anymore when the length of the fibre increases.

Equations (31) and (34) are valid for a whole category of periodic profiles of spin. To illustrate the way how to determine the phase matching conditions, we take an example of a sinusoidal spin profile. Such a profile is defined by equation (32). The integration of this profile gives $\Theta(z) = \alpha_0 \sin(\eta z)/\eta$; then we get the DGD by using equation (31):

$$DGD(z) = \gamma_\omega \left| \int_0^z \exp[-i\frac{2\alpha_0 \sin(\eta z')}{\eta}]dz' \right| \tag{35}$$

The integral can be valued analytically by using the following identity:

$$\exp[-ix\sin(\theta)] = J_0(x) + 2\sum_{n=1}^{\infty} J_{2n}(x).\cos(2n\theta) - 2i\sum_{0}^{\infty} J_{2n+1}(x)\sin[(2n+1)\theta] \tag{36}$$

Then, we get

$$DGD(z) = \gamma_\omega \left[R^2(z) + I^2(z)\right]^{1/2} \tag{37}$$

where

$$R(z) = J_0(2\alpha_0/\eta)z + \sum_{n=1}^{\infty} \frac{J_{2n}(2\alpha_0/\eta)}{\eta n}\sin(2n\eta z) \tag{38}$$

$$I(z) = \sum_{n=0}^{\infty} \frac{J_{2n+1}(2\alpha_0 / \eta)}{\eta(2n+1)} \cos[(2n+1)\eta z] \tag{39}$$

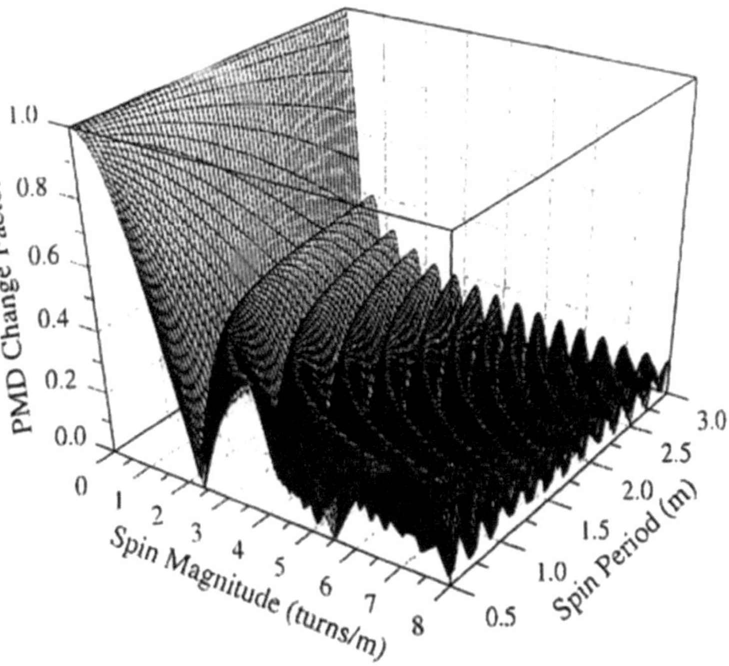


Fig. 4. The PMD change factor as a function of the spin amplitude and the spin period.

We note that when $J_0(2\alpha_0 / \eta) \neq 0$, the dominant contribution comes from the term of the linear increase of equation (38). Neglecting the oscillations term, the expression of the DGD becomes:

$$DGD(z) = \gamma_\omega J_0(2\alpha_0 / \eta) z \tag{40}$$

As in the case of spun fibres with a constant rate, in absence of random disruptions, the DGD increases linearly with the length of the fibre; in contrast with the PMDRF which takes a simpler shape:

$$PMDRF = J_0(2\alpha_0 / \eta) \tag{41}$$

Equation (41) indicates that the PMDRF is independent from the beating length in the case of spun fibres with sinusoidal profile whose beating lengths are equal to some meters or more. When $J_0(2\alpha_0 / \eta) = 0$, the linear increase term disappears, and the oscillation terms cannot be neglected any more. In this case, the DGD oscillates between 0 and a maximum value and is independent of the propagating distance. The condition where the minimum of the PMD is reached is called the condition of phase matching (figure 4).
Figure 4 illustrates the presentation in three dimensions for the graph of the PMD reduction as a function of the spin period and the spin amplitude. The phase matching condition can

be achieved for different spin parameters enabling to get an optimal reduction of PMD; though in general, amplitudes of higher spins give a better reduction of PMD. Figure (3.a) shows the evolution of DGD along the fibre with the phase matching condition. Finally, the maximum of reduction can be reached at the zeroes of the Bessel function of order zero (equation 41).

### 2.5 Reduction of the PMD for different profiles of spin

#### 2.5.1 The constant spin rate

While using equation (28), the PMDRF, as function of the spin rate, is represented for different beating lengths in figure 5; for a constant spin rate.

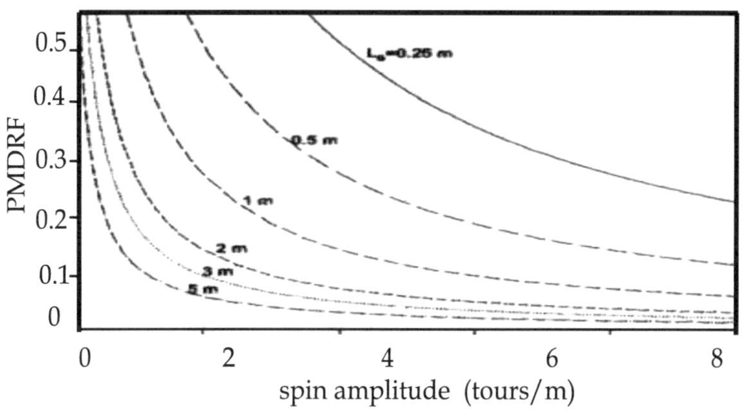


Fig. 5. The factor of reduction of the PMD with respect to the spin amplitude in the case of a constant spinning rate for different beat lengths.

We note that the PMD is reduced when the spin rate increases. For the same spin rate, PMDRF depends on the beating length. The higher the beating length is, the more reduced is the PMD. For a high PMD of the fibre (beating length <1m), a high spin rate is necessary to reduce the PMD.

#### 2.5.2 Sinusoidal spin

In figure 6, we use a beating length of 1m as example to illustrate the reduction of the PMD for sinusoidal types of spin (Ming-Jun et al., 2002).

Figure 6 shows that for sinusoidal spin types, the PMDRF oscillates with the spin amplitude, which is different from the case where the spin was constant. Furthermore, this figure shows that, for a sinusoidal spin, the phase matching condition can be gotten in order to come to a low PMD; on the other hand, in the case of constant spin, the phase matching doesn't exist. The phenomenon of phase matching can be explained by the mechanism of coupling of modes. The constant spin reduces the birefringence of the fibre, and causes no coupling of modes as well. For the sinusoidal spin, the variation in the rate of spin carries along the two modes of polarization to intercouple, reaching a compensation of the PMD. For some spin profile and birefringence of fibre, the conditions of phase matching are satisfied and the maximum of energy exchange occurs in order to provide a better reduction of PMD. The results of modeling indicate that the conditions of phase matching depend on the beating length, the period of the spin and the amplitude of the spin. We can use the same function of

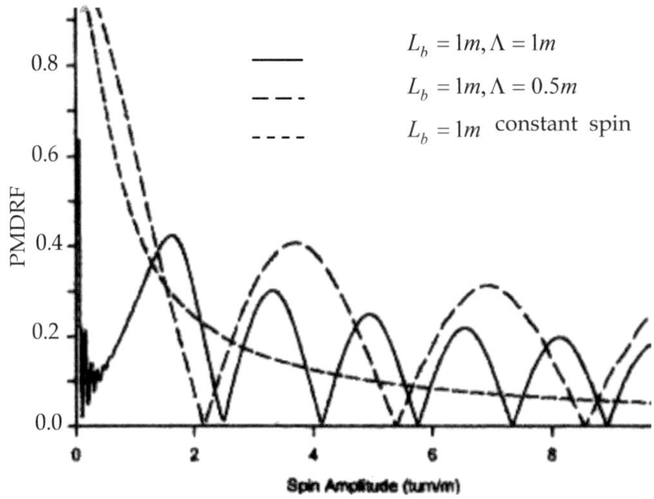


Fig. 6. The factor of reduction of the PMD versus the spin amplitude in the sinusoidal spin profiles.

spin to get a small reduction of PMD for high lengths of beating. However, for small lengths of beating; the phase matching has a strong dependence with the length of beating.
The fact that the birefringence of the real fibres is not constant and changes randomly, it is impossible to have the phase matching for the whole birefringence while using only one sinusoidal spin. This problem can be solved by admitting spin profiles with many Fourier components. To get to this point, the concept of the use of the modulated spin in amplitude and frequency has been developed by the Corning society.

### 2.6 Statistical evolution of the PMD of the spun fibres

As it was mentioned in the previous sections, the spun fibres follow a linear evolution law without the random modes coupling or in the régime of short lengths. When the random mode coupling is present, it has been found that the spun fibres follow an evolution law, a function of square root, similar to that of the unspun fibres, but with a different rate depending on the spin parameters (Chen, 2002). The random mode coupling can be characterized by a random variation of the birefringence axis and / or by the induced phase shift by the external constraints with an occurrence frequency of *1/h*, where *h* is called the coupling length of the modes. On the other side, a fibre of length «*l*» can be divided into (*l/h*) segments. Using this model, for a sinusoidal profile of spun fibre under no optimal conditions (no phase matching ), the DGD can be expressed under the following simple form:

$$\tau = \zeta \gamma_{\omega} \sqrt{hl} \tag{42}$$

We notice that the fact that the PMDRF « $\zeta$ » is independent of the beating length when the length of beating is greater than some meters, the DGD in the régime of important lengths, and in presence of the random coupling mode, is corrected by a factor $\zeta$, which is the reduction induced by the fibre spinning during the process of drawing. In this case, the property of evolution of the PMD is similar to that of the fibre possessing the linear birefringence.

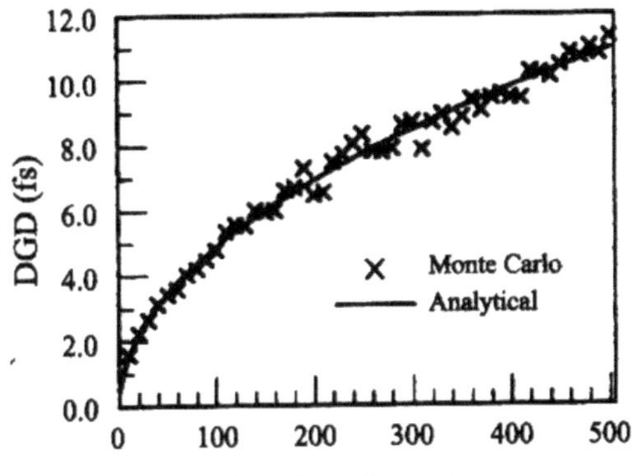


Fig. 7. The DGD of a spun fibre according to the length of the fibre. The amplitude of spin is 3.5 turn/m, the period of spin is 1m, the length of beating is 10m and the length of coupling is 10m.

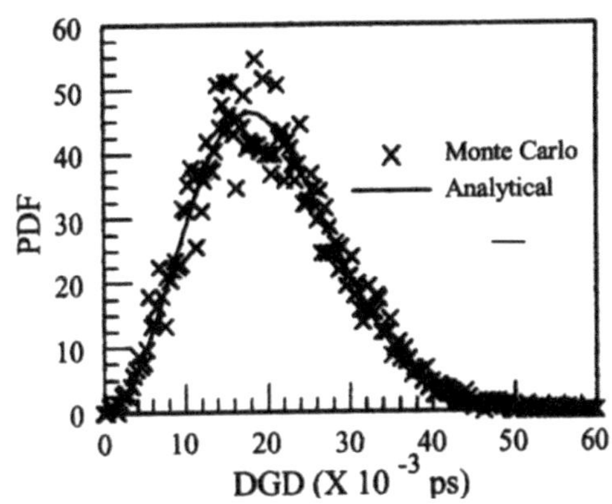


Fig. 8. Probability density function versus the DGD of the fibre when the condition of phase matching is not verified. The amplitude of spin is 3.5 turn/m, the period of spin is 1m, the length of fibre is 500m and the length of coupling is 10m.

The simplest law of evolution given by equation (42) has been verified by using a numerical modeling (Chen, 2003). Figure 7 shows the results of the numerical simulation for a sinusoidal spun fibre under the non optimal conditions. As it is shown in this figure, the numerical modeling accomodates very well with the theoretical prediction.
In the case of standard fibres (unspun fibres ) with a random coupling mode length h, the distribution of the DGD is analog to the distribution of Maxwell, where the standard deviation $\sigma$ used in the expression of PDF can be given by the following expression:

$$\sigma = (\frac{\lambda}{cL_b})\sqrt{h.l} \,/\, \sqrt{3} \tag{43}$$

We have proven that the Maxwell distribution is valid in the case of the spun fibres, except that, the parameter $\sigma$ should be corrected by the contribution of the spinning fibre. The modified parameter $\sigma$ is now under the following form

$$\sigma = \left[ J_0(2\alpha_0 / \eta) \right] (\frac{\lambda}{cL_b}) \sqrt{h.l} / \sqrt{3} \tag{44}$$

This equation has been tested and validated in (chen, 2002). Figure 8 represents the probability density function 'PDF ', according to the DGD of the spun fibre, obtained by numerical calculations and Maxwell distribution equation where we confirm according to the figures that the two results converge perfectly.

When the conditions of phase matching are satisfied, the total DGD of the fibre is a periodic function, and it oscillates between the zero value and a maximal value $\varepsilon_{\max}$. For this reason, the DGD of only one segment of a fibre is linked to the average of DGD inside one period of spin. Therefore, in the regime of high lengths ($l$ >>$h$), the total DGD can be written as follows (Ming-Jun et al., 2002):

$$\tau = \varepsilon' \varepsilon_q \sqrt{l / h} \tag{45}$$

Where $\varepsilon_q$ is the square average of the DGD in one period of spin, and $\varepsilon'$ is a coefficient that depends on the average coupling coefficient between two segments. For a condition of phase matching (for example: $\alpha_0 = 2.76\, tours / m \quad et\; \eta = 2\pi m^{-1}$ ), $\varepsilon'$ is found equal to 1.194. Besides, the DGD increases when the length of coupling of modes decreases (Ming-Jun et al., 1998). It is foreseeable, because under the conditions of phase matching, the DGD is minimum. Any disruption moves the fibre away from the optimal conditions, implying an increase of the PMD. Despite the fact that the DGD of the optimized spun fibres changes differently with the coupling length in comparison with the DGD of the non optimized spun fibres, the DGD always follows a Maxwell distribution, but with a modified parameter $\sigma$ (Chen, 2002).

$$\sigma = (\varepsilon' \varepsilon_q \sqrt{l / h}\,) / \sqrt{3} \tag{46}$$

## 3. Application of the JME method for the measurement of the PMD of the spun fibres

We used the JME method (Derickson, 1998) in order to verify the reduction of the PMD in the spun fibres (Cherbi et al., 2006). This applied method, between 1510 to 1615 nanometers, consists in determining the DGD directly between the two main states of polarization by measuring the Jones matrix of the device under test to a set of wavelengths. In order to determine the PMD of the spun fibres, we take the following steps:

- Measure Jones' matrixes $JM(\lambda_i)$ for a set of wavelengths $\lambda_1, \lambda_2, \ldots \lambda_n$ of the work range (1510 nm-1615nm)
- Do the product $JM(\lambda_i + \Delta\lambda).JM^{-1}(\lambda_i - \Delta\lambda)$
- Determine the eigen values $\rho_1$ and $\rho_2$ of the calculated product of matrix
- The DGD $(\lambda_i)$ is gotten then by (Heffner, 1992):

$$DGD(\lambda_i) = \Delta\tau = \left| \frac{\arg\left(\frac{\rho_1}{\rho_2}\right)}{\Delta\omega} \right| \tag{47}$$

The PMD of the fibre under test is determined by the arithmetic mean of the 'n' measured DGD:

$$PMD = \frac{\sum_{i=0}^{n} DGD(\lambda_i)}{n}$$

We applied the above procedure to two types of the spun fibres in order to compare their performances. The first fibre is unidirectional of length 212m in which we noted that the rotation of the spins was only in one sense once removed it from the spool. On the other hand, for the second bi-directional of length 1Km, the rotation of the spins was in the two senses. The results gotten in the figure 9.a and the figure 9.b, show that this method has a good resolution because it permitted to measure DGD of the order of femtoseconds, and to show that this type of fibre presents effectively low DGD compared to those measured in standard fibres that are of order of the picoseconds. Besides, we noted that the bi-directional fibre possesses a lower DGD than that of the unidirectional one indicating thus the efficiency of the bi-directional spun fibres in the reduction of the PMD.

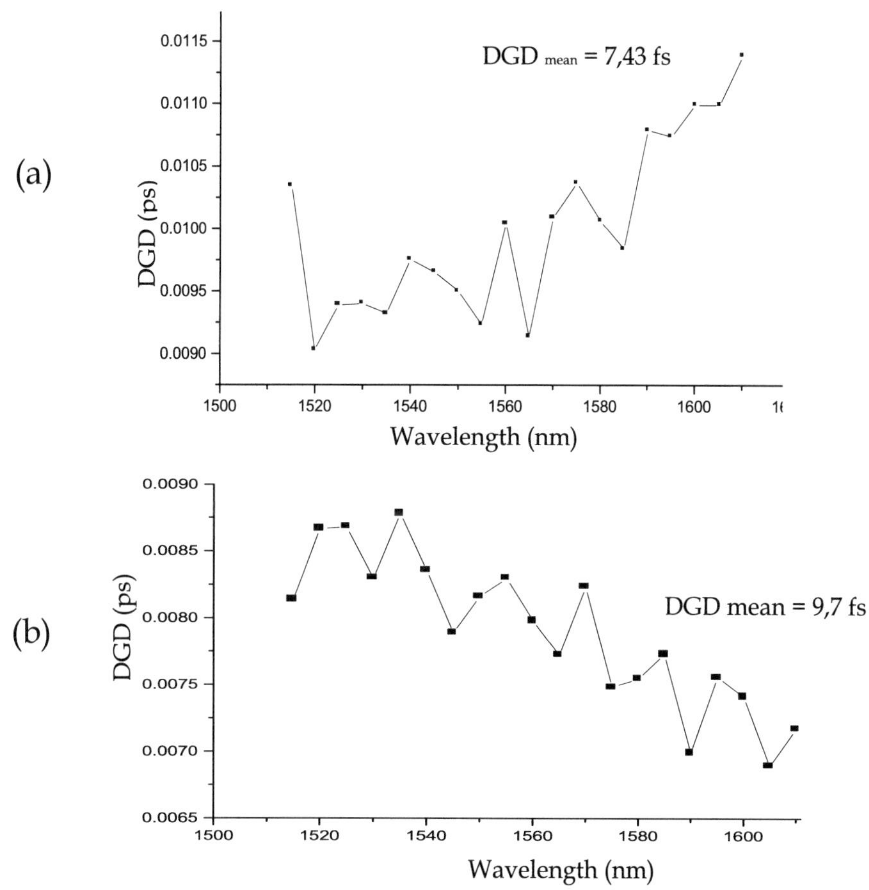


Fig. 9. Representation of DGD measured of spun fibre according to the wavelength for the length (a ) L = 290m, (b) L = 212m.

The researchers and inventors of the optical fibre of telecommunication systems predict the impact of PMD from the distribution of $\Delta\tau$ , because this results from the variation of $\Delta\tau$ as a function of wavelength and the conditions of the environment. On the other hand, due to this type of variation, the PMD of an optical path is expressed then statistically, as either the average or the root mean square (RMS) of $\Delta\tau(\lambda_i)$ (Derickson, 1998 ). It is interesting to determine the total PMD of a link made of a series of different spun fibres.

For this reason, we took three different lengths of spun fibres (fibre2, 3 and 4). We started by measuring, with the JME method, their PMD separately, for a given temperature while using a reference fibre (fibre1) used in calibration (table 1). Then, we connected the three fibres, and done the measurement of the total PMD in the same experimental conditions. The same procedure has been applied for the two fibres (fibres 3 and 4). We sought for the best relation of computation to determine the total PMD of a link of spun fibres, by testing the two following relations:

$$PMD_{totale} = PMD_1 + PMD_2 + ..............PMD_n \quad (48)$$

$$\text{Or} \quad PMD_{totale} = \sqrt{PMD_1^2 + PMD_2^2 + ..........PMD_n^2} \quad (49)$$

With $n$ the number of fibres used in the link

Our experimental results regrouped in table 1 are in very good agreement with the first relation [Cherbi et al., 2006].

| | DGD(fs) measured with the step (10nm) | Total DGD (fs) calculated with relation (48) | Total DGD (fs) calculated with relation (49) |
|---|---|---|---|
| reference fibre of (1 km) | 98,721 | | |
| Fibre 2 (212 m) | 4,8223 | | |
| Fibre 3 (290m) | 7,4315 | | |
| Fibre 4 (1 km) | 9,7399 | | |
| connected Fibres (2+3+4) | 22,9437 | 21,99 | 13,16 |
| connected fibres (3+4) | 17,1985 | 17,17 | 12,25 |

Table 1. The PMD relation of the spun fibres link.

## 4. Determination of the polarization's properties of the spun fibres using the reflectometers

The beat length of the fibre can be measured directly by the extraction of the spatial period of the backscattered signals (Wegmuller, 2002, 2004), which permits to estimate the PMD in the single-mode fibres (Ellison et al., 1998; Chen, 2002). The OFDR method is not exploited again especially for investigating of the spun fibres for the determination of its parameters and of their PMD according to the distance. In this section, we will present the relation already demonstrated experimentally by COTDR in our anterior works (Cherbi et al., 2009) existing between the spatial period of the backscattered signal and the PMD of the spun fibre and given by (Chen, 2003). Even more, the COTDR method allowed us to compare the results found with those of the JME method. Afterward we will present the POFDR method which used for spun fibres (Cherbi et al., 2009; Wegmuller et al., 2005) to obtain the beat lengths of the two types of spun fibres and the spin period of the bi-directional fibre.

Chen( Chen, 2003) has demonstrated that the spatial period of the backscattered signals obtained from a POTDR (polarization-sensitive optical time-domain reflectometer) of the spun fibres varies linearly with the beat length of the fibre. This means that for a given beat length, the spatial period $T_s$ can be used as calibration for the reduction of the PMD. A simple relation linking the spatial frequency F ($F = 1/T_s$) to the beat length and the spin parameters, is given by:

$$F = \left| J_0\left(2\alpha_0 / \eta\right) \right| / (L_b / 2) \quad (50)$$

The PMD of the bi-directional spun fibre is linked to the spatial frequency in the form (Chen, 2003):

$$PMD = \left(\lambda / 2c\right) F \quad (51)$$

Thus, the PMD of the bi-directional spun fibres can be determined directly through the measure of the spatial period as in the case of the standard fibres, while measuring the spatial period of the backscattered obtained from the reflectometers. The equation (50) shows that when the spin is zero, the spatial period converges to the one of the standard fibres.

### 4.1 Measure of the DGD in the spun fibres by the C-OTDR method

The technique (COTDR) (Wegmuller et al., 2004) is appropriate to detect the defaults in a given fibre (sites of reflection, losses) with a spatial resolution of the decimetre order. The main difference of this reflectometer (Cherbi et al., 2009) compared to a classic OTDR (Ellison & Siddiqui, 1998) resides in the use of photon counting detector (InGaAs avalanche photodiode). It is used in the so-called gated Geiger mode, which means that the detector is only active during a short time slot. During this period, only a single photon falls in the detector and triggers an avalanche, which is then detected by electronics discriminator. Contrary to the operation of a classic detector APD in linear regime, this avalanche is no longer proportional to optical input signal power, but independent of it. The detection is therefore a binary one, either there is an avalanche, or not. In order to evaluate the incident optical power (or mean photon number) on the detector during its activation, the detection process (gate opening) must be repeated many times in order to determine the detection probability of photons with a good precision. This probability is proportional to the incident signal power that is smaller than about 40% (no detector saturation) and larger than the detector thermal noise (dark counts). This condition is satisfied in our set-up by using the variable attenuator before the excitation of the fibre.

From the detection probability for a certain gate position, set by the delay generator, the reflectivity at the corresponding location in the fibre is readily gotten with a spatial 2- point resolution determined by gate duration. Thus, to have some information on the different positions in the fibre, the gate delay must be adjusted. In our set-up, the user can specify the zoom interval ( $L_{start}, L_{stop}$ ) for which the reflectivity is automatically measured with a step size (sampling resolution ).

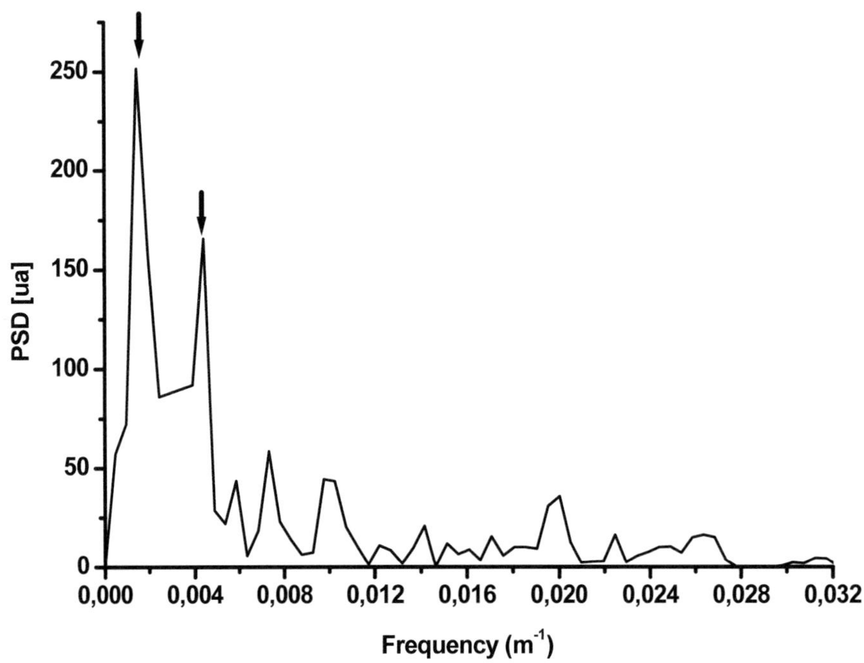


Fig. 10. The Power spectral density of C-OTDR signal for the bi-directional spun fibre of 1 Km.

Figure 10 presents the PSD of the backscattered signal power measured by the C-OTDR reflectometer with its spatial frequencies for a bi-directional spun fibre of 1 Km length. Two peaks appear, respectively, at spatial frequencies F and F/2, in the COTDR trace. A spatial frequency of $0.005\,m^{-1}$ is gotten from the backscattered signal PSD. The relation (51) gives a

DGD of the used spun fibre equal to 11.4 femtosecondes. We found for the same fibre a mean value of the DGD equal to 11,53 femtosecondes while using the JME method. These two results are in very good agreement. In conclusion, this demonstrates that the DGD of a spun fibre can be calculated from the spatial frequency of the COTDR signal in accordance with relation (51) permitting to calculate the spatial frequency of a backscattered signal COTDR in a spun fibre from its parameters": $\alpha_0 = 3.5\pi\ rad$, $\Lambda = 1.5\ m$ and $L_B = 20\ m$ . The calculated spatial frequency is equal to $F = 0.005\ m^{-1}$, which is equal to the same one measured from the C-OTDR trace. Based on that, we validate the equation, linking the spatial frequency, the spin parameters and the intrinsic birefringence of the bi-directional spun fibre, given in (Chen, 2003)

### 4.2 Measure of the beat length of the spun fibres by P-OFDR

This reflectometer implements the technique of coherent detection sensitive to the polarization in order to get information about the evolution of the polarization states along the fibre under test. In our case, a POFDR is used, implementing the detection of polarization diversity ( Cherbi et al, 2009] and a polarized beam splitter which plays the role of a fixed analyser. The former permits to remove the Rayleigh reflections independent of the polarization by subtracting output 1 from output 2, thereby removing the frequencies of the back scattered signal that are not related to the fibre birefringence.
The used laser in this reflectometer is a DFB (distributed feedback) characterised by a spectral width of the order of 1MHz on the whole tuning range, a spatial range of 80 m. Due to the coherent detection, a very good sensitivity of 100 dB is gotten with this reflectometer. The only factor limiting the resolution of this method is the tuning of the laser. The laser that we used is limited by the continuous tuning of 20 GHz that gives approximately a resolution of 9 mm.
In (cherbi et al., 2009), we have analyzed three types of fibres having the same length of 200 m: a bi-directional spun fibre, a unidirectional spun fibre and a standard fibre. They were wrapped on a table in order to minimize the external constraints. Figure 11 shows the example of the different POFDR traces for different used resolutions of the unidirectional spun fibre (dark line is the mean of different traces). The beat lengths of the two types of spun fibres and the one of the standard fibre are calculated by the following relation (Wegmuller et al., 2002):

$$< L_b > = \frac{1}{std(DSP)}\sqrt{\frac{12}{\pi}} \qquad (52)$$

Where PSD is the power spectral density of POFDR signal.
The calculated values of beat lengths derived for the PSD signals of the different fibres: unidirectional spun fibre (figure 11), standard fibre and bi-directional spun fibre (figure 12) are respectively: 50 m, 38 m, and 150 m. We note that the beat length of the bi-directional spun fibre is more important than those of the others, which means that the PMD of the bi-directional spun fibre is lower than that of the two other types of fibres, result that we found with the JME method. It also confirms that the bi-directional spun fibre reduce efficiently the PMD compared to the unidirectional spun and the standard fibres.

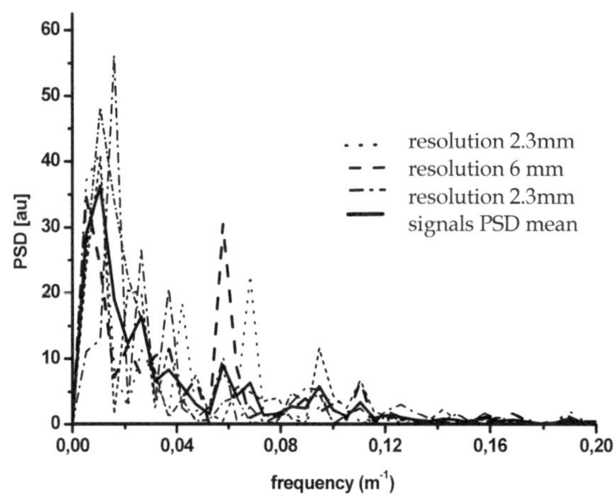


Fig. 11. The mean power spectral density of the backscattered signals POFDR, obtained for different resolutions, of the unidirectional spun fibre of 200 m length.

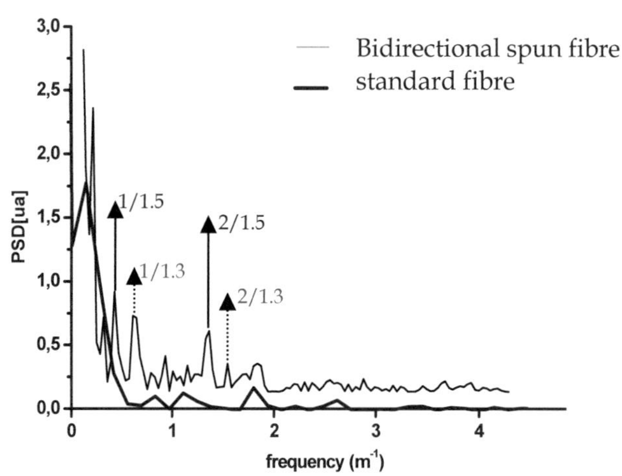


Fig. 12. The mean power spectral density of the backscattered signals POFDR obtained for bi-directional spun fibres and standards of lengths 200m.

## 5. Conclusion

In this chapter, we presented the principle of the spun fibres with their technology and their role in the reduction of the PMD in a transmission link of optical fibre. Several types of spun fibres have been given. The theory of these fibres based on the equations of the coupled modes has been detailed.
The reduction of the PMD in these fibres is verified while applying the JME method and the COTDR method used to measure the DGD of the order of femtoseconds. We also confirmed this result while measuring the beat length of these fibres with the POFDR method and compare it with that of a standard fibre. This comparison proved the efficiency of this type of fibres in the reduction of the PMD. Finally, according to the use of the COTDR and POFDR, we concluded that the validity of the data analysis, obtained from the reflectometers and used nowadays for standard fibres, has been demonstrated for the spun fibres and more precisely for the bi-directional spun fibres. Besides, the high spatial resolution of the POFDR enables again the observation of the spatial frequencies directly linked to the spin period, so a precise characterization of the spun fibres can be accomplished.

## 6. References


Barlow, A.J; Ramskov-Hansen; Payne,D.N. birefringence and polarization mode dispersion in spun single mode fibers", *applied optics*, vol.20 (1981), pp 2963.

Blaszyk, P.E. & Christoffand, W.R. method and apparatus for introducing controlled spin in optical fibers", *U.S patent 63224872 B1* (December 2001).

Chen, X.; Ming, J. Li. & Nolan, D. A. Polarization mode dispersion of spun fibers : an analytical solution . *optics Letters*, vol.27(2002),pp. 294 -296.

Chen, X. Scaling properties of polarization mode dispersion of spun fibers in the presence of random mode coupling. *optics letters*, vol. 27 (2002),pp. 1595.

Chen,X. properties of polarization evolution in spun fibers, optics letters, vol. 28 ( 2003), pp. 2028.

Cherbi,L.; Azrar,A.; Mehenni,M. &Aksas, R. Characterization of the Polarization in the Spun Fibers. *Microwave and Optical Technology Letters*, Vol.51, N°.2, (February 2009), pp. 341–347, ISSN 0895-2477

Cherbi, L.; Mehenni, M. & Wegmuller, M. Mesure de la dispersion des modes de la polarisation dans les fibres optiques spun par la méthode d'analyse des valeurs propres de la matrice de Jones . *25 iémes Journées Nationales d'Optique Guidées JNOG*, pp 176-178, Metz, France, Novembre 2006.

Dandliker, R (1992). Rotational effects of polarization in optical fibers in anisotropic and nonlinear optical waveguides, *C, G. Someda, Elsevier, New york*, pp. 39 - 76.

Derickson. D (1998). *Fiber optic, test and measurement.* Hp professionals books, Prentice hall.

Ellison, G. & Siddiqui, A. S. A fully polarimetric optical time domain reflectometer", *IEEE Photonics technology letters*, vol.10 (1998), pp. 246 -248.

Galtarossa, A.; Palmieri, L. & Pizzinat, A. optimized spinning design for low PMD fibers: an analytical approach. *Journal of lightwave technology*, vol. 19 (2001), pp. 1502.

Hart, A.; GHuf, R. & Walker. method of making a fiber having low polarization mode dispersion due to a permanent spin", *U.S patent 5*, Vol. 47 (1994), pp. 298, March 29.

Heffner,B. Automated measurement of polarization mode dispersion using Jones matrix eigenanalysis. *IEEE Photonics technology Letters*, vol.4 (1992), n°9, pp.1066.

Ming-Jun,Li &Nolan, D. fiber spin profile designs for producing fibers with low polarization mode dispersion. *optics letters*, vol. 23(1998), pp. 1659 - 1661.

Ming-Jun, li; Chen, X. & Nolas, A. fiber spinning for reducing Polarization mode dispersion in single mode fibers: theory and applications, science and technology division, *Corning Incorporated*, Sp - AR 02-2 (2002), Corning NY14831.

Noé et al. Polarization mode dispersion compensation at 10, 20, and 40 Gb/s with various optical equalizers. *IEEE journal of lightwave Technology* , vol. 17 (1999), pp 1602 - 1610.

Wegmuller, M.; Scholder, F. & Gisin,N. Photon counting OTDR for local birefringence and fault analysis in the metro environment. *Journal of lightwave technology*, vol. 22 (2004), N°. 2.

Wegmuller, M.; Legré, M. & Gisin, N. Distributed Beatlength Measurement in single mode fibers with optical frequency domain reflectometry, *journal of light wave technology*, vol 20 (2002), N° 5, pp. 829.

Wegmuller, M.; Cherbi,L. & Gisin, N. Investigation of spun fibers using high resolution reflectometry. *Proceeding OFC IEEE of Optical Fiber Conference*, America 2005.

# Part 3

## Photonic Crystal Fibers

# 9

# Photonic Crystal Fibers with Optimized Dispersion for Telecommunication Systems

Michal Lucki
*Czech Technical University in Prague,*
*Faculty of Electrical Engineering,*
*Czech Republic*

## 1. Introduction

The use of Photonic Crystal Fibers (PCF) is understood within their unique chromatic dispersion characteristics and nonlinear behavior, which is suitable for dispersion compensation or transmission of information without pulse spreading, leading to an intersymbol interference. Pulse spreading being the result of chromatic dispersion in optical fibers is considered as one of the critical issues in the design of optical fibers. Since the dispersion can result in worse system performance, it is necessary to prevent its occurrence or to compensate it.

A systematic study of dispersion properties in PCFs is presented. The investigation includes a description of fiber chromatic dispersion dependence on structural and material parameters. Potential zero or anomalous dispersion in doped PCFs is achieved. An overview of current innovations on the studied problem is presented.

Moreover, the new PCF with nearly zero ultra-flattened chromatic dispersion is introduced. It is shown from the numerical results that the dispersion of -0.025 ps/nm/km is available from the wavelength of 1200 nm to 1700 nm.

## 2. Photonic crystal fibers

PCFs, also known as microstructured or holey fibers, are investigated in view of their unique properties of light guidance. Unlike conventional step-index fibers, PCFs guide light through confining field within microstructure periodic air holes. PCFs are characterized by the periodicity of refractive index, implemented as an array of air holes around the core. The guidance mechanism in some aspects resembles the operation of semiconductor materials. In other words, the photons in PCFs have a function, which is similar to the operating principle of electrons in semiconductors.

### 2.1 Types of photonic crystal fibers

PCFs are classified in two categories: solid core high-index guiding (or simply an index guiding) fibers and hollow core low-index guiding fibers. The Index Guiding Photonic Crystal Fiber (IGPCF) guides light in a solid core by Modified Total Internal Reflection (M-TIR). This principle is similar to the guidance in conventional optical fibers. The other category of PCFs, Hollow Core Photonic Crystal Fiber (HCPCF) guides light by the Photonic Band Gap (PBG) effect. Light is confined in the low-index core, since the distribution of energy levels in the structure makes the propagation in the cladding region impossible.

The M-TIR principle of light guidance relies on a high-index core region, typically pure silica, surrounded by a lower effective index material, provided by air holes in the cladding.

### 2.2 New properties achievable in photonic crystal fibers

The effective index of such a fiber can be approximated by a standard step-index fiber, with a high-index core and a low-index cladding. However, the refractive index of a microstructured cladding in PCFs exhibits strong wavelength dependence very different from pure silica, which allows PCFs to be designed with a new set of features unattainable within the classical approach. For example, endlessly single mode PCF can be designed through the strong wavelength dependence of the effective index (reducing thus the value of normalized frequency, a parameter important for modal regimes). This is fundamentally different from the conventional fibers where, at huge core diameter to wavelength ratios, a multi-mode operation is unavoidable at shorter wavelengths, because the cladding index is constant and normalized frequency arises with wavelength, once exceeding the value critical for single-mode operation. In addition, the presence of air holes in the cladding can change the spectral characteristics of microstructured fibers.

Among PCFs with modified spectral properties, zero dispersion or anomalous dispersion fibers are very promising for group velocity dispersion compensation. The latest designs show optimal dispersion for broadband applications, in contrast to the commercially available compensating fibers, which can usually operate at a specific wavelength.

## 3. Photonic crystal fibers for dispersion compensation or zero-dispersion transmission

Chromatic dispersion directly affects the pulse width and the phase-matching conditions important for most telecommunications applications. Chromatic dispersion in lightwave systems is related to the variation in group velocity of optical signals in a fiber. The adjective "chromatic" emphasizes its wavelength-dependent nature. Chromatic dispersion limits the maximum distance, to which a signal can be transmitted without the necessity of regeneration of its shape, timing, and amplitude. The pulse spreading must be compensated or avoided, for example, by specific fiber design.

As far as basic terminology is concerned, when the chromatic dispersion coefficient is less than zero, the dispersion regime is said to be anomalous, and shorter wavelengths propagate faster than longer wavelengths. The pulse is said to be negatively chirped. In the opposite case of dispersion coefficient being greater than zero, the dispersion regime is said to be normal. Long waves are guided faster than the short ones.

### 3.1 Engineered chromatic dispersion in photonic crystal fibers

The mechanism of light dispersion depends on various reasons, therefore the techniques of suppressing particular dispersion components vary from each other. One can distinguish between a number of types of dispersion, such as modal, waveguide or material dispersion. Chromatic dispersion consists of two components. The first one comes from bulk material dispersion $D_{mat}$. The second one comes from waveguide dispersion $D_w$, where the material and the waveguide dispersion are expressed, as follows:

$$D_{mat} = \frac{-\lambda}{c}\frac{d^2 n_M}{d\lambda^2} \tag{1}$$

$$D_w = (\frac{-\lambda}{c})\frac{\partial^2[\mathrm{Re}(n_{eff})|_{n_M(\lambda)=Cons.}]}{d\lambda^2} \tag{2}$$

where $n_m$ is the matrix index. Since waveguide dispersion can be anomalous and material dispersion normal, optimal dispersion design can be achieved by the suitable balance of particular dispersion components contributing to the total dispersion. To design a fiber with zero dispersion, it is necessary to optimize both: material properties, as well as the shape of the waveguide. There exists, therefore, a wavelength, at which total dispersion is equal to zero. Beyond this, the fiber exhibits a region of anomalous dispersion, which can be used for the compression of pulses in optical fibers.

To achieve a specific value of total dispersion, one must compensate material dispersion $D_{mat}$ with waveguide dispersion $D_w$. The slope of $D_w$ should be adjusted by optimizing the fiber's geometry in order to make it parallel to $-D_{mat}$. If the goal is to obtain flattened dispersion in a target wavelength interval, one must control $D_w$ to make it follow a trajectory parallel to that of $-D_{mat}$. If material dispersion is linear in a target interval, a systematic approach can be used. Generally, this is the classical method of how to treat chromatic dispersion profiles using geometrical parameters in PCFs with successive iterations of structural parameters to improve the quality of the results.

### 3.2 Current state of the art

Due to unique dispersion flexibility, PCFs are considered as useful for achieving anomalous dispersion. They are used for the robust compensation of chromatic dispersion or dispersion-free transmission. There are several practical solutions to limit chromatic dispersion and to keep the initial width of optical pulses. One of the methods is to design fibers with zero dispersion. Resultant zero dispersion can be achieved by compensating material dispersion with waveguide dispersion. This operation is generally possible at a specific wavelength, so that the signal must be transmitted within a very narrow range of optical frequencies.

#### 3.2.1 Dispersion compensating fibers

Zero dispersion is useful for low-speed systems, but can be undesired in high-speed transmission systems, since the phase match of all the frequency components can result in nonlinear effects. Another method of keeping a constant pulse width is to retain small normal dispersion in optical fibers and compensate it by using Dispersion Compensation Fiber (DCF), added at signal repeater. In general, chromatic dispersion compensators optically restore signals that have become degraded by chromatic dispersion, significantly reducing bit error rates at the receiving end of a fiber's span. A DCF is characterized by strong anomalous dispersion, which exactly compensates normal dispersion arising between repeaters. Many studies have been published about the design and optimization of chromatic dispersion in PCFs. They tend to shift zero-dispersion wavelengths or minimum anomalous dispersion wavelength towards at the conventional band around 1550 nm, (known as C-band). Conventional dispersion compensating fibers are designed to operate at a specific wavelength, for example at 1550 nm, achieving negative value of at least hundreds ps/km/nm at the operating wavelength. Recently, the extension of operating bandwidth towards longer wavelengths is the area of interests, since short optical frequencies are more used in high-speed transmission systems.

PCFs are highly flexible for engineered dispersion. By manipulating the geometry design of the PCF (core diameter, normalized hole diameter, number of rings, hole defects), it is possible to achieve desired dispersion and losses required for specific applications. The interplay between chromatic dispersion and geometrical structure allows establishing a well-defined procedure to design specific predetermined dispersion profiles. This topic is described in many studies about the dispersion controllability.

One of the very first works with significant contribution to this topic is a work by Birks et al. (1999). The latest studies report new aspects related to the topic (for example the work of Haxha et al., the work by Liu et al. or finally the one by Razzak et al). The main topic addressed in those works is the ultra-flattened dispersion at a wide wavelength interval and at low confinement losses.

Premium DCF is demonstrated by Wu et al. (2008), where the negative dispersion value of –1350 ps/km/km at 1550 nm is achieved. Other designs aim to achieve dispersion in the wavelength range of about 1500-1625 nm. This could open a door for broadband dispersion compensation using PCFs.

### 3.2.2 Dispersion flattened photonic crystal fibers

Achieving a flattened dispersion curve is required for many telecommunication applications, in which we desire to have the same dispersion values for broad band utilization. For this purpose, some studies are focused on investigating various techniques of adjusting the PCF's geometry to obtain flattened dispersion characteristic. With this regard, a study presented by Liu et al. (2007) shows, how ultra-flattened dispersion curve could be achieved using elliptical holes. An optimized design of a PCF over ultra-wide band by replacing two rings of inner circular air holes with elliptical air holes is presented. The permitted dispersion fluctuation is 0.6–1.0 ps/nm/km within a broad band from 1000 nm to 1900 nm, which means over all: S, C, and L bands. Moreover, periodic structures having small core with large equal-sized air holes managed to shift zero-dispersion wavelengths towards shorter wavelengths.

Summarizing, the design process requires high attention to all important properties, such as flattened chromatic dispersion curve, effective mode area, confinement loss over broad bandwidth. In addition, designers should consider the complexity of new structure's fabrication process.

### 3.2.3 Doping technique for enhanced dispersion properties

The standard solid core PCF with hexagonal lattice and medium air-fraction volume exhibits chromatic dispersion characteristics far from the preferable ones for practical implementations. Doped cores can be used to enhance dispersion properties of IGPCF. The technique is based on doping of the central part of the $SiO_2$ core by the $GeO_2$ material. The germanium dioxide raises the refractive index of the doped region and hence modifies the waveguide properties of the PCF.

The dispersion behavior has been investigated for Highly Non-linear PCF (HNPCF), where the core's refractive index is increased by doping with high-index material, such as rare earth ions. The idea of doping the PCF's core with rare earth elements has been investigated. For example, an ytterbium-doped PCF can be used to achieve enhanced nonlinearities. The most widely used dopant in PCF is Germania - $GeO_2$, due to its intensified nonlinearities and enhanced photo-sensitivity. In fiber fabrication process, the refractive index of the doped core is determined mainly by the concentration of the $GeO_2$ ions embedded in the core. The accurate characterization of the dopant's location and its concentration in optical fibers is studied by Zhong et al. The dispersion dependence on the concentration of $GeO_2$ in the fiber's core is explained accurately by Hoo et al. (2004). Notice is hereby given that $GeO_2$ is a dopant commonly used for doping the core region for raising the refractive index, on the other hand, $B_2O_3$ or F are doping substances suitable for doping the cladding region that in turn lower the refractive index.

### 3.3 Shortcomings of existing solutions

The narrow bandwidth of operating wavelengths is considered as a limitation, in particular for systems with Wavelength Division Multiplexing. Therefore, recent studies focus on the ultra-low, ultra-flattened broadband dispersion over a wide spectrum of telecommunication wavelengths. PCFs can be exploited into this aim, since the large refractive index variation between silica and air permits to achieve significant waveguide dispersion over a wide wavelength range. PCFs with large air-holes have already been proposed in some studies about dispersion compensation.

Many DCFs uses the technique of doping their core with high-index material. This can result in high confinement losses, reaching even more than 1 dB/km, as indicated in catalogues of commercially available fibers. In addition, those fibers suffer from small effective mode area, since some DCFs have an extremely small core and concurrently high air fraction to enhance nonlinear evolution of spectral characteristics.

## 4. Simulation method

Huge possibilities of geometry manipulation and air-holes shapes arrangements have increased the complexity of numerical analysis of PCFs. The main objective of simulations is to study chromatic dispersion characteristics of IGPCF and HNPCF. Such structures demand efficient numerical methods to analyze them accurately. Thus, many modeling methods have been applied in this perspective, such as the plane wave expansion method, localized function method, finite element method, finite difference time domain method, finite difference frequency domain method, Fourier composition method or multipole method. The results presented in this work have been achieved by using the full-vectorial Finite Difference Frequency Domain method (FDFD), which was described in details by Zhu et al. (2002). This tool practically employs the same algorithm as the Finite Difference Time Domain (FDTD) method, the only difference between the two algorithms is that FDFD is a 2D solution, whereas FDTD is a 3D solution, which means that FDFD is easier for software implementation and meanwhile leads to the same numerical dispersion equation as that of the 3D-FDTD method.

For a given frequency, the numerical propagation constants and mode patterns can be calculated. The main geometrical quantities concerned: hole diameter $d$, the hole pitch $\Lambda$, and the core diameter, used in the implementation are displayed in Fig. 1.

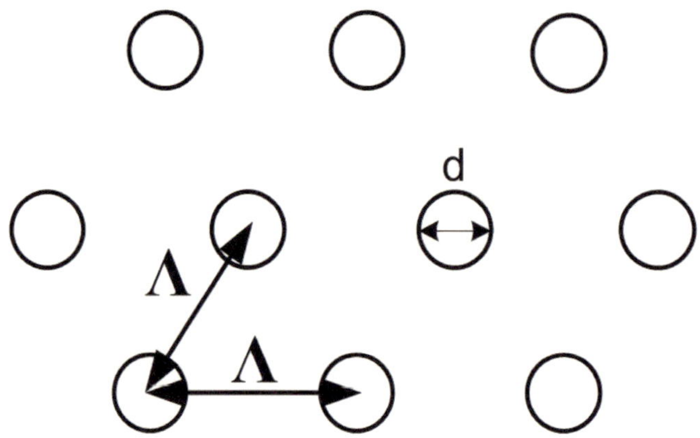


Fig. 1. Geometrical quantities describing PCFs.

In order to investigate the optical behavior of PCFs, the structure presented in Fig. 2 is used. It represents a HNPCF structure, where the core is doped with high-index material.

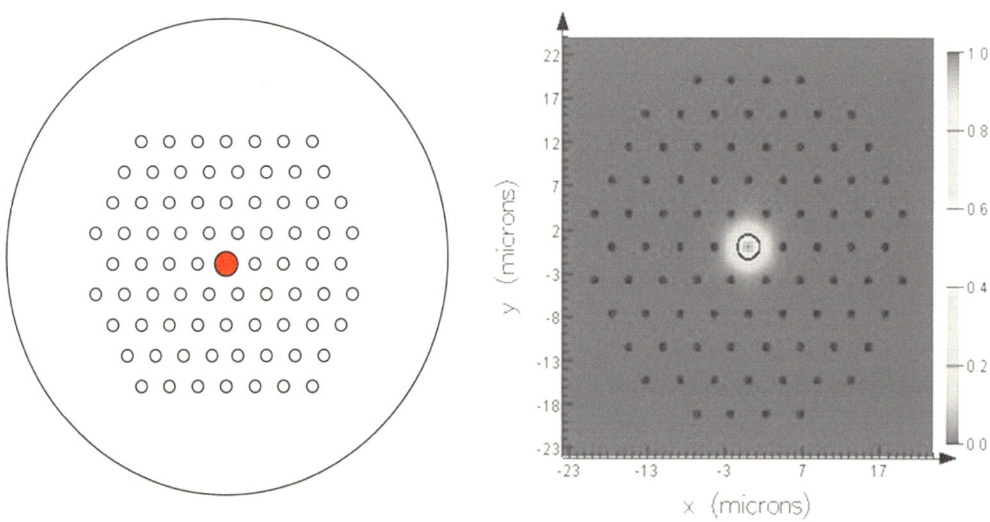


Fig. 2. Doped structure evaluated in terms of dispersion compensation (left) and the fundamental mode of the modeled PCF (right).

The basic flow of simulation is executed with several iterations to calculate the number of parameters and to obtain precise results. The simulation algorithm for parameters sweeping contains few steps: once the physical structure is created, the simulation parameters and mesh are set, as well as the monitors are defined, the simulation is run. The frequency domain information is available at any point of the cross-section of a modeled fiber.
In order to perform a series of simulations to investigate the change in measured intensity as a function of geometry or to perform any other systematic study, the built-in scripting environment is used. This scripting environment has many advantages, where one can

extract specific values of parameters or implement a required sweep in the structure and observe how chromatic dispersion or bending loss parameters are changed.

## 5. Simulation results

In order to understand the behavior of chromatic dispersion and loss in PCFs, an analysis has been proceeded to study the HNPCF with high-index doped core.

### 5.1 Dispersion in doped PCFs

The investigated HNPCF structure is specified in Table 1, where the cladding includes five rings of air holes and the core, which doped with high-index material, of which the refractive index is equal to 1.475. Relatively small air holes are preferred.

| Parameter [unit] | Value |
|---|---|
| Pitch $\Lambda$ [μm] | 4.4 |
| Hole's diameter $d$ [μm] | Varied 0.6–2.2 |
| Normalized hole diameter $d/\Lambda$ [-] | Varied 0.1–0.5 |
| Air-fraction refractive index [-] | 1 |
| Dopant's (core's) refractive index [-] | 1.475 |
| Silica glass refractive index (high-index cladding region) [-] | 1.458 |
| Propagating wavelength [nm] | 1550 |
| Core diameter [μm] | 1.4 |
| Effective index of cladding at 1550 nm | 1.4582 |
| Number of rings at the cladding $Nr$ | 5 |

Table 1. Structural parameters for the doped PCF presented in Fig. 2

Dispersion in nonlinear doped microstructured optical fiber, specified in Table 1, is shown in Fig. 3.

The doped PCF has a parabolic dispersion curve (in contrast to standard IGPCF, where dispersion shows linear increase with wavelength, which is presented in studies describing dispersion in PCFs). Usually, dispersion in fibers with a hexagonal lattice has a Zero Dispersion Wavelength (ZDW) at the O-band.

A general tendency in microstructured fibers is that both ZDWs are found at shorter wavelengths, when the fraction of air filling is increased or when the central defect is decreased.

Adjusting the geometrical parameters can be a tool to control the curvature of a dispersion profile. This can eventually lead to two closely laying ZDWs and very low minimum dispersion or, vice versa, to ZDWs far from each other, and flat dispersion curve. This mechanism shows a good agreement with the results achieved in this numerical analysis. Though, the second ZDW is located rather at longer wavelengths.

For the studied structure, the dispersion curve of HNPCF presented in Fig. 3 crosses the x-axis at two zero points, the first one appears at the shorter wavelength, usually at the O-band or the E-band, whereas the second point is located at the longer wavelength, usually at the C-band or the L-band.

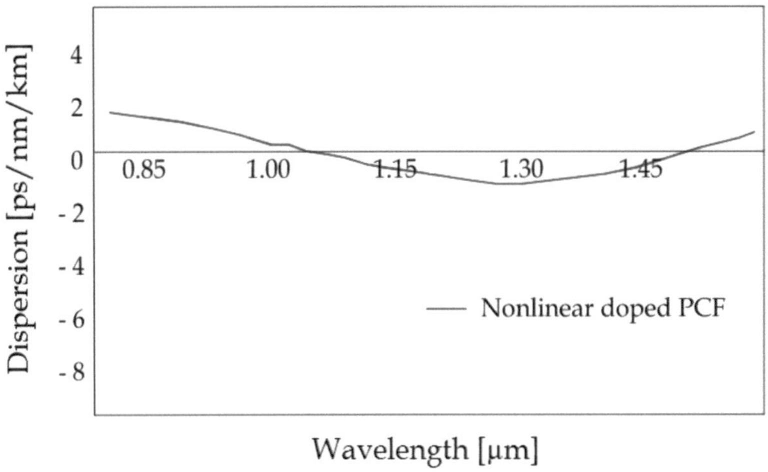


Fig. 3. Chromatic dispersion in regular solid-core PCF and modeled doped PCF.

The investigation focuses on the properties resulting from a doped core to control the dispersion in PCFs. The technique is based on doping the central part of the $SiO_2$ core by the $GeO_2$ material. The ZDWs are found at shorter wavelengths, when the fraction of air filling is increased and the central defect is decreased. Adjusting the geometrical parameters can rather result in different dispersion properties; the most mature designs assume the second ZDW being rather at longer wavelengths, since shorter optical frequencies are more used in high speed transmission systems.

The advantage of the studied structure is the flexibility of adjusting both: minimum anomalous wavelength and ZDW locations. As it is demonstrated below, such type of fibers is highly sensitive to geometrical parameters, as well as to the change of material index values. It also keeps an endlessly single mode characteristic of a solid core PCF.

### 5.2 Chromatic dispersion dependence on air-fraction volume

Results shown in Fig. 4 indicate a negative behavior of chromatic dispersion; the second ZDW is affected by the air fraction percentage.

With a decrease in hole diameter, it is possible to move the position of the second ZDW to higher wavelengths, reaching the C and L-band, with regard to current trends in systems using Wavelength Division Multiplex.

### 5.3 Chromatic dispersion dependence on core diameter

Similar results are achieved for the core diameter optimization and for varied refractive index. Parameters for the core diameter sweeping are presented in Table 2. For this purpose, all the parameters are fixed, as given in Table 2, while the core diameter is chosen to vary from 2.8 to 4.4 μm As far as core diameter is concerned, all the remaining parameters are fixed (with $d/\Lambda$ being 0.3). As depicted in Fig. 5, the minimum dispersion value arises with the increase in core diameter.

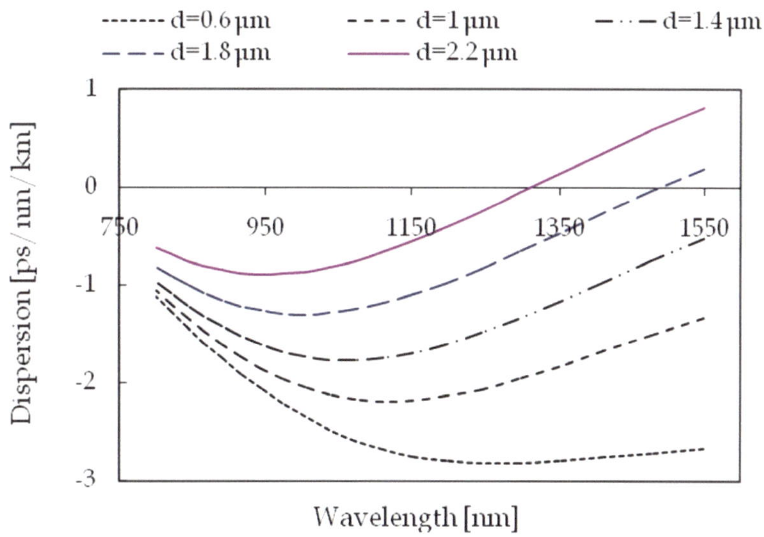


Fig. 4. Chromatic dispersion dependence on hole diameter.

Another conclusion, which reveals at Fig. 5, refers to the behavior of ZDW. We observe that greater value of a core diameter is responsible for ZDW achieved at longer wavelength. At the specific value of a core diameter (3.6 μm), the values of studied dispersion start to be positive.

| Parameter [unit] | Value |
|---|---|
| Pitch *Λ* [μm] | 4.4 |
| Hole's diameter *d* [μm] | 1.32 |
| Normalized hole diameter *d/Λ* [-] | 0.3 |
| Air-fraction index [-] | 1 |
| Dopant's (core's) refractive index [-] | 1.475 |
| Silica glass refractive index (high-index cladding region) [-] | 1.458 |
| Propagating wavelength [nm] | 1550 |
| Core diameter [μm] | Varied 2.8- 4.4 |
| Effective index of cladding at 1550 nm | 1.4586 |
| Number of rings at the cladding *Nr* | 5 |

Table 2. Structural parameters for PCF doped in a small core.

### 5.4 Chromatic dispersion dependence on doping level

In order to precisely control chromatic dispersion, the effect of changing the dopant's refractive index (that can be practically achieved by changing the concentration of $GeO_2$

from 16 to 30 %) is further investigated. By the increase in refractive index, lower minimum dispersion in the area of negative values is produced.

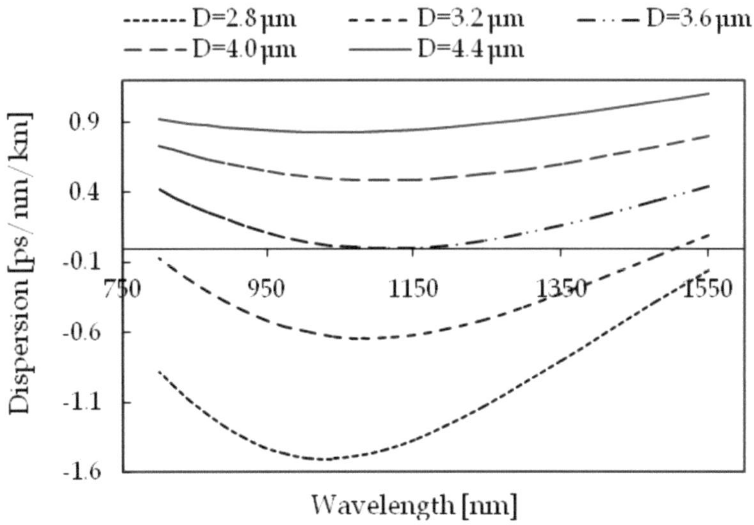


Fig. 5. Chromatic dispersion dependence on core diameter.

Fig. 6 combines the effect of the refractive index values varied from 1.472 to 1.49, in which a summarized impact over all: O, E, S, C, L bands is shown. Considering a specific wavelength, for instance 1550 nm, dispersion increases with refractive index of the doped core.

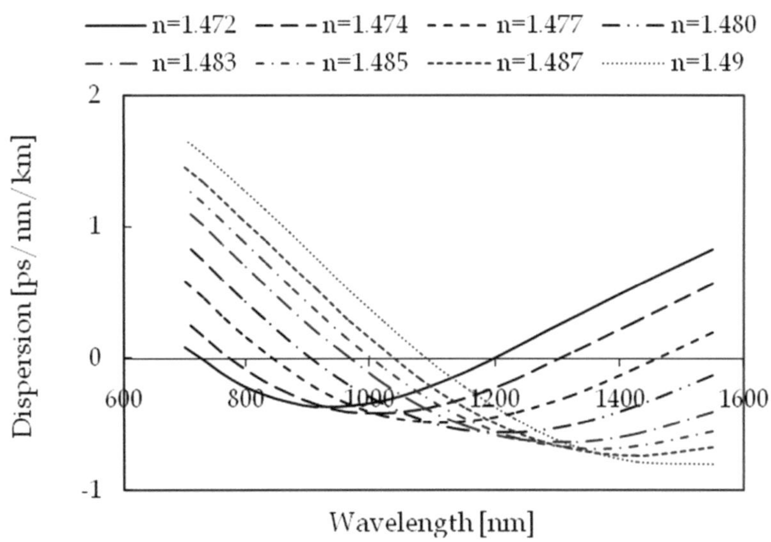


Fig. 6. Chromatic dispersion dependence on dopant material refractive index.

Extracted values of ZDW obtained for varied material refractive index are presented in Table 3.

| Refractive index [-] | First ZDW [nm] | Second ZDW [nm] |
|---|---|---|
| 1.472 | 728 | 1190 |
| 1.474 | 780 | 1320 |
| 1.477 | 835 | 1450 |
| 1.48 | 913 | 1556 |

Table 3. Location of first and second ZDW in the modeled PCF.

## 6. Design of PCF with ultra-flat chromatic dispersion

The combination of studied parameters could interplay with their effects to achieve optimal dispersion for telecommunication applications. This is generally considered as one of the major advantages of PCFs.

A PCF with flattened dispersion curve is required for telecommunication applications, in which we desire to have the same dispersion values for broadband utilization, in this case long-distance propagation with nearly zero dispersion in systems with Wavelength Division Multiplexing. The final goal is to optimize the structure to achieve flattened dispersion curve and dispersion values near zero. This could be done by finding the suitable configuration of the following parameters: hole diameter, core diameter, and selective doping.

The proposed structure is doped by using $GeO_2$. The fiber has three air rings of holes in the cladding. The doped core radius is 7.4 μm, which is relatively big compared to all above studied structures. Detailed description of the proposed structure is summarized in Table 4.

| Parameter [unit] | Value |
|---|---|
| Pitch $\Lambda$ [μm] | 4.4 |
| Hole's diameter $d$ [μm] | 1.32 |
| Normalized hole diameter $d/\Lambda$ [-] | 0.3 |
| Air-fraction index [-] | 1 |
| Dopant's (core's) refractive index [-] | 1.48 |
| Silica glass refractive index (high-index cladding region) [-] | 1.458 |
| Propagating wavelength [nm] | 1550 |
| Core diameter [μm] | 7.4 |
| Effective index of cladding at 1550 nm | 1.465 |
| Number of rings at the cladding Nr | 3 |

Table 4. Structural parameters of HNPCF for achieving ultra-flattened dispersion diagram.

As it is observed in Fig. 7, the fundamental mode is trapped in the core. The fiber operates as a single-mode PCF. A special attention should be taken during the fabrication of the core, which is much greater than the doping region, as depicted in Fig. 7.

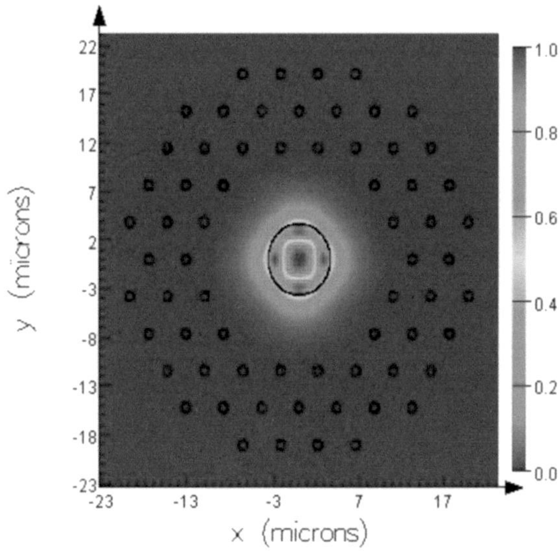


Fig. 7. The fundamental mode of the proposed near-zero ultra-flattened PCF.

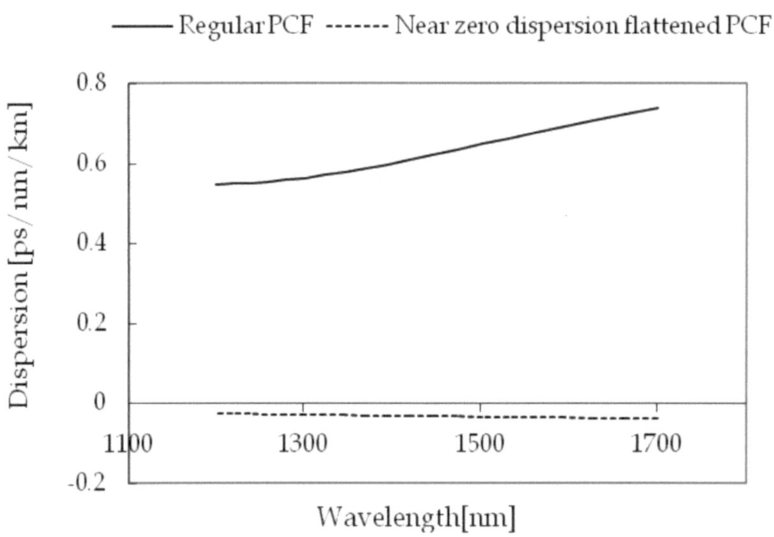


Fig. 8. Chromatic dispersion of the designed PCF compared to the regular IGPCF.

The achieved dispersion is ultra-flat with small negative values around –0.025 ps/nm/km. It can be observed that the chromatic dispersion is almost constant at a wide telecommunication wavelength range. The result is compared with the regular solid-core IGPCF. (As a reference, a standard structure made with medium-sized, pure silica core and medium air-filling fraction is concerned). In Fig. 8, a comparison between the dispersion values of the standard IGPCF and the designed structure is presented.

## 7. Conclusion

New fiber structure with near-zero ultra-flattened is proposed. It is suitable for broadband utilization in transmission systems. Before this, many fibers have been examined and many improvements have been applied to the studied structures. It is described how to control the location and shape of the chromatic dispersion curves. An investigation is carried out to study the PCF with high-index core material, in which a parabolic curve is evaluated in terms of potential ZDWs.

Investigated PCFs showed higher flexibility in fiber design. A new fiber structure is introduced and investigated. The bandwidth, in which anomalous dispersion is achieved, is getting wider with decreasing air fraction. By the increase in hole diameter, the second ZDW is extended till the U-band. Lower minimum dispersion values are achieved by the increase in doping region diameter.

Utilizing all the previous results of the interplay between chromatic dispersion on one side, and geometrical parameters as well as refractive index on the other side, has provided a well-defined procedure to design ultra-flattened and ultra-low chromatic dispersion profile.

HNPCF is doped with high-index material (dopant $GeO_2$) with the refractive index of 1.48 and only three air rings in the cladding. The achieved dispersion results were ultra-flattened with very small negative dispersion values: -0.025 [ps/nm/km] over the telecommunication band. The fiber is suitable for broadband zero-dispersion propagation of optical signals in high-speed transmission systems.

The future study will focus on achieving flattened and high anomalous chromatic dispersion for telecommunication applications. For example, the insertion of liquids in PCFs is promising for achieving optimal chromatic dispersion and nonlinear effects. Another goal is to optimize the studied structures without doping. Structures matching the characteristic of ITU-T standard fibers will be studied.

Last but not least, the future research should be highlighted on the recurrent optimization of algorithms to be developed.

## 8. Acknowledgment

This work has been supported by the Czech Science Foundation under project No. 102/09/P143.

## 9. References

Antos A. & Smith, D. (1994). Design and characterization of dispersion compensating fiber based on the $LP_{01}$ mode. *J. Lightw. Technol.*, Vol. 12, No. 10, pp. 1739-1745

Begum, F. et al. (2009). Design and analysis of novel highly nonlinear photonic crystal fibers with ultra-flattened chromatic dispersion. *Optics Communications,* Vol. 282, No. 7, pp. 1416-1421

Begum, F. et al. (2009). Design of broadband dispersion compensating photonic crystal fibers for high speed transmission systems, *Proceeding of the Conference on Optical Fiber Communication 2009 (OFC 2009)* 22-26 March 2009, San Diego, USA, pp. 1-3

Birks, T. et al. (1999). Localized function method for modeling defect modes in 2-d photonic crystals. *J. Lightwave Technol.*, Vol. 17, pp. 2078-2081

Birks, T. et al. (1999). Dispersion compensation using single-material fibers. *IEEE Photonics Technology Letters*, Vol. 11, pp. 674-676

Chen, M. & Xie, S. (2008). New nonlinear and dispersion flattened photonic crystal fiber with low confinement loss. *Optics Communications*, Vol. 281, pp. 2073-2076

Chen, W. et al. (2009). Dispersion-flattened Bragg photonic crystal fiber for large capacity optical communication system. *Front. Optoelectron. China*, Vol. 2(3), pp. 289-292

Cucinotta, A. et al. (2002). Holey fiber analysis through the finite-element method. *IEEE Photonics Technology Letters*, Vol. 14, pp. 1530-1532

Diddams S. & Diels, C. (1996). Dispersion measurements with white-light interferometry. *J. Opt. Soc. Am. B*, Vol. 13, pp. 1120-1129

Ferrando, A. et al. (2000). Vector description of higher-order modes in photonic crystal fibers. *J. Opt. Soc. Am. A*, Vol. 17, pp. 1333-1340

Ferrando, A. & Silvestre, E. (2000). Nearly zero ultraflattened dispersion in photonic crystal fibers. *Opt. Lett.* , Vol. 25, pp. 790-792

Fu, H. et al. (2008). Pressure sensor realized with polarization-maintaining photonic crystal fiber-based Sagnac interferometer. *Appl. Opt.*, Vol. 47, pp. 2835-2839

Hai, N. et al. (2007). A Novel Ultra-flattened Chromatic Dispersion Using Defected Elliptical Pores Photonic Crystal Fiber with Low Confinement Losses. *Proceedings of Antennas and Propagation Society International Symposium*, pp. 2233-2236, ISBN: 978-1-4244-0877-1, 9-15 June 2007

Hansen, K. (2003). Dispersion flattened hybrid-core nonlinear photonic crystal fiber. *Opt. Express*, Vol. 11, No. 13

Hansen, K. (2005). Introduction to nonlinear photonic crystal fibers. *J. Opt. Fiber Commun. Rep.*,Vol. 2, pp. 226-254

Haxha, S. & Ademgil, H. (2008). Novel design of photonic crystal fibers with low confinement losses, nearly zero ultra-flatted chromatic dispersion, negative chromatic dispersion and improved effective mode area. *Optics Communications*, Vol. 281, No. 2, pp. 278-286

Hoo, Y. et al. (2004). Design of photonic crystal fibers with ultra-low, ultra-flattened chromatic dispersion. *Optics Communications*, Vol. 242, No. 4-6, pp. 327-3

Huttunen, A. & Torma, P. (2005). Optimization of dual-core and microstructure fiber geometries for dispersion compensation and large mode area. *Opt. Express*, Vol. 13, No. 2, pp. 627-635

Huntington, S. et al. (1997). Atomic force microscopy for the determination of refractive index profiles of optical fibers and waveguides: A Quantitative study. *J.Appl. Phys.*, Vol. 82, pp. 2730-2734

Issa, N. & Poladian, L. (2003). "Vector wave expansion method for leaky modes of microstructured optical fibers", *J. Lightwave Technol.*, Vol. 21, pp. 1005-1012

Koshiba M. (2002). Full-vector analysis of photonic crystal fibers using the finite element method. *IEICE Trans. Electron.*, Vol. E-85C, pp. 881-888

Leong, J. et al. (2006). High-Nonlinearity Dispersion-Shifted Lead-Silicate Holey Fibers for Efficient 1-μm Pumped Supercontinuum Generation. *J. Lightwave Technol.*, Vol. 24, p. 183

Lou, S. et al. (2009). Photonic crystal fiber with novel dispersion properties. *Front. Optoelectron. China*, Vol. 2, pp. 170-177

Liu, J. et al. (2006). Enhanced nonlinearity in a simultaneously tapered and $Yb^{3+}$-doped photonic crystal fiber. *J. Opt. Soc. Am. B*, Vol. 23, pp. 2448-2453

Liu, J. et al. (2006). Modal cutoff properties in germanium-doped photonic crystal fiber. *Appl. Opt.*, Vol. 45, pp. 2035-2038

Liu, Z. et al. (2007). A broadband ultra flattened chromatic dispersion microstructured fiber for optical communications. *Optics Communications*, Vol. 272, No. 1, pp. 92-96

Matsui, T. et al. (2007). Dispersion Compensation Over All the Telecommunication Bands With Double-Cladding Photonic-Crystal Fiber. *J. Lightwave Technol.*, Vol. 25, pp. 757-762

Monro, T. et al. (2000). Modeling large air fraction holey optical fibers. *Journal of Lightwave Technology*, Vol. 18, pp. 50-56

Ni Y. et al. (2004). Dual-Core Photonic Crystal Fiber for Dispersion Compensation. *IEEE Photonics Technol. Letters*, Vol. 16, No. 6

Poli, F. et al. (2003). Dispersion and nonlinear properties of triangular photonic crystal fibers with large air-holes and small pitch. *Proc. European Conference on Optical Communication ECOC 2003*, Rimini, Italy, Sept. 21–25 2003

Razzak, S. et al. (2007). Ultra-flattened dispersion photonic crystal fibre, *Electron. Lett.* Vol. 43, No. 11, pp. 615-617

Russel, P. (2006). Photonic-Crystal Fibers. *J. lightw Technol.*, Vol. 24, No. 12, pp. 4729-4749

Saitoh, K. & Koshiba, M. (2004). Highly nonlinear dispersion-flattened photonic crystal fibers for supercontinuum generation in a telecommunication Windows. *Opt. Express*, Vol. 12, No. 10, pp. 2027-2032

Veng, M. et al. (2000). Dispersion compensating fibers. *Opt. Fiber Technol.* Vol. 6, 164–80

Wadsworth, W. et al. (2002). Supercontinuum generation in photonic crystal fibers and optical fiber tapers: a novel light source. *J. Opt. Soc. Am. B*, Vol. 19, pp. 2148-2155

Wang, Y. et al. (2009). Ultra-flattened chromatic dispersion photonic crystal fiber with high nonlinearity for supercontinuum generation *SPIE-OSA-IEEE*, Vol. 7630, 76301F-1

Wu, M. et al. (2008). Broadband dispersion compensating fiber using index-guiding photonic crystal fiber with defected core. *Chin. Opt. Lett.*, Vol. 6, pp. 22-24

Yang, S. et al. (2006). Theoretical study and experimental fabrication of high negative dispersion photonic crystal fiber with large area mode field. *Opt. Express*, Vol. 14, No. 7

Yu, Ch. et al. (2008). Tunable dual-core liquid-filled photonic crystal fibers for dispersion compensation. *Opt. Express 4443*, Vol. 16, No. 17

Zhong, Q. & Inniss, D. (1994). Characterisation of lightguiding structure of optical fibers by atomic force microscopy. *J. Lightwave. Technol.*, Vol. 12, pp. 1517-1523

Zhu, Z. & Brown, T. (2002). Full-vectorial finite-difference analysis of microstructured optical fibers. *Opt. Express*, Vol. 10, pp. 854–864

Zsigri, B. et al. (2004). A novel photonic crystal fiber design for dispersion compensation. *J. Opt. A: Pure Appl. Opt.*, Vol. 6, pp. 717-720

# 10

# Photonics Crystal Fiber Loop Mirrors and Their Applications

Chun-Liu Zhao[1], Xinyong Dong[1], H. Y. Fu[2] and H. Y. Tam[2]
*[1]Institute of Optoelectronic Technology, China Jiliang University, Hangzhou*
*[2]Photonics Research Centre, Department of Electrical Engineering,*
*The Hong Kong Polytechnic University, Hung Hom, Kowloon,*
*Hong Kong SAR*
*China*

## 1. Introduction

Fiber loop mirrors (FLMs), also called Sagnac interferometers, are interesting and very useful components for use in optical devices and systems [1, 2]. Many components based on FLMs have been demonstrated for applications in wavelength-division-multiplexing filters and in sensors, among others [3-7]. In FLM, the two interfering waves counter-propagate through the same fiber and are exposed to the same environment. This makes it less sensitive to noise from the environment. In general, a conventional fiber loop mirror made of high-birefringent fibers (HiBi fibers) or polarization-maintaining fibers (PMFs) has several advantages compared with a Mach–Zehnder interferometer, such as insensitivity, high extinction ratio, in-dependence of input polarization, easy to manufacture and low cost [1, 2]. However, conventional PMFs (e.g., Panda and bow-tie PMFs) have a high thermal sensitivity due to the large thermal expansion coefficient difference between boron-doped stress-applying parts and the cladding (normally pure silica). Consequently, conventional PMFs exhibit temperature-sensitive birefringence [8]. Therefore, conventional PMF based Sagnac interferometers exhibit relatively high temperature sensitivity, which is about 1 and 2 orders of magnitude higher than that of long-period fiber grating (LPG) and fiber Bragg grating (FBG) sensors [9, 10]. This can limit the practical use of the devices in some applications.

Various kinds of sensors based on HiBi-FLMs have been proposed and realized since HiBi-FLMs are sensitive to many parameters and have a high sensitivity, such as temperature sensors, level liquid sensors, refractive index sensors, strain sensors and biochemical sensors [7, 9-12]. However, when a HiBi-FLM is used to measure strain or other parameters, its cross-sensitivity to temperature may degrade sensor performance since the optical path length of the HiBi-FLM shows temperature dependence caused by thermal refractive-index change and thermal expansion effect. Thus, the temperature effect must be discriminated or eliminated when they are used for sensing [13-15].

The photonic crystal fiber (PCF) is a new class of optical fiber that emerged in recent years. Typically, these fibers incorporate a number of air holes that run along the length of the fiber and have a variety of different shapes, sizes, and distributions [16-17]. Of the many unusual properties exhibited by a PCF, a particularly exciting feature is that the PCF can be made HiBi by arranging the core and the air-hole cladding geometry, thereby introducing asymmetry [18-19]. Their birefringence can be of the order of $10^{-3}$, which is about one order of magnitude larger than that of conventional HiBi fibers. Unlike conventional PMFs (bow-tie, elliptical core, or Panda), which contain at least two different glasses each with a different thermal expansion coefficient, thereby causing the polarization of the propagation wave to vary with changing temperature, the PCF birefringence is highly insensitive to temperature because it is made of only one material (and air holes). Recently, some of FLMs used PCFs have been developed and applied on various devices [20-22] and optical fiber sensors [24-35], including strain sensors, pressure sensors, temperature sensors and curvature sensors, and so on.

In this chapter, we will first introduce the basic operation principle of FLMs, secondly, will demonstrate a temperature-insensitive interferometer based on a HiBi-PCF FLM. We will then move on to various applications in optical sensors such as strain sensors, pressure sensors, and temperature sensors. Following, we will discuss a demodulation technology of HiBi-PCF FLM based sensors. Finally, we will describe several multiplexing schemes for HiBi-PCF based FLM sensors.

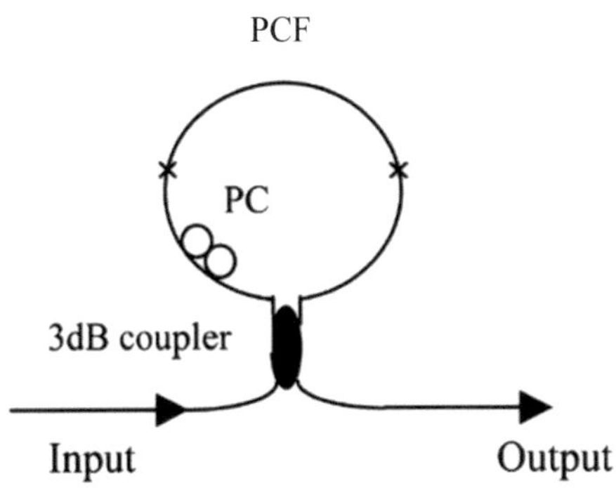


Fig. 1. Configuration of FLM made of a PCF.

## 2. Basic principle of FLMs

As shown in Fig. 1, the 3-dB coupler splits the input signal equally into two counter-propagating waves which subsequently recombine (at the coupler) after propagating around the loop. The interference of the counter-propagating waves will be constructive or destructive, depending on the birefringence of the cavity, and thus, the loop transmission response is wavelength dependent. The phase difference between the fast and slow beams that propagate in the PCF is given by [1, 2]:

$$\theta = 2\pi BL / \lambda \tag{1}$$

where B, L, and λ are the birefringence of the PCF, the length of the PCF and the wavelength, respectively. When the variation of B following the wavelength is small, there is $B=/n_x-n_y/$, where $n_x$ and $n_y$ are the effective refractive index for each polarization mode. Ignoring insertion loss of the 3-dB coupler and the attenuation of the PCF and the single-mode fiber in the loop, the transmission spectrum of the fiber loop is approximately a periodic function of the wavelength, namely,

$$T = (1 - \cos\theta) / 2 \tag{2}$$

The transmission dip wavelengths are the resonant wavelengths satisfying $2\pi BL/\lambda_{dip} = 2k\pi$, where k is any integer. Thus, the resonant dip wavelengths can be described as

$$\lambda_{dip} = BL / k \tag{3}$$

And the wavelength spacing between transmission dips can be expressed as

$$S = \lambda^2 / BL \tag{4}$$

When some varies (strain or temperature) applied on the PCF sensing element, they will cause the birefringence change $\Delta B$ and length change $\Delta L$ of the PCF. So the $\lambda_{dip}$ has a change and it can be expressed as:

$$\Delta\lambda_{dip} = (\Delta BL + B\Delta L) / k \tag{5}$$

So the change of varies can be obtained by measuring the wavelength shift of the dip in the output spectrum. The setting of the polarization controller (PC) can affect the contrast of the transmission function. By adjusting the state of the PC, transmission bands with large extinction ratio can be obtained.

## 3. Temperature-insensitive interferometer using a HiBi-PCF FLM [21]

In general, the optical path length of a conventional HiBi-FLM shows temperature dependence caused by thermal refractive-index change and thermal expansion of the devices [8]. This can limit the practical use of the device. In this part, utilizing the high birefringence and the low temperature coefficient of birefringence, a temperature-insensitive interferometer based on a HiBi-PCF FLM is realized.

In this experiment, a 6.5-cm-long HiBi-PCF was used, which was fabricated by Blaze-Photonics Com., and the cross-sectional scanning electron micrograph is shown in Fig. 2. Mode field diameters at the two orthogonal polarizations are 3.6 and 3.1 μm. The HiBi-PCF has a group birefringence $\Delta n_g$ of $8.65\times10^{-4}$ at 1550 nm, and a nominal beat length of 1.8 mm. Both ends of the HiBi-PCF are spliced to conventional single-mode fiber (SMF) by using a $CO_2$ laser splicing system. The PCF-SMF splicing loss is large (about 3.5 dB) because of mismatching of mode field and numerical apertures between the PCF and the SMF. The splicing loss will be reduced when a pre-tapering technology is used. The PCF-SMF splicing losses will increase the total insertion loss of the HiBi-PCF-FLM. The device characteristics are measured with a tunable laser source (Agilent 81689 A) which can be tuned from 1.5 to 1.6 μm and a power sensor (Agilent 81634 A).

Fig. 3 shows the transmission spectra of the HiBi-PCF-FLM at different temperatures. The temperature of the HiBi-PCF-FLM is controlled by a temperature chamber during measurement. The transmission spectrum is approximately a periodic function of wavelength, as given by equation (2). The corresponding wavelength spacing between transmission peaks is about 0.43 nm, which is consistent with equation (4). The extinction ratio is nearly 26 dB and the total insertion loss of the HiBi-PCF-FLM is 10 dB.

Since the phase difference is given by equation (1), a change of the phase matching condition caused by the environment leads to a wavelength spacing variation and a resonance wavelength shift. As shown in Fig. 3 and Fig. 4, when the ambient temperature of the HiBi-PCF-FLM is increased, the transmission peaks shift a little to shorter wavelength. We choose

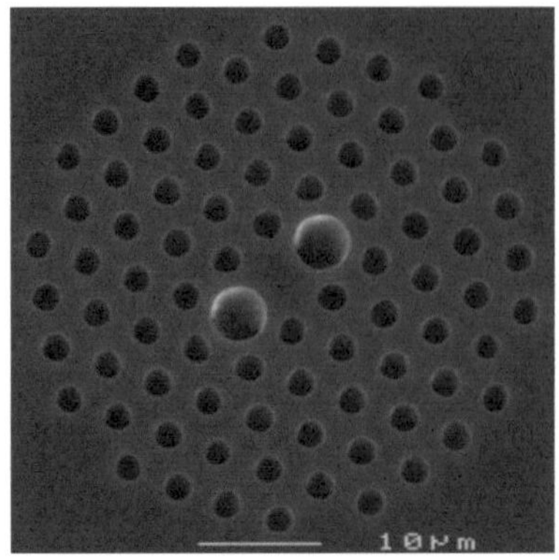


Fig. 2. Scanning electron micrograph of the cross section of the HiBi-PCF.

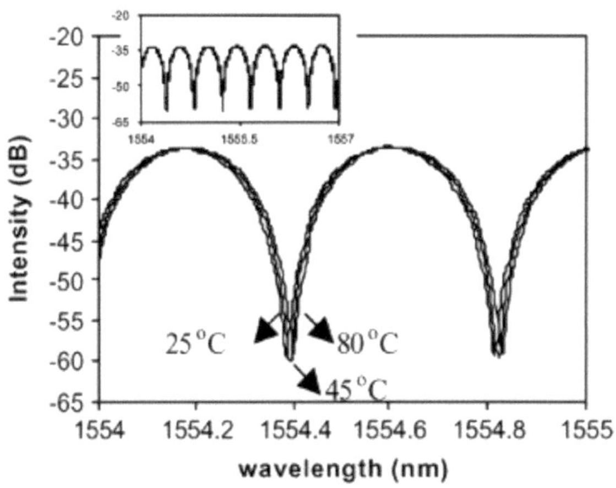


Fig. 3. Transmission spectra as a function of temperature for the HiBi-PCF-FLM, insertion: the transmission spectra in the range of 1554 -1557 nm.

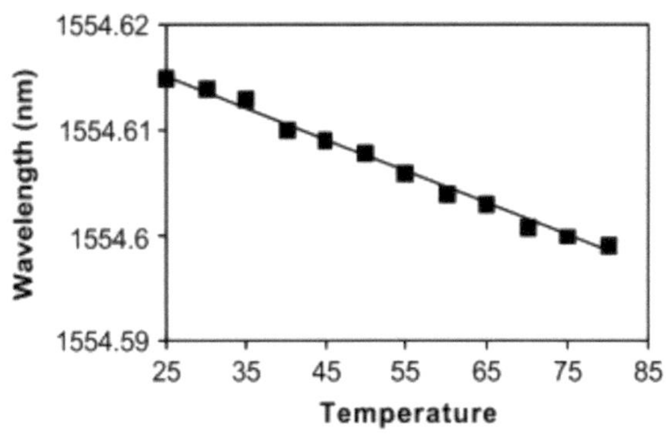


Fig. 4. Variation of the transmission peak wavelength at 1554.6 nm with temperature for the HiBi-PCF-FLM.

the transmission peak at 1554.6 nm as an example. The wavelength shift of the transmission peak with temperature is 0.3 pm /°C. The line (a) in Fig. 5, which is for the HiBi-PCF-FLM, shows the wavelength spacing change with temperature. The variation of wavelength spacing is very small: only 0.05 pm /°C.

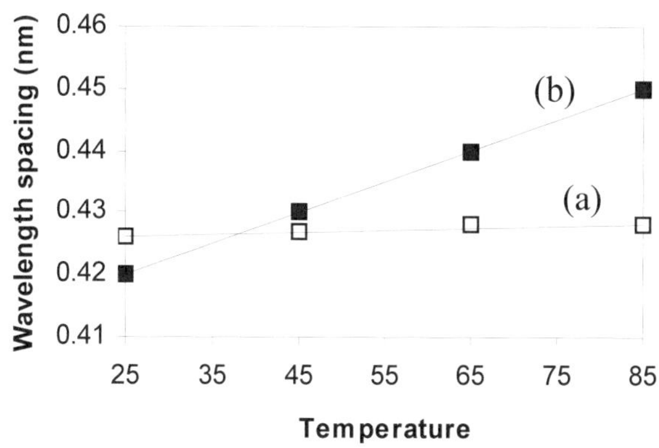


Fig. 5. Variation of the wavelength spacing with temperature (a) the HiBi-PCF-FLM; (b) the PMF-FLM.

In order to compare the new HiBi-PCF-FLM with the conventional FLM, we used a Panda polarization maintaining fiber (PMF) as the HiBi fiber. The Panda PMF is from Fujikura (SM-13P) with a measured birefringence of $\Delta n_g$ = $3.85\times10^{-4}$ at 1550 nm. The length of the Panda PMF is about 14.8 m. The wavelength spacing of the PMF-FLM is about 0.42 nm at temperature 25 °C. The extinction ratio is about 25 dB. As shown in Fig. 6, the transmission peaks shift very significantly at different temperatures. The line (b) in Fig. 5 shows the temperature dependence of the wavelength spacing for the conventional PMF-FLM. The variation of the wavelength spacing with temperature is about 0.5 pm /°C, which is nearly

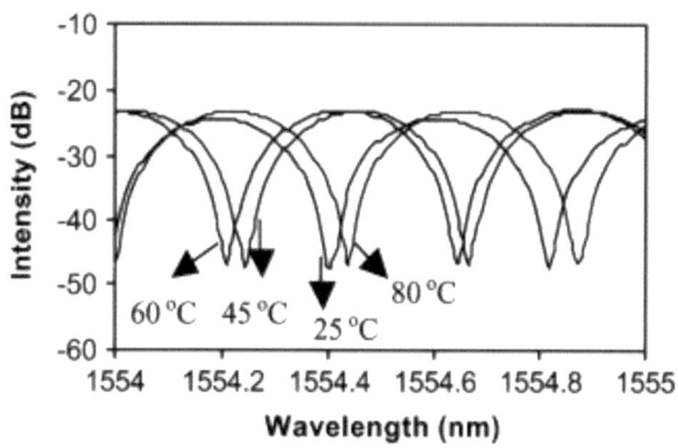


Fig. 6. Transmission spectra for the HiBi-PMF-FLM at different temperatures.

ten times of that for the HiBi-PCF-FLM. Furthermore, Fig. 7 shows the transmission peak shift as a function of temperature for the PMF-FLM. In theory, the wavelength shift of transmission peaks with temperature is nearly 16.6 pm /°C. In the experiment, however, the polarization of the propagation wave may vary with temperature, because different glasses of the PMF have different thermal expansion coefficient. This also effects the stability of the PMF-FLM. Such a large variation of the properties of the FLM made of conventional PMF with temperature makes it unsuitable for many applications in optical communication or sensor systems. However, by using HiBi-PCF, temperature-insensitivity of the FLM is improved by about 55 times.

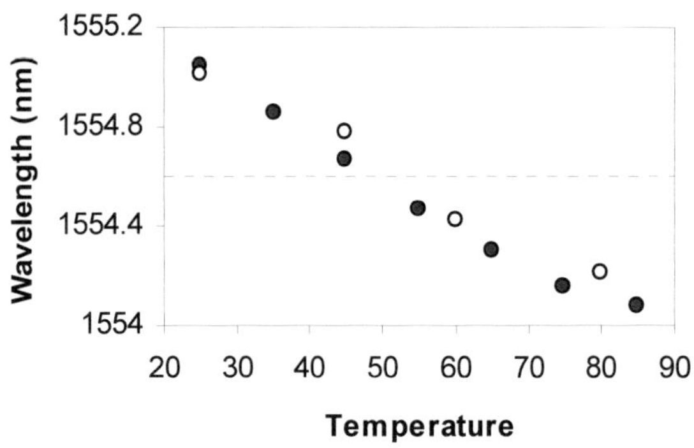


Fig. 7. Variation of the transmission peak wavelength near 1554.6 nm with temperature for the HiBi-PMF-FLM (•: theoretical and ○: experimental results).

## 4. Optical fiber sensors based on a HiBi-PCF FLM

### 4.1 A temperature independent strain sensor based on a HiBi-PCF FLM [24]

Strain sensors based on the strain-induced variation in birefringence of the HiBi fibers used in FLMs were also proposed and characterized. These sensors possess lots of advantages including simple design, easy to manufacture, high sensitivity, and low cost. However, previously reported FLM sensors are all based on conventional HiBi fibers whose birefringence is dependent on temperature. When they are used for sensing other measurands such as strain, the high thermal response of conventional PMFs may cause serious cross-sensitivity effects and reduce the measurement accuracy. In this part, a HiBi-PCF FLM strain sensor is demonstrated. The strain measurement is inherently temperature insensitive due to the great thermal stability of HiBi-PCF based FLM.

The proposed FLM strain sensor is as shown in Fig.8. When a strain is applied on the HiBi-PCF, the phase change induced by an elongation $\Delta L$ (i.e., a strain $\varepsilon = \Delta L / L$) to the PM-PCF can be given approximately by

$$\Delta\theta = \frac{2\pi}{\lambda}[\Delta LB + L\Delta B] \tag{6}$$

where $\Delta B = \Delta n_x - \Delta n_y$, is the variation of birefringence of the PM-PCF caused by photoelastic effect. Based on the analysis of photoelastic effect in single-mode fibers [35], the change of effective refractive index in the fiber core is related to the applied strain with a coefficient named effective photoelastic constant. It is therefore assumed that $\triangle n_x$ and $\triangle n_y$ have similar descriptions but different effective photoelastic constants, expressed as follows:

$$\Delta n_x = p_e^x n_x \varepsilon \text{ , and} \tag{7a}$$

$$\Delta n_y = p_e^y n_y \varepsilon \tag{7b}$$

Where $p_e^x$ and $p_e^y$ are the effective photoelastic constant for the slow and fast axes, respectively. By substituting Eqs. (7a) and (7b) into Eq.(6) and considering the relationship between spectrum (or peak wavelength) shift and phase change, i.e., $\Delta\lambda = S\Delta\theta / (2\pi)$, the following relationship can be obtained:

$$\Delta\lambda = \lambda(1 + \mathrm{p_e}')\varepsilon \tag{8}$$

where $P_e' = (n_y p_e^y - n_x p_e^x) / B$, is a constant that describes the strain-induced variation of the birefringence of the PM-PCF.

From Eq. (8), it can be seen that $\Delta\lambda$ is directly proportional to $\varepsilon$; therefore, linear spectrum (or peak wavelength) shift is expected with change of the applied strain.

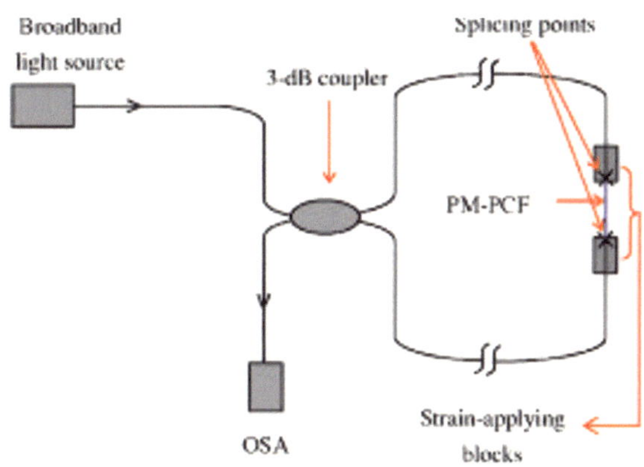


Fig. 8. Experimental setup of the proposed strain sensor based on a FLM made of a highly birefringent PCF.

In the experiment, the HiBi-PCF is 86 mm long, whose structure is the same as that in the section 3. Fig. 9 shows the transmission spectrum of the HiBi-PCF based FLM within a wide wavelength range of 70 nm. The wavelength spacing between the two transmission minima is 32.5 nm, and a good extinction ratio of 32 dB was achieved at the first transmission minimum located at 1547 nm. Since the light source we used is not polarized and there is no polarization-dependent element used in the sensor system, the stability of the sensor output against environmental variations, such as small vibrations, is good.

We fixed one end of the HiBi-PCF and stretched the other end by using a precision translation stage. Fig. 10 shows several measured transmission spectra around the transmission minimum at 1547 nm under different applied strains. The spectrum shifted 7.5 nm to the longer wavelength direction when the strain was increased from 0 to 32 m. The measured data are shown in Fig.11. A linear fitting to the experimental data gives a

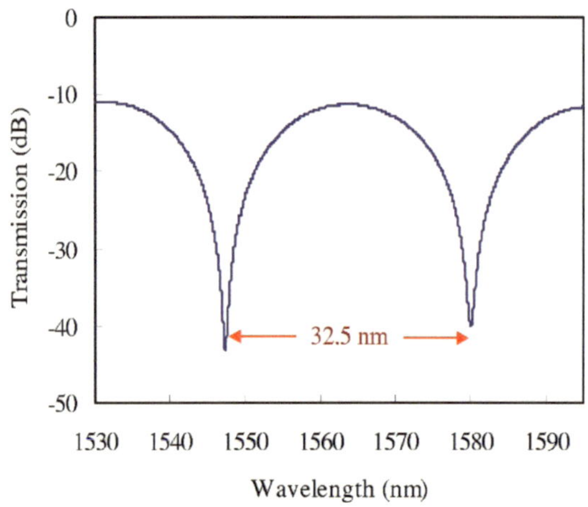


Fig. 9. The transmission spectrum of the HiBi-PCF FLM.

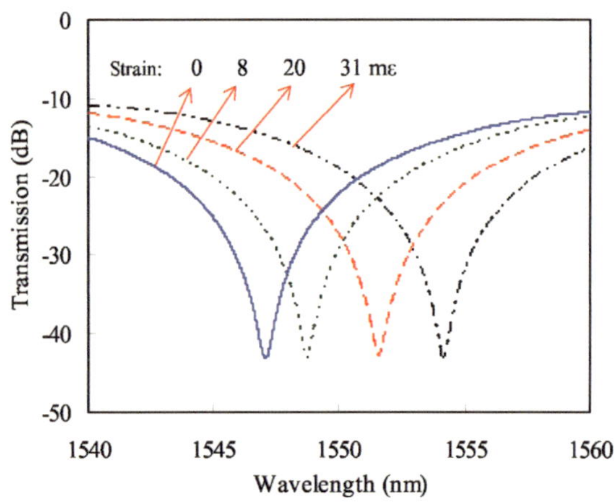


Fig. 10. Measured transmission spectra under different strains.

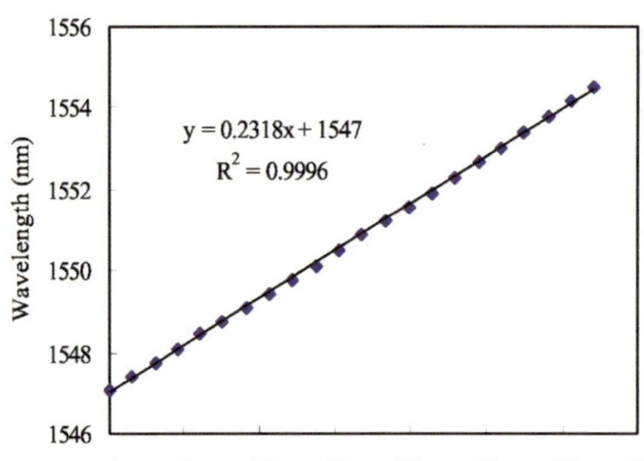


Fig. 11. Wavelength shift of the transmission minimum at 1547 nm against the applied strain.

wavelength-strain sensitivity of 0.23 pm/$\mu\varepsilon$ and a high $R^2$ value of 0.9996, which shows that the linearity of the wavelength to strain response is excellent. Therefore, the experimental data agree well with the theoretical prediction, and the constant $p_e^{'}$ in Eq. (8), calculated from the wavelength-strain sensitivity value, is -0.82.

The resolution of the strain measurement, limited by the 10 pm wavelength resolution of the used OSA, is 43 $\mu\varepsilon$, which is actually quite high when taking into account the large measurement range. The maximum value of the applied strain is mostly determined by the maximum strain that the HiBi-PCF can endure, not the strength of the fusion splicing points because the two splicing points between the HiBi-PCF and SMFs were prevented from being stretched as they were glued to the strain-applying blocks. As a result, the measurement range is several times larger than that of fiber Bragg grating and long-period grating sensors, where the fiber strength is significantly weaken during the grating inscription by high power ultraviolet laser beams [37]. This may be regarded as one of the several advantages of the proposed HiBi-PCF based the strain sensor over the two kinds of fiber grating sensors.

Temperature stability of the HiBi-PCF FLM strain sensor was also tested by setting the sensor head into a temperature-controlled container. The transmission minimum at 1547 nm was moved to shorter wavelength by only 22 pm when the temperature was increased up to 80°C. Measurement results are shown in Fig.12. The temperature sensitivity is only 0.29 pm/°C, which, compared with the reported value of 0.99 nm/°C of the FLM temperature sensor based on conventional PMF [7], is about 3000 times lower. The temperature sensitivity is also in good agreement with the previously reported value in Ref. 21 where the same HiBi-PCF was used. If a temperature variation of 30°C is assumed, the corresponding wavelength shift of the strain sensor is only 8.7 pm, which is even smaller than the wavelength resolution of the OSA. Therefore, such a low temperature sensitivity can be totally neglected when the sensor is operated in normal environmental condition without very large temperature variations.

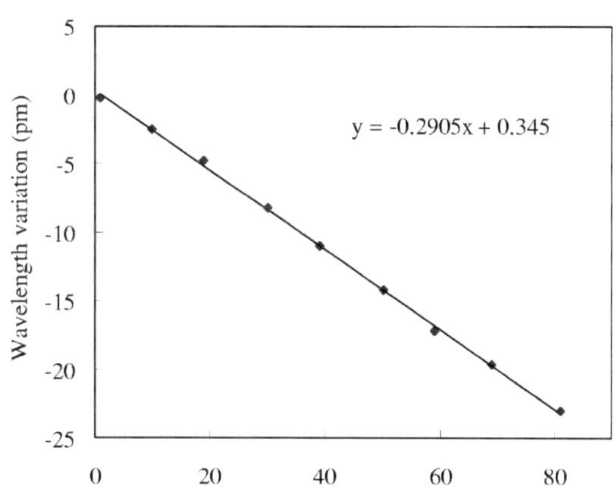


Fig. 12. Wavelength variation of the transmission minimum at 1547 nm against temperature.

Compared with the conventional HiBi fiber based FLM sensors and fiber Bragg grating or long-period grating sensors, the HiBi-PCF FLM strain sensor is inherently insensitive to temperature, eliminating the requirement for temperature compensation. It is also simple, easy to manufacture, potentially low cost, and possesses a much larger measurement range.

**4.2 Pressure sensor realized with a HiBi-PCF FLM [29]**

In this part, we demonstrate a pressure sensor based on a HiBi-PCF FLM. The FLM itself acts as a sensitive pressure sensing element, making it an ideal candidate for pressure sensor. Other reported fiber optic pressure sensors generally required some sort of modification to the fiber to increase their sensitivity [38]. The HiBi-PCF FLM pressure sensor does not require polarimetric detection and the pressure information is wavelength encoded.

Fig. 13 shows the experimental setup of the pressure sensor with the HiBi-PCF based FLM interferometer. The used HiBi-PCF is 58.4 cm and is laid in an open metal box and the box is placed inside a sealed air tank. The tank is connected to an air compressor with adjustable air pressure that was measured with a pressure meter. The input and output ends of the FLM are placed outside the air tank.

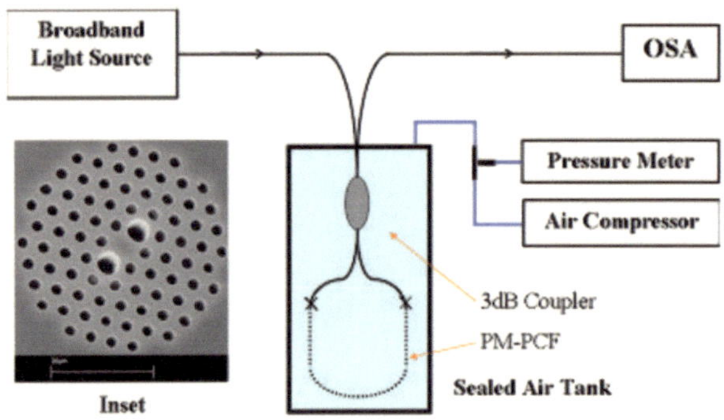


Fig. 13. The experimental setup of our proposed pressure sensor.

Ignoring the loss of the FLM, the transmission spectrum of the fiber loop is approximately a periodic function of the wavelength and is given as Eq. (1). The total phase difference θ introduced by the HiBi-PCF can be expressed as

$$\theta = \theta_0 + \theta_{\mathrm{P}} \tag{9}$$

where $\theta_0$ and $\theta_P$ are the phase differences due to the intrinsic and pressure-induced birefringence over the length L of the HiBi-PCF and are given by

$$\theta_0 = \frac{2\pi \cdot B \cdot L}{\lambda} \tag{10}$$

$$\theta_{\mathrm{P}} = \frac{2\pi \cdot \left(\mathrm{K_P \Delta p}\right) \cdot L}{\lambda} \tag{11}$$

$\Delta P$ is the applied pressure and the birefringence-pressure coefficient of HiBi-PCF can be described as [39]

$$K_P = \frac{\partial n_s}{\partial P} - \frac{\partial n_f}{\partial P}. \tag{12}$$

The pressure-induce wavelength shift of the transmission minimum is $\Delta\lambda = S \cdot \theta_P / 2\pi$. Thus the relationship between wavelength shift and applied pressure can be obtained as

$$\Delta\lambda = (\frac{K_P \cdot \lambda}{B}) \cdot \Delta P \,. \tag{13}$$

Eq. (13) shows that for a small wavelength shift, the spectral shift is linearly proportional to the applied pressure.

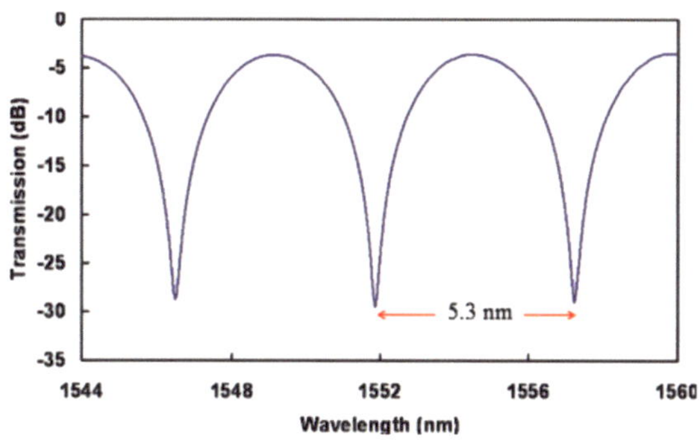


Fig. 14. Transmission spectrum of the HiBi-PCF based Sagnac interferometer.

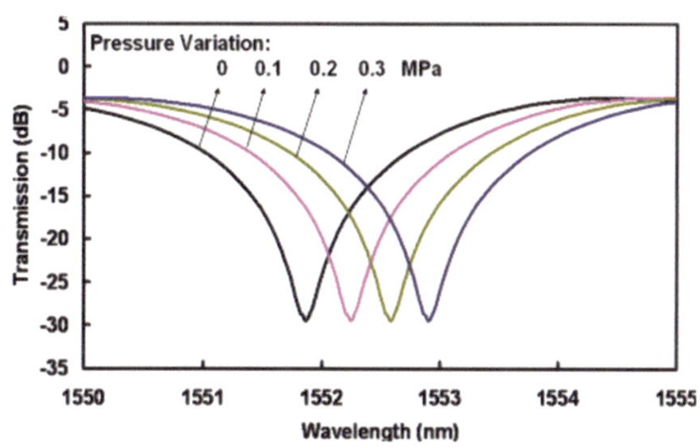


Fig. 15. Measured transmission spectra under different pressures.

Fig 14 shows the transmission spectrum of the HiBi-PCF FLM at atmospheric pressure, i.e., at zero applied pressure. The spacing between two adjacent transmission minimums is ~5.3 nm and an extinction ratio of better than 20 dB was achieved. The intrinsic birefringence of the HiBi-PCF used in our experiment is $7.8 \times 10^{-4}$ at 1550 nm.

The air compressor is initially at one atmospheric pressure (about 0.1MPa). In the experiment, we can increase air pressure up to 0.3 MPa; thus, the maximum pressure that can be applied to the HiBi-PCF-based FLM sensor is ~0.4 MPa. At one atmospheric pressure one of the transmission minimums occurs at 1551.86 nm and shifts to a longer wavelength with applied pressure. When the applied pressure was increased by 0.3MPa, a 1.04 nm wavelength shift of the transmission minimum was measured, as shown in Fig 15. Fig. 16 shows the experimental data of the wavelength-pressure variation and the linear curve fitting. The measured wavelength-pressure coefficient is 3.42 nm/MPa with a good $R^2$ value of 0.999, which agrees well with the theoretical prediction. From Eq. (13), the birefringence-pressure coefficient is ~$1.7 \times 10^{-6}$ $MPa^{-1}$. The resolution of the pressure measurement is ~2.9 kPa when using an OSA with a 10 pm wavelength resolution. Because of the limitations of our equipment, we have not studied the performance of this pressure sensor for high pressure at this stage. However, we found that the HiBi-PCF can stand pressure of 10 MPa without damage to its structure. This part of the work is ongoing and will be reported in our further studies.

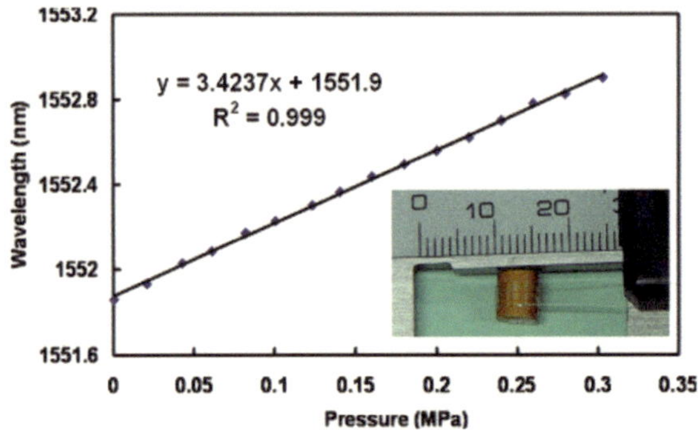


Fig. 16. Wavelength shift of the transmission minimum at 1551.86 nm against applied pressure with variation up to 0.3Mpa.

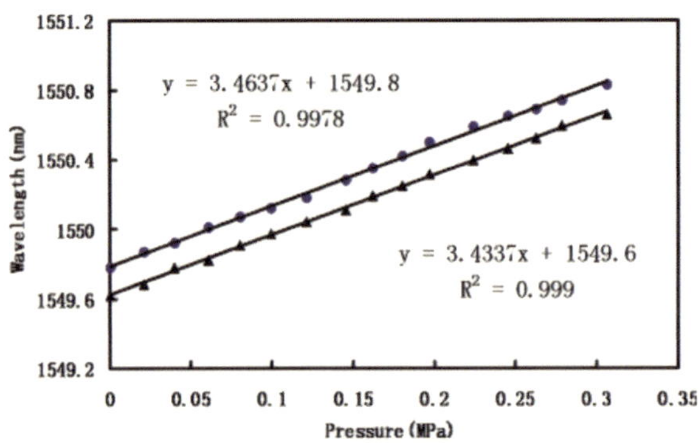


Fig. 17. Wavelength shift of the transmission minimum against applied pressure for HiBi-PCFs with length of 40 (circles) and 79.6 cm (triangles); the wavelength pressure coefficients are 3.46 and 3.43 nm/MPa, respectively.

Although the length of HiBi-PCF used in our experiment is 58.4 cm, it is important to note that the HiBi-PCF can be coiled into a very small diameter circle with virtually no additional bending loss so that a compact pressure sensor design can be achieved. The induced bending loss by coiling the HiBi-PCF into 10 turns of a 5mm diameter circle, shown in the inset of Fig. 16, is measured to be less than 0.01 dB with a power meter (FSM-8210, ILX Lightwave Corporation). The exceptionally low bending loss will simplify sensor design and packaging and fulfils the strict requirements of some applications where small size is needed, such as in down-hole oil well applications. To investigate the effects of coiling, we have studied two extreme cases in which the HiBi-PCF was wound with its fast axis and then its slow axis on the same plane of the coil. There were no measurable changes for either the birefringence or the wavelength–pressure coefficient when the fiber was coiled into 15 and 6mm diameter circles with both of the orientations coiling. The coiling of the HiBi-PCF

into small diameter circles makes the entire sensor very compact and could reduce any unwanted environmental distortions, such as vibrations.
The wavelength-pressure coefficient is independent of the length of the HiBi-PCF, as described in Eq. (13). Fig. 17 shows the wavelength-pressure coefficients are 3.46 and 3.43 nm/MPa for HiBi-PCFs with lengths of 40 and 79.6 cm, respectively. After comparing the two wavelength–pressure coefficients with that of the pressure sensor with a 58.4 cm HiBi-PCF (Fig. 17), we observed that the wavelength–pressure coefficient is constant around 1550 nm; this agrees well with our theoretical prediction. However, the length of the PM-PCF cannot be reduced too much because this would result in broad attenuation peaks in the transmission spectrum and that would reduce the reading accuracy of the transmission minimums.

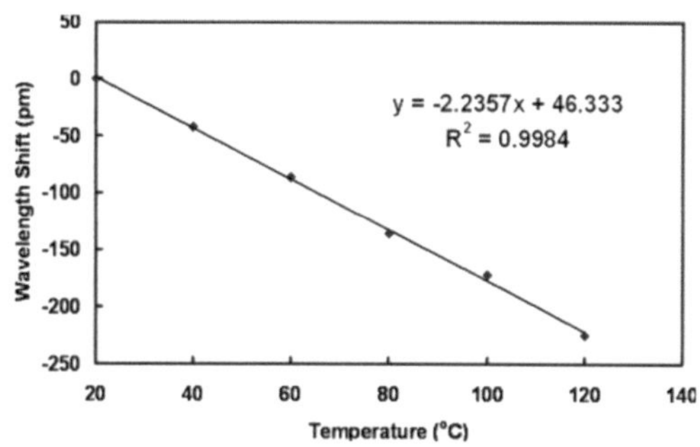


Fig. 18. Wavelength shift of the transmission minimum at 1551.86 nm against temperature.

Temperature sensitivity of the proposed pressure sensor is also investigated by placing the sensor into an oven and varying its temperature. Fig 18 shows the wavelength shift of a transmission minimum versus temperature linearly with a good $R^2$ value of 0.9984. The measured temperature coefficient is -2.2 pm/°C, which is much smaller than the 10 pm/°C of fiber Bragg grating. The temperature may be neglected for applications that operate over a normal temperature variation range.
Based on the small size, the high wavelength pressure coefficient, the reduced temperature sensitivity characteristic, and other intrinsic advantages of fiber optic sensors, such as light weight and electro-magnetically passive operation, the proposed pressure sensor is a promising candidate for pressure sensing even in harsh environments. Considering the whole pressure sensing system, we can also replace the light source with laser and use a photodiode for intensity detection at the sensing signal receiving end. Since the power fluctuation is very small even when the HiBi-PCF is bent, intensity detection is practical for real applications. Because of the compact size of the laser and photodiode, the entire system can be made into a very portable system. Furthermore, the use of intensity detection instead of wavelength measurement would greatly enhance interrogation speed and consequently makes the system much more attractive.

### 4.3 A high sensitive temperature sensor based on a FLM made of an alcohol-filled PCF [33]

HiBi-PCFs have a low thermo-optic and thermo-expansion coefficient HiBi-PCF, so HiBi-PCF FLMs can not be used to measure temperature directly. However, by inserting a short alcohol-filled HiBi-PCF into a FLM, a temperature sensor with an extremely high sensitivity can be realized by measuring the wavelength shift of the resonant dips of the alcohol-filled HiBi-PCF FLM.

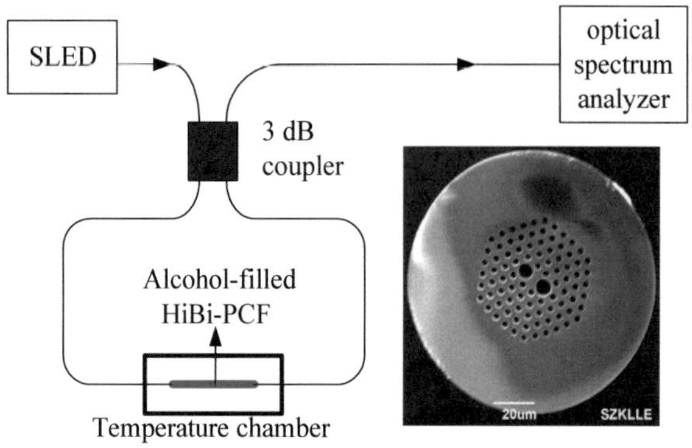


Fig. 19. Experimental setup of the temperature sensor based on a FLM inserted an alcohol-filled highly birefringent PCF. Insertion: SEM of the cross section of the used PCF.

The temperature sensor, as shown in Fig. 19, consists of a 3dB coupler and a short alcohol-filled PCF. Alcohol is chosen to fill into HiBi-PCF since it is an easy-filled liquid with a high temperature sensitivity. Here, an alcohol-filled HiBi-PCF is inserted into a FLM as a temperature sensing head. Birefringence change ΔB and length change ΔL of the alcohol-filled HiBi-PCF caused by temperature, leads a wavelength shifting of the resonant dips according to Eq. (3). The relationship between the dip wavelength change $\Delta\lambda_{dip}$, ΔB and ΔL is simply expressed as Eq. (5), $\Delta\lambda_{dip} = (\Delta BL + B\Delta L) / k$ , where ΔB is the birefringence change caused by the thermo-optic effect, including that of the original HiBi-PCF and that of the filled alcohol, and ΔL is the length change caused by the thermo-expansion effect, which also includes the elongation of the original HiBi-PCF and the expansion of the filled alcohol. We neglect ΔB and ΔL caused by the HiBi-PCF itself because of a good thermal independence of the HiBi-PCF. Further, ΔL caused by the thermo-expansion of the filled alcohol is also ignored since the volume of alcohol filled into the air-holes of the HiBi-PCF is small. Thus, $\Delta\lambda_{dip}$ mainly depends on ΔB of the alcohol-filled HiBi-PCF. The birefringence-temperature dependence of the alcohol-filled HiBi-PCF is analyzed by using a full-vector finite element method (FEM). The diameters of the bigger and smaller holes are 7 and 3.2 μm, respectively, and the pitch length between centers of two adjacent holes is 5.46 μm, according to the HiBi-PCF used in experiment. The refractive index of pure silica and the filled alcohol is taken as 1.4457 and the empirical value which is calculated by an empirical equation according to [40]. Fig. 20 shows the empirical temperature dependence of the refractive index of alcohol and the theoretical temperature dependence of the birefringence of the alcohol-filled HiBi-PCF. With the temperature rising, the refractive index of alcohol decreases linearly, while the birefringence of the alcohol-filled HiBi-PCF increases linearly. The mode fields of the two orthogonal polarizations at 20 °C are shown in the insertion of Fig. 21. The birefringence of the alcohol-filled HiBi-PCF is calculated at $3.5\times10^{-4}$ at 20 °C. $P_t$ is defined as a thermo-optic constant on the birefringence of the alcohol-filled HiBi-PCF, which equals to the slope of the temperature dependence curve of birefringence and is calculated at $1.5\times10^{-6}$ /°C. According to Eq. (2) and Eq. (5), the relationship between the resonant dip wavelengths shift $\Delta\lambda_{dip}$ and the temperature change ΔT can be deduced as

$$\Delta\lambda_{dip} = \Delta BL / k = \frac{LP_t}{k}\Delta T = \frac{\lambda_{dip}}{B}P_t\Delta T \qquad (14)$$

Based on the above equation, the temperature sensitivity of the alcohol-filled HiBi-PCF FLM is related to $\lambda_{dip}$, $P_t$ and B. A high temperature sensitivity depends on a long wavelength $\lambda_{dip}$ of the measured resonant dip, a high thermo-optic constant $P_t$ and a small birefringence B of the filled HiBi-PCF.

The HiBi-PCF used in the experiment is provided by Yangtze Optical Fibre and Cable Company. The HiBi-PCF has a birefringence of $10.2\times10^{-4}$ at 1550 nm, and the length is 6.1 cm. After the HiBi-PCF filling with alcohol by air-holes capillary force, the birefringence of the PCF reduces significantly, which bring advantages on a larger wavelength space between two resonant dips and on a wider measurement range. Both ends of the alcohol-filled HiBi-PCF are spliced to conventional single-mode fiber (SMF) by using a regular arc splicing machine (Fujikura FSM 60).

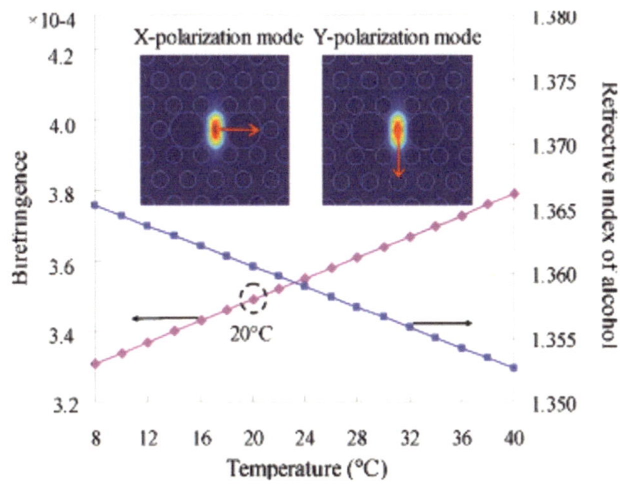


Fig. 20. Temperature dependence of the refractive index of alcohol and the birefringence of polarization mode fields of the alcohol-filled HiBi-PCF at 20 °C.

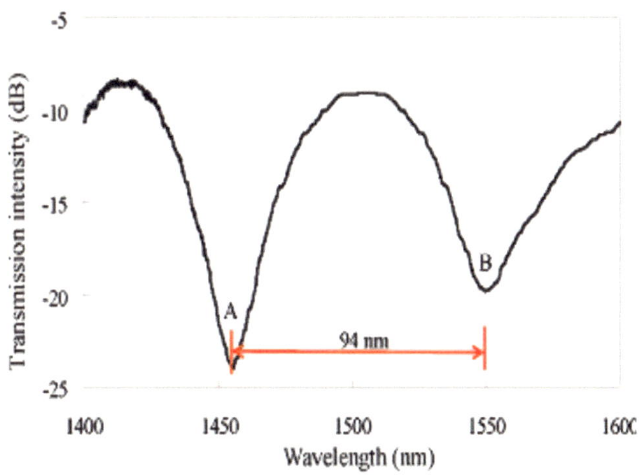


Fig. 21. Transmission spectrum of the alcohol-filled HiBi-PCF FLM at 20 °C.

Fig. 21 shows the transmission spectrum of the alcohol-filled HiBi-PCF FLM at room temperature (20 °C). Two resonant dips of the FLM display in the wavelength range from 1400 to 1600 nm. One is at the wavelength of 1455.8 nm (dip A) with 15.5 dB extinction ratio; the other is at about 1549.8 nm (dip B) with 10.5 dB extinction ratio. The wavelength spacing between these two dips is ~94 nm and the corresponding birefringence of the alcohol-filled HiBi-PCF is ~$3.9\times10^{-4}$ at 20 °C, which is close to the theoretical value (~$3.5\times10^{-4}$). The little difference between the experimental and theoretical values may be caused by the error of air-holes geometry size of HiBi-PCF according the SEM.

In the experiment, the temperature characteristic of the alcohol-filled HiBi-PCF FLM is tested by placing the alcohol-filled HiBi-PCF of the FLM at a temperature-controlled container. Fig. 22 (a) and (b) show the transmission spectra of the alcohol-filled HiBi-PCF FLM at temperature range of 20 to 34 °C and 8 to 20 °C, respectively. Dip A red-shifts from 1455.8 to 1543.7 nm with temperature increasing gradually from 20 to 34 °C, at the same time, the extinction ratio of dip A decreases. While, dip B blue-shifts from 1549.8 to 1470.4 nm with the temperature decreasing gradually from 20 to 8 °C.

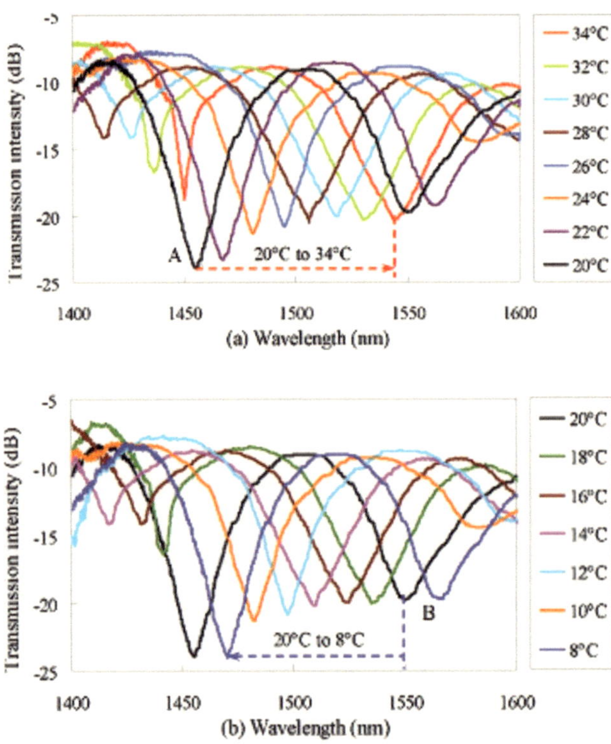


Fig. 22. Transmission spectra of the alcohol-filled HiBi-PCF FLM (a) when temperature increases from 20 and 34 °C and (b) when temperature decreases from 20 and 8°C.

Fig. 23 shows the experimental relationship between temperature and the resonant wavelength of dip A and dip B. The fitting curves can be expressed as y = 6.2176x+1331.7 for dip A and y = 6.6335x+1416.7 for dip B, and the high fitting degrees 0.9997 and 0.9995 mean the linearity of the resonant wavelength to temperature is excellent. The experimental

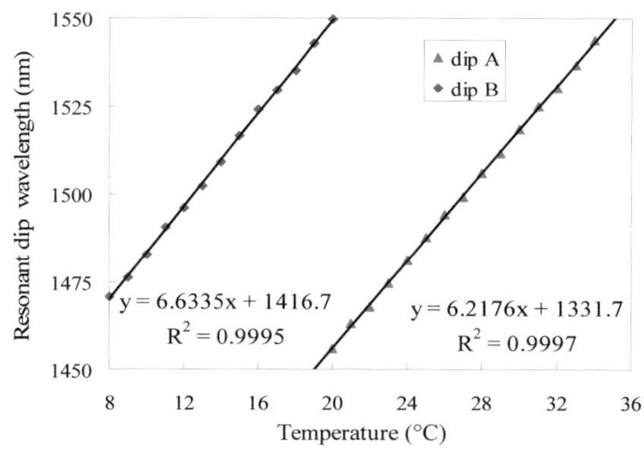


Fig. 23. The relationship between temperature and the resonant wavelength of dip A and dip B.

temperature sensitivities of dip A and dip B are ~6.2 nm/°C and ~6.6 nm/°C, respectively. And the theoretical sensitivities are ~6.1 nm/°C and ~6.5 nm/°C from Eq. (14). It is clear that the theoretical and the experimental results are in accordance. The temperature sensitivity of the alcohol-filled HiBi-PCF FLM is very high, and reach up to about 660 and 7 times higher than that of a FBG (~0.01 nm/°C) and that of the FLM made of a conventional HiBi fiber with a 72 cm length (~0.94 nm/°C) [10].

In practical uses, for a wider measurement range of temperature, the length L of the HiBi-PCF can be shortened in order to widen the spacing between two resonant dips based on $S = \lambda^2/BL$. For example, when the alcohol-filled HiBi-PCF is 1 cm, the spacing of the proposed FLM sensor is ~564 nm. It can be provided the measurement range of ~84 °C with the same temperature sensitivity ~6.6 nm /°C according to Eq. (14), in which the length of the sensing fiber is the same as the length of FBG sensing head and is shortened 72 times than that of the conventional HiBi-FLM temperature sensor.

## 5. Demodulation of sensors based on HiBi-PCF FLM [34]

All HiBi-PCF FLM sensors demonstrated above are based on monitoring the resonant wavelength variation of the FLM. In these configurations, a broadband light source and an optical spectrum analyzer (OSA) are needed, which cause the sensors expensive. In this part, we introduce a simple demodulation technology for a strain sensor based on HiBi-PCF FLM, which can also be used in other FLM based sensors. By utilizing the fact that the transmission intensity of a FLM at a fixed wavelength is strongly affected by the strain applied on a piece of HiBi-PCF in the FLM since the transmission spectrum of the FLM shifts with the applied strain, but the resonant dip (both wavelength and intensity) is insensitive to temperature, a low-cost temperature-insensitive strain sensor based on a HiBi-PCF FLM is achieved. The sensor uses a distributed-feedback (DFB) laser as the light source. Since the output intensity of the FLM is directly proportional to the applied strain, only an optical power meter is sufficient to detect strain variation, avoiding the need for an expensive OSA.

Since the HiBi-PCF is insensitive to temperature, the strain applied on the HiBi-PCF is an only influence factor on the transmission spectrum of the FLM. When an axial strain is applied on the HiBi-PCF, the phase difference of the FLM is changed, which is induced by an elongation of the HiBi-PCF and the variation of birefringence of the HiBi-PCF caused by photoelastic effect. The relationship between the FLM phase change $\Delta\theta$ and the axial strain applied on the HiBi-PCF can be expressed as

$$\Delta\theta = \frac{2\pi}{\lambda}(LB + LP_e)\varepsilon \tag{15}$$

where $P_e = P_e^x n_x - P_e^y n_y$, and $p_e^x$ and $p_e^y$ are the effective photoelastic constant for the slow and fast axes, respectively.
So, when an axial strain is applied on the HiBi-PCF, the transmission spectrum of the FLM can be described as

$$T' = [1 - \cos(\theta + \Delta\theta)]/2 \tag{16}$$

Fig. 24 shows the theoretical transmission spectra of the FLM at a free state and at the state of an axial strain (6000με) applied on the HiBi-PCF, which is gotten based on the equation (16). In theoretical calculation, the length and the birefringence of the HiBi-PCF are taken as L = 79.5 mm and B = $8.5\times10^{-4}$, respectively, in accordance to the experimental data. $P_e$ of the HiBi-PCF is assumed to $P_e$ = $-2.24\times10^{-4}$ [16], which best fits the experiment. As shown in Fig. 25, the transmission spectrum of the FLM shifts to longer wavelength since the phase matching condition is changed when an axial strain is applied on the HiBi-PCF. Therefore, the applied strain can be gotten by monitoring the resonant wavelength shift of the FLM through using a broadband light source and an OSA in a high cost. The theoretical sensitivity of strain based on monitoring the resonant wavelength shift is obtained at 1.1 pm/με.

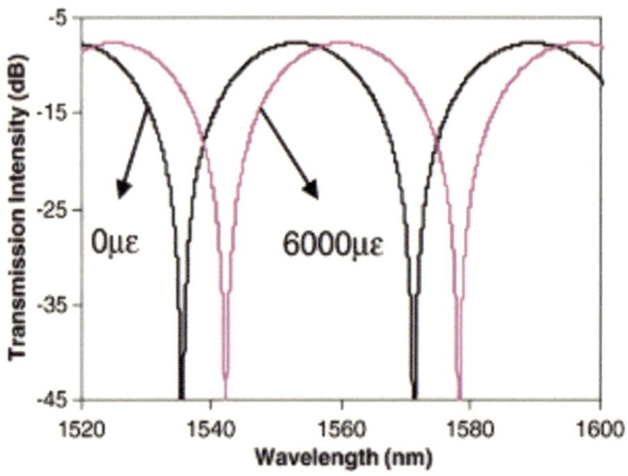


Fig. 24. Theoretical transmission spectra of the FLM at a free state and at the state of an axial strain (6000 με) applied on the HiBi-PCF.

Meanwhile, the transmission intensity at a fixed wavelength changes when the transmission spectrum of the FLM shifts with the strain applied on the HiBi-PCF. Thus, the information of the applied strain can be also gotten by monitoring the transmission intensity. The HiBi-PCF FLM sensor based on intensity measurement can be achieved in a low cost by using a DFB laser and an optical power meter, instead of an expansive broadband source and an

OSA. When an axial strain is applied on the HiBi-PCF, the FLM transmission intensity at a fixed wavelength can be described as

$$T_0^{'} = [1 - \cos(\theta_0 + C_0\varepsilon)]/2 \tag{17}$$

where $\theta_0 = \frac{2\pi}{\lambda_0}LB$, $C_0 = \frac{2\pi}{\lambda_0}(LB + LP_e)$. It is clear that, the transmission intensity of the FLM at a fixed wavelength varies accordingly with the applied strain. The transmission intensity variation of the FLM with the change of the axial strain applied on the HiBi-PCF can be deduced as

$$\frac{dT_0^{'}}{d\varepsilon} = \frac{1}{2}C_0 \sin(\theta_0 + C_0\varepsilon) \tag{18}$$

Fig. 25 is the theoretical relationship between the transmission intensity of the FLM sensor and the applied axial strain at three different wavelengths, which are gotten from the equation (17). As shown in the Fig. 24, the transmission intensity of the FLM at a fixed wavelength is approximately a periodic function of the axial strain applied on the HiBi-PCF. The strain spacing is about 33000 με. This means the maximal measurement range is about 16500 με, in which the relationship between the applied strain and the transmission intensity of the FLM is a proportional dependence. When the strain is measured from 0 με, the measurement range of the applied strain is different for the different fixed wavelength. When the fixed wavelength is chosen at the resonant wavelength (1535.6 nm), the measurement range of the strain is maximum, which is from 0 to 16500 με.
Fig. 26 is the enlarged drawing of the circle part in Fig. 25. Fig. 26 shows that all of the transmission intensity of the FLM at four different wavelengths (1530 nm, 1532 nm, 1545 nm and 1547 nm) are proportional to the applied strain, when the strain is in the range of 0 ~ 6000 με. The strain sensitivity is in positive when the fixed wavelength (1530 nm and 1530 nm) is shorter than the resonant wavelength (1535.6 nm) of the FLM; on the other hand, the strain sensitivity is in negative when the fixed wavelength (1545 nm and 1547 nm) is longer than the resonant wavelength (1535.6 nm) of the FLM.

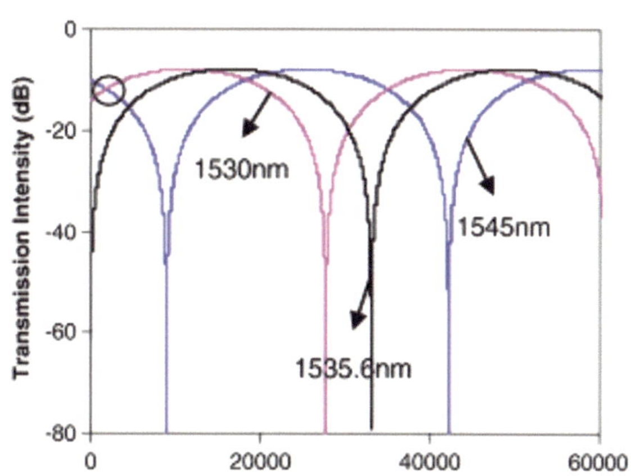


Fig. 25. Theoretical strain dependence of the transmission intensity of the FLM at different wavelengths.

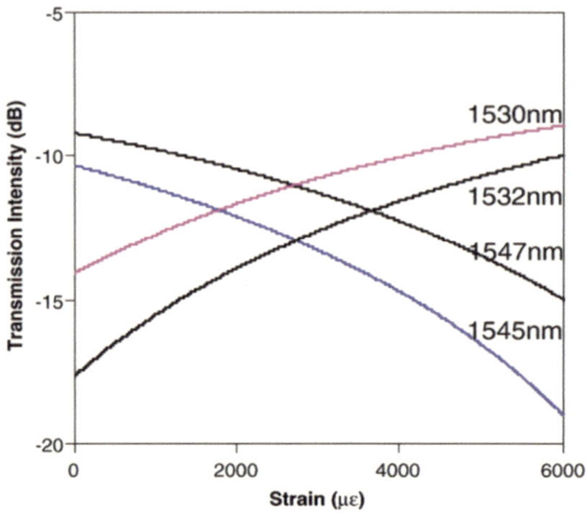


Fig. 26. Theoretical strain dependence of the transmission intensity of the FLM at different wavelengths in the strain range of 0~6000 με.

Fig. 27 shows the transmission spectrum of the HiBi-PCF FLM. The HiBi-PCF has a birefringence B of ~$8.5\times10^{-4}$ at 1550 nm, and the length L of 79.5 mm. The corresponding wavelength spacing between transmission peaks (or transmission dips) is about 35.6 nm, and the extinction ratio is nearly 26 dB. Fig. 28 shows the strain characteristics of the FLM at different strain. The whole transmission spectrum shifts toward longer wavelength with the applied strain increasing because the length of the HiBi-PCF increases with the axial stretching and the birefringence of the HiBi-PCF decreases due to the photoelastic effect of the fiber. When the strain sensor is based on the resonant wavelength monitoring, the strain sensitivity with wavelength which is the slope of the curve, is estimated to be 1.1 pm/με as shown in Fig. 39. Experimental results are identical with the theoretical analysis. When an OSA with a wavelength resolution of 10 pm is used, the strain resolution is about 9.1 με.

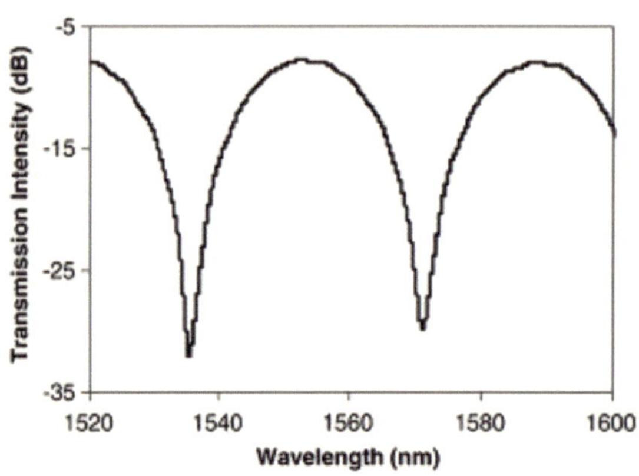


Fig. 27. Experimental transmission spectrum of the HiBi-PCF FLM.

When the strain sensor is based on the transmission intensity measurement, a single wavelength source such as a wavelength tunable laser or a DFB laser is used as a light source. The HiBi-PCF FLM sensor based on optical intensity measurement is measured with

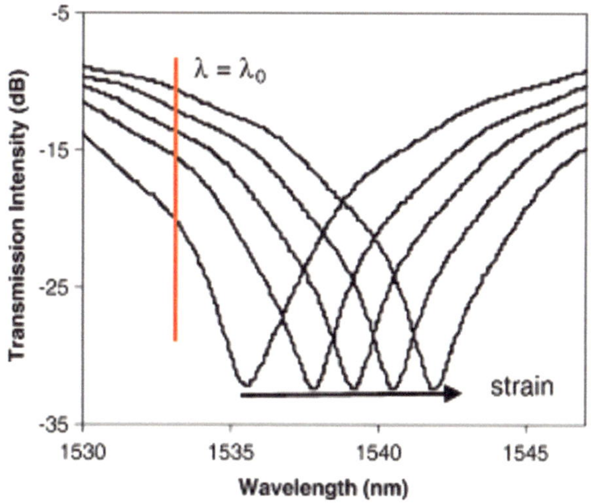


Fig. 28. Experimental transmission spectra of the FLM at different strain applied on the HiBi-PCF (from left to right, the strain: 0, 2137, 3357, 4565 and 5770 με, respectively).

a tunable laser source (Agilent 81689 A) and a power meter (Agilent 81634 A). The wavelength of the tunable laser is near the resonant wavelength of the FLM and hence the output light intensity from the FLM is directly related to the FLM's transmission at the wavelength of the DFB laser. Since the FLM's transmission is insensitive to temperature, the output power is only affected by the transmission spectrum change caused by the strain applied on the HiBi-PCF. Fig. 30 shows the measured and theoretical relationship between the output intensity of the FLM sensor and the applied axial strain for various laser wavelengths. It's clear that the strain sensitivity with intensity is related to the wavelength of the used laser source. In our experiment, a tunable laser is used for easiness of wavelength adjustment. In practice, a DFB laser with appropriate wavelength would be better for the purpose of reducing cost.

As shown in Fig. 30, for laser wavelengths of 1530 nm and 1532 nm, which are shorter than the resonant wavelength (1535.6 nm) of the FLM, the output intensity increases with applied strain and the strain sensitivity is positive. Meanwhile, for laser wavelength longer than the resonant wavelength (1535.6 nm), the output intensity decreases with the applied strain and the intensity sensitivities are negative. Fig. 30 also shows the theoretical curves of the relation between the output intensity and the applied strain. The experimental results are in a good agreement with the theoretical analysis.

From the equation (17), the output intensity in response of the strain can be expressed in $T = [1-\cos(\theta_0+C_0\varepsilon)]/2$. When laser wavelengths are 1530 nm and 1545 nm, the theoretical relationships between the output intensity and the applied strain are $T_{1530} = [1-\cos(277+203\varepsilon)]/2$ and $T_{1545} = [1-\cos(274+201\varepsilon)]/2$, respectively. The coefficients $\theta_0$ and $C_0$ of the above equations are different since the wavelength is different. In the above theoretical equations, dispersion effect on the $B$ is ignored. The experimental and theoretical results are identical, and the fitting degrees between them are obtained highly as $R^2 = 0.997$ at the wavelength 1530 nm and $R^2 = 0.994$ at the wavelength 1545 nm, respectively. Furthermore, the strain sensitivity is various with the applied strain. When the applied strain is 3000 με, the strain sensitivity is 2.7 dB/1000με at 1530 nm and -3.2dB/1000με at 1545 nm. When an optical power meter with an intensity resolution of 0.01 dB is used, a strain resolution of 3.7 με at 1530 nm and 3.1 με at 1545 nm is achieved, which is about 2.5 times higher than that of the strain sensor based on the resonant wavelength measurement.

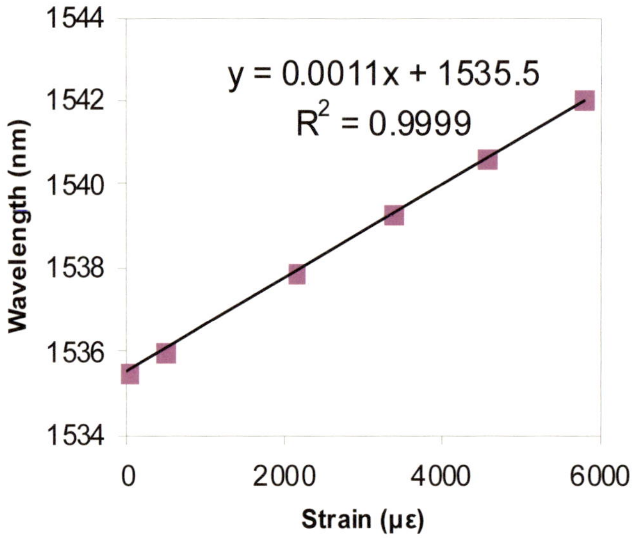


Fig. 29. The experimental relationship between the of the FLM at different wavelengths. Lines: theoretical curves. Pointes: experimental data.

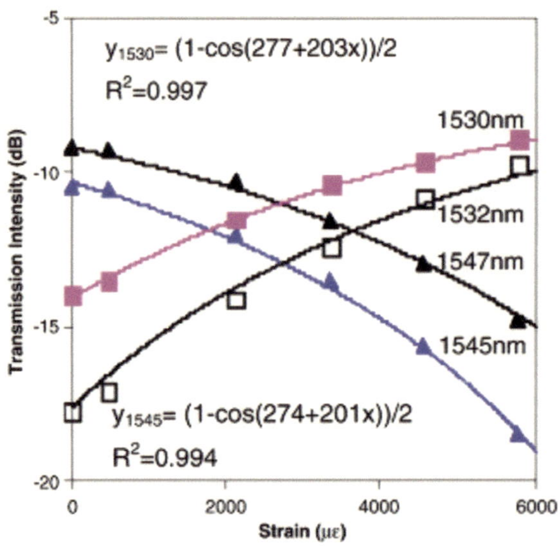


Fig. 30. Strain dependence of the transmission intensity wavelength of the transmission peak near 1535.6 nm of the FLM and the strain applied on the HiBi-PCF.

## 6. Multiplexing of HiBi-PCF based Sagnac interferometric sensors [35]

In this part, three multiplexing schemes for PM-PCF based Sagnac interferometric sensors are presented. The first scheme is to multiplex sensors in the wavelength domain using coarse wavelength division multiplexers (CWDMs). The sensing signal from each sensor can be measured within a specific wavelength channel of the CWDM. The second scheme is to multiplex sensors by connecting them in series along a single fiber. It is simple in terms of system architecture as no additional fiber-optic components are needed. The third scheme is to multiplex sensors in parallel by using fiber-optic couplers. The sensing information of the first multiplexing technique can be obtained by direct measurement such as with an optical spectrum analyzer. For the serial and parallel multiplexing, signal processing methods are required to demultiplex the complex sensing signal. Two mathematical transformations, namely the discrete wavelet transform (DWT) and the Fourier transform (FT), are used independently to convert the multiplexed sensing signal back to their constituent sensor signals. These two transform methods are experimentally demonstrated via two multiplexed Sagnac interferometric sensors. Their operating principles, experimental setup, and overall performance are discussed. In the part of 4.2, we have demonstrated the utilization of PM-PCF based Sagnac interferometers for pressure sensing [29]. Similar pressure sensing experiments were performed here for the purposes of demonstration and verification of the multiplexing schemes as well as the demultiplexing methods.

### 6.1 Multiplexing technique base on CWDM

Wavelength division multiplexing is a direct multiplexing technique that can be readily implemented into Sagnac interferometric sensors. Since the output interference spectra of all the sensors cover the whole bandwidth of the light source, individual sensor signals can be physically separated by CWDMs into different wavelength channels. The experimental setup of two multiplexed sensors using CWDMs is illustrated in Fig.31. It includes a

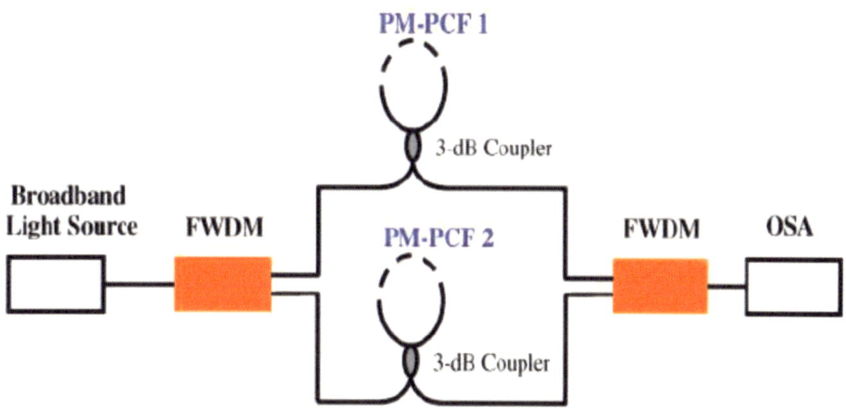


Fig. 31. Experimental setup of FWDM multiplexing technique for two PM-PCF based Sagnac interferometric sensors.

broadband light source, an OSA, two identical filter wavelength division multiplexers (FWDMs) with the two output ports having respective operation range in the C and L bands (1500~1562 nm/1570~1640 nm). The two Sagnac interferometric sensors, PM-PCF1 and PM-PCF2, have effective PM-PCF lengths of 40 cm and 80 cm, respectively. After the broadband light was launched into the first FWDM, the light was split into C and L bands. These two bands of light then illuminated the two sensors separately and were recombined by the second FWDM.

Figure 32 shows the output spectrum of the two Sagnac interferometric sensors multiplexed by FWDM. From the figure, sensors PM-PCF 1 and PM-PCF 2 are found in the L band and C band, respectively. The FWDMs are shown to have good flatness in their operating wavelength range. There is an abrupt discontinuity at the edges of the two FWDMs at around 1562 nm–1570 nm, where such range should be excluded from measurements. By measuring the shifts of individual transmission minima (or maxima) of the two Sagnac interferometric sensors within their corresponding wavelength ranges, sensing information of both sensors can be obtained.

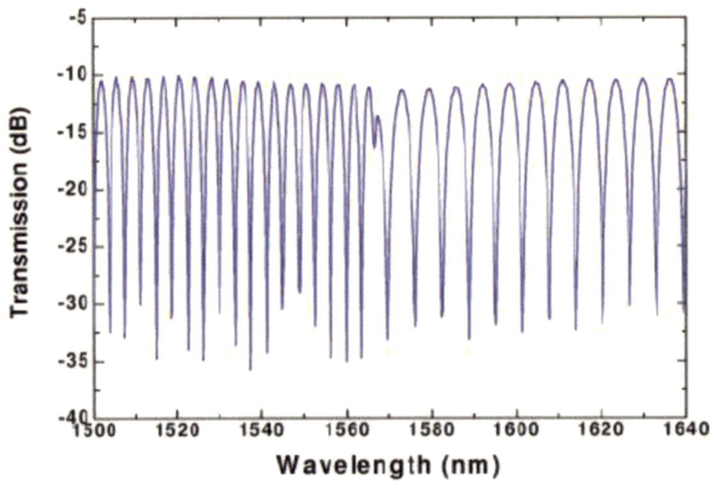


Fig. 32. Output Spectrum of the CWDM multiplexing technique for PM-PCF based Sagnac interferometric sensor.

### 6.2 Multiplexed in series along a single fiber with transmitted signals

The second multiplexing scheme is to multiplex Sagnac interferometric sensors in series along a single fiber. Similar concatenated sensor configuration has been employed previously in optical filtering [41], and in strain and temperature discrimination [42]. However, in both cases, multiplexing was not the main focus, and so the techniques of multiplexing were not studied. Figure 33 illustrates such a scheme by simply cascading the sensors together. For K Sagnac interferometric sensors multiplexed in series, the output spectrum is given by,

$$\frac{P_{output}}{P_{input}} = 10Log\prod_{k=1}^{k}\{\frac{1}{2}L_k[1-\cos(\frac{2\pi}{S_k}\lambda+\theta_k)]\}[dB] \quad (19)$$

where $L_k$, $S_k$, $\theta_k$ are the loss, the period of the output spectrum and the initial phase of the k-th sensor, respectively. Note that the output spectrum is the multiplication of all individual sensor signals.

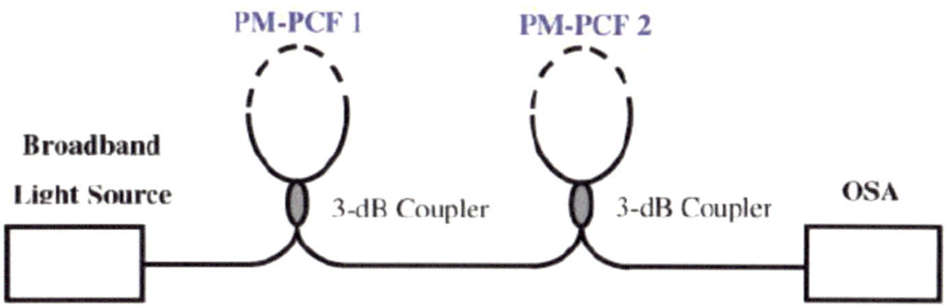


Fig. 33. Experimental setup of in series multiplexing technique for PM-PCF based Sagnac interferometric sensor.

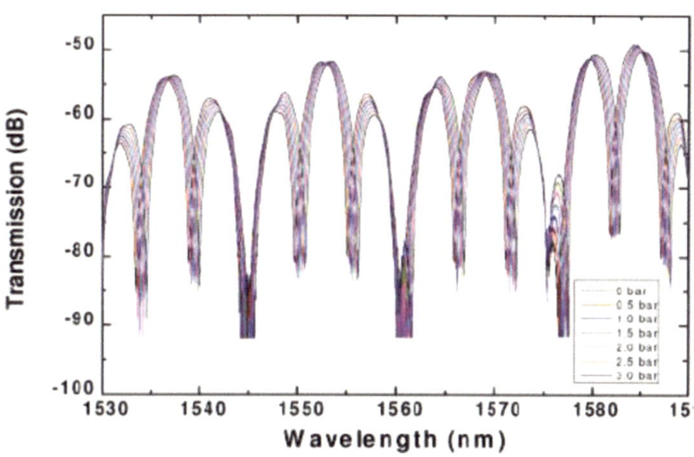


Fig. 34. Output transmission spectra of the two multiplexed Sagnac interferometric sensors in series with one sensor under applied pressure variations.

In the experimental demonstration, two sensors were spliced together adjacent to each other in series. The effective lengths of PM-PCF1 and PM-PCF2 were 20 cm and 60 cm, respectively. PM-PCF1 was placed freely on a table, while PM-PCF2 was placed inside a sealed pressure chamber. Pressure was applied to PM-PCF2 from 0–3 bars in steps of 0.5 bar, and was measured by a pressure gauge (COMARK C9557). Figure 34 shows the output spectra of various pressure values measured by the OSA. In principle, to obtain the sensing information, the wavelength shift of the transmission minima of each sensor needs to be determined. However, as can be seen, the multiplexed sensor signal is more complex, and so simply tracing the initial phase may not yield accurate results. Thus, in order to separate the multiplexed signals, the DWT and FT methods were used independently to demultiplex the sensing signals. They worked by transforming the signals into another domain, such that each individual sensor signal can be easily identified, and their phase shifts measured.

### 6.3 DWT demultiplexing method

The principle of the DWT demultiplexing method has been outlined in Ref. [43].When DWT is applied to a signal, it is decomposed and halved into high and low frequency components, represented as detail and approximation coefficients, respectively. This is similar to applying both a high-pass and a low-pass filter simultaneously to a signal. Then, the approximation coefficients (i.e., low frequency components) of the signal can be further decomposed into 2nd-level detail and approximation coefficients. This iterative process continues until all individual sensor signals are separated and appear on different wavelet levels. In other words, it continues until the spatial frequency of the sensing signals matches with the frequency range at which the wavelet level represents. Figure 35 shows the extracted detail coefficients of the two sensors at different wavelet levels. By tracking their phase shifts, the response of the two sensors under various pressure levels can be detected. Figure 36(a) shows the phase shifts of the two sensors as a function of applied pressure. It is clear that PM-PCF2 shifted linearly with applied pressure, while PM-PCF1 remained about the zero shift position. The crosstalk between the two multiplexed sensor signals was also measured. The crosstalk given here is the ratio of the phase shift of PM-PCF1 (no pressure applied) to that of PM-PCF2 (pressured applied), and is shown in Fig. 36(b). It should be noted that, the crosstalk measurement represented here includes other sources of errors, such as measurement error and ambient noise. As can be seen from the figure, the crosstalk between the two sensors is less than 5% and decreases progressively at higher pressure values. This means the absolute crosstalk values are quite stable for the measured pressure range, and implies that the errors are mainly due to sources other than the actual crosstalk between the two sensors. On the other hand, if the crosstalk measurement shows a trend that correlates with the applied pressure, this would mean there is actual crosstalk present in the multiplexed sensor system.

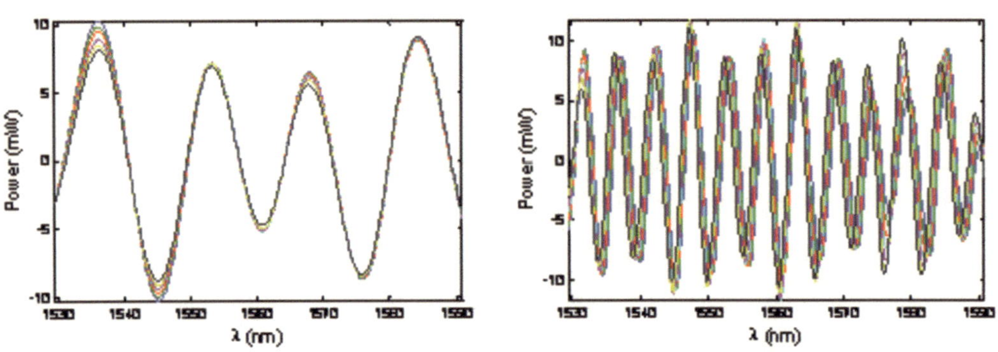


Fig. 35. Sensing signals of the two Sagnac interferometric sensors extracted using the wavelet method.

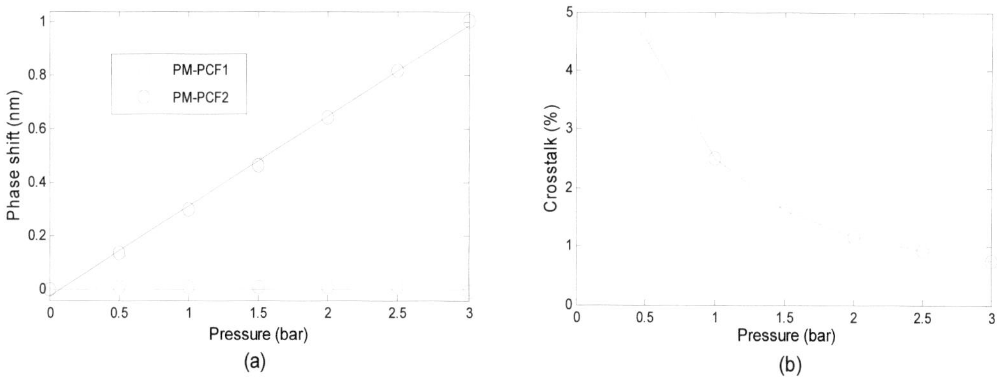


Fig. 36. (a) The wavelength shift as a function of pressure variation for the two Sagnac interferometric sensors,(b) sensing signal crosstalk of the two Sagnac interferometric sensors.

### 6.4 FT demultiplexing method

Besides the DWT, we also employed the FT method and the operating principle can be found in Ref. [44].The FT method works by transforming the multiplexed sensing signal from the original (wavelength) domain, into its dual (spatial frequency) domain, and is represented in the FT magnitude and phase spectra. Since the multiplexed signal is periodic, each individual sensor appeared as an finite amplitude peak in the FT magnitude spectrum; residing at a position dependent on the spatial frequency of the original sinusoidal signals. Thus, provided no two sensors have the same spatial frequency, each sensor can be distinctly identified. Normally, there are two ways of tracing the measurand-induced changes of individual sensors: (i) if the spatial-frequencies of the sensors change, measurands can be detected by the amount the amplitude peaks shift in the magnitude spectrum; and (ii) if the phase of the sensors change (and not the spatial-frequencies), measurands can be detected by the change of slope of the phase spectrum over the region corresponding to the amplitude peaks of the sensors in the magnitude spectrum. For the PM-PCF Sagnac interferometric sensors, when pressure was applied, the phase of the signals shifted proportionally while the spatial frequencies have no noticeable change, and so the second method applies. Figure 37 gives the FT magnitude and phase spectra of the multiplexed sensing signals after taking the FT. The corresponding regions of phase for the two sensors are shown in Fig.38. From the figure, one can see that PM-PCF1 is held constant (no noticeable change in the phase slope), while PM-PCF2 is under a varying amount of applied pressure which resulted in a gradual change of the phase slope. The calculated equivalent wavelength shift and crosstalk between the two sensors are shown in Figs. 39(a) and 39(b), respectively. From the figure, the maximum crosstalk is~5%, which is considered small.

### 6.5 Multiplexed in parallel by using coupler with reflected signals

The third multiplexing scheme is to multiplex Sagnac interferometric sensors in parallel, and is illustrated in Fig.40.The effective lengths of PM-PCF1 and PM-PCF2 are 20 cm and 60 cm, respectively. The source light is split equally by the 3-dB coupler into two paths to illuminate the two sensors separately. The sensing signals reflected back from the two

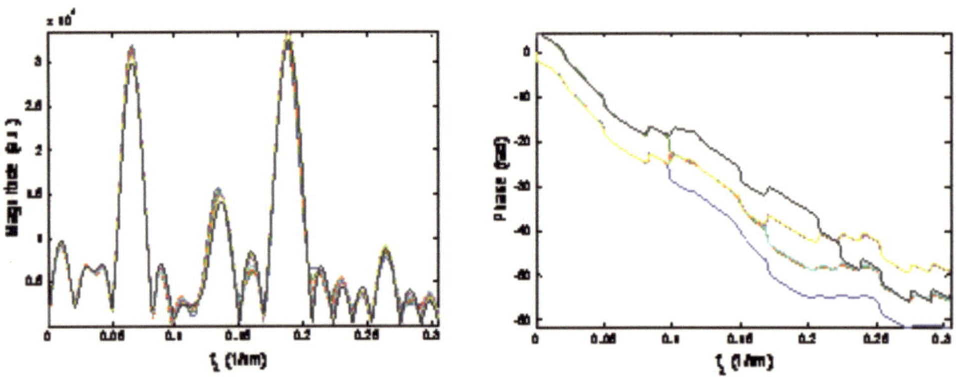

Fig. 37. Magnitude spectra and phase spectra of the sensing signal under Fourier transformation.

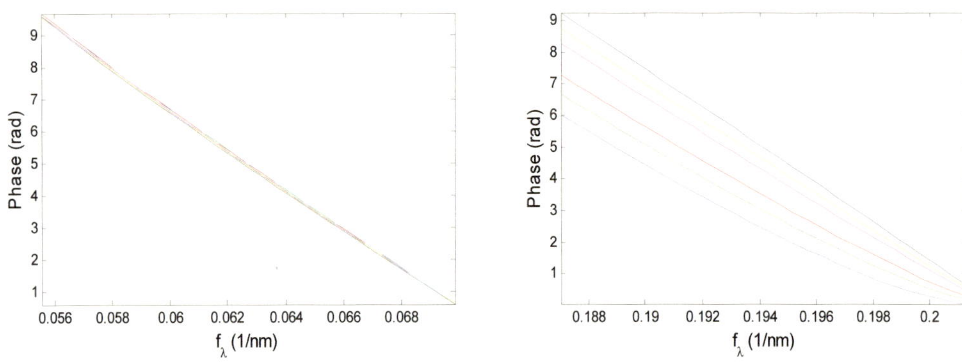


Fig. 38. Phase shift of the sensing signal from the two Sagnac interferometric sensors.

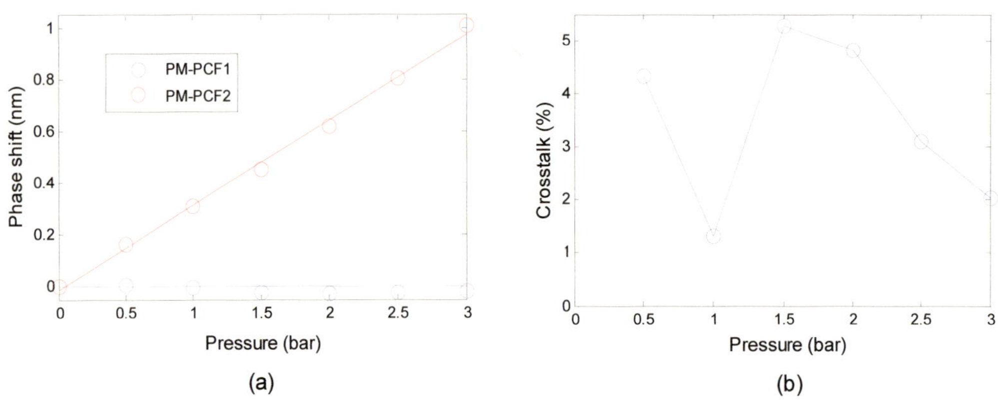


Fig. 39. (a) The wavelength shifts as a function of pressure variation for the two Sagnac interferometric sensors, (b) sensing signal crosstalk of the two Sagnac interferometric sensors.

sensors are then coupled together by the same 3-dB coupler, and were measured with an OSA. The unused ends of the sensors were coiled in small loops to minimize Fresnel reflections. As compared to the serial multiplexing scheme, it required an additional 3-dB coupler. Note that the reflected sensing signals were taken instead of the transmitted signals, and there were two reasons for it. First, it helped to use one less 3-dB coupler to combine individual sensor signals at the output side and so reduced the system cost and complexity. Second, the reflected signal spectrum is, mathematically, the complement of the transmitted spectrum; and since the spectrum is of the form of sinusoidal pattern, the only difference is the phase angle of π. For K Sagnac interferometric sensors multiplexed in parallel, the output spectrum is given by,

$$\frac{P_{output}}{P_{input}} = 10Log_{10}\sum_{k=1}^{k}\{\frac{1}{2}L_k R_k[1+\cos(\frac{2\pi}{S_k}+\theta_k)]\}[dB] \qquad (20)$$

where $R_k$, $L_k$, $S_k$, $\theta_k$ are the coupling ratio, the loss, the period of the output spectrum and the initial phase of the k-th sensor, respectively. Note that the output spectrum is the arithmetic sum of all individual sensor signals, as opposed to multiplication in the serial multiplexing case.

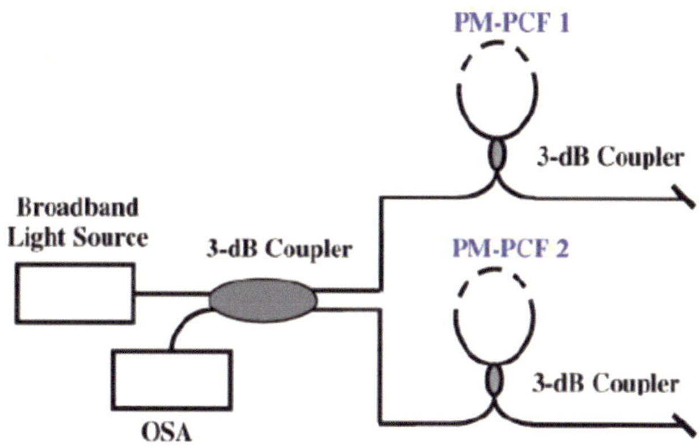


Fig. 40. Experimental setup of in parallel multiplexing technique for PM-PCF based Sagnac interferometric sensors.

As an experimental demonstration, a similar pressure sensing experiment to the previous multiplexing scheme was performed. Figure 41 shows the output spectra, with PM-PCF1 placed freely on the table and PM-PCF2 placed inside the pressure chamber. Again, we employed both the DWT and FT methods independently to demultiplex the sensing signal.

### 6.6 DWT demultiplexing method

After taking the DWT of the multiplexed sensing signal, Fig.42 shows the detail coefficients of the two sensors at different wavelet levels. It is apparent from the figure that PM-PCF 1 remained almost constant, while PM-PCF2 can visibly be seen to have had the whole signal shifted. The phase shifts of the two sensors and the corresponding crosstalk measurement are shown in Figs. 43(a) and 43(b), respectively. The crosstalk between the two sensing signals is indeed very small, with a maximum value of less than 2%.

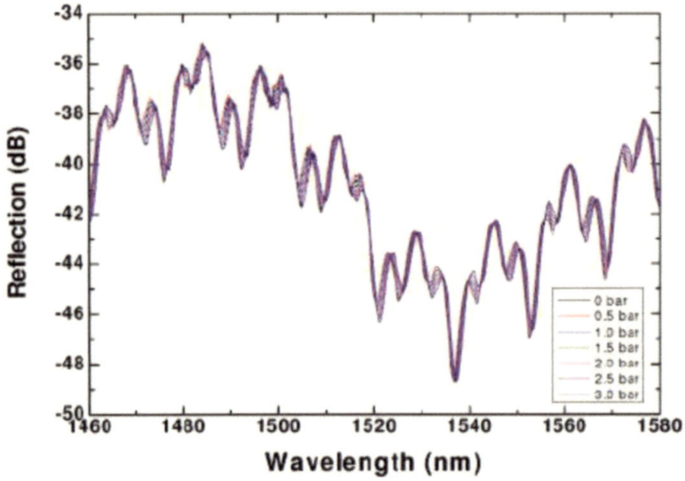


Fig. 41. Output transmission spectra of the two multiplexed Sagnac interferometric sensors in parallel with one sensor under applied pressure variations.

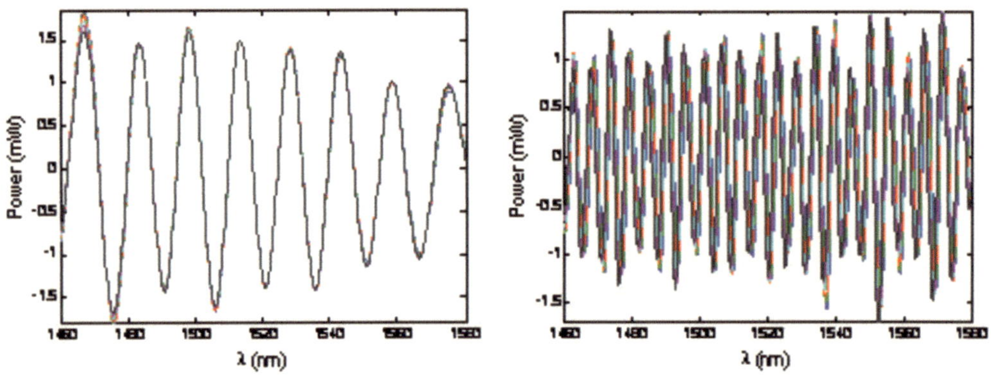


Fig. 42. Sensing signals of the two Sagnac interferometric sensors extracted using the wavelet method.

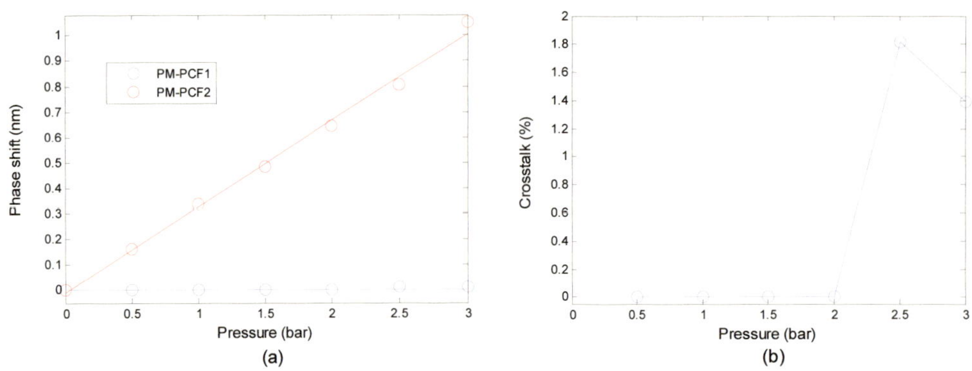


Fig. 43. (a) The wavelength shifts as a function of pressure variation for the two Sagnac interferometric sensors, (b) sensing signal crosstalk of the two Sagnac interferometric sensors.

### 6.7 FT demultiplexing method

With the FT method applied, Fig. 44 gives the FT magnitude and phase spectra of the multiplexed sensing signals. The corresponding regions of phase for the two sensors are illustrated in Fig.45. From the figure, one can notice that PM-PCF1 has no noticeable change in the phase slope, while PM-PCF2 experienced pressure changes which resulted in a gradual change in the phase slope. The calculated equivalent wavelength shifts and the corresponding crosstalk measurement are shown in Figs. 46(a) and 46(b), respectively. Again, the crosstalk is very small, with a maximum of less than 3%.

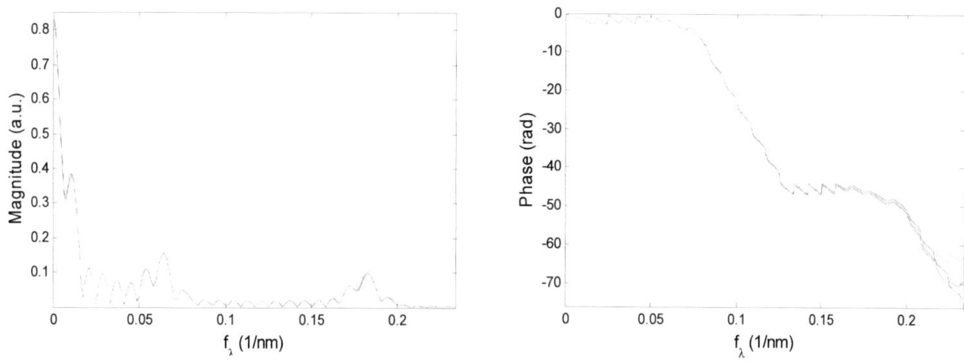


Fig. 44. Magnitude spectrum and phase spectrum of the sensing signal under Fourier transformation.

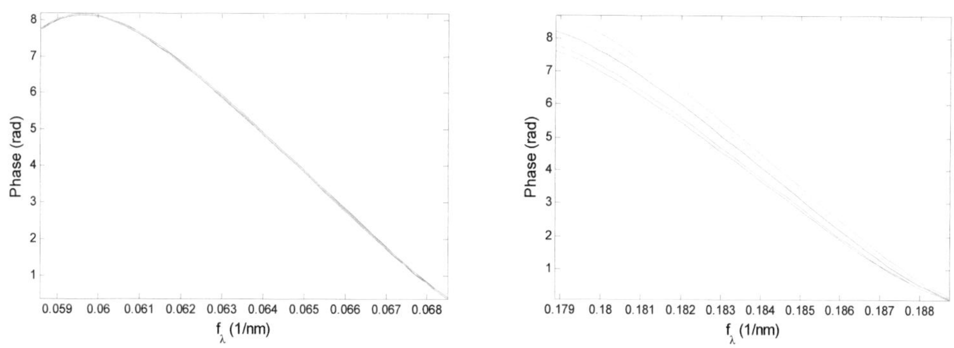


Fig. 45. Phase shift of the sensing signal from the two Sagnac interferometric sensors.

### 6.8 Discussions

Each of the three multiplexing schemes has its own characteristics and is suitable for different applications. The CWDM scheme enables easy real-time system implementation. It provides a direct measurement without the need for dealing with crosstalk between signals from different channels. The number of sensors that can be multiplexed is limited by the available channels of the CWDM at a fixed light source bandwidth. Although with more channels, more sensors can be multiplexed; the bandwidth of each channel becomes narrower. In principle, the minimum bandwidth of each channel has to be larger than the period of the sensor signal, plus a bit of guard band between channel edges to avoid erroneous results due to signal discontinuities.

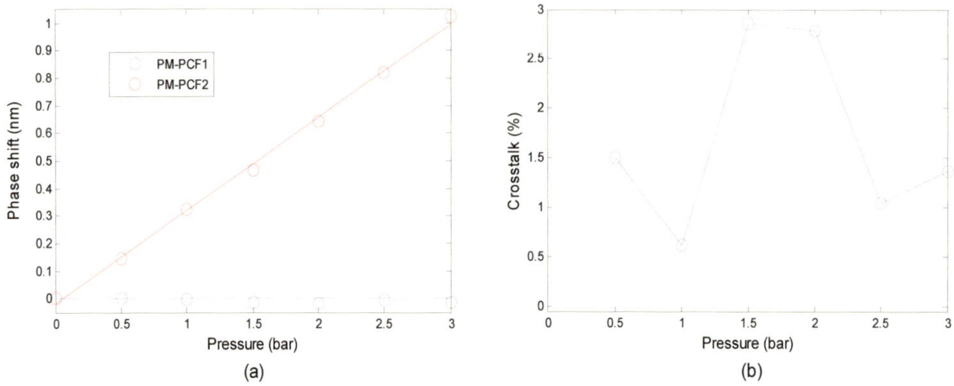


Fig. 46. (a) The wavelength shifts as a function of pressure variation for the two Sagnac interferometric sensors, (b) sensing signal crosstalk of the two Sagnac interferometric sensors.

For the serial multiplexing scheme, no additional fiber-optic components are needed. The sensors are multiplexed easily by connecting them together one by one, which makes this scheme the simplest in terms of sensor system architecture. The number of sensors that can be multiplexed is mainly limited by the splicing loss between PM-PCFs and SMFs. On the other hand, for the parallel multiplexing scheme, it requires the addition of fiber couplers, which makes the system architecture relatively more complex and increases the total system cost. In addition, it increases the insertion loss due to splicing and fiber couplers. Nevertheless, the errors and adverse effects are also less because individual sensor signals are added rather than multiplied, and so they do not suffer from spectral shadowing and nonlinear mapping as is found in the serial multiplexing scheme [42]. It is evident from our experiments that parallel multiplexing has less crosstalk (with other sources of errors included) than that of serial multiplexing. It should be pointed out that the measurement errors due to fluctuations in the applied pressure played a role in our results, which can be noticed in their deviation from ideal values. This implies the intrinsic crosstalk is believed to be quite low.

There is a consideration when using the DWT and FT methods to demultiplex the sensor signals obtained from the serial and parallel multiplexing schemes. The effective length of PM-PCFs must be properly chosen not to be too close to each other in order to avoid overlap after performing the transformations. However, it is not an issue for the CWDM scheme because signals from sensors are well distinguished by each channel. These three multiplexing schemes can be implemented together to further increase the number of sensors.

For example, within each channel in the CWDM, sensors can be multiplexed in series or in parallel. This combined configuration cannot only increase the number of sensors by several times, but also maximizes the full use of the light source bandwidth. To sum up, from practicability point of view, the CWDM scheme is among the easiest and simplest, whereas serial multiplexing is more practical in real applications. On the other hand, parallel multiplexing offers slightly better performance in terms of crosstalk and measurement errors. At present stage, the main limitations on the last two multiplexing schemes are the insertion loss. The presented multiplexing schemes, together with the two demultiplexing methods, are not only limited to use for PM-PCF Sagnac interferometric sensors. Indeed, they can be applied in any PCF sensor that has sinusoidal patterns. This will be one step closer towards a more practical sensing system using PCF based sensors.

## 7. Summary

In this chapter, we have introduced and demonstrated the basic operation principle of FLMs, and their applications in optical devices and in optical sensors, which include:

i. Temperature-insensitive interferometer based on HiBi-PCF FLM. The temperature-insensitivity of the FLM is improved 55 times by using the HiBi-PCF, mainly because the temperature coefficient of birefringence in PCF is measured to be 30 times lower than that of conventional PMF;
ii. Temperature-insensitive strain sensor based on HiBi-PCF FLM. Strain measurement with a sensitivity of 0.23 pm/$\mu\varepsilon$ is achieved, and the measurement range, by stretching the PM-PCF only, is up to 32 m$\varepsilon$. The strain measurement is inherently temperature insensitive due to the great thermal stability of PM-PCF based Sagnac interferometers. That improves the accuracy of strain measurement and eliminates the requirement for temperature compensation;
iii. Pressure sensor realized with HiBi-PCF based Sagnac interferometer. The Sagnac loop itself acts as a sensitive pressure sensing element, making it an ideal candidate for pressure sensor. Pressure measurement results show a sensing sensitivity of 3.42 nm/MPa, which is achieved by using a 58.4 cm PM-PCF-based Sagnac interferometer. Important features of the pressure sensor are the low thermal coefficient and the exceptionally low bending loss of the PM-PCF, which permits the fiber to be coiled into a 5mm diameter circle. This allows the realization of a very small pressure sensor;
iv. Compact and highly sensitive temperature sensor based on an alcohol-filled HiBi-PCF FLM. Due to the high temperature sensitivity of the filled alcohol, an alcohol-filled HiBi-PCF FLM with an extremely high sensitivity on temperature are presented Experimental results show that the sensitivity is as high as 6.6 nm/°C, which is 660 and 7 times higher than that of a FBG and that of the FLM made of a conventional HiBi fiber;
v. Demodulation of sensors based on HiBi-PCF FLM. The sensor demodulation is based on the intensity measurement, in which a distributed-feedback (DFB) laser is used as the light source. Since the output intensity of the FLM is directly proportional to the applied strain, only an optical power meter is sufficient to detect strain variation, avoiding the need for an expensive OSA;
vi. Multiplexing of HiBi-PCF based Sagnac interferometric sensors. Three multiplexing schemes are presented for HiBi-PCF based Sagnac interferometric sensors. The first technique is wavelength division multiplexing using coarse wavelength division multiplexers (CWDMs) to distinguish signals from each multiplexed sensor in different wavelength channels. The other two schemes are to multiplex sensors in series along a single fiber link and in parallel by using fiber-optic couplers. While for the CWDM scheme, the multiplexed sensing signal can be obtained by direct measurement; for the other two multiplexing techniques, the sensing signal is more complex and cannot be easily demultiplexed. Thus, some signal processing methods are required. In this regard, two mathematical transformations, namely the discrete wavelet transform and Fourier transform, have been independently and successfully implemented into these two schemes. The operating principles, experimental setup, and overall performance are discussed.

## 8. References

[1] V. Vali and R. W. Shorthill, "Fiber ring interferometer," Appl. Opt. vol.15, pp.1099–1103 (1976).
[2] D. B. Mortimore, "Fiber loop reflectors", J. Lightwave Technol., vol. 6, pp. 1217-1224, (1988).
[3] X. Fang and R. O. Claus, "Polarization-independent all-fiber wavelength-division multiplexer based on a Sagnac interferometer", Opt. Lett., vol. 20, pp. 2146-2148, (1995).
[4] S. Li, K. S. Chiang, and W. A. "Gambling, Gain flattening of an erbium-doped fiber amplifier using a high-birefringence fiber loop mirror", IEEE Photon. Technol. Lett., vol. 13, pp. 942-944, (2001).
[5] L. Yuan, W. Jin, L. Zhou, Y. L. Hoo, M. S. Demokan, "Enhancement of multiplexing capability of low-coherence interferometric fiber sensor array by use of a loop topology", J. Lightwave Technol., Vol. 21, pp. 1313-1319, (2003).
[6] S. Yang, Z. Li, X. Dong, G. Kai, and Q. Zhao, "Generation of wavelength-switched optical pulse from a fiber ring laser with an F-P semiconductor modulator and a HiBi fiber loop mirror," IEEE Photon.Technol. Lett., vol. 14, pp. 774–776, (2002).
[7] N. Starodumov, L. A. Zenteno, D. Monzon, and E. De La Rose, "Fiber Sagnac interferometer temperature sensor," Appl. Phys. Lett. Vol. 70, pp. 19–21, (1997).
[8] S. Knudsen, A. B. Tveten, and A. Dandridge, "Measurements of fundamental thermal induced phase fluctuations in the fiber of a sagnac interferometer," IEEE Photon. Technol. Lett., vol. 7, pp. 90–92, (1995).
[9] E. De La Rose, L. A. Zenteno, A. N. Starodumov, and D. Monzon, "All-fiber absolute temperature sensor using an unbalanced high-birefringence Sagnac loop," Opt. Lett. vol. 22, pp. 481–483, (1997).
[10] Y. Liu, B. Liu, X. Feng, W. Zhang, G. Zhou, S. Yuan, G. Kai, and X. Dong, "High-birefringence fiber loop mirrors and their applications as sensors, " Appl. Opt., vol. 44, pp. 2382-2390, (2005).
[11] B. Dong, Q. Zhao, J. Lv, T. Guo, L. Xue, S. Li, and H. Gu, "Liquid-level sensor with a high-birefringence-fiber loop mirror, " Appl. Opt. , vol. 45, pp. 7767-7771, (2006).
[12] O. Frazão, B.V. Marquesa, P. Jorge, J.M. Baptista, J.L. Santos, "High birefringence D-type fibre loop mirror used as refractometer," Sensors and Actuators B: Chemical, vol. 135, pp. 108-111, (2008).
[13] O. Frazão, J.L. Santos, J.M. Baptista, "Simultaneous measurement for strain and temperature based on a long-period grating combined with a high-birefringence fiber loop mirror," IEEE Photon. Technol. Lett. vol. 18, pp. 2407-2409 (2006).
[14] G. Sun, D.S. Moon, Y. Chung, "Simultaneous temperature and strain measurement using two types of high-birefringence fibers in Sagnac loop mirror," IEEE Photon. Technol. Lett., vol. 19, pp. 2027-2029, (2007).
[15] O. Frazão, S. O. Silva, J. M. Baptista, J. L. Santos, G. Statkiewicz-Barabach, W. Urbanczyk, and J. Wojcik, "Simultaneous measurement of multiparameters using a Sagnac interferometer with polarization maintaining side-hole fiber" Appl. Opt. vol. 47, pp. 4841-4847, (2008).
[16] T. A. Birks, J. C. Knight, and P. S. J. Russell, "Endlessly single-mode photonic crystal fiber," Opt. Lett., vol. 22, pp. 961–963, (1997).
[17] P. St. J. Russell, "Photonic crystal fibers," Science, vol. 299, pp. 358–362, (2003).

[18] T. P. Hansen, J. Broeng, Stig E. B. Libori, E. Knudsen, A. Bjarklev, J. R. Jensen, and H. Simonsen, "highly birefringent index-guiding photonic crystal fibers", IEEE Photon. Technol. Lett., vol. 13, pp. 588-590, (2001).
[19] Ortigosa-Blanch, J. C. Knight, W. J. Wadsworth, J. arriaga, B. J. Mangan, T. A. Birks, and P. St. J. Russell, "Highly birefringent photonic crystal fibers", Optics Lett., vol. 25, pp. 1325-1327, (2000).
[20] K. Suzuki, H. Kubota, S. Kawanishi, H. Kubota, S. Kawanishi, "Optical properties of a low-loss polarization-maintaining photonic crystal fiber", Optics Express, vol. 9, pp. 676-680, (2001).
[21] C.-L. Zhao, X. Yang, C. Lu, W. Jin, M. S. Demokan, "Temperature- insensitive interferometer using a highly birefringent photonic crystal fiber loop mirror," IEEE Photon. Technol. Lett., vol. 16, pp. 2535-2537, (2004).
[22] D.-H. Kim, and J. U. Kang, "Sagnac loop interferometer based on polarization maintaining photonic crystal fiber with reduced temperature sensitivity," Opt. Express, vol. 12, pp. 4490–4495, (2004).
[23] Xiufeng Yang , Chun-Liu Zhao, Qizhen Peng, Xiaoqun Zhou, Chao Lu, "FBG sensor interrogation with high temperature insensitivity by using a HiBi-PCF Sagnac loop filter", Optics Communications, vol. 250， pp. 63-68 , (2005).
[24] X. Dong, H.Y. Tam, and P. Shum, "Temperature-insensitive strain sensor with polarization-maintaining photonic crystal fiber based Sagnac interferometer", Appl. Physics Lett., vol. 90, pp. 151113, (2007).
[25] O. Frazão, J. M. Baptista, and J. L. Santos, "Temperature-independent strain sensor based on a Hi-Bi photonic crystal fiber loop mirror," IEEE Sens. J. vol. 7, pp. 1453–1455, (2007).
[26] O. Frazão, J. M. Baptista, J. L. Santos, and P. Roy, "Curvature sensor using a highly birefringent photonic crystal fiber with two asymmetric hole regions in a Sagnac interferometer," Appl. Opt. vol. 47, pp. 2520–2523, (2008).
[27] G. Kim, T. Cho, K. Hwang, K. Lee, K. S. Lee, Y.-G. Han, and S. B. Lee, "Strain and temperature sensitivities of an elliptical hollow-core photonic bandgap fiber based on Sagnac interferometer," Opt. Express, vol. 17, pp. 2481–2486, (2009).
[28] H. M. Kim, T. H. Kim, B. Kim, and Y. Chung, "Temperature-insensitive torsion sensor with enhanced sensitivity by use of a highly birefringent photonic crystal fiber," IEEE Photonics Technology Letters, vol. 22, pp.1539-1541, (2010).
[29] H.Y. Fu, H.Y. Tam, L.-Y. Shao, X. Dong, P.K.A. Wai, C. Lu, and Sunil K. Khijwania, "Pressure sensor realized with polarization-maintaining photonic crystal fiber-based Sagnac interferometer", Appl. Opt., vol. 47, pp. 2835-1839, (2008).
[30] H. Gong, C. C. Chan, L. Chen, and X. Dong, "Strain sensor realized by using low-birefringence photonic-crystal-fiber-based Sagnac loop", IEEE Photon. Technol. Lett., vol. 22, pp. 1238-1240, (2010).
[31] H. Gong, C. C. Chan, P. Zu, L. Chen, and X. Dong, "Curvature measurement by using low-birefringence photonic crystal fiber based Sagnac loop", Opt. commun., vol. 283, pp. 3142-3144, (2010).
[32] P. Zu, C. C. Chan, Yongxing Jin, Tianxun Gong, Yifan Zhang, Li Han Chen and Xinyong Dong "A temperature-insensitive twist sensor by using low-birefringence photonic crystal fiber based sagnac interferometer", IEEE Photonics Technology Letters, vol. 25, pp. 1041-1135, (2011).

[33] W. Qian, C.-L. Zhao, S. He, X. Dong, S. Zhang, Z. Zhang, S. Jin, J. Guo and H. Wei, "High-sensitivity temperature sensor based on an alcohol-filled photonic crystal fiber loop mirror", Opt. Lett., vol. 36, pp. 1548-1550, (2011).
[34] W. Qian, C.-L. Zhao, X. Dong, and W. Jin, "Temperature independent strain sensor based on intensity measurement using a highly birefringent photonic crystal fiber loop mirror," Opt. commun., vol. 283, pp. 5250-5254, (2010).
[35] H. Y. Fu, A. C. L. Wong, P. A. Childs, H. Y. Tam, Y. B. Liao, C. Lu, and P. K. A. Wai, "Multiplexing of polarization-maintaining photonic crystal fiber based Sagnac interferometric sensors", Optics Express, vol. 17, pp. 18501-18512, (2009).
[36] Bertholds and R. Dandliker, "Determination of the individual strain-optic coefficients in single-mode optical fibres", J. Lightwave Technol. vol. 6, pp. 17-20, (1988).
[37] D. Kersey, M. A. Davis, H. J. Patrick, M. LeBlanc, K. P. Koo, C. G. Askins, M. A. Putnam, and E. J. Friebele, "Fiber grating sensors", J. Lightwave Technol. vol. 15, pp. 1442-1463, (1997).
[38] Y. Zhang, D. Feng, Z. Liu, Z. Guo, X. Dong, K. S. Chiang, and B.C. B. Chu, "High-sensitivity pressure sensor using a shielded polymer-coated fiber Bragg grating," IEEE Photon. Technol. Lett. vol. 13, pp. 618–619 (2001).
[39] H. K. Gahir and D. Khanna, "Design and development of a temperature-compensated fiber optic polarimetric pressure sensor based on photonic crystal fiber at 1550 nm," Appl. Opt. vol. 46, pp. 1184–1189 (2007).
[40] Y. Yu, X. Li, X. Hong, Y. Deng, K. Song, Y. Geng, H. Wei, and W. Tong, "Some features of the photonic crystal fiber temperature sensor with liquid ethanol filling", Opt. Express, vol. 18, pp. 15383-15388, (2010).
[41] L. Liu, Q. Zhao, G. Zhou, H. Zhang, S. Chen, L. Zhao, Y. Yao, P. Guo, and X. Dong, "Study on an optical filter constituted by concatenated Hi-Bi fiber loop mirrors," Microw. Opt. Technol. Lett. vol. 43, pp. 23–26, (2004).
[42] O. Frazão, J. L. Santos, and J. M. Baptista, "Strain and temperature discrimination using concatenated high-birefringence fiber loop mirrors," IEEE Photon. Technol. Lett. vol. 19, pp. 1260–1262, (2007).
[43] C. L. Wong, P. A. Childs, and G. D. Peng, "Multiplexed fibre Fizeau interferometer and fibre Bragg grating sensor system for simultaneous measurement of quasi-static strain and temperature using discrete wavelet transform," Meas. Sci. Technol. vol. 17, pp. 384–392, (2006).
[44] P. A. Childs, "An FBG sensing system utilizing both WDM and a novel harmonic division scheme," J. Lightwave Technol. vol. 23, pp. 348–354, (2005), "Erratum" 23, 931 (2005).

# 11

# Photonic Crystal Fiber for Medical Applications

Feroza Begum and Yoshinori Namihira
*Graduate School of Engineering and Science, University of the Ryukyus, Okinawa, Japan*

## 1. Introduction

Optical coherence tomography (OCT) is a new technology for noninvasive cross-sectional imaging of tissue structure in biological system by directing a focused beam of light at the tissue to be image [Bouma et al., 1995; Jiang et al., 2005; Ryu et al., 2005]. The technique measures the optical pulse time delay and intensity of backscattered light using interferometry with broadband light sources or with frequency swept lasers. It is analogous to ultrasound imaging or radar, except that it uses light rather than sound or radio waves. In addition, unlike ultrasound, OCT does not require direct contact with the tissue being imaged. OCT depends on optical ranging; in other words, distances are measured by shining a beam of light onto the object, then recording the optical pulse time delay of light. Since the velocity of light is so high, it is not possible to directly measure the optical pulse time delay of reflections; therefore, a technique known as low-coherence interferometry compares reflected light from the biological tissue to that reflected from a reference path of known length. Different internal structures produce different time delays, and cross-sectional images of the structures can be generated by scanning the incident optical beam. Earlier OCT systems typically required many seconds or minutes to generate a single OCT image of tissue structure, raising the likelihood of suffering from motion artifacts and patient discomfort during *in vivo* imaging. To counter such problems, techniques have been developed for scanning the reference arm mirror at sufficiently high speeds to enable real-time OCT imaging [Tearnery et al., 1997]. OCT can be used where excisional biopsy would be hazardous or impossible, such as imaging the retina, coronary arteries or nervous tissue. OCT has had the largest impact in ophthalmology where it can be used to create cross-sectional images of retinal pathology with higher resolution than any other noninvasive imaging technique. Now a days OCT is a prospective technology which is used not only for ophthalmology but also for dermatology, dental as well as for the early detection of cancer in digestive organs. The wavelength range of the OCT light source is spread from the 0.8 to 1.6 μm band. This spectral region is of particular interest for OCT because it penetrates deeply into biological tissue and permits spectrally resolved imaging of water absorption bands. In this spectral region, attenuation is minimum due to absorption and scattering. It should be noted that scattering decreases at longer wavelengths in proportion to $1/\lambda^4$, indicating that the scattering magnitude at 0.8 ~ 1.6 μm wavelengths is lower than at the visible wavelengths [Agrawal, 1995]. Ultrahigh-resolution OCT imaging in the spectral region from 0.8 to 1.6 μm requires extremely broad bandwidths because longitudinal resolution depends on the coherence length. The coherence length is inversely proportional to the bandwidth and proportional to square of the light source center wavelength. This can be achieved by supercontinuum (SC) light using photonic crystal fibers. The ophthalmology and dermatology OCT imaging are done predominantly at near 0.8 μm center wavelength [Bouma et al., 1995; Drexler et al., 1999; Ohmi et al., 2004; Pan et al., 1998; Welzel et al., 1997]. The dentistry OCT imaging is performed at 1.3 μm wavelength [Boppart et al., 1998; Colston et al., 1998; Hartl et al., 2001; Herz et al., 2004]. Currently, it is reported that the OCT imaging at 1.5 ~ 1.6 μm broadband light source can be readily applied to take images of human tooth samples [Lee et al., 2009]. On the other hand, telecommunication window (around 1.55 μm) is the most attractive window in optical communication systems, dispersion compensation and nonlinear optics because of the minimum transmission loss of the fiber [Begum et al., 2007a, 2007b, 2009a].

Photonic crystal fibers (PCFs) [Russel, 2003], a pure silica core optical fibers with tiny air holes embedded in the host silica matrix running along the propagation axis, have boosted the fiber optic research due to their remarkable modal properties such as provide single-mode operation for very short operating wavelengths [Knight et al., 1996], remain single-mode for large scale fibers [Knight et al., 1998], achieve high birefringence [Kaijage et al., 2000], and controllable dispersion characteristics [Begum et al., 2009b] which cannot be achieved with conventional optical fibers. These fibers are also termed as microstructured fibers (MSFs) or microstructured optical fibers (MOFs). PCFs are dived into two categories according to the light confinement mechanisms: one is index-guiding or solid core fibers [Knight et al., 1996] and the other is photonic bandgap (PBG) or hollow core fibers [Couny et al., 2008]. Those with a solid core light can confine in a high-index core by modified total internal reflection which is same index guiding principle as conventional optical fibers. However, they can have a much higher effective-index contrast between core and cladding, and therefore can have much stronger confinement for applications in nonlinear optical devices, polarization maintaining fibers, etc. Alternatively, in PBG fibers where the light is confined in a lower index core by a photonic bandgap created by the microstructured cladding. The presence of air holes in the cladding gives rise to strong wavelength dependence of the cladding index which is primarily responsible for its magnificent characteristics. The extra degrees of freedom in PCFs facilitate a complete control on its properties such as ultraflattened dispersion and high negative dispersion. The precise control of geometrical parameters can provide ultraflattened dispersion in PCFs. PCFs are very attractive and efficient to produce high power light source in OCT system. Because PCFs can generate SC spectrum due to their design degree of freedom which make it possible to enhance the nonlinear effects by reducing effective area and tailor chromatic dispersion. As it is well known, the optical attenuation sources in PCFs include intrinsic losses due to Rayleigh scattering, imperfection losses due to the fabrication, and confinement losses caused by finite number of air holes in the cladding. Since the core has the same refractive index as the cladding, the guided mode is intrinsically leaky and experiences confinement losses. In fact, confinement losses occur even in the absence of the other two losses. By careful design, it is possible to reduce confinement losses to negligible values compared with the intrinsic losses. Control of chromatic dispersion keeping a low confinement loss to a level below the Rayleigh scattering limit is a very important for any optical system supporting ultrashort soliton pulse propagation [Agrawal, 1995]. In all cases, almost flattened fiber dispersion and low confinement loss behavior becomes a crucial issue.

Although the resolution power of the currently available OCT machines are remarkable, they are not sufficiently high to unequivocally identify all retinal sublayers and make 'biopsy'-like diagnoses. Resolution is limited mainly by the bandwidth of the light source, usually a superluminescent diode (SLD) [Colston et al., 1998; Ryu et al., 2005] and increased resolution will require wider bandwidth light sources. The emergence of ultrabroad bandwidth femtosecond laser technology has allowed the development of an ultra-high resolution OCT [Boppart et al., 1998; Bouma et al., 1995; Drexler et al., 1999; Hartl et al., 2001; Herz et al., 2004; Jiang et al., 2005; Lee et al., 2009; Ohmi et al., 2004; Pan et al., 1998; Tearnery et al., 1997; Welzel et al., 1997]. The ultrahigh resolution OCT will in effect be a microscope capable of revealing certain histopathological aspects of macular disease in the living tissue. Femtosecond laser source is expensive than picosecond laser source and low incident power. Consequently, currently researchers are paying attention to develop picosecond light sources for using ultrahigh-resolution OCT system. Picosecond pulse laser source gives more narrow spectra than femtosecond laser source but since the laser source is cheaper in this case it attracts practical implementation. The ultrahigh resolution OCT will in effect be a microscope capable of revealing certain histopathological aspects of macular disease in the living tissue.

In this work, we report a broadband SC generation in highly nonlinear photonic crystal fiber (HN-PCF) at center wavelength 0.8 μm, 1.3 μm and 1.55 μm using high power picosecond pulses which can be applicable in ultrahigh-resolution OCT system for ophthalmology, dermatology and dental imaging. The proposed HN-PCF is investigated through a full-vector finite difference method with anisotropic perfectly matched layer. Through numerical simulation, it is demonstrated that it is possible to achieve different properties of the proposed HN-PCF. Based on the nonlinear Schrödinger equation, we find that the proposed HN-PCF, having four rings and two different sizes of air holes, can achieve SC spectrum with input picosecond pulses. We have further investigated the full width of half maximum of the generated SC spectrum of HN-PCF that can gives significant information on the longitudinal resolution in biological tissue by assuming coherent length. The achieved longitudinal resolutions in tissue are 0.97 μm at 0.8 μm for ophthalmology and dermatology, 0.85 μm at 1.3 μm for dental imaging and 1.1 μm at 1.55 μm also for dental imaging. To our knowledge, these are the highest resolution achieved in biological tissue to date at 0.8 μm, 1.3 μm and 1.55 μm wavelength. Furthermore, numerical simulation result shown that it is possible to obtain ultra-flattened chromatic dispersion, low dispersion slope, high nonlinear coefficient and very low confinement loss, simultaneously from the proposed HN-PCF.

## 2. Proposed HN-PCF structure

Fig. 1 (a) shows the schematic cross section of the conventional PCF structure. This PCF consists of a triangular lattice of air holes where the core is defined by a missing air hole. The core diameter is 2a, where 'a' equals $\Lambda$-$d/2$. The air hole pitch is labeled $\Lambda$, and measures the period of the air hole structure (the distance between the centers of neighboring air holes). The air hole size is labeled $d$, and measures the diameter of the holes. The background material is regular silica with a cladding refractive index $n$ = 1.45. Fig. 1 (b) shows the proposed HN-PCF structure. It has a pitch $\Lambda$, two air holes with diameters $d_1$ and $d$. The pitch constant is chosen to be $\Lambda$ = 0.87 μm, while the diameter of the air holes in the cladding of the fiber are $d_1$ = 0.46 μm, $d$ = 0.80 μm, with a total number of 4 hole layers in the cladding. Designing HN-PCF for the OCT and telecommunication window using a conventional PCF structure is difficult: therefore, the dimensions of the first rings of the proposed HN-PCF are scaled downed to shape the dispersion characteristics. The dimensions of the other rings are retained sufficiently large for better field confinement.

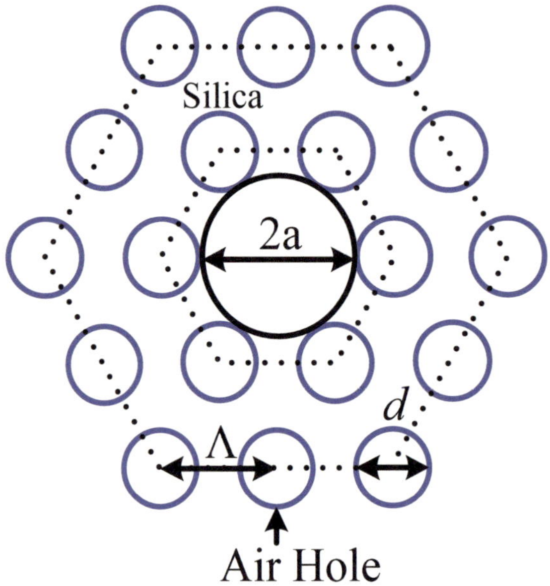


Fig. 1(a). Schematic cross section of the conventional PCF structure.

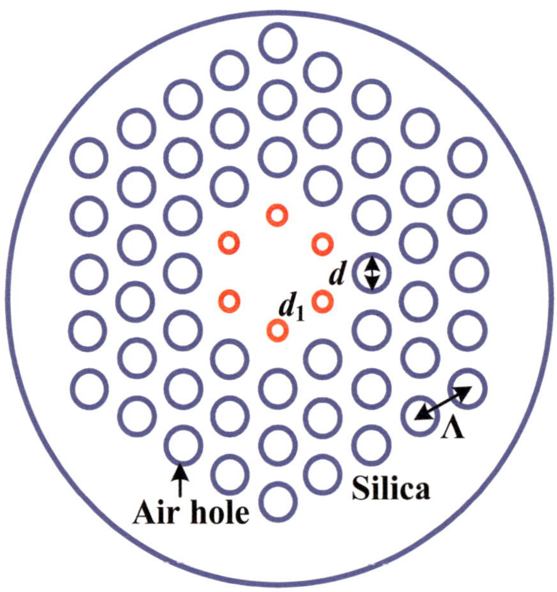


Fig. 1(b). The proposed HN-PCF structure.

This HN-PCF structure can provide ultra-flattened chromatic dispersion characteristics with very high nonlinearity, and low confinement loss for the OCT and telecommunication window. We analyzed the proposed HN-PCF with low confinement losses by modulating only dimension of the first rings, in order to simplify the structure and decrease the fabrication difficulties. In telecommunication widow, the parameters $\Lambda$ = 0.79 μm, $d_1$ = 0.28

μm, $d$ = 0.69 μm, with a total number of 7 hole layers in the cladding are selected for achieving ultra-flattened chromatic dispersion characteristics, small effective area, and low confinement loss. In this case, 7 air hole layers are selected only for reducing confinement loss below 0.2 dB/km.

## 3. Numerical model

The situation in photonics is especially favorable for computation because the Maxwell equations are practically exact, the relevant material properties are well known, and the length scales are not too small. The results of such computations have consistently agreed with experiments. This makes it possible and preferable to optimize the design of photonic crystals on a computer, and then manufacture them. For this proposed HN-PCF structure, by using an accurate modal analysis based on a full-vector finite difference method (FDM) [Begum et al., 2011; Shen et al., 2003] with anisotropic perfectly matched boundary layers (PML), we evaluate the different properties of HN-PCF. The PML in fact is not a boundary condition, but an additional domain that absorbs the incident radiation waves without producing reflections. Once the effective refractive index $n_{eff}$ is obtained by solving an eigenvalue problem drawn from the Maxwell's equations using the FDM, the parameter chromatic dispersion $D(\lambda)$, confinement loss $L_c$, effective area $A_{eff}$ and nonlinear coefficient $\gamma$ can be calculated [Begum et al., 2011; Shen et al., 2003].

### 3.1 Chromatic dispersion

The group-velocity dispersion $D(\lambda)$ is defined as the change in pulse width per unit distance of propagation (i.e., ps/(nm.km). It means that $D(\lambda)$ causes a short pulse of light to spread in time as a result of different frequency components of the pulse traveling at different velocities. This can be calculated from following equation.

$$D(\lambda)=\frac{d\beta_1}{d\lambda}=\frac{d}{d\lambda}\left(\frac{1}{v_g(\lambda)}\right)=-\frac{2\pi c}{\lambda^2}\beta_2=-\frac{\lambda}{c}\frac{d^2\,\mathrm{Re}[n_{eff}]}{d\lambda^2} \quad (1)$$

where, $\beta_1$ and $\beta_2$ are the propagation constant parameters, $v_g$ is the group velocity, $\lambda$ is the operating wavelength in μm, $c$ is the velocity of the light in a vacuum, Re[$n_{eff}$] is the real part of the effective index.

The corresponding dispersion slope $S(\lambda)$ is defined as

$$S(\lambda)=\frac{dD(\lambda)}{d\lambda} \quad (2)$$

Since the total chromatic dispersion is the summation of material dispersion $D_m(\lambda)$ and waveguide dispersion $D_w(\lambda)$. The material dispersion quantified from the Sellmeier equation is directly included in the FDM calculation process. The reason for this is that $D_m(\lambda)$ is mostly determined by the wavelength dependence of the fiber material and for this reason it cannot be altered significantly in the engineering process. On the other hand, $D_w(\lambda)$, which is strongly dependent to the silica-air structure. Therefore, in our calculation chromatic dispersion $D(\lambda)$ [Begum et al., 2011; Shen et al., 2003] corresponds to the total dispersion of the PCFs.

### 3.2 Confinement loss

The attenuation caused by the waveguide geometry is called confinement loss $L_c$. This is an additional form of loss that occurs in single-material fibers particularly in PCFs because they are usually made of pure silica and given by [Begum et al., 2011; Shen et al., 2003]

$$L_c = -20\log_{10} e^{-k_0 \operatorname{Im}[n_{eff}]} = 8.686 k_0 \operatorname{Im}[n_{eff}] \tag{3}$$

where, $k_0$ is the propagation constant in free space, $\lambda$ is the operating wavelength in μm, and $\operatorname{Im}(n_{eff})$ is the imaginary part of the complex effective index $n_{eff}$.

### 3.3 Effective area

The effective area $A_{eff}$ is defined as follows [Begum et al., 2011; Shen et al., 2003]

$$A_{eff} = \frac{\left(\int_{-\infty}^{\infty}\int_{-\infty}^{\infty} |E|^2\, dxdy\right)^2}{\int_{-\infty}^{\infty}\int_{-\infty}^{\infty} |E|^4\, dxdy} \tag{4}$$

where, $E$ is the electric field derived by solving Maxwell's equations. From this equation, it is seen that effective area $A_{eff}$ depends on the fiber parameters such as the mode field diameter and core-cladding index difference.

### 3.4 Nonlinear coefficient

In this research, silica is used as a background material for designing PCFs. Since silica can be treated as a homogeneous material, the lowest-order nonlinear coefficient is the third-order susceptibility $\chi^{(3)}$. Most of the nonlinear effects in optical fibers therefore originate from nonlinear refraction, a phenomenon that refers to the intensity dependence of the refractive index resulting from the contribution of $\chi^{(3)}$, i.e., the refractive index of the fiber becomes [Agrawal, 1995]

$$n = n_1 + n_2 E^2 \tag{5}$$

where, $n_1$ is the linear refractive index which is responsible for material dispersion, $E^2$ is the optical intensity inside the fiber, $n_2$ is the nonlinear refractive index related to $\chi^{(3)}$ by the following relation

$$n_2 = \frac{3}{8n_1} \operatorname{Re}\left(\chi^{(3)}_{xxxx}\right) \tag{6}$$

where, Re stands for the real part. Another way to represents the refractive index is

$$n = n_1 + n_2 \frac{P}{A_{eff}} \tag{7}$$

where, $P$ is the incident light power and $A_{eff}$ is the effective area of the fiber. From nonlinear part of Eq. (5) and Eq. (7), we can write

$$E^2 = \frac{P}{A_{eff}} \tag{8}$$

From Eq. (8), it is clear that optical intensity inside the fiber $E$ can be increased by two ways. One is by focusing the light tightly to reduced $A_{\mathrm{eff}}$ and by increasing incident optical power. The nonlinear coefficient of PCFs depends on the value of nonlinear refractive index and the effective area of the PCFs. The nonlinear coefficient is calculated according to following equation [Agrawal, 1995].

$$\gamma = \left(\frac{\omega}{c}\right)\left(\frac{n_2}{A_{eff}}\right) = \left(\frac{2\pi}{\lambda}\right)\left(\frac{n_2}{A_{eff}}\right) \tag{9}$$

where, $\gamma$ is the nonlinear coefficient, $\omega$ is the angular frequency, $n_2$ is the nonlinear refractive index, $\lambda$ is the wavelength of the light, $(n_2/A_{\mathrm{eff}})$ is the nonlinear constant. It is possible to enhance the nonlinearity by reducing the effective area $A_{\mathrm{eff}}$ through a smaller core diameter and increasing nonlinear refractive index of a material $n_2$. This $n_2$ is constant and depending on the material of the fibers while is variable and varied from $2.2\sim3.4\times10^{-20}$ m$^2$/W.

### 3.5 Nonlinear Schrödinger equation

Nonlinear Schrödinger equation (NLSE) is used for numerical calculation of SC spectrum [Agrawal, 1995]. The propagation equation Eq. (10) is a nonlinear partial differential equation that does not generally lend itself to analytic solutions when both the nonlinearity and the dispersion effect are present. A numerical approach is therefore often necessary for an understanding of the nonlinear effects in optical fibers. The split-step Fourier method is one of these, and is the most popular algorithm because of its good accuracy and relatively modest computing time [Agrawal, 1995].

$$\frac{\partial A}{\partial Z} + \frac{\alpha}{2}A + \frac{i}{2}\beta_2\frac{\partial^2 A}{\partial T^2} - \frac{1}{6}\beta_3\frac{\partial^3 A}{\partial T^3} = i\gamma\left[|A|^2 A + i\frac{\lambda_c}{2\pi c}\frac{\partial}{\partial T}\left(|A|^2 A\right) - T_R A\frac{\partial |A|^2}{\partial T}\right] \tag{10}$$

where, $A$ is the complex amplitude of the optical field, $z$ is the propagation distance, $a$ is the attenuation constant of the fiber, $T = t - z/v_g$ ($t$ is the physical time, $v_g$ is the group velocity at the center wavelength), $\gamma$ is the nonlinear coefficient, $\lambda_c$ is the center wavelength, and $T_R$ is the slope of the Raman gain, $\beta_n$ (n =1 to 3) are the n-th order propagation constant. This propagation constant $\beta(\omega)$ is approximated by a few first terms of a Taylor series expansion about the carrier frequency $\omega_0$, that is

$$\beta(\omega) = \beta_0 + (\omega - \omega_0)\beta_1 + \frac{1}{2}(\omega - \omega_0)^2\beta_2 + \frac{1}{6}(\omega - \omega_0)^3\beta_3 + \cdots \tag{11}$$

where,

$$\beta_n = \left(\frac{d^n\beta}{d\omega^n}\right)_{\omega=\omega_0} \tag{12}$$

The second order propagation constant $\beta_2$ [ps$^2$/km], accounts for the dispersion effects in fiber-optic communication systems. Depending on the sign of $\beta_2$, the dispersion region can be classified into two regions, normal dispersion region ($\beta_2$ >0) and anomalous dispersion region ($\beta_2$ < 0).

### 3.6 Coherence length

Coherence length $l_c$ is one of the important parameter in estimating the longitudinal resolution of the OCT source. The shorter the coherence length of the source, the more closely the sample and reference arm group delays must be matched for the constructive interference to occur. On the other word, we can say the combination of the reflected light from the sample arm (containing the item of the interest) and the reference light from the reference arm (usually a mirror) gives rise to an interference pattern but only if light from both arms traveled the same optical distance. The same optical distance means a difference of less than a coherence length. For a Gaussian spectrum the FWHM-duration of the coherence time $t_c$ is

$$t_c = \frac{4\ln 2}{\pi \Delta v} \tag{13}$$

where, the half-power bandwidth $\Delta v$ represents the spectral bandwidth of the source in the optical frequency domain.

Because of the backscattering configuration of OCT that the light travels back and forth in the interferometer, the coherence length $l_c$ (in air) is expressed by the formula [Bouma et al., 1995]

$$l_c = \frac{ct_c}{2} = \frac{2c\ln(2)}{\pi} \cdot \frac{1}{\Delta v} = \frac{2\ln(2)}{\pi} \cdot \frac{\lambda_c^2}{\Delta\lambda} \tag{14}$$

$$l_c \approx 0.44 \frac{\lambda_c^2}{\Delta\lambda} \tag{15}$$

where, $c$ is the velocity of light in free-space, $\lambda_c$ is the center wavelength of the spectrum and $\Delta\lambda$ is the FWHM-wavelength width, $\Delta v = \frac{c\Delta\lambda}{\lambda_c^2}$ is the spectral bandwidth. This $l_c$ is very important for estimating the longitudinal resolution $l_r$ in air and biological tissue.

### 3.7 Longitudinal resolution

The axial or longitudinal and lateral or transverse resolutions of OCT are decoupled from one another; the former being an equivalent to the coherence length $l_c$ of the light source and the latter being a function of the optics. After calculating coherence length $l_c$, longitudinal resolution in air and biological tissue can be estimated by [Bouma et al., 1995]

$$l_r - \frac{l_c}{n_{tissue}} \tag{16}$$

where, $n_{\text{tissue}}$ is the refractive index of the biological tissue. For ultrahigh-resolution OCT imaging $l_c$ should be low value because $l_r$ is proportional with $l_c$.

## 4. Simulation results

Fig. 2 (a), (b) and (c) shows the wavelength dependence properties of chromatic dispersion, dispersion slope, effective area, nonlinear coefficient and confinement loss for the four-rings

HN-PCF in Fig. 1. As shown in Fig. 1, only the diameter of the first air hole ring is varied and the diameters of the remaining air holes remain the same, where $d_1$ = 0.46 µm, $d$ = 0.80 µm, for a fixed pitch $\Lambda$ = 0.87 µm. From Fig. 2, it is found that the proposed HN-PCF owning ultra-flattened chromatic dispersion and dispersion slope at 0.8 µm are 0.55 ps/(nm.km) and 0.2 ps/(nm$^2$.km), respectively. The nonlinear coefficient is larger than 208.0 [W·km]$^{-1}$ at 0.8 µm wavelength. Besides, the confinement loss is calculated and it is found that confinement loss is less than $10^{-2}$ dB/km in the wavelength range of 0.75 µm to 1.0 µm which is lower than Rayleigh scattering loss in conventional fiber.

Fig. 3 (a), (b) and (c) demonstrates the wavelength dependence properties of chromatic dispersion, confinement loss and effective area for the seven-rings HN-PCF in Fig. 1, where $d_1$ = 0.28 µm, $d$ = 0.69 µm, for a fixed pitch $\Lambda$ = 0.79 µm. In this case, it has been selected 7 air hole rings for reducing confinement loss lower than Rayleigh scattering loss in conventional fiber at 1.55 µm. Numerical simulation results show that the 7-rings HN-PCF have nonlinear coefficients more than 54.0 [W·km]$^{-1}$ and confinement loss lower than 0.1 dB/km at 1.55 µm, ultra-flattened chromatic dispersion of -2.3 ps/(nm.km) at 1.55 µm wavelength.

SC generation in the proposed HN-PCF is numerically calculated which is shown in Fig. 4 (a), (b) and (c). In Fig. 4 consider the propagation of the sech$^2$ (square of the hyperbolic-secant) waveform with the full width at half maximum (FWHM), $T_{FWHM}$ and Raman scattering parameter are 1.0 ps and 3.0 fs, respectively, through the proposed HN-PCF. The input power $P_{in}$ of the incident pulses are 18.0 W, 55.0 W and 58.0 W at 0.8 µm, 1.3 µm and 1.55 µm, respectively. The propagation constant around the carrier frequency $\beta_2$ and $\beta_3$ are 1.88 ps$^2$/km and 0.02 ps$^3$/km, respectively for Fig. 4 (a). Again, the propagation constant around the carrier frequency $\beta_2$ and $\beta_3$ are 2.55 ps$^2$/km and -0.03 ps$^3$/km, respectively for Fig. 4 (b). Moreover, the propagation constant around the carrier frequency $\beta_2$ and $\beta_3$ are 1.51 ps$^2$/km and 0.01 ps$^3$/km, respectively for Fig. 4 (c). The achieved fiber length is 10.0 m in all cases. The calculated SC spectrum FWHM bandwidth is 200 nm, 530 nm and 590 nm at center wavelength 0.8 µm, 1.3 µm and 1.55 µm, respectively. From these results, it is evident that high quality SC spectrum is readily generated with relatively short fiber length and good incident power compared to the previously reported ones [Boppart et al., 1998; Bouma et al., 1995; Colston et al., 1998; Drexler et al., 1999; Hartl et al., 2001; Herz et al., 2004; Jiang et al., 2005; Lee et al., 2009; Ohmi et al., 2004; Pan et al., 1998; Ryu et al., 2005; Tearnery et al., 1997; Welzel et al., 1997].

Fig. 5 (a), (b) and (c) represents the intensity spectra of the proposed HN-PCF at center wavelengths 0.8 µm, 1.3 µm and 1.55 µm, respectively when changing incident optical powers. It should be noted that in this time, the fiber lengths are remain unchanged in all of the center wavelengths. From these figures, it is seen that intensity spectra are gradually broadening with increasing the input power, $P_{in}$ at the particular wavelength. Therefore, it is clearly seen that the SC spectral width is dependent to the incident power.

Fig. 6 (a), (b) and (c) represents the intensity spectra of the proposed HN-PCF at center wavelengths 0.8 µm, 1.3 µm and 1.55 µm, respectively in different fiber lengths while incident optical powers are remain unchanged. From these figures, it is observed that intensity spectra are gradually broadening with increasing the fiber length, $L_F$ at the particular wavelength. So, it is noted that the SC spectral width is dependent to the fiber length. From Fig. 5 and 6, it is clear that the SC spectral width is dependent to the incident power and fiber length as well.

Fig. 7 (a), (b) and (c) demonstrates the output powers of the proposed HN-PCF at center wavelengths 0.8 µm, 1.3 µm and 1.55 µm, respectively when the fiber length is 10 m in all center wavelengths. From these figures, it is found that output powers are increased with increasing incident input powers at particular wavelength.

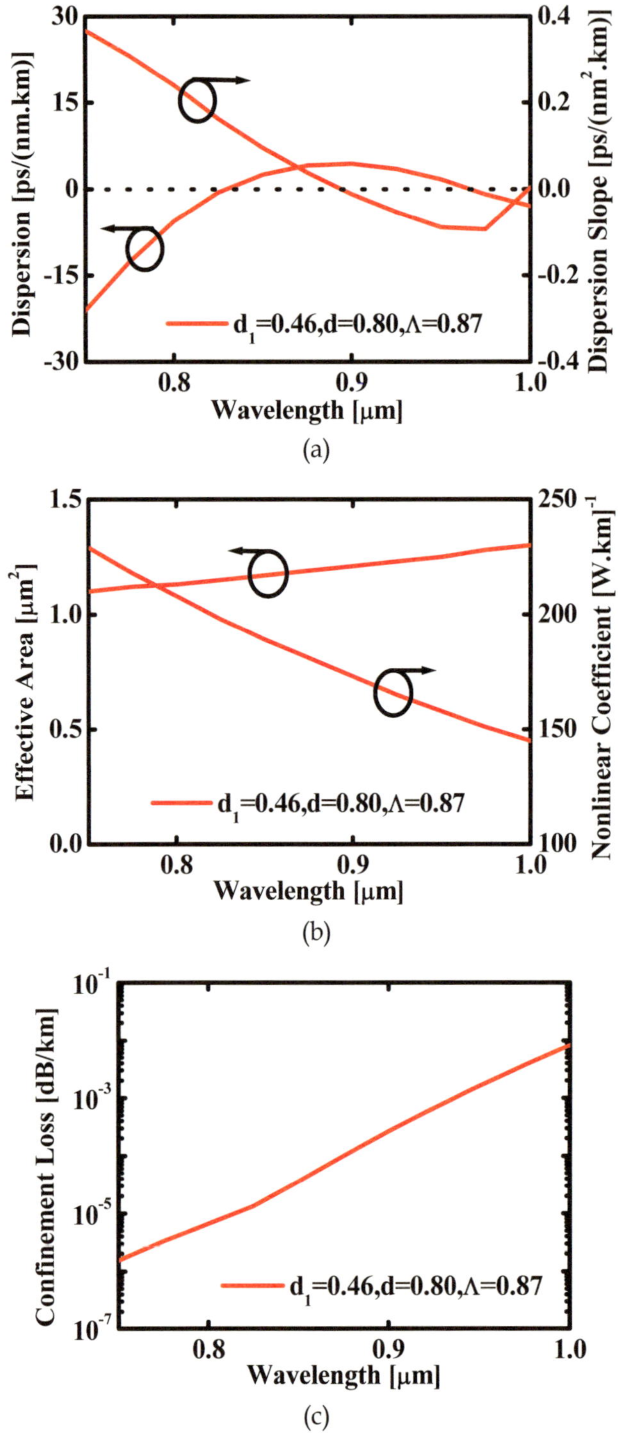


Fig. 2. (a) Chromatic dispersion and dispersion slope, (b) Effective area and nonlinear coefficient and (c) Confinement loss of the 4-rings proposed HN-PCF.

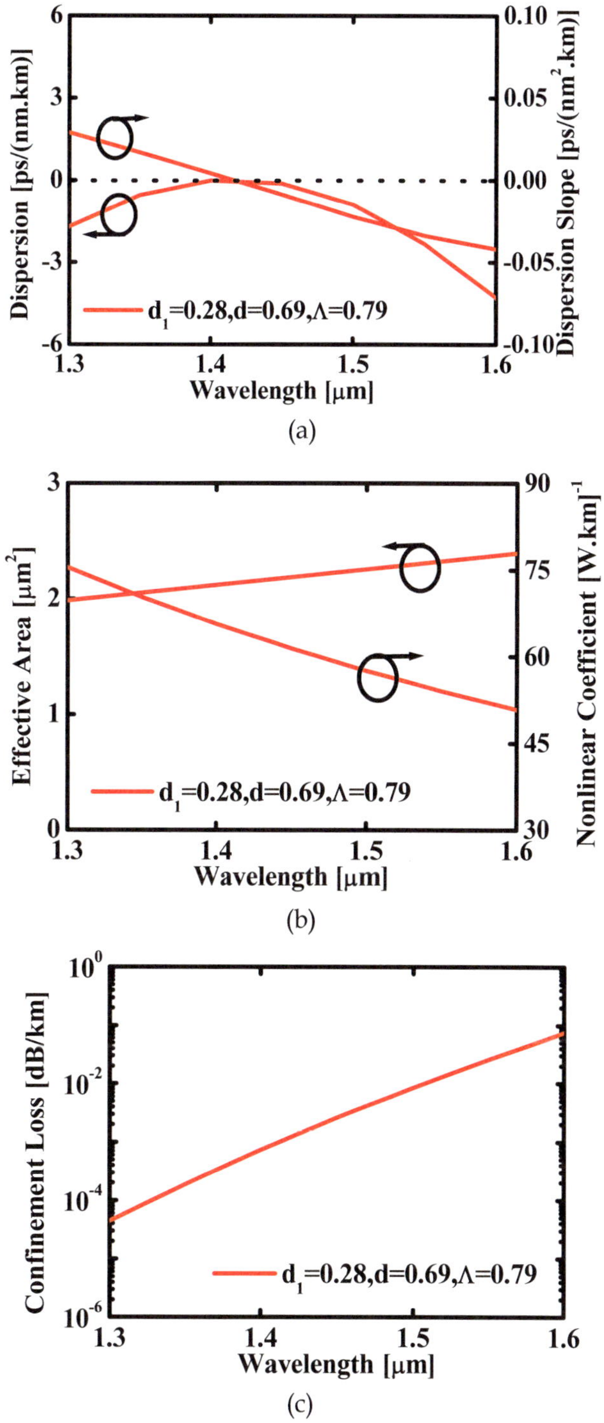


Fig. 3. (a) Chromatic dispersion and dispersion slope, (b) Effective area and nonlinear coefficient and (c) Confinement loss of the 7-rings HN-PCF.

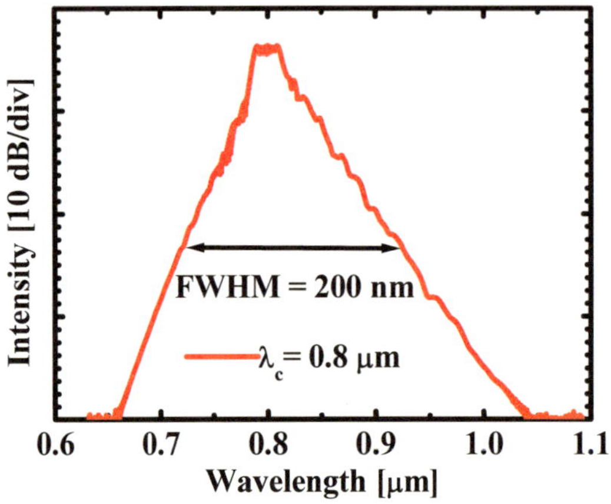


(a)

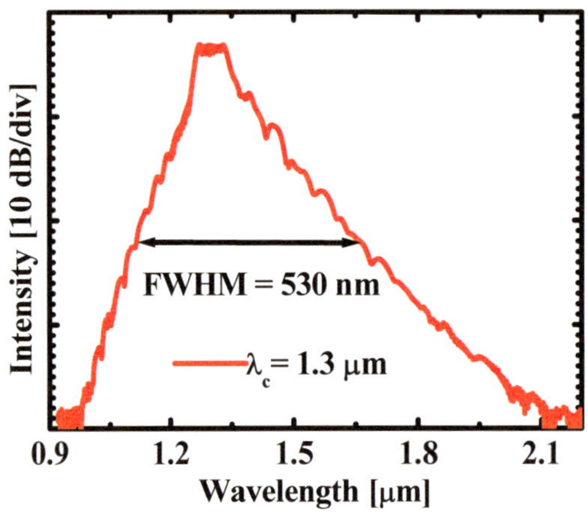


(b)

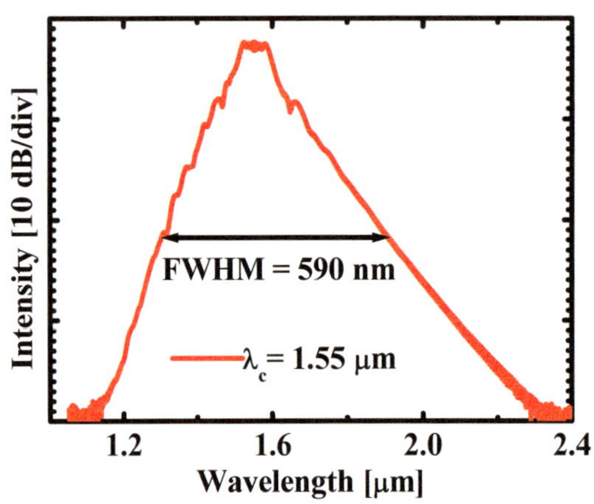


(c)

Fig. 4. Spectral intensity at (a) 0.8 μm (b) 1.3 μm and (c) at 1.55 μm of the proposed HN-PCF which is shown in Fig. 1.

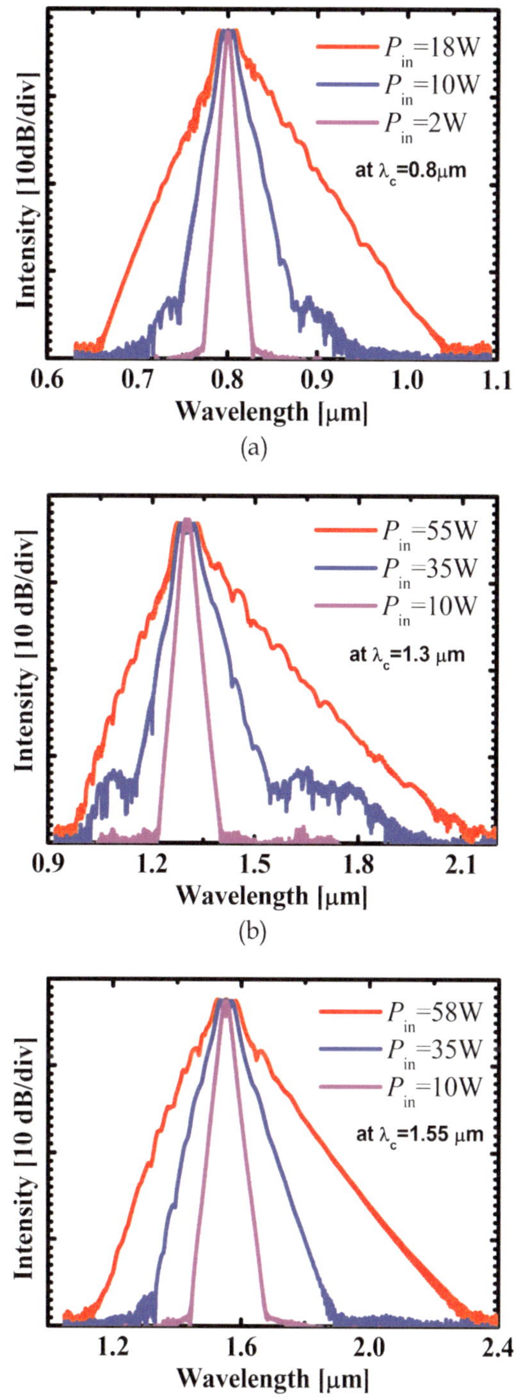


Fig. 5. Intensity spectra at center wavelengths (a) 0.8 μm, (b) 1.3 μm and (c) 1.55 μm of the proposed HN-PCF which is shown in Fig. 1 when changing incident optical powers.

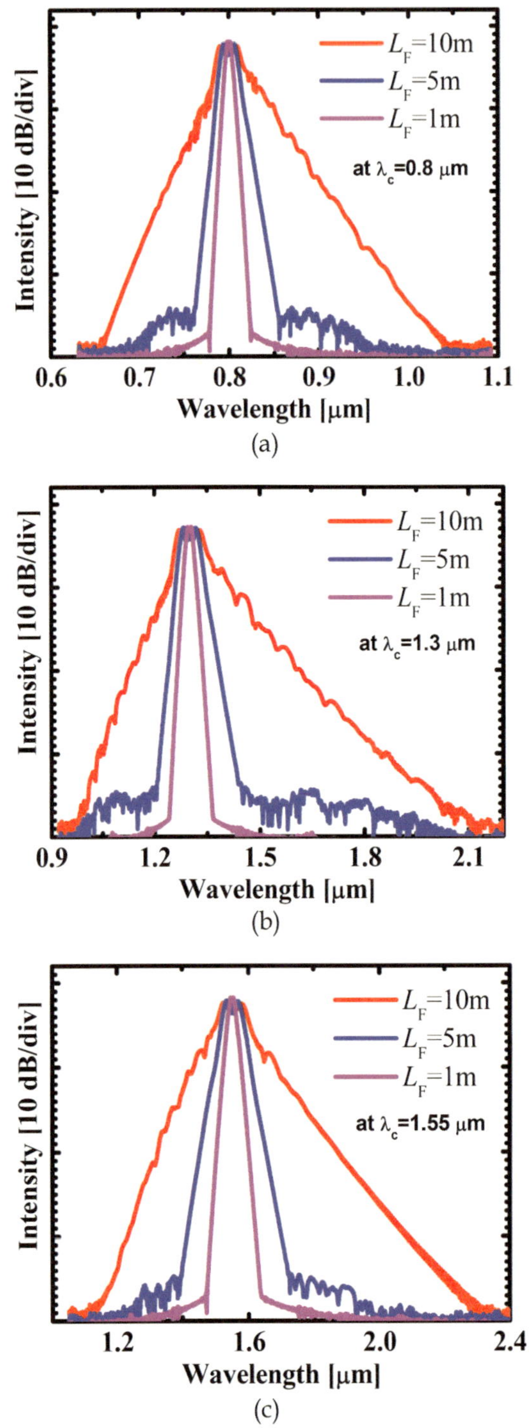


Fig. 6. Intensity spectra at center wavelengths (a) 0.8 μm, (b) 1.3 μm and (c) 1.55 μm of the proposed HN-PCF which is shown in Fig. 1 in different fiber lengths.

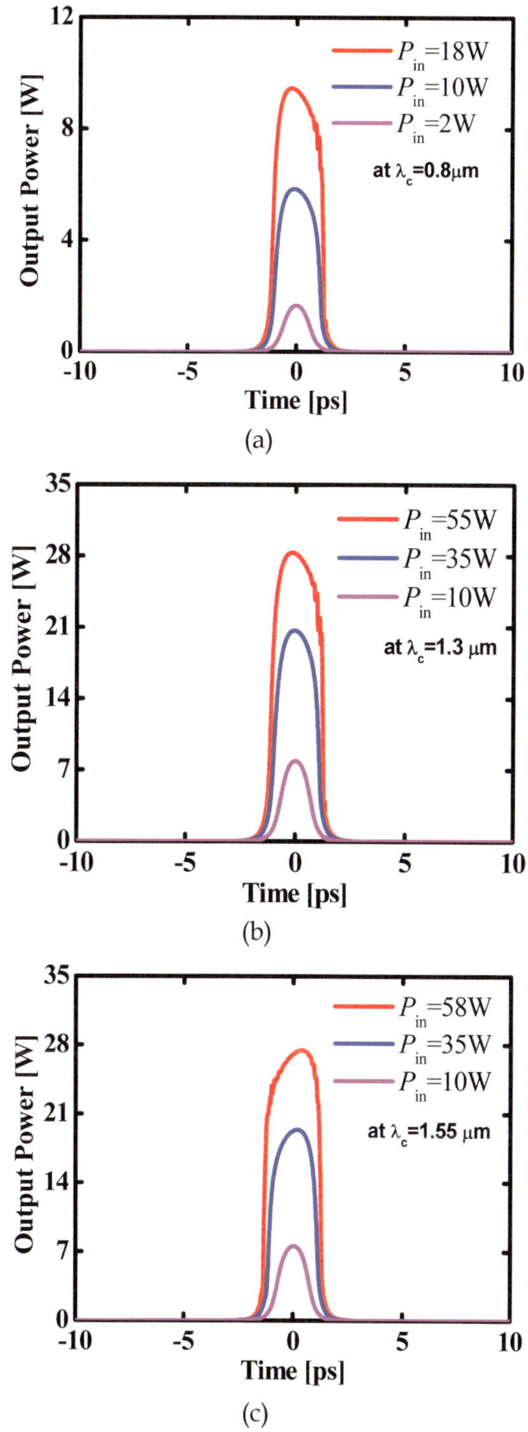


Fig. 7. Output power at center wavelengths (a) 0.8 μm, (b) 1.3 μm and (c) 1.55 μm of the proposed HN-PCF which is shown in Fig. 1.

The spectral bandwidths, FWHM are 200 nm, 530 nm and 590 nm at center wavelength 0.8 µm, 1.3 µm and 1.55 µm, respectively. The calculated $l_c$ values are 1.4 µm, 1.4 µm and 1.8 µm at center wavelength 0.8 µm, 1.3 µm and 1.55 µm, respectively. The calculated $l_r$ values are 0.97 µm, 0.85 µm and 1.1 µm when typical $n_{tissue}$ is 1.44, 1.65 and 1.65 at center wavelengths 0.8 µm, 1.3 µm and 1.55 µm, respectively [Ohmi et al., 2000]. These calculated $l_r$ value is better than that of Ref. [Boppart et al., 1998; Bouma et al., 1995; Colston et al., 1998; Drexler et al., 1999; Hartl et al., 2001; Herz et al., 2004; Jiang et al., 2005; Lee et al., 2009; Ohmi et al., 2004; Pan et al., 1998; Ryu et al., 2005; Tearnery et al., 1997; Welzel et al., 1997] and SLDs with OCT imaging longitudinal resolution of ≈ 10 – 15 µm. Some calculated parameters of the proposed HN-PCF are shown in table 1. From this Table 1, it is seen that the highest longitudinal resolution and wider FWHM is obtained at 1.3 µm and 1.55 µm wavelengths, respectively.

| **Paramters** | **$\lambda_c$ = 0.8 µm** | **$\lambda_c$ = 1.3 µm** | **$\lambda_c$ = 1.55 µm** |
|---|---|---|---|
| $\beta_2$ [ps²/km] | 1.88 | 2.55 | 1.51 |
| $\beta_3$ [ps³/km] | 0.02 | -0.03 | 0.01 |
| $T_R$ [fs] | 3.0 | 3.0 | 3.0 |
| $T_{FWHM}$ [ps] | 1.0 | 1.0 | 1.0 |
| $P_{in}$ [W] | 18.0 | 55.0 | 58.0 |
| $L_F$ [m] | 10.0 | 10.0 | 10.0 |
| FWHM [nm] | 200.0 | 530.0 | 590.0 |
| $l_c$ [µm] | 1.4 | 1.4 | 1.8 |
| $l_r$ [µm] | 0.97 | 0.85 | 1.1 |

Table 1. Some calculated parameters of the proposed HN-PCF.

The apparent advantages of our HN-PCF design are the facts that it simultaneously exhibits numerous optical properties such as flattened dispersion, low confinement loss, high nonlinearity at three central wavelengths 0.8 µm, 1.3 µm and 1.55 µm. Moreover, one can take advantage of the different dispersion characteristics of the two different geometrical parameters to get one more degree of freedom for tailoring the generated SC spectrum. Furthermore, the proposed fiber can be used to make a fiber-based light source to generate SC in three different central wavelengths for ophthalmology, dermatology and dentistry OCT imaging application. Hence, the same fiber with three center wavelengths can be used in several OCT imaging and optical communication applications while exhibiting relatively good longitudinal resolution performance, high power, and in turn can pave the way for the compact, robust and cheap fiber-based OCT light sources. Therefore, picosecond pulse based PCFs are among the most specialized optical lightguides in the new optical fiber technology which is highly competitive compared to traditional laser designs.

## 5. Conclusions

We have proposed broadband SC generated HN-PCF which can be used as a high power picoseconds pulses light source in ultrahigh-resolution OCT system for ophthalmology, dermatology and dental imaging. Moreover, it has been sent that this proposed HN-PCF would be applicable in optical communication. We achieved longitudinal resolutions in tissue are 0.97 µm, 0.85 µm and 1.1 µm at center wavelength of 0.8 µm, 1.3 µm and 1.55 µm, respectively. Furthermore, from numerical simulation results it was found that the proposed HN-PCFs have high nonlinear coefficients with ultra-flattened chromatic dispersion, low dispersion slopes, and very low confinement losses, simultaneously. The broad bandwidth of the light source permits high resolution for bright OCT imaging in the wavelength ranges from 0.8 μm to 1.6 μm. For the less number of geometrical parameters, this light source has the potential to be made compact, robust and cheap fiber-based OCT light sources and suitable for clinical applications. Consequently, the same proposed fiber can be used in different optical communication applications such as dispersion controlling, wavelength conversion, SC generation, optical parametric amplification, and so on.

## 6. Acknowledgement

The authors are indebted and grateful to the Japan Society for Promotion of Science (JSPS) for their support in carrying out this research work, JSPS ID number P 09078. Dr. Feroza Begum is a JSPS Postdoctoral Research Fellow.

## 7. References

Agrawal G.P. (1995). *Nonlinear Fiber Optics*, Academic Press, ISBN 0-12-045142-5

Begum F., Namihira Y., Razzak S.M.A., and Zou N. (2007a). Novel Square Photonic Crystal Fibers with Ultra-flattened Chromatic Dispersion and Low Confinement Losses. *IEICE Trans. on Elec.*, Vol. E90-C, No. 3, pp. 607-612, ISSN 0916-8524

Begum F., Namihira Y., Razzak S.M.A., Kaijage S., Miyagi K., Hai N.H., and Zou N. (2007b). Highly Nonlinear Dispersion-Flattened Square Photonic Crystal Fibers with Low Confinement Losses. *Opt. Review*, Vol. 14, No. 3, pp. 120-124, ISSN 1340-6000

Begum F., Namihira Y., Razzak S.M.A., Kaijage S., Hai N.H., Kinjo T., Miyagi K., and Zou N. (2009a). Novel Broadband Dispersion Compensating Photonic Crystal Fibers: Applications in High Speed Transmission Systems. *Jour. of Opt. & Laser Tech.*, Vol. 41, No. 5, pp. 679-686, ISSN 0030-3992

Begum F., Namihira Y., Kaijage S., Razzak S.M.A., Hai N.H., Kinjo T., Miyagi K., and Zou N. (2009b). Design and analysis of novel highly nonlinear hexagonal photonic crystal fibers with ultra-flattened chromatic dispersion. *Opt. Comm.*, Vol. 282, No. 7, pp. 1416-1421, ISSN 0030-4018

Begum F., Namihira Y., Kinjo T., and Kaijage S. (2011). Supercontinuum generation in square photonic crystal fiber with nearly zero ultra-flattened chromatic dispersion and fabrication tolerance analysis. *Opt. Commun.*, Vol. 284, No. 4, pp. 965-970, ISSN 0030-4018

Boppart S.A., Bouma B.E., Pitris C., Southern J.F., Brezinski M.E., and Fujimoto J.G. (1998). In vivo cellular optical coherence tomography imaging. *Nature Medicine*, Vol. 4, No. 7, pp. 861-865, ISSN 1078-8956

Bouma B.; Tearney G.J.; Boppart S.A.; Hee M.R.; Brezinski M.E., and Fujimoto J.G. (1995). High-resolution optical coherence tomographic imaging using a mode-locked Ti:Al2O3 laser source. *Opt. Lett.*, Vol. 20, No. 13, pp. 1486-1488, ISSN 0146-9592

Colston B.W., Jr., Sathyam U.S., DaSilva L.B., Everett M.J., Stroeve P., and Otis L.L. (1998). Dental OCT. *Opt. Express*, Vol. 3, No. 6, pp. 230-238, ISSN 1094-4087

Couny F., Roberts P.J., Birks T.A., and Benabid F. (2008). Square-lattice large-pitch hollow-core photonic crystal fiber. *Optics Express*, Vol. 16, No. 25, pp. 20626-20636, ISSN 1094-4087

Drexler W., Morgner U., Kärtner F.X., Pitris C., Boppart S.A., Li X.D., Ippen E.P., and Fujimoto J.G. (1999). In vivo ultrahigh-resolution optical coherence tomography. *Opt. Lett.*, Vol. 24, No. 17, pp. 1221-1223, ISSN 0146-9592

Hartl I., Li X.D., Chudoba C., Ghanta R.K., Ko T.H., Fujimoto J.G., Ranka J.K., and Windeler R.S. (2001). Ultrahigh-resolution optical coherence tomography using continuum generation in an air-silica microstructure optical fiber. *Opt. Lett.*, Vol. 26, No. 9, pp. 608-610, ISSN 0146-9592

Herz P.R., Chen Y., Aguirre A.D., Fujimoto J.G., Mashimo H., Schmitt J., Koski A., Goodnow J., and Peterson C. (2004). Ultrahigh resolution optical biopsy with endoscopic optical coherence tomography. *Optics Express*, Vol. 12, No. 15, pp. 3532-3542, ISSN 1094-4087

Jiang Y., Tomov I.V., Wang Y., and Chen Z. (2005). High-resolution second-harmonic optical coherence tomography of collagen in rat-tail tendon. *Applied Physics Lett.*, Vol. 86, No. 13, pp. 133901-133903, ISSN 0003-6951

Kaijage S.F., Namihira Y., Hai N.H., Begum F., Razzak S.M.A., Kinjo T., Higa H., and Zou N. (2008). Multiple Defect-core Hexagonal Photonic Crystal Fiber with Flattened Dispersion and Polarization Maintaining Properties. *Opt. Review,* Vol. 15, No. 1, pp. 31-37, ISSN 1340-6000

Knight J.C., Birks T.A., Russell P.St.J., and Atkin D.M. (1996). All-silica single-mode optical fiber with photonic crystal cladding. *Opt. Lett.*, Vol. 21, No. 19, pp. 1547-1549, ISSN 0146-9592

Knight J.C., Birks T.A., Cregan R.F., Russell P.St.J., and de Sandro J.-P. (1998). Large mode area photonic crystal fiber. *Electron. Lett.*, Vol. 34, No. 13, pp. 1347-1348, ISSN 0013-5194

Lee J.H., Jung E.J., and Kim C.-S. (2009). Incoherent, CW supercontinuum source based on erbium fiber ASE for optical coherence tomography imaging. *Proceedings of OptoEelectronics and Communication Conference*, Paper number FD3, Hongkong, 13-17 July 2009, ISBN 978-1-4244-4102-0

Ohmi M., Ohnishi Y., Yoden K., and Haruna M. (2000). In vitro simultaneous measurement of refractive index and thickness of biological tissue by the low coherence interferometry. *IEEE Trans. on Biomed. Eng.*, Vol. 47, No. 9, pp. 1266-1270, ISSN 0018-9294

Ohmi M., Yamazaki R., Kunizawa N., Takahashi M., and Haruna M. (2004). In vivo observation of micro-tissue structures by high-resolution optical coherence tomography with a femtosecond laser. *Japanese Society for Medical and Biological Engineering (Japanese paper)*, Vol. 42, No. 4, pp. 204-210, ISSN 1881-4379

Pan Y., and Farkas DL. (1998). Noninvasive imaging of living human skin with dual-wavelength optical coherence tomography in two and three dimensions. *J Biomed Opt.*, Vol. 3, No. 4, pp. 446–455, ISSN 1083-3668

Russel P.St.J. (2003). Photonic crystal fibers. *Science*, Vol. 299, No. 5605, pp. 358-362, ISSN 0036-8075

Ryu S.Y., Choi H.Y., Choi J.N.E., Yang G.-H. and Lee B.H. (2005). Optical Coherence tomography implemented by photonic crystal fiber. *Opt. and Quantum Elec.*, Vol. 37, No. 13-15, pp. 1191-1198, ISSN 0306-8919

Shen L.-P., Huang W.-P., and Jian S.-S. (2003). Design of photonic crystal fibers for dispersion-related applications. *J. Lightwave Technol.*, Vol. 21, No. 7, pp. 1644-1651, ISSN 0733-8724

Tearney G.J., Bouma B.E., and Fujimoto J.G. (1997). High-speed phase- and group-delay scanning with a grating-based phase control delay line. *Opt Lett.*, Vol. 22, No. 23, pp. 1181–1183, ISSN 0146-9592

Welzel J., Lankenau E., Birngruber R., and Engelhardt R. (1997). Optical coherence tomography of the human skin. *J Am Acad Dermatol.*, Vol. 37, No. 6, pp. 958–963, ISSN 0190-9622

# 12

# Optic Fiber on the Basis of Photonic Crystal

Yurij V. Sorokin
*MIREA(TU),*
*Russia*

## 1. Introduction

### 1.1 Phylums of optic fibers

The optic fiber represents internal dielectric medium (crystal, glass etc.), which one is contained a main body of a quantity of light transmitted on a fiber, and which one is called as a core. The core can be surrounded by a layer with lower refractive index, which one is called as a shell. For protection against exposures and for increase of a mechanical strength the core with a shell can be coated with a padding layer of plastic.

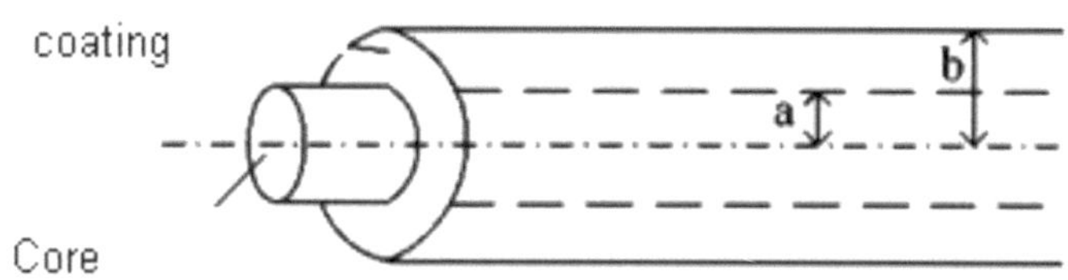


Fig. 1. An optic fiber.

There are different phylums of fibers. The optic fibers without a shell represent simply glass or quartz fiber. They are friable and are ineffective. For them large losses, as on border of two mediums the electrical field is not equal to zero point and the border is rather incomplete. Besides, that such fiber was monomode; his diameter should be less than 1 micron. Such fibers now practically are not applied.

Optic fibers with a shell. The core in such optic fibers is coated with a shell with lower refractive index. The losses in fibers with a shell are much less than losses in fibers without a shell. As we shall see hereinafter, the illumination in such fibers depends on reduced frequency. And essentialist: the manufacturing of such fibers is technologically possible, in which one one mode of propagation will be diffused only. Hereinafter we shall esteem basically only fibers with a shell.

On a structure of refractive index of a fiber it is possible to secure two most often meeting of a type: stepwise and gradient.

In a stepwise fiber the refractive index in a core remains to a constant (see fig. 2a):

In a gradient fiber the refractive index of a core varies depending on r - spacing interval from an axis of a fiber (see fig. 2b).

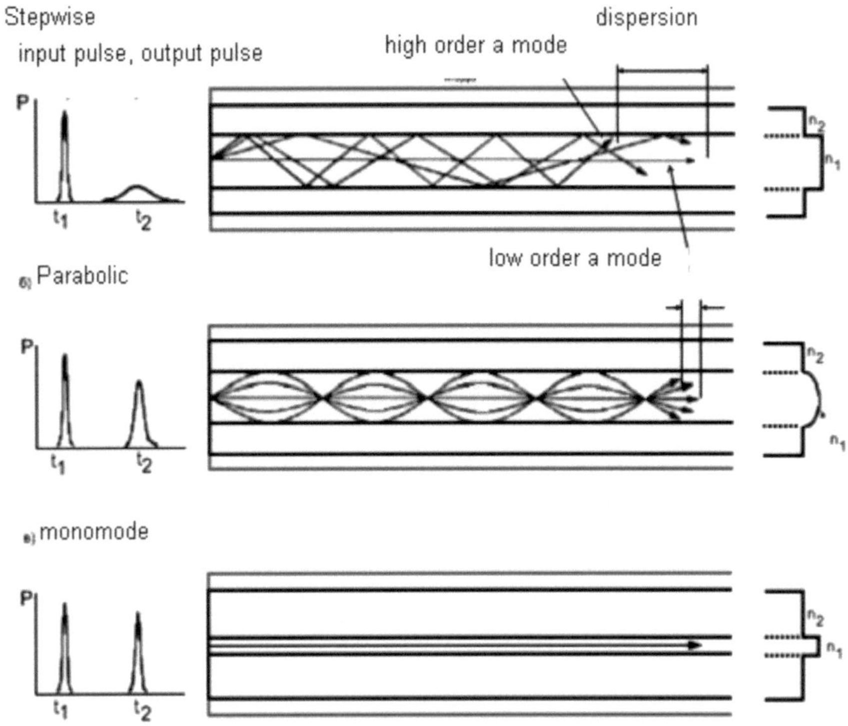


Fig. 2. Phylums of optic fibers, their structures of refractive index and broadering of an optical impulse: and - stepwise; - parabolic; in - monomode.

As we shall see later, in a gradient fiber, in which one the refractive index varies under the parabolic law, the optical pathes of different beams will be practically identical, that essentially reduces a dispersion of a fiber. The gradient fiber as contrasted to stepwise has the best characteristics on dispersion and consequently has large throughput capacity.

The selected law of change of refractive index can be more or less composite. The directional illumination is possible as well in a homogeneous material, if to him to give the definite form. Gears of an illumination in most often used stepwise and gradient fibers.

### 1.2 Stepwise fiber - A numbered aperture

Let's consider a stepwise optic fiber (fig. 2a). Let and - radius of a core, b - radius of a shell. If diameter of a fiber about several tens micrometers, and difference of refractive indexes about 10-2, it is possible to use concepts of a ray optics and to speak about propagation of light rays.

Let's consider the gear of an illumination in a fiber, neglecting absorption in stuff, it is necessary to allow which one, generally speaking. Let light beam in a core is diffused bevel way $\theta$ to an axis Oz, the axis Oz is directed on an axis of a fiber (fig. 3).

Longitudinal wave number or propagation coefficients:

$$\beta = k\cos\theta = \frac{\omega}{c} n_1 \cos\theta = \text{const}\,.$$

The surge, gated in in a core of a fiber, will be retained in her at the expense of full internal reflection at fulfilment of a condition $\theta < \theta_{kr}$ , where $\theta_{kr}$ - critical angle. At fulfilment of a condition of full internal reflection the surge in a shell is an only imaginary and fast damp on exponential law at deleting from a demarcation a core a shell. At increase of a angle? The condition of total reflection ceases to be executed, and the surge in a shell becomes real. Pursuant to above mentioned it is possible to secure three kinds of rays:

1. Routed rays (rays distributing in a fiber),
2. Beams distributing with outflow (loss),
3. Refracted beams? If is satisfied condition of full internal reflection,

And alone area, where the beam is real, is the core, the beam is considered as routed (fig. 3).
If the beam appears by real in some part of a shell, he is diffused with outflow ( (fig. 4).
If the beam appears by real in all volume of a shell, we deal with a refracted beam.

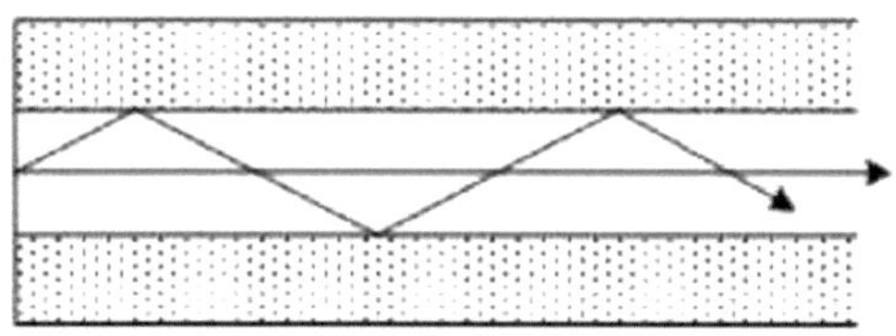

Fig. 3. Routed beam. There is a total reflection from a shell.

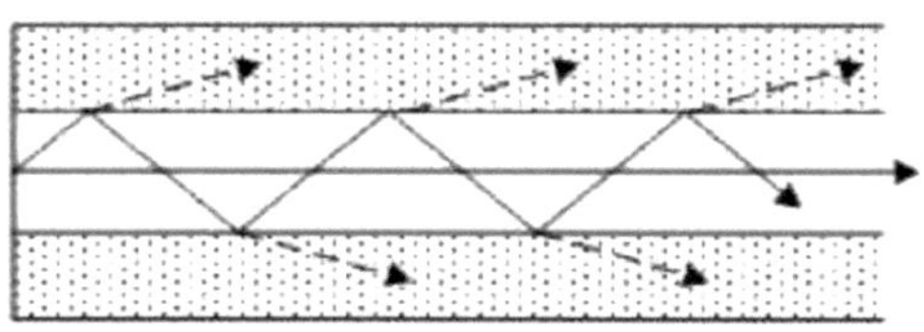

Fig. 4. Beam distributing with outflow. The part of a beam inpours into a shell.

Let's consider in more detail beams distributing in a fiber. Let beam drops from air on butt end of a fiber bevel way $\Omega$. Let's find a maximum angle $\Omega_m$ , under which one it is possible to enter this beam into a fiber, that the beam was hereinafter diffused in a fiber. Thus the ray in a core will be diffused bevel way $\theta_{kr}$, conforming to a case of total reflection from a demarcation with a shell (see fig. 5).

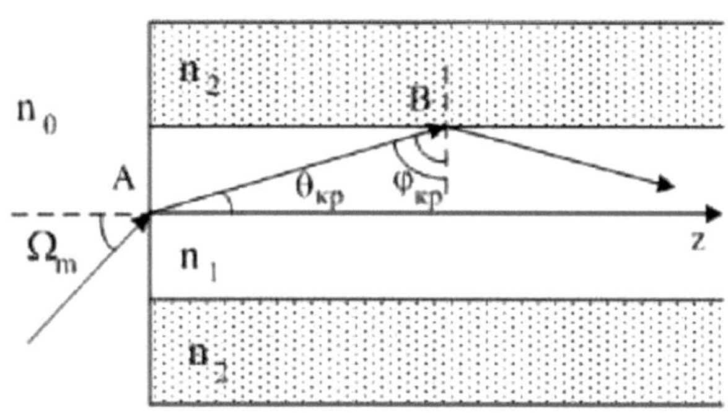


Fig. 5. An illumination in a stepwise fiber.

For a demarcation an air - core of a fiber (point A):

$$\frac{\sin\Omega_m}{\sin\theta_{кр}} = \frac{n_1}{n_0}.$$

Here n0 - refractive index of air. Let's count $n_0 = 1$.
Let's find $\sin\Omega_m$

$$\cos\theta_{кр} = n_2 / n_1 .$$

Angle $\theta_{кр}$ is discovered:

$$\sin\Omega_m = n_1 \sin\theta_{кр} = n_1\sqrt{1-\cos^2\theta_{кр}} = n_1\sqrt{1-\frac{n_2^2}{n_1^2}} = \sqrt{n_1^2 - n_2^2} .$$

Value $\sin\Omega_m$ call as a numbered aperture of a fiber. The numbered aperture has notation NA. Thus, the numbered aperture is peer

$$NA = \sin\Omega_m = \sqrt{n_1^2 - n_2^2} .$$

The numbered aperture of a fiber determines a maximum corner(angle) of input in a fiber of a beam, which one will test total reflection and to be diffused in a fiber.
If the condition of total reflection is defaulted, the beams with outflow or refracted beams will be diffused.

### 1.3 Gradiant fiber - A numbered aperture

Let's consider a gradient optic fiber (see fig. 2б). His(its) refractive index, as against a stepwise fiber, varies at change r:

$$\begin{cases} n = n(r), \text{ если } r < a, \\ n = n_2, \text{ если } r \geq a. \end{cases}$$

To similarly stepwise fiber, it is possible to find a maximum angle of input of radiation in a fiber, only he will depend on spacing interval r:? m =? m (r). Value $\sin\Omega_m$ (r) we shall call as a local numbered aperture of a fiber:

$$NA(r) = \sin\Omega_m(r) = \sqrt{n^2(r) - n_2^2} .$$

Any beams dropping on butt end of a fiber apart r from an axis and falling inside of an aperture tumulus with an apex angle $\Omega_m(r)$ , tests after input total reflection and is diffused in a fiber. The local numbered aperture is max on an axis of a fiber and up to zero point on border a core and shells drops.
As numbered aperture of a gradient fiber we shall call maximum value of the local numeric aperture.
For a gradient fiber with a quadratic structure of refractive index the effective numbered aperture is determined, which one is peer:

$$NA_{eff} = \frac{\sqrt{n^2(0) - n_2^2}}{\sqrt{2}} .$$

### 1.4 Power entered into a fiber

Let's show, that only part of light which is radiated a small diffuse source, placed on an axis of sighting of a fiber near to his butt end, can be entered into a fiber. Let's consider a small diffuse light source, the brightness which one is identical in all directions figured in figure 6. Let I0- power which is radiated in unit of solid angle on a normal to a source, I (0) = $I_0$cos0 - power which is radiated bevel way 0. Then power which is radiated in small solid angle δθ, is peer:

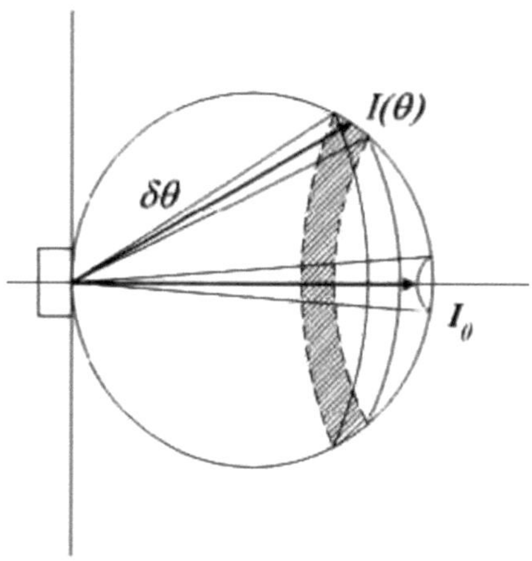


Fig. 6.

$$I_0 \cos\theta\,\delta\theta = I_0 \cos\theta\, 2\pi \sin\theta\,\delta\theta .$$

The total power which is radiated such source, is by integrating of this expression on all directions:

$$\Phi_0 = \int_0^{\pi/2} (I_0 \cos\theta)(2\pi)(\sin\theta) d\theta = 2\pi I_0 \int_0^{\pi/2} \sin\theta\, d(\sin\theta) =$$

$$= 2\pi I_0 \left( \frac{\sin^2\theta}{2} \right) \Bigg|_{\theta=0}^{\pi/2} = \pi I_0 .$$

The power, gated in in a fiber, diameter of a core which one is more than diameter of a source, is determined by a following integral:

$$\Phi = \int_0^{\Omega_m} (I_0 \cos\theta)(2\pi)(\sin\theta) d\theta = 2\pi I_0 \left( \frac{\sin^2\theta}{2} \right) \Bigg|_{\theta=0}^{\Omega_m} =$$

$$= \pi I_0 \sin^2 \Omega_m = \Phi_0 (NA)^2 .$$

$$\frac{\Phi}{\Phi_0} = \sin^2 \Omega_m = (NA)^2 .$$

$$NA = \sqrt{n_1^2 - n_2^2} = \sqrt{\frac{(n_1 - n_2)(n_1 + n_2) n_1}{n_1}} \approx$$

$$\approx \sqrt{\frac{2 n_1^2 (n_1 - n_2)}{n_1}} = n_1 \sqrt{2\Delta} ,$$

$$\text{где } \Delta = \frac{n_1^2 - n_2^2}{2 n_1^2} .$$

The power entered into a fiber, depends on a numbered aperture of a fiber NA.

To enter into a fiber maximal light, it is necessary to supply large values of values $n_1$ and $\Delta$. Apparently, that best, that it can be made to use for manufacturing of a fiber glass with large refractive index and to not cover with his shell. However thus alongside with increase of power entered into a fiber, there are two problems:

1. The part of a surge even at full internal reflection inpours out through an echoing area. And the foregone availability of irregularities and heterogeneities on her will convert a surge, fading in air, in distributing, that results in large losses.
2. At increase $\Delta$ the intermodal dispersion is augmented, that results in signal degradation.

### 1.5 Pathway of light rayss

a. Stepwise fiber.

The refractive index of a core of a stepwise fiber n1 is a constant. A angle $\theta$, under which one the beam is diffused in a fiber, is a constant. The beam is diffused, testing total reflection on a demarcation a core - shell. Between two series total reflections a ray path straight-line.

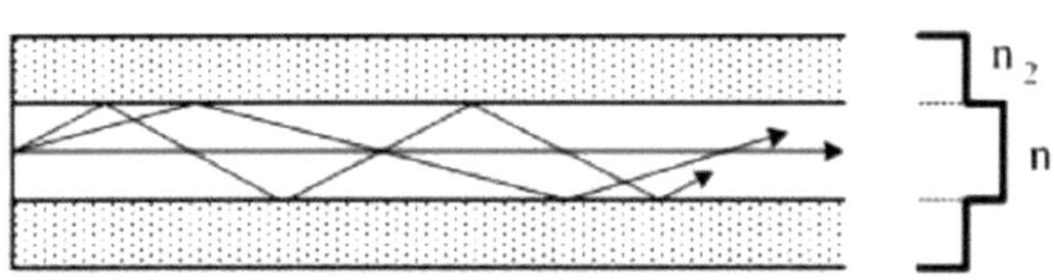


Fig. 7. A pathway of rays in a stepwise multimode fiber.

The pathway consists of equal sections received one of other by a mixing length lengthwise axis Oz on definite spacing interval and turn on a angle. Around of an axis Oz. In a transverse projection they concern the same circumference of radius R.

b. Gradient fiber.

Because of that the refractive index of a core varies, the pathway of beams in a gradient fiber has composite nature and depends on a concrete view of relation n (r). In that specific case fibers with quadratic refractive index

$$n = n_0 (1 - \frac{n_2}{2n_0} r^2)$$

The ray path in a transverse projection represents a closed curve. In a longitudinal section of a pathway are smoothly varying lines (fig. 8).

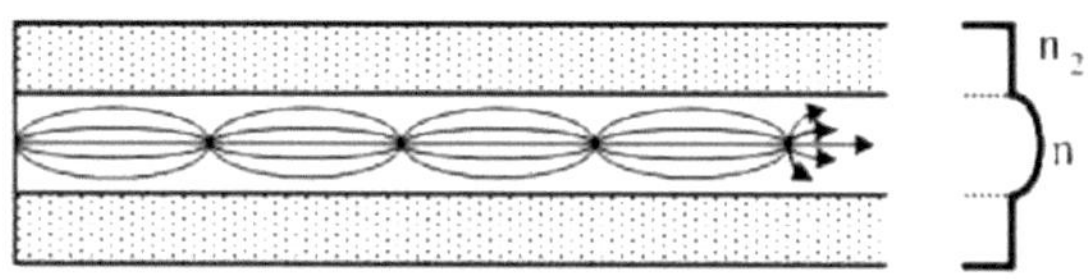


Fig. 8. Feature of such gradient fiber is that the optical lengths of paths are identical to all beams, that corresponds to absence of an intermodal dispersion.

## 2. Special optic fibers

### 2.1 Total characteristic of special optic fibers

Overwhelming majority of optic fibers for telecommunications acting on the world market - of a fiber conforming international standards: ITU T Recommendation G.652 - G.656. It, so-called main optic fibers, main problem which one - delivery of a maximum amount of information with maximum speed on maximum spacing intervals with minimum losses.

Main problem of special optic fibers - fulfilment of miscellaneous operations with light signals and flows (strengthening, modulation, filtration etc.), and also activity of fibers in special modes and conditions (for example, at high mechanical loads - impact or static, heat, irradiation, humidity, in YF mean IR and distant IR ranges). Therefore requirements to optical losses in such fibers depart on the second schedule. Representative length of special optic fibers not kilometres, as in case of main fibers, and from units up to several tens meters.

Many sires of special optic fibers dilate the clients in an orb of a biomedicine, aircraft and in military branches. Other sires see more capabilities for special optic fibers in application in sensors and fiber optic gyros. Already it becomes now clear, that in any version of further development the special optic fibers will be used in the equipment of communications networks of following breed.

Now it is possible to call about twenty phylums of special optic fibers distinguished by the design characteristics and the basic properties. Is resulted the basic items of information on some eurysynusic special optic fibers conditionally categorized on the most relevant areas of their application in optical communication below. Further in sections are resulted more in-depth information on four phylums of special optic fibers: activated, photosensing, anisotropic, photonic crystal.

### 2.1.1 Fiber, as fissile medium, for fiber lasers and amplifiers

The optic fibers, doped by erbium, are designed for erbium of fiber amplifiers (EDFA) with a broad band of the requirements to characteristics, predestined for DWDM, CATV and other applications of a telecommunication. EDFA amplifiers actuates power amplifiers, preamplifiers both linear amplifiers for C- and L-ranges.

In representative erbium the fiber amplifier doped by erbium a fiber 980 nm (or 1480 nm) is pumped by a laser diode with a wavelength to supply amplification in range 1550 nm. Erbium the fiber should be executed by such to supply a peak efficiency of absorption of pumping with a wavelength 980 nm, and also optimum signal amplification in range 1550 nm. It is executed by creation of a fiber with a high numbered aperture with representative value from 0.23 to reach reasonable overlapping of areas of a field of pumping and field of a signal. The wavelength of a cut-off of a fiber has also critical value in his design, as it determines a wavelength, on which one the fiber should work in a single mode. Representative erbium the fiber has such wavelength of a cut-off, which one guarantees, that the pumping will be diffused in a single mode ensuring maximum overlap between area of a field and area erbium of ions in a core of a fiber.

Ytterbium a fiber and ytterbium a fiber with a double shell will be used in high-power stimulus sources and amplifiers. These fibers were designed to meet the requirements to optical high-power amplifiers, industrial and military lasers, and also infrared sources. The fibers were specially designed effectively to aggregate a monomode signal and high power of pumping from the multimode diode in a passive fiber with a double shell. Integrating cheap with large output power multimode diodes of pumping on a wavelength 915 or 976 nm with these fibers is possible easily to reach high-watt power levels with effective attitude of electrical power to optical. Using stepwise fibers in a mode of continuous radiation, the output power reaches kilowatt with an angle of divergence restricted only by diffraction. In a pulse mode the mean power about 100 watt even for femtosecond can be reached fiber laser. Amplifiers with ytterbium by a fiber with a double shell - attractive technology for the phased high-power gratings. They have many advantages, including large strengthening and ease in control of a thermal way.

### 2.1.2 Fiber for pumping fiber lasers

These fibers have a multimode core conforming on the sizes to diameter of an inner shell ytterbium of a fiber used as a fissile member for fiber lasers and amplifiers. They will be used for power transmission of radiation from an optical source of pumping of a fiber laser (or amplifier) to his fissile member and delivery of an output laser emission for different applications. They can be utilised as connectors - pigtails for laser diodes of pumping and as shoulders for fiber couplers and summators. Summarizes output power from several laser diodes of pumping in one fiber, augmenting thereby power of pumping.

The data of a fiber have following features: they multimode, have a large numbered aperture (~0,45), damping on a wavelength 915 nm about 3 db/kms. Some fibers for power transmission of radiation from an optical source of pumping can reallocate back distributing light, reflected from an active fiber of the laser, which one is the main cause of failures of multimode laser diodes of pumping.

### 2.1.3 Fiber for optical multiplexers and demultiplexers

The optical multiplexers and demultiplexers of an input / conclusion usually form with usage of photosensing fibers. Capacity of an optic fiber under operating of light to change refractive index of a core is called as a photosensitivity of a fiber. When the ultraviolet radiation illuminates a core of a fiber doped by germanium, the ultra-violet photons lacerate electron-pair bindings, the refractive index of a core changes and after irradiation remains invariable. The photosensing fibers will be used for creation of fiber Bragg gratings, which one, are a main component of multiplexers and demultiplexers of an input / conclusion of radiation. The fiber Bragg grating represents an optic fiber with an alternation of refractive index along his core. Irradiating a photosensing fiber by the laser through a phase mask, it is possible to create a fiber Bragg grating.

The main property of this grating is the reflection of light, distributing on a fiber, in a narrow bandwidth, which one is centered about a Bragg wavelength. The fiber Bragg grating has a high reflection coefficient on a definite wavelength, small insertion losses, sharp selectivity of a wavelength and small crosstalks. Therefore she is the rather attractive device for the installation in multiplexers and demultiplexers of an input / conclusion. To carve out an input signal from opposite to a distributing radio echo, the optical not mutual circulator will be used. Each multiplexer of an input / conclusion has two circulators: one for input of a definite wavelength, other - for a conclusion. The circulator usually introduces losses from 0,5 up to 1 db. The insertion losses grow the more, than more gratings and circulators in the multiplexer (demultiplexer).

### 2.1.4 Fiber for optical choppers

There are two types of optical wave-guide modulators: planar and fiber. Both types are be by phase modulators more often. The planar modulator is constructed as an optical waveguide on a substrate (integrally - optical chopper). He provides modulation and coordination with fibers established on an input and an output of the planar chip, which one can be either customary monomode fibers or polarization fibers .

Alternative version of external modulators is completely fiber acousto-optical modulator. Most often completely fiber acousto-optical modulator represents devices executing frequency shift on the basis of surface acoustic waves. In them the phenomenon of communication of polarized modes in polarization fibers or spatial communication of modes in customary monomode fibers will be used.

Thus, in optical fiber modulators will be used both polarization fibers, and customary optic fibers. The monomode fibers with birefringence transmit optical radiation by two disconnected modes, which one are linearly polarized, are orthogonally related and have different phase velocities of propagation. The polarization fibers are constructed so that to transmit input light only to one linear polarization. The desirable direction of a polarization plane receives on the basis of a principle of creation of mechanical pressure, using in a fiber an elliptical shell an ambient round core or round shell an ambient elliptical core, and also other frames of a fiber.

### 2.1.5 Fibers for optical filters

Now there are many phylums of optical fiber filters: filters on diffraction or Bragg gratings, filters Fabre-Pero, etc. Fabre-Pero the filter represents a resonator consisting from two bound among themselves of optical waveguides with particulateing reflect mirrors on test leadss. The filters of Mach are constructed with usage of two directional couplers and two customary fibers, one of which is a reference shoulder, and in the friend the refractive index is varied pursuant to a control signal. The Bragg fiber filter represents a photosensing fiber, on a part which one is formed the Bragg grating. The characteristics of such fiber are submitted in section 5.3. If to change (to operate)((control)) the season of a grating of the Bragg filter, he becomes a tunable filter. The season of a grating can be changed at the expense of heating or mechanical pressure.

#### 2.1.6 Fibers for compensation of a dispersion

Indemnification of dispersion can be executed by several methods. For example, the special fibers or devices dispersions, named by compensators, (dispersion compensating modules) can be applied. These fibers have a large negative dispersion (80-100 ps/nm), and also negative slope of a dispersion curve. With the help of fibers compensatory dispersion, it is possible to execute the broad audience of operations.

The second example of indemnification of dispersion can serve fiber Bragg of a grating with the variable season. In these fibers the season changes along a fiber linearly. Thereof, the surges of miscellaneous length are mirrored from gratings arranged on miscellaneous spacing intervals from an input, that results in miscellaneous time of their propagation and accordingly to indemnification of a chromatic dispersion. All compensators with the linear season of a grating are not rebuilt devices. In rebuilt compensators the change of the season of a grating along a fiber should be non-linear. The variation of indemnification of dispersion is reached by stretching of a fiber by a mechanical or thermal way.

Thus, for indemnification of a dispersion the optic fibers with a negative dispersion and photosensing fibers will be used, from which one receive Bragg fiber gratings with the variable season.

#### 2.1.7 Fiber for sources of supercontinuum

The special examples of special optic fibers are Photonic crystal fibers. Due to a development of a series of unique properties they find a use not only in optical communication, but also in transfer of large powers, sensing sensors, nonlinear circuits and other areas. In photonic crystal fibers the area of a shell of a fiber with longitudinal air passages will be used, which one encircles a core, where the radiation is massed. Their internal periodic frame made of filled air capillary tubes represent in cross section hexagonal or square grating. The handling phylum of a grating, its step, form of air passages and refractive index of a glass allows to receive properties, which one do not exist for customary fibers. So, for example, the brightly expressed non-linear properties do photonic crystal fibers capable to generate supercontinuum, i.e. to convert light of a definite wavelength to the public with more by lengthy and more by short waves. Thus, the creation of broadband light sources on new principles is possible.

### 2.2 Activated fibers for the optical amplifiers and lasers

#### 2.2.1 Stuffs for the Erbium fiber amplifiers

In fact amplifying medium of the amplifier is an Erbium fiber - the optical fiber with impurity of the Erbium ions. Such optical waveguides are produced by the same methods, as optical waveguides for a transmission of information, with attachment of intermediate operation of impregnation not foundered stuff of a core by solution of salts of erbium or operation of doping by ions of erbium from a gas phase directly in a precipitation process of a core. The wave-guide parameters of the erbium optical fiber do similar to the parameters of optical wave-guides used for a transmission of information, with the purposes of reduction losses on connections. Principled is the selection of addition elements reshaping a core of the fissile optical wave-guide, and also guard ropes of an ion concentration of erbium. The different components in a quartz glass change nature of Stark scission of energy levels of erbium ions (fig. 1.2.4). By-turn it results in change of absorption spectrums and radiation. In a fig.. 1.2.8 the radiation spectrums of ions of erbium are submitted in a quartz glass doped most often used in technology of optical fibers by the used of in technology, phosphorum and aluminium. From the introduced data it is visible, that the most broad luminescent spectrum (so, and spectrum of strengthening), amounting about 40 nm on half-height, is reached at usage as the component of aluminium. Therefore this member became indispensable component of a stuff of a core erbium of optical fibers.

The ion concentration of erbium in a core of an optic fiber actually determines his length used in the amplifier at given signal levels and pumping. The high limit of concentration of fissile ions is determined by originating of effect of cooperative up-conversion. This phenomenon is that is possible at large concentration of fissile ions the formation of clusters consisting of two and more ions of erbium. When these ions appear in an exited state, there is an exchange of energies, as a result of which one of them passes in a condition with higher energy, and second - is nonradioactive relaxation on an index plane. Thus, the part of ions of erbium occludes radiation of a reinforced signal, reducing efficiency of the amplifier.

Other direction of researches in an expansion region of a band of strengthening erbium of amplifiers, and also increase of an ion concentration of erbium is connected to looking for other (not silicate) glass-forming matrixes for a core of a fiber. Recently so has appeared considerable concern to phosphate, telluride and fluoride glasses.

Width of a luminescent spectrum for phosphate glasses is close to by silicate (fig. 1.2.14). Here of scoring for these stuffs as contrasted to by silicate matrixes no. Nevertheless, increase of concentration of erbium in phosphate glasses does not result in noticeable formation of erbium clusters, as it takes place in silicate glasses. Therefore phosphate glasses have lower coefficients of non-linear up-conversion luminescence damping in comparison with silicate glasses. It allows realizing in phosphate glasses much higher erbium ion concentrations without noticeable concentration damping, in comparison with silicate glasses.

High concentration phosphate glasses doped by erbium and ytterbium, have found the application at mining planar wave-guide amplifiers.

In spite of attractiveness telluride and fluoride matrixes, they do not find yet broad usage in optical fiber amplifiers in a kind of composite technology of an extract of a fiber.

### 2.2.2 Activated fibers with a double shell

In fiber lasers the optical pumping will be used, that is creations of inverse in fissile medium need external radiation of optical range. For example, pumpings Nd of lasers need radiation with a wavelength in region 810 nm, for Yb of lasers in the field of 910-980 nm, though it is possible to use and other lengths of surges falling in a band of absorption.

The pumping of the maiden fiber lasers implemented through a lateral area with the help of radiation of lamps - flashes. Such scheme of pumping allowed to reach efficiency of generating, that is attitude of power of generating to power of sources of pumping, no more than 5 %. It is connected, first of all, that the large part of power of pumping was not occluded. The pumping of a fiber laser through butt end of the optical waveguide utilised directly in a core. Such scheme allowed to occlude all radiated power of pumping, so and it is essential to increase efficiency of generating. However, apparently, that in this case it is impossible to use a lamp pumping because of its small brightness, and is unique by a possible source of pumping under such scheme there are lasers. Thus, for effective pumping of fiber lasers it is possible to use solid-state or semiconducting lasers, and the brightness last allows till now to enter into a monomode core power more several watt.

To raise output power of fiber lasers and to simplify input of radiation of semiconducting laser diodes in an optical fiber, have offered to use the optical waveguide with a double clad fiber - DCF. The optical waveguide of such design represents (fig. 9) monomode a core - 1 inside the multimode optical waveguide (maiden shell) - 2, surrounded second shell (polymer or from a quartz glass) with lower refractive index - 3. Out of door such design sometimes covers

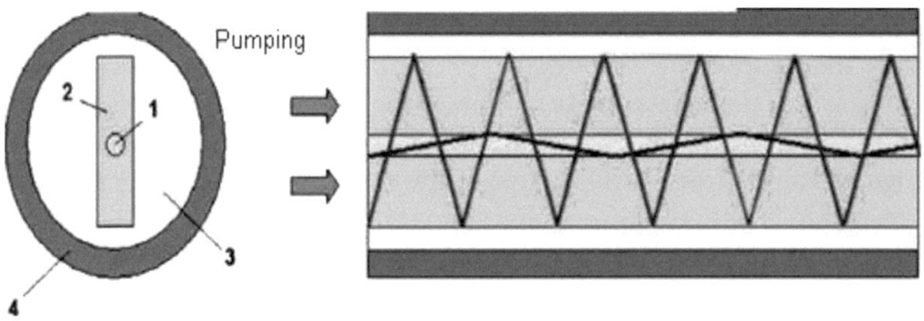


Fig. 9.

By containment shell - 4. In such frame the radiation of pumping at the expense of full internal reflection from the second shell is diffused on the maiden shell, being step-by-step occluded in a core, doped by fissile ions, on which one the radiation of generating is diffused. The area of the maiden shell can be much more area of a core, that allows to enter into such frame much more powers of pumping, than in a core. Despite of such advancing usage a lamp pumping for such fiber lasers practically is eliminated, as the maximum sectional area of the maiden shell does not exceed 1 mm $^2$, and as a rule lies within the limits of 0,01-0,1 MM$^2$. The increase of the area of the maiden shell is limited first of all to necessity to have sufficient absorption of radiation of pumping from the maiden shell. The section(cross-section) of the maiden shell can be made rectangular, and thus it is possible by a maximum mode to agree the aperture and frame of fields of the channel of pumping laser diode used for pumping and, accordingly to increase efficiency of pumping.

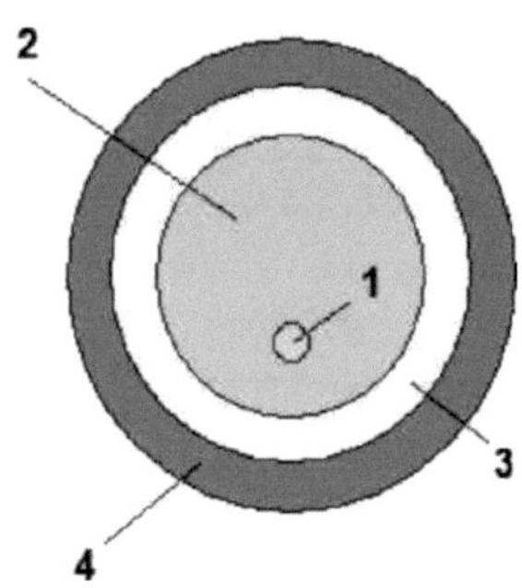


Fig. 10. An active fiber with a double shell (DCF): 1 - core, activated ytterbium, with refractive index n1, 2 - maiden shell for distribution of pumping with n2 (n2 < n1), 3 - second shell with n3 (n3 < n2), 4 protective coatings.

The absorption on a core is limited to maximum technologically accessible concentration of fissile ions, and the area of a core is limited to conditions her one mode or monomode and other parameters. Depending on a cross-section profile of the optical waveguide the lobe of modes which are not blocked with a core varies. Apparently, that best absorption of radiation of pumping needs such form of the optical waveguide, which one minimized or would eliminate existence of such modes. Besides for increase of absorption in optical waveguides accepting distribution of such modes is possible to place a core of the optical waveguide not in center (fig. 10), or to use a bending of the optical waveguide, that enriches exchange between modes intersecting a core and modes, having in centre a minimum.

### 2.2.3 Photonic crystal activated fibers

Recently rough development was received by lasers on the basis of photonic crystal fibers. Photonic crystal fiber have following distinctive features as contrasted to by customary fibers:

- High numbered aperture 0.6 (limiting idealized values 0.9);
- Large diameter of a core (up to 40 microns), which one can support a single mode. As a result of it in photonic crystal fibers it is possible to realize high powers of pumping and generating without noticeable heating;
- Absence of non-linear effects;
- A high anisotropy of frame of a fiber permitting to skip radiation with a high scale of polarization.

### 2.2.4 Photonic crystal (microstructured) fibers

Photonic crystal the fiber, in has a solid core and also has the expressed non-linear - optical behaviour. As opposed to him, for a hollow-by-a-core fiber, the non-linear - optical behaviour show is gentle. Last two types of fibers are applied as spectral selectors and to indemnification of a dispersion in fiber communication circuits.

Some advantages and lacks photonic crystal of fibers, as contrasted to customary, are adduced in tab. 1

| The characteristics | a customary fiber | FC-fiber |
|---|---|---|
| The numbered aperture, NA | 0.06 | > 0.6 is reached<br>0.9 - limit |
| Diameter of a fiber for a single mode, micron | 7<br>$\lambda$= 1540 nm | > 40<br>$\lambda$= 300 – 2000 nm |
| The area of a core, mkm$^2$ | 50 | from 3 to 1000 |
| The non-linear effects | a full set | miss or are brightly expressed |
| The losses, db/km | 0.2 are close to an idealized limit | 10- reached idealized limit 0.0005 |

Table 1. The comparative characteristics customary and photonic crystal of fibers.

From the table it is visible, that photonic crystal of a fiber can have a large numbered aperture, that easies input of radiation in them. The non-linear - optical effects in a photonic crystal fiber can be overwhelmed or, to the contrary, are increased. The losses in photonic crystal fibers, now considerably exceed losses in fibers regular style. In a fig. 11 is given frame photonic crystal of a fiber and channelling inside it the beams.

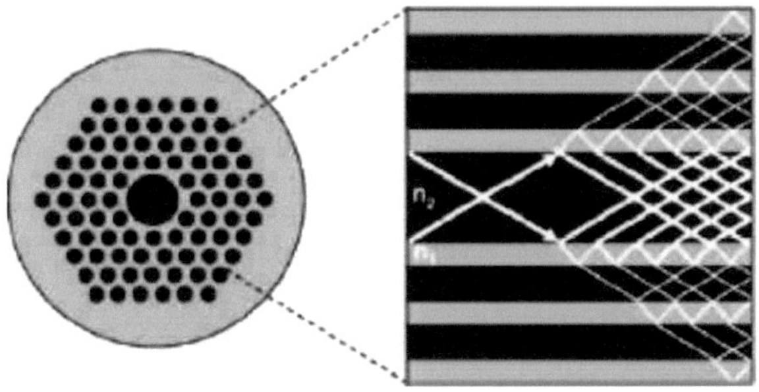


Fig. 11.

As a result of it in a core of photonic crystal are reshaped wave modes, similar to modes of common fibers (fig. 12)

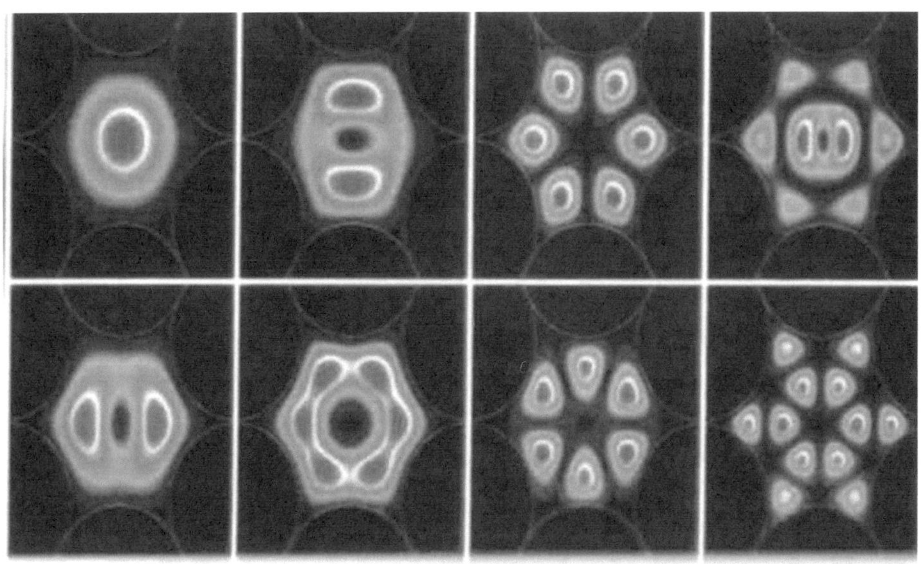

Fig. 12. Modes of photonic crystal fibers.

In a fig. 12 are shown the cross sections of some phylums photonic crystal of fibers having the special properties. The maiden phylum of a fiber a multimode fiber with the solid heart and large numbered aperture NA. Such fibers can be applied to pumping fiber lasers. In monomode photonic crystal fibers by selection of diameter of channels it is possible over a wide range to change dispersion. The similar fibers have the expressed non-linear - optical behaviour and are applied in fiber lasers, and also to control of optical signals.

In too time, theoretically photonic crystal of a fiber with empty by a core the losses at a level of 0.0005 db/kms can have.

Unique property of optical photonic fibers is strong dependence of dispersion properties from geometrical parameters of a fiber. The selection of geometry of a fiber allows to realise.

Positive, negative and zero dispersions, and also to vary a slope dispersion by a curve. Therefore photonic crystals of a fiber are perspective for usage in multiway fiber communication circuits for indemnification of a broadering of optical impulses. Photonic crystal of a fiber with a small chromatic dispersion can be utilised in rebuilt lasers, and also optical multiplexers and demultiplexers.

### 2.2.4.1 Common views about photon chips and their properties

Photonic crystal waveguides and the fibers are new phylum of optical waveguides. Their occurrence is connected to creation and research of new optical objects - photon chips.
Three types of optical fibers with frame of photon chips are now known. It is optical fibers with solid light-guided habitation, optical fibers with hollow light-guided habitation and optical fibers with coaxial frame (fig. 13). Between them there is a relevant distinction in gears ensuring wave properties of optical waveguides (fig. 14).

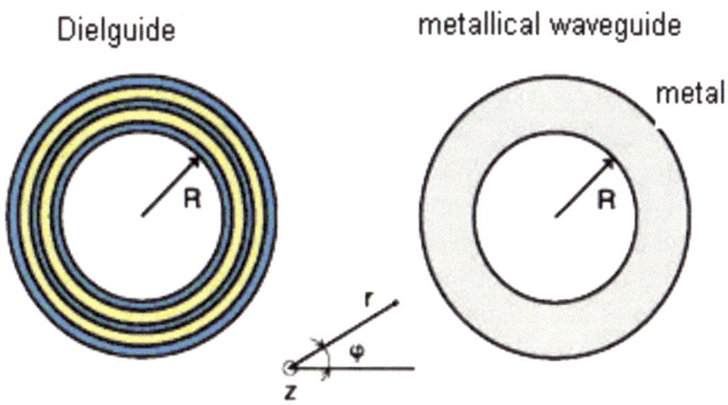


Fig. 13. A coaxial fiber of a Bragg type.

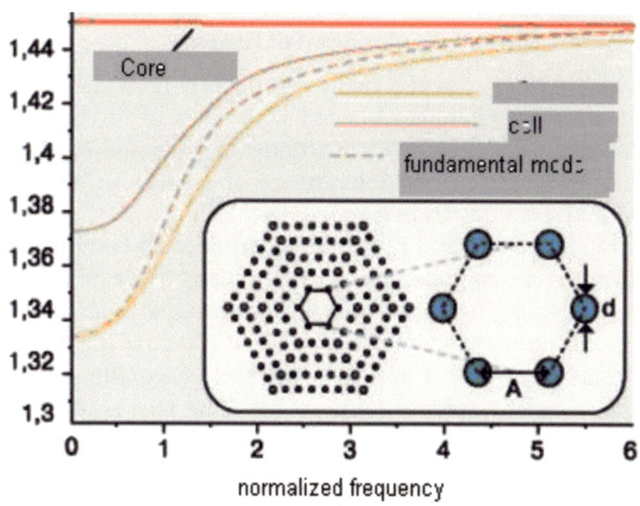


Fig. 14. Relation of effective refractive indexes and fundamental mode from the normalized frequency **A/λ**.

The perforated optical waveguide with solid light-guiding habitation represents a core from a quartz glass in a shell from a photon chip (quartz glass with air-vessels by channels), having lower mean factor of an interception in relation to a vein. Therefore wave guiding of property of such optical waveguides are provided simultaneously with two effects: full internal reflection, as in customary optical waveguides, and zonal properties of a photon chip. The availability of a shell by the way of photon chip essentially distinguishes perforated fibers from customary optical fibers.

The photon chips represent periodic frames from dielectrics with distinguished refractive index. The season of these frames - about a wavelength. Unidimensional (1D) the photon chip represents interleaving dielectric layers with high and low refractive indexes. As a rule, the optical distance of these layers is aliquot. From here is apparent, that the Bragg reflector and Bragg waveguide are at the same time unidimensional photon chips. Elementary bivariate (2D) the photon chip represents a dielectric lamina with the in batches arranged foramens. Three-dimensional (3D) the photon chip can be formed, for example, from dielectric spheres. The similar photon chip is called as a simulated opal, as his frame and the optical behaviour are close to frame and properties natural precious of a rock of an opal.

The title of photon chips is called by that the properties of photons in such periodic frames are look-alike to properties of electrons in a periodic electrical field of atoms of customary chips. It is known, that the electron has wave properties. In a customary chip there is an interference between «surge "- electron" and periodic electrical field of atoms. It results in occurrence of allowed and forbidden wave bands or energies of electrons in a chip. So there are a valence band and conduction band - ranges of energies, allowed for an electron, and forbidden region - area of energies, which one an electron in a chip receive can not. In a photon chip takes place the similar situation. The photon, which one simultaneously is an electromagnetic wave, interferes with periodic frame of a photon chip. In outcome there are ranges of allowed and forbidden energies of photons (or lengths of surges of an electromagnetic wave) in a photon chip. The photons with forbidden energies are mirrored from a photon chip; and the photons with allowed energies in him in pour. For such photons he is transparent.

#### 2.2.4.2 Property and application photonic crystal of fibers

In photonic crystal a fiber properties of a photon chip has only medium ambient his core. In dielguides regular style channelling is provided with effect of full internal reflection from border of a core of the waveguide with medium. In photonic crystal waveguides the channelling descends as a result of an interference of a surge in medium with photonic crystal by properties and reflection from it.

The representative frame of an optical fiber with a double shell is submitted in a fig. 15. He consists of three layers: a monomode core 1, doped both fissile impurity of a rare earth member, and impurity reshaping a structure PP; an internal quartz shell 2; an external polymer shell 3 with PP, under as contrasted to PP of a quartz glass. The internal quartz shell has the representative size 0.1 - 1 mm, that provides a capability of input of radiation of pumping from semiconducting sources with power some tens watt. At distribution on a quartz shell the radiation of pumping is occluded by fissile ions of a rare earth member, invoking luminescence, which one if there is a resonator formed VBR 4, develops in a lasing localized in a core of the optical waveguide, diameter by which one makes 5-10 microns. For more effective absorption of pumping the quartz shell, as a rule, has rectangular or Δ. Figurative cross section.

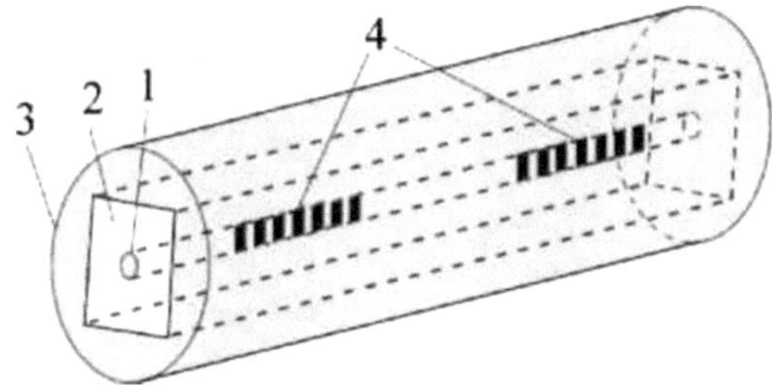


Fig. 15.

Already now on the basis of optical waveguides with a double shell are designed the laser systems have output power ~1 kW. Such systems are applied to processing of different stuffs, and also as sources of pumping for fiber lasers operating a phenomenon of an enforced Raman effect of light (VKR-LASERS).

**2.2.4.3 Photonic crystal coaxial optic fiber**

The transfer of a potent laser radiation for the technological purposes at the help fiber of optical waveguides is an actual problem of modern optoelectronics. An interrupting in implementation of fiber optic transmission systems of a potent laser radiation is the occurrence of undesirable non-linear effects: enforced dissipation VRMB and VKR and four-photon mixture. The solution of the given problem results in mining optical fibers with the increased section of a field of a dominant mode.

The photon chip, in particular unidimensional (fig. 16), is a periodic dielectric frame, the season by which one consists, as a minimum, of two layers. Let's consider the elementary example of unidimensional infinite periodic frame. The refractive index of such frame (fig. 16) is determined with the help of a periodic function:

$$\mathbf{n\ (x) = n1,\ 0 < x < h1} \tag{1}$$

$$\mathbf{n\ (x) = n2,\ h1 < x < h} \tag{2}$$

Where $\mathbf{\Lambda = h_1 + h_2}$ the season of a grating.

Electrical and magnetic permeability depend on a parameter interceptions by a conventional mode:

$$\mathbf{n_m = (\varepsilon_m\ \mu_m)^{-0.5},\ m = 1, 2.} \tag{3}$$

Thus, the uni-dimensional photon chip is anything diverse as a Bragg's mirror, consisting from alternate layers with low and high refractive index. Such frame precludes with an illumination in a definite wave band dependent on a pitch angle of a plane wave on frame. In other words for photonic crystal frames there is an area of frequencies, where the illumination is forbidden inside a stuff particulate or completely. This area is called as a forbidden region, by analogy with a solid (chip), where the areas of possible energy of electrons "are sorted out" by forbidden regions.

The effect reflection of light in such frame will be used in multilayer dielectric mirrors. Thus the optical distance of layers should be comparable to a wavelength, and also at an angular variation of dip the area of forbidden lengths of frequencies displaces.

On the other hand, the radiation can be diffused in bridge to layers and such frame represents the multiway planar waveguide, in which one there is an essential

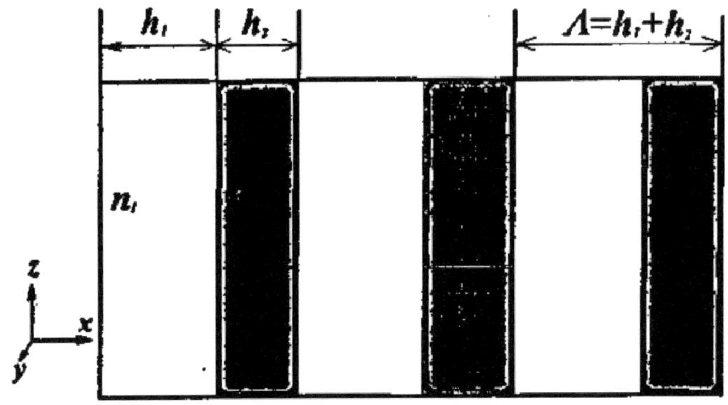


Fig. 16. An arbitrary segment of a unidimensional photon chip.
$n_1$ Refractive indexes and h of depth of the conforming layers, $\mathbf{\Lambda = h_1 + h_2}$ the season of frame. [5].

communication of channels. This communication conducts to "spreading" of dispersion curves separately taken the planar waveguide and formation of a zone, similar zones of a solid (zone of passing). The similar situations descend and in bivariate photon chips (FC), Bragg optic fibers consisting, for example, from the regularly arranged parallel dielectric barrels. The illumination perpendicularly to axes of barrels nor is always possible, while along barrels the routed surges can be diffused at any frequency. Thus, at inclined dip the excitation and routed modes of multiway waveguides and partial passing is possible.
For a qualitative analysis of properties of natural modes of a photonic crystal fiber the model of the coaxial waveguide can be utilised. The physical gear of wave-guide propagation of electromagnetic radiation in waveguides of the given type is similar to the gear of wave-guide propagation in hollow FK-fibers and is connected to availability of photon forbidden regions in a transmission spectrum of a shell of the waveguide. The bivariate periodic frame of a shell of a FC-fiber is changed within the framework of this model with a system of coaxial glass barrels (fig. 17) by depth b c by an inner radius N-th of the waveguide

$$\mathbf{r\,N = r0 + N\,(b+c),} \tag{4}$$

Where r0 - radius hollow of a core, with - depth of a backlash between walls of barrels.
The geometrical sizes of layers which are generatrix the coaxial waveguide, are selected with allowance for of space factor by air of a FK-envelope of the microstructured fiber (MS). Space factor of a shell of a fiber by air is under the formula:

$$\mathbf{\eta = \pi a2/4\Lambda.} \tag{5}$$

With allowance for of this factor are selected the parameters of the coaxial waveguide b and from (fig. 15).
The similar model allows using visual physical submissions for an estimation of dispersion properties and obtaining of a qualitative picture of distribution of intensity of electromagnetic radiation in waveguide modes localized in hollow to a core of a fiber. The relevant characteristic of waveguide modes of MS-fibers is the degree of localization of a light field in a core of a fiber. For increase of efficiency of non-linear - optical interplays in a central lode of a fiber it is required to reach, probably, higher localization of a field by reduction an effective area of a mode. This problem can be resolved by the conforming

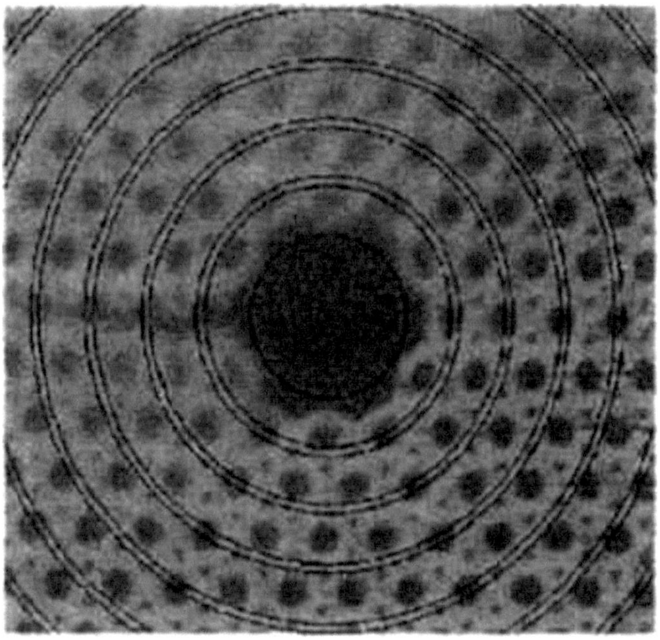

Fig. 17. The possible scheme of selection of parameters of the coaxial waveguide for simulation of a transmission spectrum of a FC-fiber. The black circles figure borders of periodic layers with different refractive indexes.

selection of attitude of the size of a core of a fiber to a wavelength of radiation and usage of a fiber with the greater difference of refractive index of a shell and core.

In a fig. 18 the schedule of a dispersion of group velocities (DGV) is submitted. The MS-fibers with two cycles of air foramens around of a central waveguide channel about a diameter about 3 microns provide an abnormal mode of a dispersion for radiation with a wavelength less than 900 nm. The wave band (from 1,05 up to 1,35 microns) with near-zero by a dispersion actuates some lengths of surges of standard radiation 1,06 microns and 1.3 microns. The affinity of a wavelength of a laser radiation to a wavelength conforming to zero value of a dispersion of a group velocity, allows to reduce to minimum influencing effects of dispersion bleed at distribution femto second of momentums impulses in a fiber and to supply fulfilment of conditions of the phase coordination for parametric processes of four-wave interplay.

To the similar requirements effectively there corresponds coaxial frame of an optic fiber with a condition of synchronization of modes [2,3]. According to the solution of characteristic equation the condition of self-conformity of a light field gives the solution for distribution of waveguide modes. The distribution optical waveguide of modes descends at a strictly definite ratio of the sizes of a layer. In the designed metalized optic fiber the light guide mode is channelled at the expense of distribution between in layers with a factor of an interception $n_c$ and spacing interval from a stimulus source - and, instituted from a ratio:

$$n_c = (n_N^2 - r_N^2 / (a^2 + r_N^2))^{0,5} \tag{6}$$

At such ratio there is a surface wave-guide radiative transfer. For normalization of distribution of radiation and exception of non-linear local effects the principle of equality of

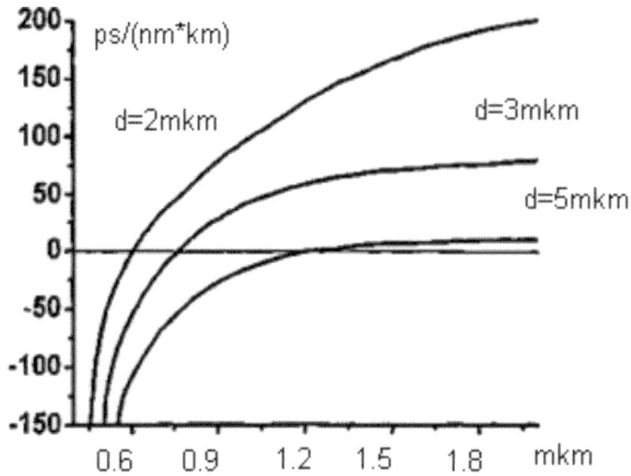


Fig. 18. A dispersion of a group velocity of a dominant mode of a MS-fiber with two cycles of ring structures around of a central wave-guide lode (d - diameter of a core).

the areas between light guide zones will be used, and the area corresponds theoretically to defined value:

$$S_N = 0,5 *3,14*\lambda*(0,5\ M)^{0,5} *(r_{N1}+r_{N2}), \tag{7}$$

Where λ - wavelength of radiation, N - of a ring-type zone, M - number of modes of radiation, rN1, rN2 - external and inner radius N- й of a ring-type zone. The layer between zones is executed from a condition of a ratio for full internal reflection with a factor of an interception nc. At a mode of distribution of refractive index conforming to the law of Gauss, there is a coordination between irradiance both refractive index and the directivity of modes are increased. As such frame was earlier represented the multiway planar waveguide, in which one there is an essential communication of channels. At which one there is "spreading" of dispersion curves separately of taken planar waveguide by the way of ring-type zone and formation of a zone, similar zones of a solid (zone of passing). Due to full filling the light guide ring-type zones of all cross section of an optic fiber descend sharp increase of skipped power up to more than in 100 times.

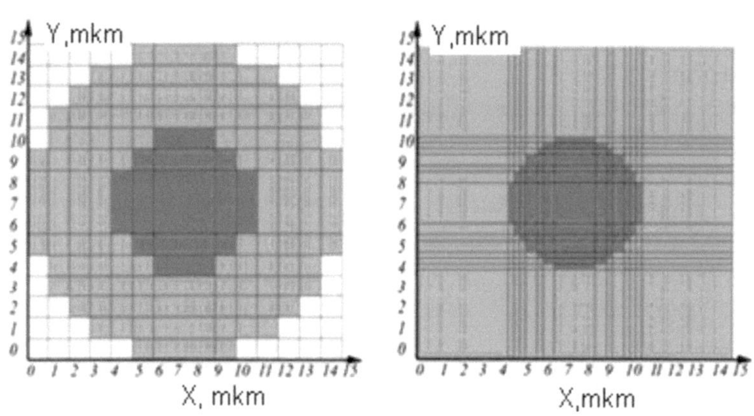


Fig. 19. Model of a standard fiber with a stepwise structure of refractive index.

The simulation of frame of an optical field conducted on a method of matched sine waves has shown the different configuration of a field pattern for a standard and coaxial optic fiber (fig. 19-22).

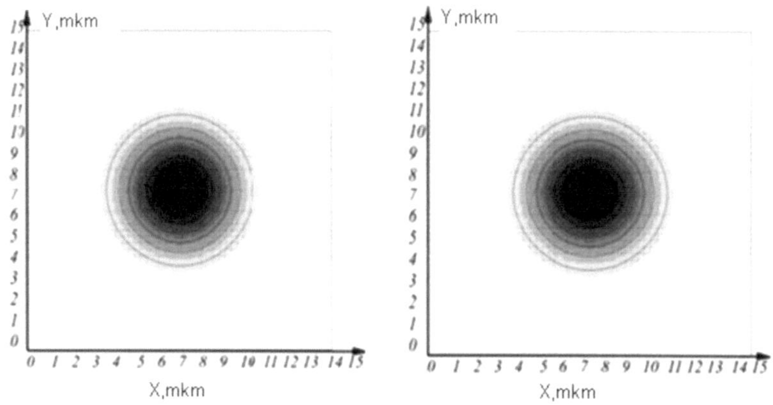


Fig. 20. Distribution of an optical field on an output of a standard fiber.

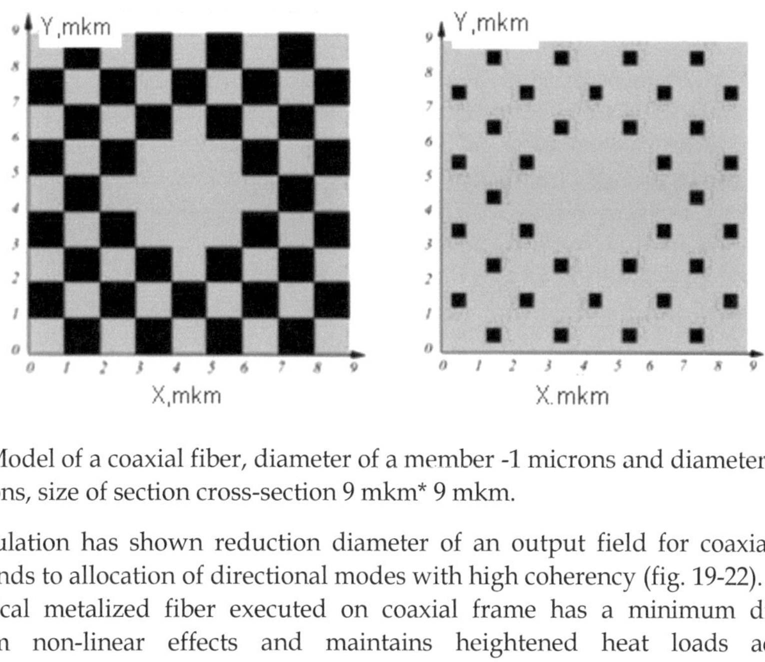


Fig. 21. Model of a coaxial fiber, diameter of a member -1 microns and diameter of a member 0.4 microns, size of section cross-section 9 mkm* 9 mkm.

The simulation has shown reduction diameter of an output field for coaxial frame, that corresponds to allocation of directional modes with high coherency (fig. 19-22).
The optical metalized fiber executed on coaxial frame has a minimum dispersion, by minimum non-linear effects and maintains heightened heat loads accompanying distribution of potent optical radiation. Allows to augment a numbered aperture and to transmit a heightened radiated power. Last researches demonstrate the friend recursive approach in the analysis of a radiative transfer in a photon chip [4,5]. The development of this direction allows to use new operating characteristics of such models for phylum of Cantor fractal.

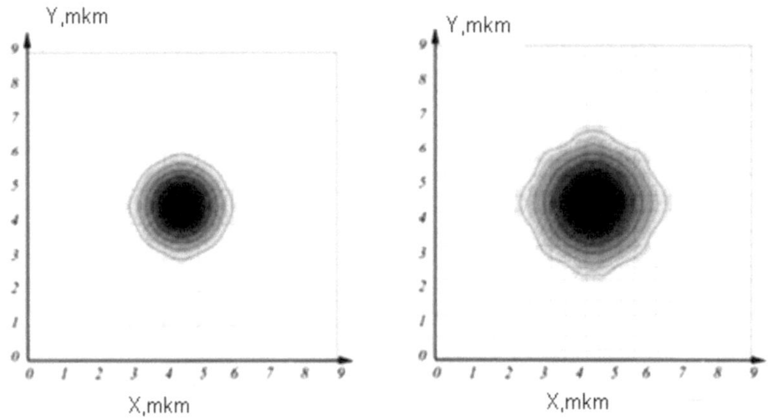


Fig. 22. Distribution of an optical field in the conforming models.

### 2.3 Anisotropic monomode optical waveguides

Alongside with trunk links of communication the optical fibers will widely be used in the diversified measuring, diagnostic and highly sensitive monitoring systems and control. On the basis of anisotropic monomode optical fibers there are sensors for measurement of different physical quantities and such unique devices as fiber optic gyros.

Most reasonable phylum anisotropic OF for a commercial production is the optical waveguide with an elliptical exerting shell.

The production process of such optic fiber lies in manufacturing of bar MSVD by a method with concentric frame of internal layers, abrasive processing of round bar and hyper thermal round off. The exerting shell contains 15-20 mol. % $B_2O_3$ and $GeO_2$, in quantity indispensable for indemnification of change of refractive index conditioned by the introducing of a boron. The isolating shell, ambient a core, is indispensable for a decrease of optical losses on a wavelength more than 1 micron conditioned by oscillation of atoms. The containment shell isolates deposited layers from hydrogen diffusing from a reference quartz tube. All shells in fiberglass have value of refractive index close to refractive index of a quartz glass.

The abrasive work on a work piece is encompass byed grooving of two flutes with diametrically opposite of the parties. At the subsequent hyper thermal round off of a flute peter, the bar becomes round, and exerting shell elliptical.

The optical waveguides with an elliptical core or exerting shell succumb on an optical behaviour OF such as "«PANDA" a little, however, is expedient differ under the cost, and also simplicity and stability of a master schedule of their manufacturing. Effecting of such optical waveguides in world practice bases, basically, on MSVD a method of manufacturing of bars. The optic fibers with elliptical members of frame can be received by one of three methods: to hyper thermal compression of a handset with marked in layers at rarefaction in some mm of a water pile; by parallel plate grinding with round off at 2100-2200°C and pressing of round bar at series heating of sites up to 1800-2000°C.

The design of an optic fiber with an elliptical shell MSVD by a method can also be produced BC such as "tie - bowtie" with application of unilateral internal etching of layers of an exerting shell on an internal surface of a handset which is heated up from the diametrically opposite parties.

Now most perspective OF for sensors are the anisotropic fibers, in which one there are pressure by definite building blocks essentially distinguished by coefficient of thermal expansion from a base material.
In the customary monomode optical waveguide with round cross section of the heart and axisymmetrical distribution of refractive index two are diffused orthogonally polarized modes $HE_{11}$, which one are accepted for meaning $HE^{x}_{11}$ and $HE^{y}_{11}$. At the introducing in a fiber of one of these modes the condition of polarization changes because of transformation to an orthogonal mode under effect of external factors: pressure, temperature, chatterings and etc. Linearly polarized radiation becomes ellipse polarized. The swapping of a quantity of light from one mode in other is conditioned by that they are vacuous, that is their propagation coefficients px and py are identical.
The condition of polarization of radiation can be kept if to break a symmetry of the form or refractive index of a core. In this case Px Py will differ, limiting a degree of metamorphosis of orthogonal modes. The optic fibers of such type are called as anisotropic monomode optical waveguides. The geometrical anisotropy forms by metamorphosis of the round form of a core in elliptical, and the anisotropy of refractive index is provided with orthogonal orientation of pressure (voltage, stresses) at usage of stuffs with miscellaneous coefficients of thermal expansion. A measure of an anisotropy of such optical waveguide is the modal birefringence:

$$B = (p_x - p_y)/(2tc\mathbf{A},) \quad (8)$$

On which one count the basis of measurement of length of beats of orthogonal modes (length, on which one the phase phase progression of polarization modes makes 2п) $L_b$:

$$B = L/L_b \quad (9)$$

Than less than length of beats, the more birefringence and, therefore, is less communication(connection) between polarization modes.
The lobe of power gated in in the optical waveguide of linear-polarized radiation Px, passed on an orthogonal (spurious) mode Py, is characterized by an extinction coefficient m:

$$m = 10 \lg (Py/Px) = 10 \lg (hL) \quad (10)$$

Where h - degree of preservation of polarization of radiation, L - length of the optical waveguide.
From this equation follows, that:

$$h = (P_y/P_x)L^{-1} \quad (11)$$

The birefringence OF with an out-of-round core having large (a) and small (b) an axes, at an ellipticity (a/b-1) more unit is proportional to a square of a difference of refractive indexes of a core and shell ($An^2$).
Conclusions: The photonic crystal fiber allows to increase the characteristics of fiber-optic links of communication and to create a new generation of telecommunication instrumentation.

## 3. References

[1] Optical fiber amplifiers. Materials, devices and applications. by ed. Shoichi Sudo // Artech House, Inc, Boston, 1997, 627 p.

[2] Becker P.C., Olsson N.A., Simpson J.R. Erbium-doped fiber amplifiers/ Fundamentals and technologies // Academic Press, SanDiego, 1999, 460 p.
[3] Canning J. Fiber laser and related technologies // Optics and lasers in engineering, 2006, №44, p.647-676
[4] Snitzer E. Proposed fiber cavities for optical masers // J. Appl. Phys., 1961, Vol.32, №1, p.36-39.
[5] Koester C. J. and Snitzer E. Amplification in a fiber laser // Appl. Opt., 1964, Vol.3 №10, p.1182.
[6] A.M. Zheltikov. Optician of the microstructured fibers // M.: science, 2004.
[7] Y. V. Sorokin. Optical fiber with plating of surface. Works collection "Optical fiber technics", 1998, p. 29-33.
[8] Y. V. Sorokin, V. V. Sorokin " Optical fiber ", Patent Russia, № 2060520, 01.04.1994
[9] Y.V. Sorokin "Metallized optical fiber", patent of Russia № 2178192 (2002.01.10)
[10] W.Takeda, S.Kirihara, Y.Miyamoto, K.Sakoda, and K.Honda, «Localization of electromagnetic waves in three-dimensional fractal cavities», Phys. Rev. Lett., (5 March 2004), DOI: 10.1103/PhysRevLett.92.093902, PACS number: 42.70. Qs, 63.20. Pw
[11] Y.Miyamoto, et all, "Smart Processing Development of Photonic Crystals and Fractals", Int. J. Applied Ceramic Technology, 1., 40-48 (2004).
[12] Y. V. Sorokin "The optical summator with Photonic Crystal". Patent Russia u.m. № 86761, 2010.
[13] Y. V. Sorokin "Optical source on the generator of supercontinuum". Patent Russia u.m. №86800, 2010.
[14] Y. V. Sorokin "The fiber optical multiplexer - demultiplexer with Photonic Crystal". Patent Russia u.m. № 86826 , 2010.
[15] Y. V. Sorokin "The fiber optic filter on Photonic Crystal". Patent Russia u.m. № 86825, 2010.

13

# Microstructured Fibre Taper with Constant Outer Diameter

Ming-Leung Vincent Tse[1], D. Chen[1,2], C. Lu[1], P. K. A. Wai[1] and H. Y. Tam[1]
[1]*The Hong Kong Polytechnic University, Hong Kong SAR,*
[2]*Zhejiang Normal University,*
*China*

## 1. Introduction

Microstructured Optical Fibres (MOFs) can be tapered in many ways depending on the length scale. Short taper of a few centimetres in length is commonly produced using the flame brush technique on a conventional fibre tapering rig (Bilodeau et al., 1988). Meso-taper of up to tens of metres in length can be produced using improved tapering rig with a ceramic microheater (Vukovic et al., 2008). It is also possible to taper optical fibre on a standard fibre draw tower, both for conventional step index fibre (Chernikov et al., 1993) and for MOFs (Tse et al., 2006b). Longitudinal variation of the MOF structure can lead to a comprehensive control of dispersion and nonlinearity, for spectral control under general conditions (Tse et al., 2008). The possibility of different long-length fibre-taper designs can lead to exciting applications in nonlinear fibre optics, such as uniform and stable supercontinuum generation for telecommunications spectral slicing (Chen et al., 2009a; Dudley & Coen, 2002; Genty et al., 2009; Vukovic & Broderick 2010), adiabatic soliton compression (Hu et al., 2006; Tse et al., 2006b, 2008), mode conversion (Town & Lizier, 2001) and pulse transformation (Broderick, 2010).

The early dispersion-decreasing microstructured fibres were fabricated by stacking of glass capillaries with radial (2D) designs consisted of uniform air hole size (Travers et al. 2007; Tse et al., 2006b). The preforms were drawn into canes (diameter in mm scale), and finally tapered down either by changing the pressure of all the air holes, or by reducing the outer diameter (OD) of the fibre during the drawing process. Both methods led to variation of hole and core sizes along the fibre. The disadvantage of such tapering schemes is that when the features reduce in size, the associated confinement loss increases (Marks et al. 2006; Nguyen et al., 2005). Therefore, a large number of rings of air holes are needed to reduce the loss to a low level; the fabrication of these MOFs is labour intensive. Moreover, decrease of the outer diameter of the fibre may reduce the mechanical strength of the fibre and handling may became difficulty, at the same time induce complication when connecting to standard fibres.

It has been reported that selective holes within the microstructure region of MOFs can be independently pressurised during the fibre drawing process (Couny et al. 2008). Together with the possibility to vary the pressure of the holes during the drawing process (Tse et al.,

2006b; Voyce et al. 2009), more complex 3D fibre designs can be achieved. A design concept is introduced for long microstructured fibre taper to be produced by stacking of silica capillaries, and to be drawn on a traditional optical fibre draw tower with multi-pressure control (Tse et al., 2009b).

In this chapter, we will investigate in detail the optical properties of these draw-tower tapers, in particular, the Effective Refractive Index ($n_{eff}$), Confinement Loss (CL), Dispersion (D) and Effective Mode Area ($A_{eff}$). In Section II, an outline of the concept of the proposed fibre tapering scheme and the simulation method will be presented. In Section III, the simulation result for the effective refractive index and confinement loss of the chosen fibre designs will be presented and discussed. Section IV investigates the dispersion profile and the effective mode area at different positions along various fibre tapers. In Section V, the proposed fabrication method and preliminary results will be presented. Final section summarises the results with conclusion.

## 2. Long microstructured fibre taper design concept

The proposed tapering scheme consists of microstructure features with large air-holes (high air-filling fraction) in every ring initially. The holes of the innermost rings are then tapered longitudinally by an independent pressure control, while keeping the outer diameter of the fibre constant by varying the draw speed at the same time. However, the variation of draw speed is not necessary for fibres with d << OD, because the variation of the glass volume is very small. Here, we investigate the simplest case, in which only the innermost ring-of-holes is varied in size. Thus, the core is made of 7-cell defects at the beginning of an index guiding taper, and the core is reduced to a 1-cell defect at the end of the taper, see Fig. 1. This tapering scheme offers low confinement loss over the entire length of the fibre, and large end-to-end core-size variation. Moreover, the mechanical strength of the fibre is the same throughout, because of the constant outer diameter. This concept has already found application in core expansion of MOFs using the flame brush method for fibre splicing, see (Chen et al., 2009b). However, the proposed draw-tower, multi-pressure MOF taper here should not be confused with tapers that are made on a tapering rig by post-processing methods after a fibre is drawn.

In this work, a full vector finite-element-method (FEM) based optical mode solver (Mode SolutionsTM 3.0 by Lumerical Solution Inc.) was used to study the fibre designs. Pure silica fibres are simulated. Dense grids consisting of large number of mesh cells (typically between 90,000 and 100,000), with emphasis at the centre region (where the guided modes were found), were used in the simulation to ensure accuracy. Perfectly matched layer (PML) boundary conditions were used. The effective index was also checked against a more commonly used simulation tool for modeling MOF (COMSOL Multiphysics), and similar results were found. The modal behaviour at different positions along various tapered MOFs for pump wavelength of 1060 nm was investigated. This wavelength coincides with that of efficient Yb-doped fibre laser sources. Dispersion profiles around 1060 nm were simulated.

MOFs with design consisting of 8 initial rings of air-holes are studied, with small hole-to-hole spacing (pitch), $\Lambda$, varied from 0.5 to 0.7 μm. The ultra-small pitch or core ensured the minimum effective mode area is obtained, and also reduced the multimodeness. The air-filling fraction, $d_{2+}/\Lambda$, of ring 2 to 8 varied from 0.6 to 0.9 for different fibres. The large

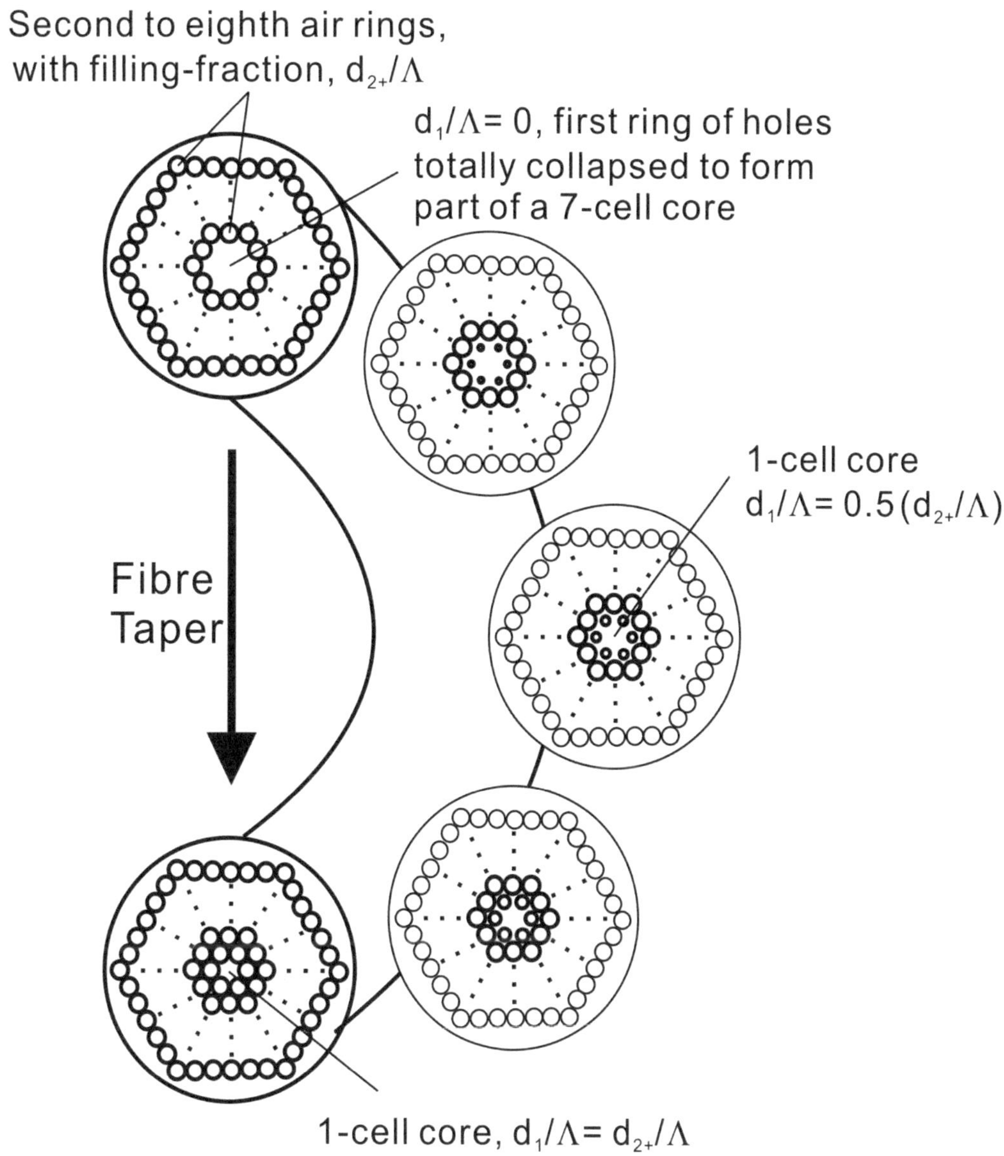


Fig. 1. A schematic of the proposed MOF taper scheme. Rings 2-8 consisted of large air-holes. The core-size is reduced by introducing the innermost ring of holes (ring 1). A 7-cell defect solid-core is found in the beginning of the fibre , which is formed partly by totally collapsing the holes in ring 1. A single-cell defect solid-core is found at the end of the fibre.

$d_{2+}/\Lambda$ ensured low confinement loss. The air-filling fraction, $d_1/\Lambda$, of the innermost ring (ring 1) was varied from 0 to $d_{2+}/\Lambda$ in each MOF taper. We studied the characteristics of each taper by simulating at least ten 2D cross-sections along the fibre. For example, when $d_{2+}/\Lambda= 0.9$, cross-sections with $d_1/\Lambda= 0, 0.1, 0.2...0.9$ were simulated.

## 3. Effective refractive index and confinement loss

In this section, we investigate the effective refractive index and confinement loss of long MOF tapers with different feature sizes. The simulated effective index of the fundamental mode and the first higher order mode along the fibres under investigation at 1060 nm are presented in Fig. 2. The result of the simulated associated confinement loss is presented in Fig. 3.

Fig. 2 shows that as the first ring-of-holes is being introduced ($d_1/\Lambda$ increases), the effective index decreases, owing to the increased amount of mode field propagating in the air-holes. Therefore, as the core-size decreases ($d_1/\Lambda$ increases), the confinement loss increases, see Fig. 3. It is more evident by looking at the mode field amplitude found from the simulated meshed structure. The mode field images of one of the taper design ($\Lambda$= 0.6 μm and $d_{2+}/\Lambda= 0.75$) are shown in Fig. 4. It shows that at the beginning of the taper with a 7-cell core ($d_1/\Lambda= 0$), the fundamental mode is well confined in the silica core. Near the middle of the taper ($d_1/\Lambda= 0.4$) and at the end of the taper ($d_1/\Lambda= 0.75$), significant amount of the mode field is propagating in the first ring of air-holes. For comparison, the mode fields of the 1$^{st}$ higher order mode are also shown in Fig. 4. The overlap of mode field and air is even more apparent for the higher order modes. The power in the air-holes will add an additional loss through scattering, as more power overlapping the air-glass boundary (White et al., 2002).

The confinement loss of the HE11 mode and higher order modes varied along the MOF tapers. We assumed for effectively single-mode guidance, the CL of the higher order mode is 10,000 times greater than that of the fundamental mode. Therefore, for each MOF taper design, part of the taper is effectively single-mode and part of the taper is slightly multi-mode. In general, the tapers are becoming more single-mode as $d_1/\Lambda$ increases. The multi-mode to single-mode transition is indicated with an asterisk in Fig. 3. For $\Lambda$= 0.6 μm and $d_{2+}/\Lambda= 0.6$, the entire MOF taper is effectively single mode, however, loss is very high (up to 100 dB/m). We have also simulated the same MOF taper with fibre bending radius of 1 mm, and found that there is negligible changes to the loss. This is expected for all the structures consider here, as the cores are very small. Bending loss is usually more significant in Large Mode MOFs (Baggett et al., 2003).

Note that, the result is presented with a general fibre length scale. In practice, depending on how the pressure is varied with time during the drawing process, the length of particular portions of the fibre can be chosen according to application requirement. A MOF taper produced on a fibre draw tower can have length of a few metres up to kilometres.

## 4. Simulated dispersion profiles and effective mode areas

In this section, the dispersion profiles and the effective mode area at different position along the fibre tapers around 1060 nm are investigated. Dispersion profiles are studied in Section 4.1 and mode field areas are presented in Section 4.2.

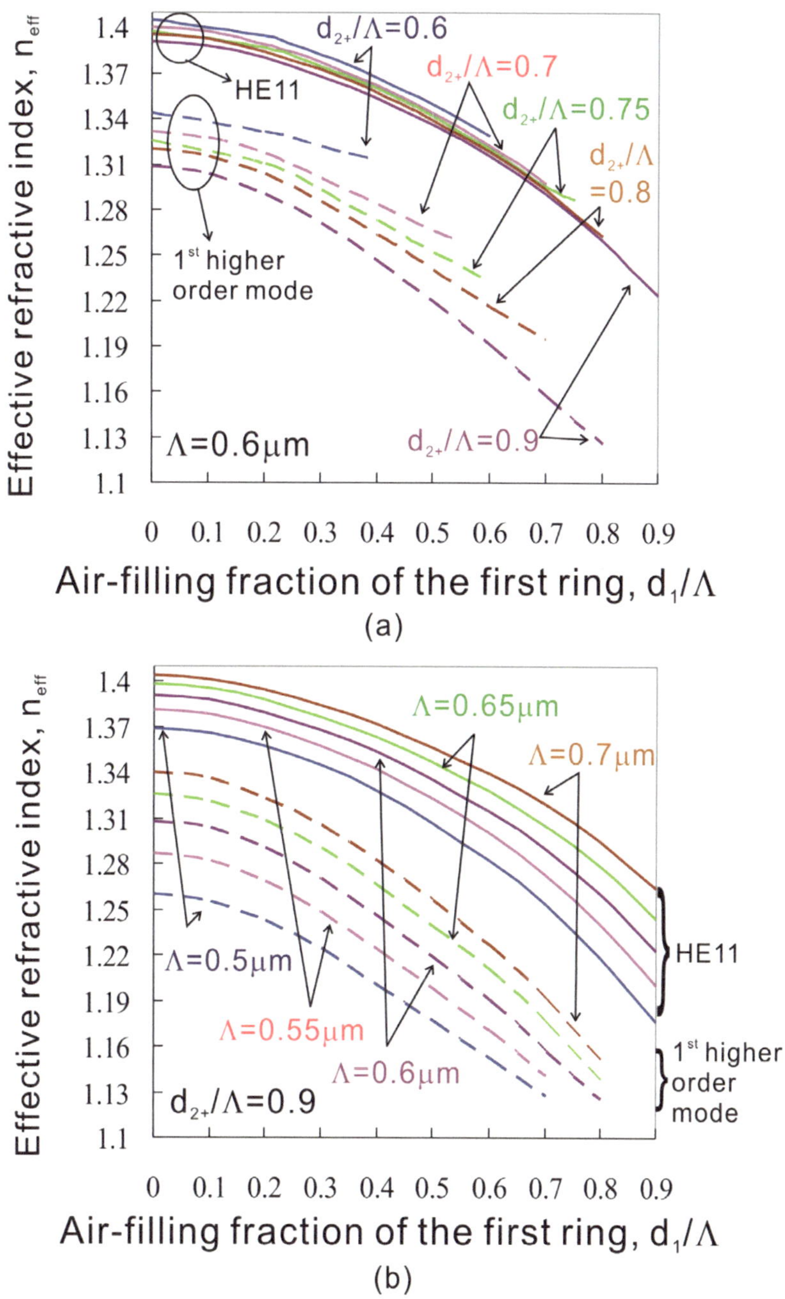


Fig. 2. Simulated effective refractive index of the HE11 mode (solid line) and the 1st higher order mode (dotted line) along tapers with different $\Lambda$ and $d_{2+}/\Lambda$: (a) $\Lambda$= 0.6 μm, $d_{2+}/\Lambda$ = 0.6, 0.7, 0.75 0.8 and 0.9 (b) $d_{2+}/\Lambda$= 0.9, $\Lambda$ = 0.5, 0.55, 0.6, 0.65 and 0.7 μm.

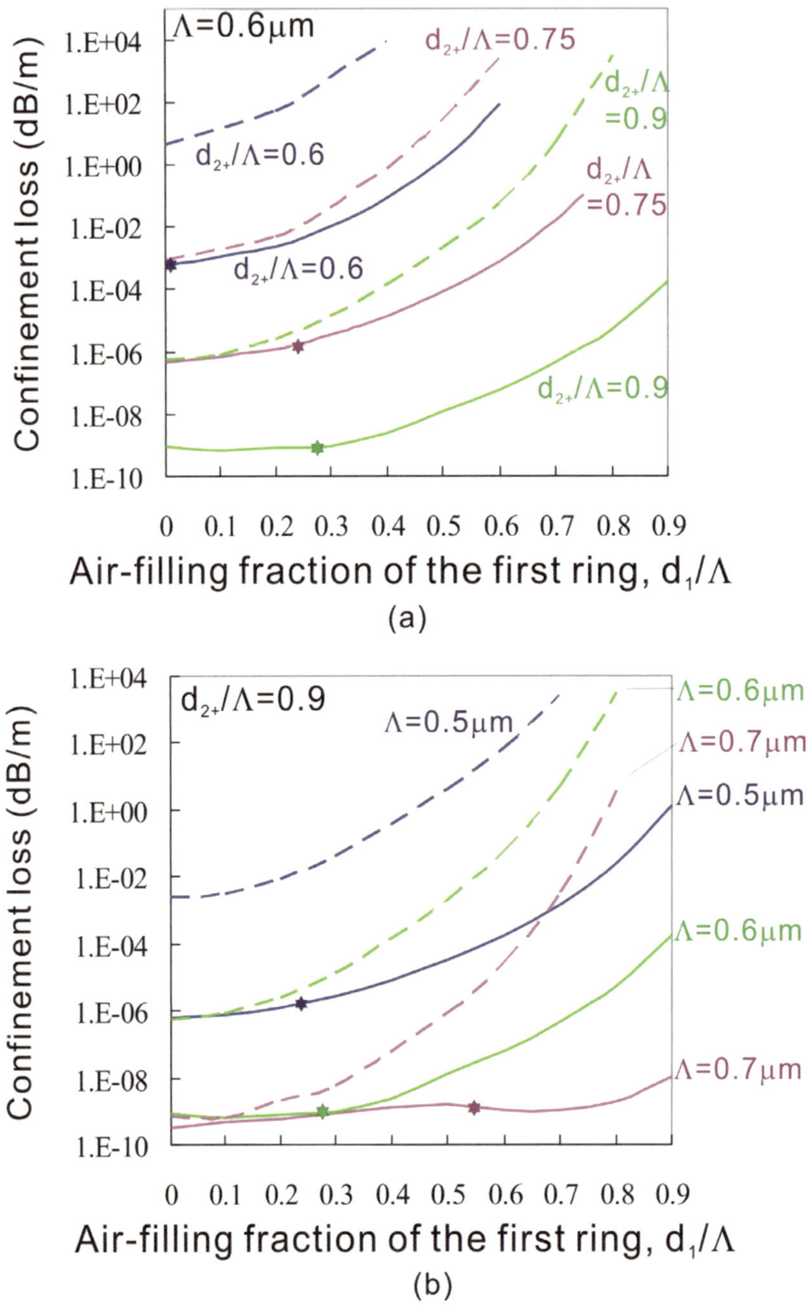


Fig. 3. Simulated confinement loss of the HE11 mode (solid line) and the 1st higher order mode (dotted line) along tapers with different Λ and $d_{2+}/\Lambda$: (a) Λ= 0.6 μm, $d_{2+}/\Lambda$= 0.6, 0.75 and 0.9 (b) $d_{2+}/\Lambda$= 0.9, Λ= 0.5, 0.6 and 0.7 μm. (Asterisks indicate where CL of the 1st higher order mode is 10,000 times larger than that of the fundamental mode)

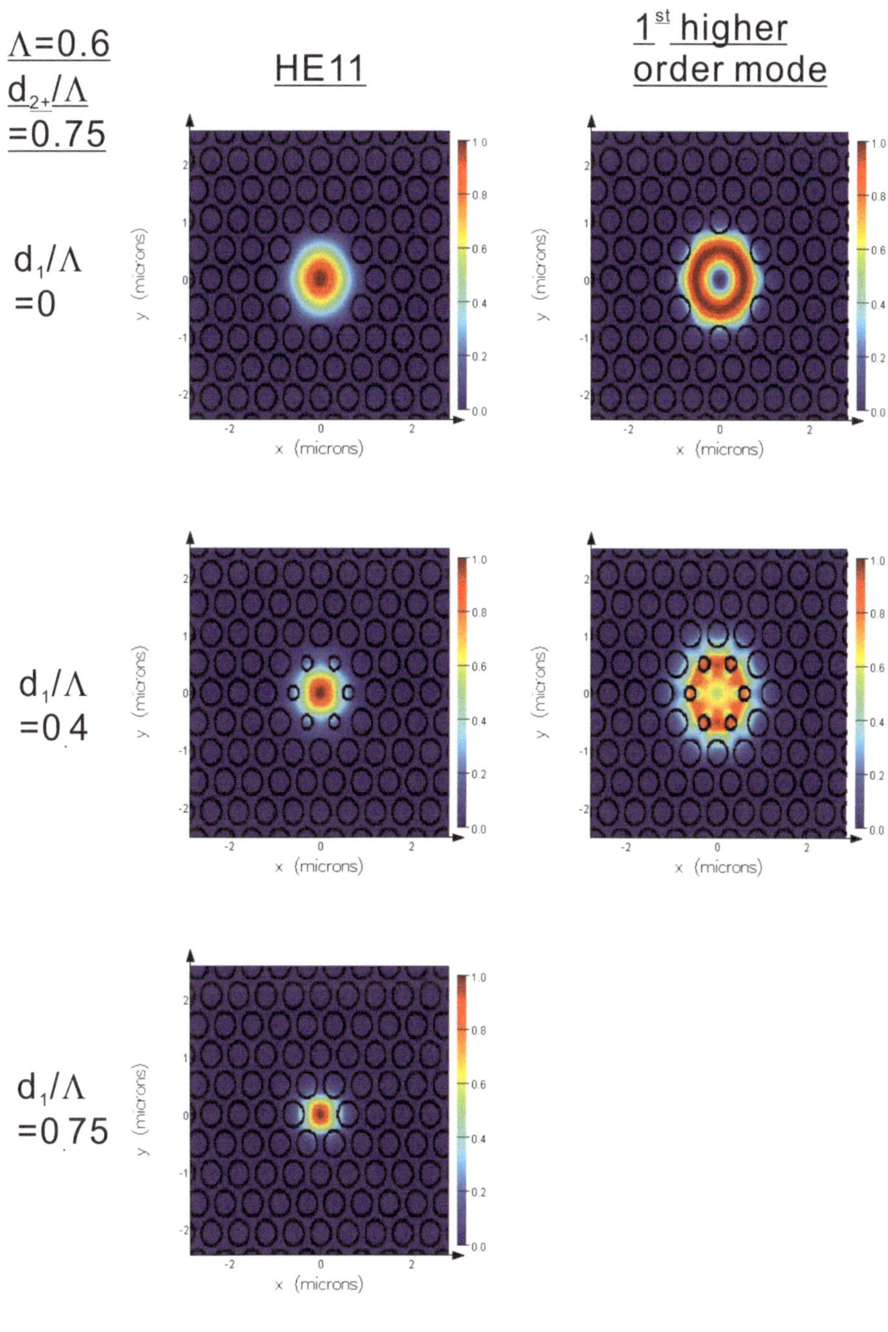


Fig. 4. Simulated modal field images of the HE11 mode and the 1st higher order mode for a taper design with $\Lambda$=0.6 μm and $d_{2+}/\Lambda$= 0.75, when $d_1/\Lambda$= 0, 0.4 and 0.75.

### 4.1 Dispersion

Dispersion profiles of the fundamental modes simulated from selected cross-sections along five MOF tapers are presented in Fig. 5. The MOF taper designs covered a large range of normal and anomalous dispersion at 1060 nm wavelength. Moreover, relatively flat dispersion slopes were found in many simulated cross-sections, especially at 1060 nm. The convex, flattened and decreasing dispersion profile would provide the ideal condition for generation of highly uniform and stable supercontinuum, which is required in telecommunications spectral slicing applications. Supercontinuum generated separately in MOFs with similar dispersion profiles pumping at 1060 nm had been studied experimentally (Tse et al., 2006a), but not in a single taper. If MOF taper technology is to be employed in optical communications networks, then the proposed constant fibre outer diameter is preferred when connecting to standard fibre via strong fusion splicing (Tse et al., 2009a; Chen et al., 2009b).

Fig. 5 shows that for various designed feature parameters, different dispersion profiles were found at different position along the tapers. In general, the dispersion curves varied with $\Lambda$ and $d_{2+}/\Lambda$ according to some predictable patterns. At around 1060 nm, as $d_1/\Lambda$ decreases, the dispersion curve moves toward the normal dispersion regime. However, the flattest point of the curve may shift away from 1060 nm wavelength, as clearly shown in fig. 5(b). For $\Lambda$= 0.6 μm, as $d_{2+}/\Lambda$ decreases, the dispersion also moves toward the normal dispersion regime for the same $d_1/\Lambda$, see for example, Fig. 5(b), (d) and (e), $d_1/\Lambda$= 0.2 and 0.3. Therefore, a greater portion (assuming constant rate of variation in $d_1/\Lambda$) of taper falls in the anomalous dispersion regime for a design with larger $d_{2+}/\Lambda$. It is less predictable for $d_{2+}/\Lambda$= 0.9 and varying $\Lambda$. The gradient of the curves varied asymmetrically in the wavelength range studied here. In practice, these results provide some valuable fabrication tolerance levels.

Individual dispersion profiles similar to or better than those shown in Fig. 5 can be achieved in fibres without longitudinal variation, however, to achieve multiple profiles in a single fibre, longitudinal variation is required (i.e. a fibre-taper).

For some applications, for example, soliton compression, a large anomalous dispersion variation may be preferred. However, if single-mode guidance is required, the dispersion profiles should be studied together with the confinement loss results presented in Fig. 3. In some cases, at the 7-cell core end, a few modes are supported in the core. For most applications, it is likely that only part of a MOF taper is useful, and not the entire length that $d_1/\Lambda$ varied from 0 to $d_{2+}/\Lambda$. The unwanted portions can either be discarded, or shorten in length during the fibre drawing process.

### 4.2 Effective mode area

The effective mode area of the fundamental mode at 1060 nm pump wavelength is studied. As shown in Fig. 6, the effective area decreases along each MOF taper as the core-size decreases with $d_1/\Lambda$ increases. For $\Lambda$= 0.5, 0.6 and 0.7 μm, part of the tapers has a sub-wavelength core size, thus greatly enhanced the nonlinearity. At the output of the tapers with $d_{2+}/\Lambda$= 0.9, the minimum effective area ($A_{eff}$) is around 0.65 μm$^2$, the nonlinearity ($\gamma$) is > 200 W$^{-1}$km$^{-1}$ at 1060 nm. For $\Lambda$= 0.7 μm, the end-to-end $A_{eff}$ ratio is about 3, $A_{eff}$= 2.02 μm$^2$ at the input, however, a large portion of the taper is multi-mode. For $\Lambda$= 0.6 μm, smaller portion of the tapers is multi-mode, but at the expense of the $A_{eff}$ ratio. For a MOF taper design with 8 rings of air holes, $\Lambda$= 0.6 μm and $d_{2+}/\Lambda$ = 0.9, $A_{eff}$= 1.58 μm$^2$ at the input, the confinement loss of the entire taper is less than 0.0002 dB/m.

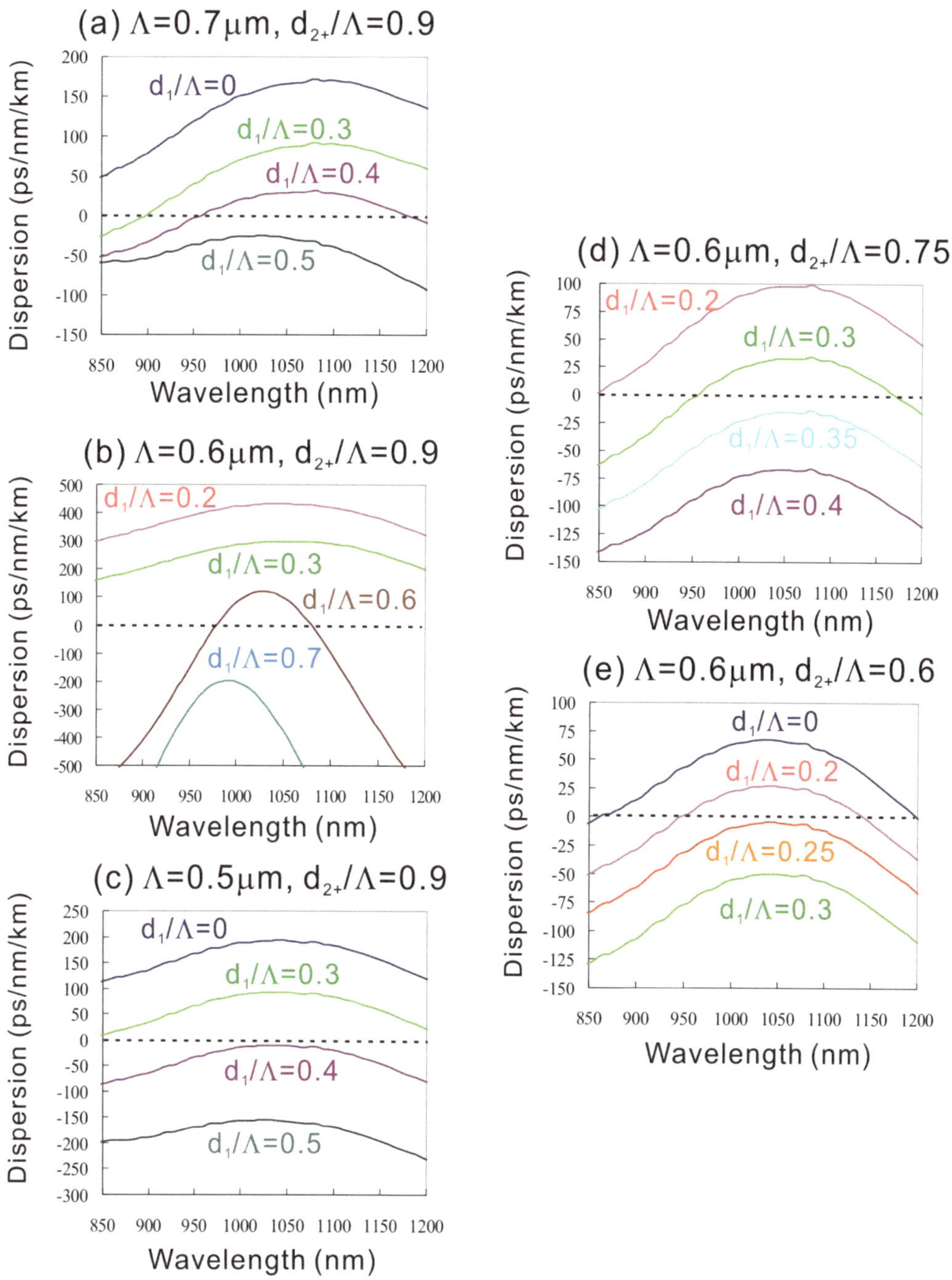


Fig. 5. Simulated dispersion profiles of the fundamental mode around 1060 nm at different position along five tapered fibres with different $\Lambda$ and $d_{2+}/\Lambda$: (a) $\Lambda$= 0.7 μm and $d_{2+}/\Lambda$= 0.9, (b) $\Lambda$= 0.6 μm and $d_{2+}/\Lambda$= 0.9, (c) $\Lambda$= 0.5 μm and $d_{2+}/\Lambda$= 0.9, (d) $\Lambda$= 0.6 μm and $d_{2+}/\Lambda$= 0.75, (e) $\Lambda$= 0.6 μm and $d_{2+}/\Lambda$= 0.6.

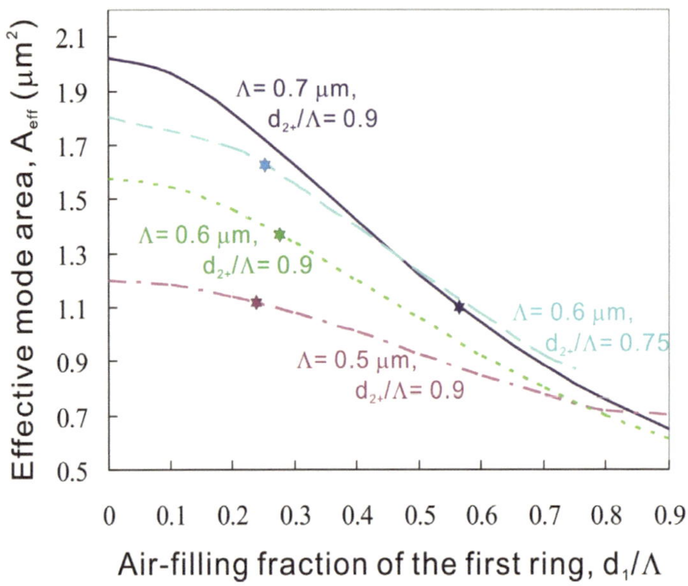


Fig. 6. Simulated effective mode area profiles along 4 tapered fibres with different $\Lambda$ and $d_{2+}/\Lambda$. (Asterisks indicate where CL of the 1st higher order mode is 10,000 times larger than that of the fundamental mode)

## 5. Fibre fabrication

In this section, a possible method for the fabrication of the proposed fibre-taper design is described by a modified capillaries stacking technique with multi-pressure hole-size control (Tse, 2007). The proposed generic method is given in Section 5.1 and the early fabrication result is presented in Section 5.2

### 5.1 Proposed fabrication method

Fig. 7 illustrates the modified capillary stacking method. The scheme consists of a traditional stack of open-ended and rigid capillaries in a holding tube, see Fig. 7(a). (Note that, the usual microstructured preform is made by stacking capillaries that are sealed at the top end.) Additional capillaries with smaller diameter are inserted into the open-ended capillaries. The outer diameter (perhaps 250 μm to 400 μm) of the inserted capillaries should be of good-fit to the inner diameter of the rigid capillaries. The flexible capillaries will provide the necessary physical elasticity for connecting to different pressure channels. Therefore, the pressure or hole-size of each hole can be controlled independently during fibre-draw. Here, we only need to consider controlling the pressure in each ring-of-holes. Since our design only require tapering one ring-of-holes (ring 1), and hole-size in rings 2 to 8 are to be kept constant, thus only two pressure regulatory channels are needed.

The modified stacking technique should work for both one-stage and two-stage fibre-draw. For one-stage draw, fibre is drawn directly from the stack. For 2-stage draw, the stack is drawn into cane first, an extra jacketing tube is added, then the jacket and cane are drawn together into fibre, see Fig. 7(b). The interstitial holes should collapse if the fibre is drawn close to 2000 $^0$C with low draw-tension.

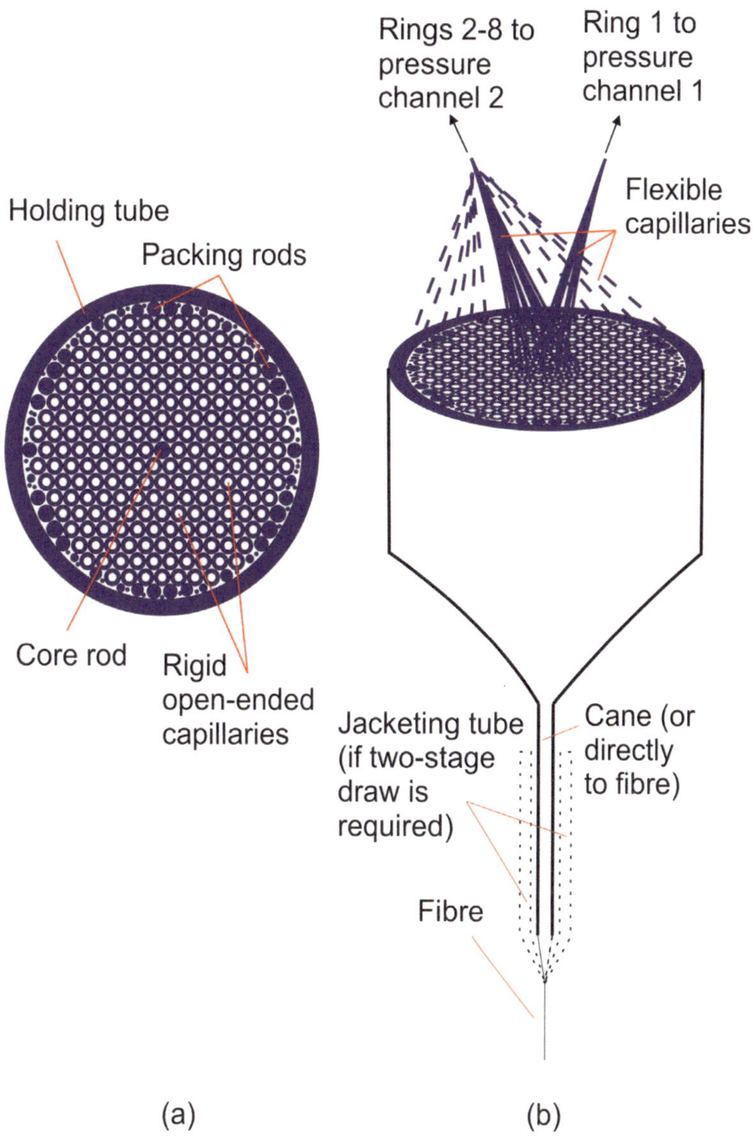


Fig. 7. A schematic of the proposed MOF taper fabrication method. (a) End-view of the stacking arrangement. (b) Shows the inserted flexible capillaries extension which can be connected to different pressure regulatory channels.

The main challenges will be on improving the pressure regulatory management system, the connection of the flexible capillaries to the pressure channels, and the feedback control of the fibre drawing system. In order to precisely control the hole-sizes, accurate active pressurisation is required. The pressures needed are often small, ranged to within a few millibars. Fibre draw tower should be equipped with fast and accurate pressure regulatory, drawing temperature and outer diameter feedback controls.

Constant pressure should be applied to all the holes when establishing the initial stable fibre drawing state. With all the feedback controls mentioned above, the pressure in the first ring-of-holes can then be decreased gradually to form the required taper structure. In previous experiment (Tse et al., 2006b), collapsing of holes using the capillaries self-pressurise method during fibre-draw can be stable and smooth. Similar stable draw can be expected for the active pressurise method proposed here, see the next section for some preliminary results.

## 5.2 Active pressure fabrication experiment

### 5.2.1 One-stage cane drawing

In this section, the effect of active pressure made to a stacked preform is experimentally investigated. A relatively loose stack with only 2 ring-of-holes is used in this experiment. The whole perform is made of high grade low OH F300 silica glass. The stack arrangement and a photo of the preform are shown in Fig. 8. The preform is 1200 mm in length, consisted of 1 core rod with diameter of 1 mm, 20 rigid capillaries with outer diameter (OD) of 0.800 mm and inner diameter (ID) of 0.520 mm, 7 flexible capillaries with OD= 0.335 mm and ID= 0.220 mm, and a holding tube with OD= 12 mm and ID= 4.3 mm.

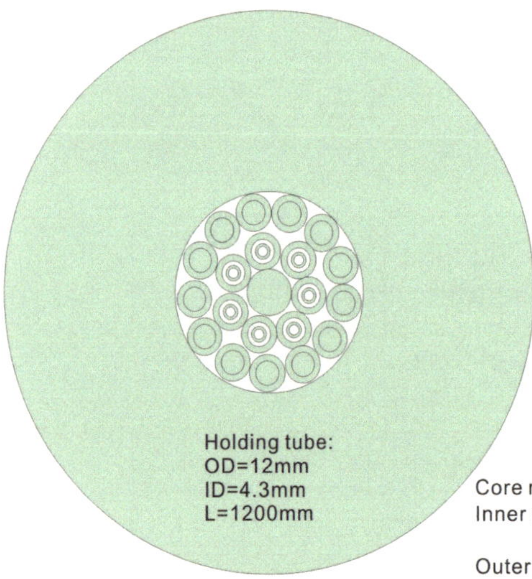


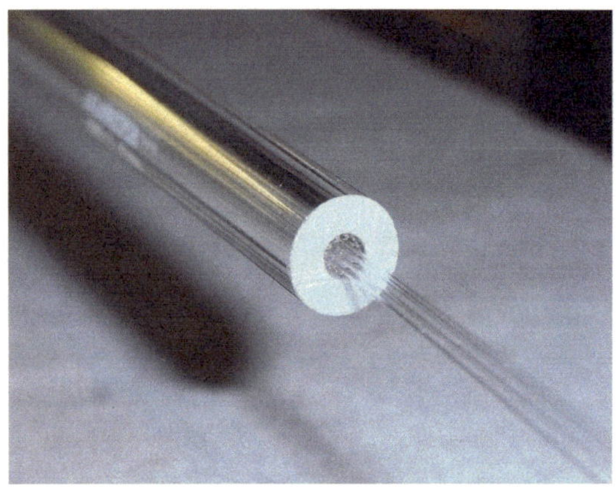

Core rod: 1000μm.
Inner ring: Open-ended capillaries with OD=800μm and ID=520μm.
Flexible capillaries with OD=335μm and ID=220μm.
Outer ring:Sealed-ends capillaries with OD=800μm and ID=520μm.

Fig. 8. (Left) A scaled schematic of the stack arrangement of the preform used in the active pressure experiment. (Right) A photo of the stacked preform with the flexible capillaries extended out. The dimensions of the stack elements are included in the figure.

The preform is arranged with the core rod surrounded by 7 open-ended capillaries, which forms the inner ring. A further 7 flexible capillaries are inserted in the inner ring, and extended out from the top of the rigid part of the prefrom. The flexible capillaries are then connected to a nitrogen pressure channel accordingly. The outer ring had 13 sealed-ends capillaries arranged closely around the wall of the holding tube. Therefore, active pressure is applied to the inner ring-of-holes only, and the outer ring is self-pressurised. According to the study carried out by Voyce et. al., the pressure inside a sealed capillary can remain relatively constant provided that long length is used and significantly extended out upward from the furnace, which is the case here (Voyce et. al., 2009). This setup will provide the necessary differential pressure for different ring-of-holes during draw to demonstrate the generic fabrication method proposed in section 5.1. The inner holes should stay open with a pressure of >2.73 kPa, suggested in the work done by Wadsworth et. al. following a simple relationship of $P_{st}$(kPa)= 600/d(μm), where Pst is the pressure required in unit of kilopascal to prevent a hole with diameter d in unit of micron from collapsing (Wadsworth et. al., 2005). The preform is drawn to cane size with OD of around 1.2 mm using a tractor belt puller. The drawing temperature was 1905 °C, the preform feeding speed was 10 mm/min, the pulling speed was 1 m/min, and the initial pressure applied to the inner ring was 8 kPa. At low temperature hole collapsing due to surface tension is reduced significantly. The furnace resistance graphite element has a diameter of >50 mm, with a capillary stack of ~4.3 mm wide loaded in the middle of the furnace, the temperature gradient across the stack is minimised. Note that, the preform is drawn without vacuum, which is often used in a traditional MOF drawing to get rid of the interstitial space and to help expanding the holes by introducing a negative pressure in the region. Thus the drawing conditions presented here led to a more accurate result of the effect caused to the capillaries by actively controlling the pressure in the holes only, with space for shrinkage and expansion.

The cane drawing result is summarised in Fig. 9, presented as microscope images of the cross-section. A total length of ~50 m of canes is drawn. Three pressure values were set during the draw, started at 8 kPa, then subsequently changed to 16 kPa → 11 kPa → 8 kPa → 11 kPa. The effect of the pressure change can be seen clearly from the pictures shown in Fig. 9 at different draw distance. By doubling the initial pressure of 8 kPa to 16 kPa, the pressure experienced by the holes was clearly over the set pressure of 16 kPa, see Fig. 9(a). This is because a lagging time is required for the pressure to stabilise by the pressure feedback loop control. It would be better to increase the pressure by small increment for a more stable hole-size increase. For the dimension of capillaries used in this preform, 16 kPa would inflate the inner holes and over power the outer ring-of-holes almost completely. At 11 kPa, the inner holes are almost at balance in pressure with similar size to the outer holes. At 8 kPa, shrinkage of the 7 inner holes relative to the outer holes is clearly visible. The outer holes were almost constant in size with pressure applied to the inner holes at 8 kPa and 11 kPa (Fig. 9(d)-(j)), which confirmed that a constant pressure is possible with long sealed-ends capillaries under self-pressurisation. It was found that decreasing pressure gave a more stable result without pressure over shooting the set point.

The study carried out in this section provide some useful data as a starting point for realising fibre-tapers on a fibre draw tower. It shows that an accurate control of selective hole-size is possible during one-stage cane-draw with active pressure. Optimal structure was achieved with a pressure of ~11 kPa, which is about 4x of the suggested pressure ($P_{st}$) required to prevent hole from collapsing.

### 5.2.2 Two-stage fibre drawing

The process of cane drawing studied in the previous section represents an one-stage draw with selective pressure control, and well controlled fibre structures were obtained. Next, a cane with the 7 inner holes connected to the same pressure control is inserted into a jacketing tube (OD= 12 mm, ID= 4.3 mm), and the outer holes were again subjected to self-pressurisation. The cane is being drawn to fibre as a two-stage draw with selective active pressure control.

The microscope image of the cane used in the fibre-draw is shown in Fig. 10(a), the diameter of the largest hole in the inner ring was assumed to be ~110 μm. The pressure applied to the inner holes was set initially at 5.5 kPa. The drawing temperature was 1905 °C, the preform

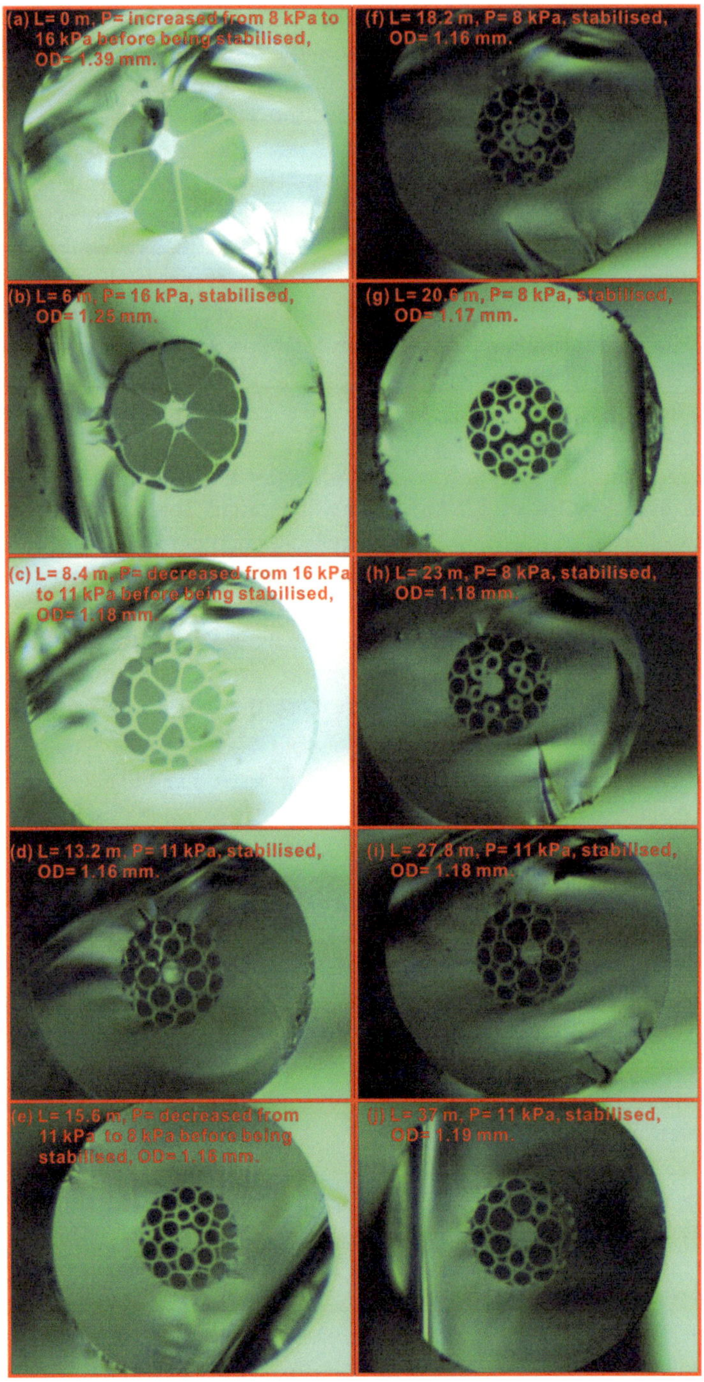


Fig. 9. Optical microscope images of the cross-section of the canes drawn from the preform with active pressure applied to the inner ring and self-pressurised for the outer ring. Draw temperature= 1905 °C, feed speed= 10 mm/min and draw speed= 1 m/min. The Draw length (L), active pressure (P) and outer diameter (OD) are included in the diagram.

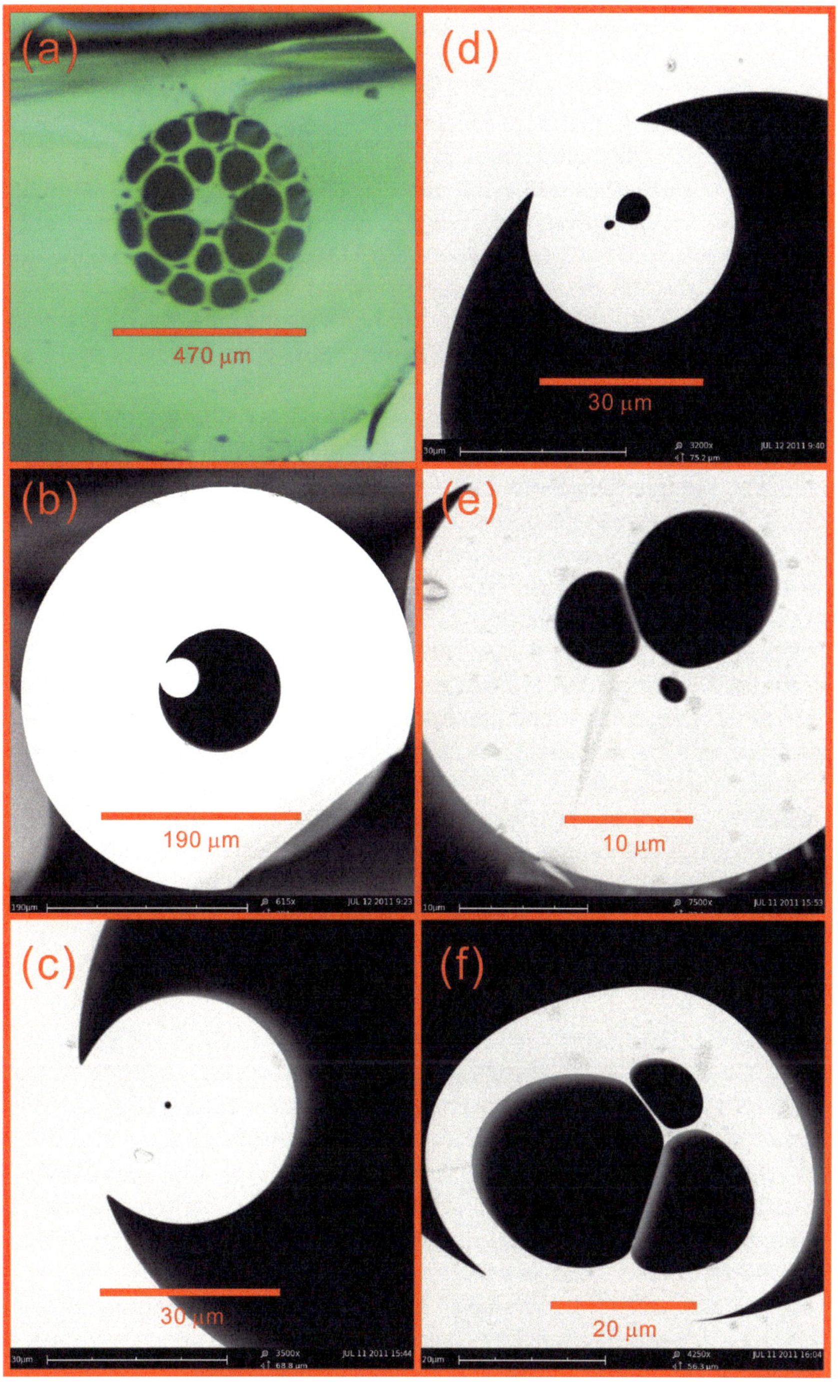


Fig. 10. (a) A photo to show the structure of the cane used in the two-stage fibre drawing experiment with active pressure control in the inner ring. (b) The SEM images of the fibre when pressure was below 18 kPa. (c)-(f) The SEM images of the fibre over the length of about 2 metres with pressure set point increased from 18 to 25.

feeding speed was 3 mm/min and the drawing speed was 3 m/min. No vacuum is applied to the space between the cane and the jacketing tube. The pressure in unit of kPa was then adjusted every few minutes to 5.8 → 6.2 → 6.7 → 7.3 → 8 → 9 → 11 → 14 → 18 → 25. The outer diameter of the fibre varied from 350 μm to 365 μm.

The SEM images of the fibre are shown in Fig. 10 (b)-(f). It was found that for pressure below 18 kPa, all the holes including the outer holes collapsed, with no structure found, SEM image of an example is presented in Fig. 10 (b). Some holes are opened and observed when the pressure was increased from 18 kPa to 25 kPa, see Fig. 10 (c)-(f). However, due to the large increment between the pressure set points, the pressure was increased too rapidly and was over shooting the set value, thus the holes in the cane massively expanded or 'burst' (picture excluded in the figure). A jacketing tube with ID closely fitted to the cane would solve the bursting problem.

Nonetheless, the effect of increasing pressure actively can be seen in a short portion of the fibre. The evolution of the fibre structure shown in Fig. 10 (c)-(f) only account for ~2 m of fibre in total. It is worth noting that not all 7 holes opened at the same time, and some may not open at all. This is because the hole-sizes of the 7 holes are different in the cane initially, and thus suggesting that it is a very important factor to consider when applying the same pressure to a ring-of-holes. The larger holes will open first and may dominate the structure.

With the drawing conditions chosen, none of the outer holes opened under self-pressurisation, and only the largest holes are observed in the active pressured inner ring. Further study need to be carried out for pressure adjusted with small increment and for pressure decreases when the optimal pressure is known. The results obtained here suggest again, a pressure of ~4$P_{st}$ is prefer for obtaining good structure, and should be chosen to establish a stable draw initially.

Selective active pressure control method for MOF fabrication is demonstrated for both one-stage and two-stage fibre drawings. Further work need to be done by employing more than one pressure/vacuum channels, and refining the pressure adjustment, capillaries and jacketing tube selections in order to achieve the required fibre design.

## 6. Conclusions

A new design concept for microstructured fibre taper that can be produced on a traditional optical fibre draw tower with multi-pressure control is proposed. A study of the simplest case with different MOF parameters is presented. The design consists of the innermost ring-of-holes varied in size along the fibre, and outer rings with large holes of constant size to provide low confinement loss. The outer diameter of the taper is preserved with effective area taper ratio of between 2 and 3 is achieved. Potentially, larger ratio is possible by tapering more than one ring-of-holes, while the outer diameter for the taper is preserved. Designs with effective mode area as small as 0.65 μm$^2$ operating at 1.06 μm are simulated. Different designs offer different advantages depending on the application requirements. In general, this tapering concept offers low confinement, dispersion and nonlinearity tailoring, and high mechanical strength and ease of handling over the entire length of the MOF taper.

Early experiments demonstrated the feasibility for the proposed fabrication method with encouraging result. Further work is required to achieve the proposed fibre taper design.

The proposed active pressure control scheme not only produces tapered fibres, but other designs in three-dimensions. Preliminary experimental results showed that hole-size can be selectively controlled longitudinally by both up-pressure and down-pressure; which suggest that fibre structure can be controlled comprehensively under general conditions. Similar scheme should also works for non-silica glass or polymer MOFs. By designing a more complex 2D fibre structure (Poletti et. al., 2005) together with the extra 3D design degree of freedom proposed here, one would expect this approach further extend the versatility of the microstructured fibre technology.

## 7. Acknowledgment

This work was supported by the University Grants Council's Matching Grant of the Hong Kong Special Administrative Region Government under the Niche Areas Project J-BB9J.

## 8. References

Baggett, J. C.; Monro, T. M.; Furusawa, K.; Finazzi V. & Richardson, D. J. (2003). Understanding bending losses in holey optical fibres, *Optics Communications*, Vol. 227, No. 4-6, (November 2003), pp. 317-335, ISSN 0030-4018

Bilodeau, F.; Hill, K. 0.; Faucher, S. & Johnson, D. C. (1988). Low-loss highly overcoupled fused couplers: fabrication and sensitivity to external pressure, *Journal of Lightwave Technology*, Vol. 6, No. 10, (October 1988), pp. 1476-1482, ISSN 0733-8724

Chen, Z.; Taylor, A. J. & Efimov, A. (2009a). Coherent mid-infrared broadband continuum generation in non-uniform ZBLAN fiber taper, *Optics Express*, Vol. 17, No. 7, (March 2009), pp. 5852-5860, ISSN 1094-4087

Chen, Z.; Xiong, C.; Xiao, L. M.; Wadsworth, W. J. & Birks, T. A. (2009b). More than threefold expansion of highly nonlinear photonic crystal fiber cores for low-loss fusion splicing, *Optics Letters*, Vol. 34, No. 14, (July 2009), pp. 2240-2242, ISSN 0146-9592

Chernikov, S. V.; Dianov, E. M.; Richardson, D. J. & Payne, D. N. (1993). Soliton pulse compression in dispersion-decreasing fibre, *Optics Letters*, Vol. 18, No. 7, (September 1993), pp. 476-478, ISSN 0146-9592

Couny, F.; Roberts, P. J.; Birks T. A. & Benabid, F. (2008). Square-lattice large-pitch hollow-core photonic crystal fibre, *Optics Express*, Vol. 16, No. 25, (November 2008), pp. 20626-20636, ISSN 1094-4087

Dudley J. & Coen, S. (2002). Numerical simulations and coherence properties of supercontinuum generation in photonic crystal and tapered optical fibres, *IEEE Journal of Selected Topics in Quantum Electronics*, Vol. 8, No.3, (May-June 2002), pp. 651-659, ISSN 1077-260X

Genty, G.; Coen S. & Dudley, J. M. (2007). Fibre supercontinuum sources, *Journal of the Optical Society of America B (Optical Physics)*, Vol. 24, No. 8, (August 2007), pp. 1771-1785, ISSN 0740-3224

Hu, J.; Marks, B. S.; Menyuk, C. R.; Kim, J.; Carruthers, T. F.; Wright, B. M.; Taunay, T. F. & Friebele, E. J. (2006). Pulse compression using a tapered microstructure optical fiber, *Optics Express*, Vol. 17, No. 9, (May 2006), pp. 4026-4036, ISSN 1094-4087

Nguyen, H. C.; Kuhlmey, B. T.; Steel, M. J.; Smith, C. L.; Mägi, E. C.; McPhedran, R. C. & Eggleton, B. J. (2005). Leakage of the fundamental mode in photonic crystal fibre tapers, *Optics Letters*, Vol. 30, No. 10, (May 2005), pp. 1123-1125, ISSN 0146-9592

Poletti, F.; Finazzi, V.; Monro, T. M.; Broderick, N. G. R.; Tse, V. & Richardson, D. J. (2005). Inverse design and fabrication tolerances of ultra-flattened dispersion holey fibers, *Optics Express*, Vol. 13, No. 10, (May 2005), pp. 3728-3736, ISSN 1094-4087

Town G. E. & Lizier J. T. (2001). Tapered holey fibers for spot-size and numerical-aperture conversion, *Optics Letters*, Vol. 26, No. 14, (July 2001), pp. 1042-1044, ISSN 0146-9592

Travers, J. C.; Stone, J. M.; Rulkov, A. B.; Cumberland, B. A.; George, A. K.; Popov, S. V.; Knight, J. C. & Taylor, J. R. (2007). Optical pulse compression in dispersion decreasing photonic crystal fibre, *Optics Express*, Vol. 15, No. 20, (October 2007), pp. 13203-13211, ISSN 1094-4087

Tse, M. L. V.; Horak, P.; Poletti, F.; Broderick, N. G. R.; Price, J. H. V.; Hayes, J. R. & Richardson, D. J. (2006a). Supercontinuum generation at 1.06 μm in holey fibres with dispersion flattened profiles, *Optics Express*, Vol. 14, No. 10, (May 2006), pp. 4445-4451, ISSN 1094-4087

Tse, M. L. V.; Horak, P.; Price, J. H. V.; Poletti, F.; He, F. & Richardson, D. J. (2006b). Pulse compression at 1.06 μm in dispersion-decreasing holey fibres, *Optics Letters*, Vol. 31, No. 23, (December 2006), pp. 3504-3506, ISSN 0146-9592

Tse, M. L. V. (2007). *Development and applications of dispersion controlled high nonlinearity microstructured fibres*, PhD thesis, University of Southampton, Southampton, U.K.

Tse, M. L. V.; Horak, P.; Poletti, F. & Richardson, D. J. (2008). Designing tapered holey fibres for soliton compression, *IEEE Journal of Quantum Electronics*, Vol. 44, No. 2, (February 2008), pp. 192-198, ISSN 0018-9197

Tse, M. L. V.; Tam, H. Y.; Fu, L. B.; Thomas, B. K.; Dong, L.; Lu, C. & Wai, P. K. A. (2009a). Fusion splicing holey fibres and single-mode fibres: a simple method to reduce loss and increase strength, *IEEE Photonics Technology Letters*, Vol. 21, No. 3, (February 2009), pp. 164-166, ISSN 1041-1135

Tse, M. L. V.; Tam, H. Y.; Lu, C. & Wai, P. K. A. (2009b). Novel design of a microstructured fibre taper, *Proceedings of 14th OptoElectronics and Communications Conference (OECC)*, ThLP32 , ISSN 978-1-4244-4102-0, Hong Kong, China, July 13-17, 2009

Voyce, C. J.; Fitt, A. D.; Hayes, J. R. & Monro, T. M. (2009). Mathematical modeling of the self-pressurising mechanism for microstructured fiber drawing, *Journal of Lightwave Technology*, Vol. 27, No. 7, (April 2009), pp. 871-878, ISSN 0733-8724

Vukovic, N.; Broderick, N. G. R.; Petrovich M. & Brambilla G. (2008). Novel method for the fabrication of long optical fibre tapers, *IEEE Photonics Technology Letters*, Vol. 20, No. 14, (July 2008), pp. 1264-1266, ISSN 1041-1135

Wadsworth, W. J.; Witkowska, A.; Leon-Saval, S. G. & Birks, T. A. (2005). Hole inflation and tapering of stock photonic crystal fibres, *Optics Express*, Vol. 13, No. 18, (September 2005), pp. 1094-4087, ISSN 1094-4087

White, T. P.; McPhedran, R. C.; Martijn de Sterke, C.; Litchinitser, N. M. & Eggleton, B. J. (2002). Resonance and scattering in microstructured optical fibres, *Optics Letters*, Vol. 27, No. 22, (November 2002), pp. 1977-1979, ISSN 0146-9592

# Part 4

## Special Characteristics and Applications

# 14

# Long Period Gratings in New Generation Optical Fibers

Agostino Iadicicco[1], Domenico Paladino[2], Pierluigi Pilla[2]
Stefania Campopiano[1], Antonello Cutolo[2] and Andrea Cusano[2]
[1]*Department for Technologies, University of Naples "Parthenope"*
[2]*Optoelectronic Division-Engineering Department, University of Sannio*
*Italy*

## 1. Introduction

The development of fiber gratings has had a significant impact on research and development in telecommunications and fiber optic sensing. Fiber gratings are intrinsic devices that allow control over the properties of light propagating within the fiber—they are used as spectral filters, as dispersion compensating components and in wavelength division multiplexing systems (Erdogan, 1997). The sensitivity of their properties to perturbation of the fiber by the surrounding environmental conditions has led to extensive study of their use as fiber sensor elements (Kersey et al., 1997). Fiber gratings consist of a periodic perturbation of the properties of the optical fiber, generally of the refractive index of the core and/or geometry, and fall into two general classifications based upon the period of the grating. Short-period fiber gratings, or fiber Bragg gratings (FBGs), have a sub-micron period and act to couple light from the forward-propagating core mode of the optical fiber to a backward, counter-propagating one (Kashyap, 1999; Canning, 2008; Cusano et al., 2009a). The long-period gratings (LPGs), instead, have period typically in the range 0.1-1 mm (James & Tatam, 2003). The LPG promotes coupling between the propagating core mode and co-propagating cladding modes. The high attenuation of the cladding modes results in the transmission spectrum of the fiber containing a series of attenuation bands centred at discrete wavelengths, each attenuation band corresponding to the coupling to a different cladding mode. The exact form of the spectrum, and the centre wavelengths of the attenuation bands, are sensitive to the period of the LPG, the length of the LPG and to the local environment: temperature, strain, bend radius and the refractive index (RI) of the medium surrounding the fiber. The peculiar spectral features of LPGs made them broadly used in many applications ranging from telecommunications to sensing (Bhatia, 1999). In particular, LPGs represent above all one of the most promising fiber grating technological platforms, to be employed in a number of chemical applications because of their intrinsic sensitivity to surrounding RI (SRI) changes (Shu et al., 2002). Up to now great efforts have been done in order to enhance the performance of LPGs in single mode fibers (SMFs) in terms of tuning capability and/or sensitivity. For instance several approaches have been proposed to achieve remarkable sensitivities such as cladding etching, LPG design for coupling to higher order modes near their dispersion turning points or in-fiber complex configuration including multi-gratings (Chung & Yin, 2004; Iadicicco et al., 2007, 2008; Pilla et al., 2008). Additionally, once the effects of depositing a thin high RI (HRI) layer onto the cladding over the grating region have been discovered, huge sensitivity enhancements in comparison to bare LPGs have been obtained due to the so-called modal transition (Del Villar et al., 2005; Cusano et al., 2005, 2006a).

On the other side, it is worth noting that new fiber designs such as D-shaped fibers and photonic crystal fibers (PCFs) (with solid and air core) capable to offer new perspective in sensing and telecommunications applications have attracted the attention of several researcher groups and scientists (Tseng & Chen, 1992; Smith et al., 2004; Gordon et al., 2007; Kaiser et al., 1974; Knight et al., 1996). However to increase the impact of the new generation fibers technology, in-fiber components such as grating filters are required. In this chapter the recent progresses of LPGs into new generation fibers will be reported. Fabrication techniques and novel applications fields offered by the hosting fiber will be discussed. In particular here the following optical fiber designs will be take into considerations:

i. *D-shaped fiber*. This category refers to a generic optical fiber showing a D-shaped transversal section. In such a fiber the core can be very close to the flat side of the "D" shape. This proximity allows access to the core electromagnetic fields more easily than in standard SMF and thus D-fiber is extremely attractive especially in sensing applications. A D-shaped fiber can be readily obtained from a standard SMF by side-polishing (Tseng & Chen, 1992) or, alternatively, is commercially available from KVH Industries, Inc. (Smith et al., 2004; Gordon et al., 2007). The possibility to combine LPGs with D-fibers has represented and still represents an open challenge for the scientific community. In this kind of fiber LPGs can be achieved impressing physical modification of the core (intra-core LPGs). Alternatively, thanks to the proximity of the core region to the flat surface, periodic modification of the effective RI of the core mode (forming the LPG) can be induced via evanescent-wave interaction if the flat surface is morphologically modified with appropriate pitch (Jang et al., 2009). Here fabrications and applications of both D-fiber LPGs are resumed.

ii. *Photonic crystal fibers*. They refer to a new class of optical fibers that have wavelength-scale morphological microstructure running down their length (Knight et al., 1996). They, according to their guiding mechanisms, may be divided into index-guiding PCFs and photonic band-gap fibers (PBFs). The former permit light to be guided in silica solid core while the second one enable the light guiding in the air core. Even if the first PCF was proposed in 1974 (Kaiser et al., 1974), the first pure silica PCF was achieved for practical use in the middle of the 1990s (Birks et al., 1995; Knight et al., 1996). Such structured optical fibers, indeed, thanks to their composite nature enable a plenty of possibilities and functionalities hitherto not possible - long range spectroscopy as well as large mode areas fiber laser just to name a few (Canning, 2008). Particular attention has been focused on hollow core PCFs (HC-PCFs) due to the lattice assisted light propagation within the hollow core (Smith et al., 2003). This particular feature, indeed, has a number of advantages such as lower Rayleigh scattering, reduced nonlinearity, novel dispersion characteristics, and potentially lower loss compared to conventional optical fibers. Fabrication of gratings in PCFs fiber still represents a challenge for the scientific community (Cusano et al, 2009; Y. Wang, 2010). Here the fabrication of LPGs in PCFs as well as the novel application fields offered by the hosting fiber will be discussed.

The next sections are organized as follows: Section 2 provides a brief review of LPGs in SMFs, Section 3 focuses on LPGs in D-shaped fibers and Section 4 reports recent progresses about LPGs in PCFs.

## 2. Long period gratings: a view back

Long Period Gratings are a periodic perturbation of the properties of the optical fiber, generally of the refractive index of the core and/or geometry, in a single mode fiber. They have periods typically ranging between 200 μm and 500 μm and lengths around 2-3 cm. The perturbation acts on the fundamental core mode enabling power transfer to a discrete set of co-propagating cladding modes that are excited at different wavelengths where a phase matching condition is satisfied. This modal coupling process determines a loss in the core mode that is reflected into a series of attenuation bands in the transmission spectrum of the optical fiber. Although they were primarily introduced as devices for optical communications (Vengsarkar et al., 1996), for which they have been used to develop band rejection filters, gain equalizers, optical amplifiers, fiber couplers, dispersion compensators (Chiang & Liu, 2006), however, they have immediately found vast application in the sensing field (Bathia et al., 1996). In fact any physical entity able to affect the difference of the core and cladding effective indices and/or the grating period and length, results in a change of the transmission spectrum in terms of central wavelength, depth and bandwidth of attenuation bands. Therefore LPGs have been investigated as sensors for a number of environmental parameters such as temperature, strain, bending and ambient RI (James & Tatam, 2003). LPGs are classically realized by exposing an optical fiber to UV lasers through an amplitude mask and exploiting the photosensitivity of silica glass. In this regard, despite extensive research on the physical mechanisms underlying the fiber photosensitivity in the past decades, there are some aspects that are not fully understood. The reason is that a number of mechanisms take part in this optical phenomenon, sometimes simultaneously, whose relative weight depends on the specific chemical composition of the fiber and drawing process, the photosensitization (hydrogen loading, flame brushing, co-doping, strain) and writing processes (irradiation power , wavelength, duration) (Vasiliev et al., 2005). The amplitude mask is usually made of a chrome- plated silica substrate that is patterned in order to have light transmitting slits alternating with reflective regions. The fiber is placed within a few millimetres behind the amplitude mask with its axis oriented perpendicular to the mask slits. A cylindrical lens focuses the Gaussian spot of the laser into a line parallel to the fiber axis. The UV light passing through the amplitude mask imprints a RI modulation onto the photosensitive fiber core thus yielding a grating with the same period as that of the mask pattern. The shortcomings of the amplitude mask technique are the restrictions on the grating period and length that are fixed by the geometrical features of the mask itself. Moreover amplitude masks can be easily damaged if they are exposed to UV light whose intensity exceeds their damage threshold thus requiring long exposure times at limited source intensities. Another widespread grating inscription method is the point-by-point writing technique, in which the grating is obtained by focusing the laser source in a single spot on the fiber and successively displacing the fiber of the required grating periodicity to induce the next index change. This method is far more flexible than the amplitude mask because length and grating index profiles are fully reconfigurable. On the other side, the former allows the grating to be written all at once and offers more precision in the spectral response which is critical for some devices such as cascaded long-period gratings.

The most frequently employed UV laser wavelengths for LPGs fabrication are 248nm and 193nm (from KrF and ArF excimer lasers, respectively) where are located strong absorption peaks due to defects of the $GeO_2$-$SiO_2$ network. Ultra-short wavelength of 152 nm from an $F_2$ laser was used to produce LPGs in SMF-28 without prior hydrogen loading. The inscribed gratings, being immune to the problem of post-writing hydrogen out-diffusion, showed higher thermal stability compared to those realized by the standard photosensitizing technique. Recently, high intensity femtosecond laser pulses at longer wavelengths (211nm, 264nm and even 800nm) are becoming a fairly widespread method to induce RI modulation through a multi-photon absorption process that does not necessarily require photosensitization (Kalachev et al., 2005). Even if the UV writing methods by means of an amplitude mask or through a point-by-point process are the most commonly and readily used writing methods in research and industry, they have certain general shortcomings: a large number of masks is required to fabricate gratings with different periods; photosensitizing pre-treatments are necessary to facilitate the RI change; UV written gratings generally suffer poor thermal stability; last but not least, UV laser sources are expensive. For these reasons several non-photosensitive techniques for grating fabrication have been investigated. In this regard, refractive index modulation produced by high temperature thermal treatments exploiting $CO_2$ lasers (10.6 µm wavelength) or an electric arc discharge has received great attention in last years. Both methods, to obtain the localized heating of the fiber, rely on a point-by-point writing approach and therefore they inherit the advantageous flexibility already mentioned. Also for these techniques there are several mechanisms that contribute to the refractive index modulation: relaxation of frozen stresses during fiber drawing, physical deformation, glass compaction or expansion, core dopants diffusion, among which the predominant cause depends on the heating treatment, the fiber type and any mechanical stress applied (Rego et al., 2005a; Y. Wang, 2010). For example, the arc discharge technique was used to form LPGs into pure-silica PCFs without any physical deformation (e.g. air holes collapse) by exploiting the glass structure change. The refractive index modulation was attributed to a glass density reduction due to the rapid heating-cooling process (Morishita & Miyake, 2004). It is worth to observe that an additional benefit of the electric arc technique lie in the fact that it is based on a very simple fabrication procedure needing inexpensive equipment. However it should be also pointed out as a major pitfall of these techniques that the intrinsic asymmetry in the heating process leads to birefringence with consequent polarization dependent losses or coupling to azimuthally asymmetric cladding modes (Rego et al., 2006).

Coming back to the applications of LPGs as sensors we can identify four physical parameters of interest: applied tensile stress can modify the effective indices of core and cladding modes through the elasto-optic effect and the grating period because of elongation ; thermo-optic effect is responsible for the effective index change while thermal expansion for period modification in the case of the temperature changes (Shu et al., 2002); bending breaks the cylindrical symmetry of the waveguide promoting coupling to azhymuthally asymmetric cladding modes that are differently affected in their effective indices depending on the region of the fiber where they are confined (Block et al., 2006); finally the effective indices of cladding modes directly depend on the index contrast between the cladding and the surrounding medium being a boundary condition in the solution of the waveguide equation (Patrick et al. 1998). An interesting feature of LPGs is that the sensitivity to a particular measurand depends drastically on the order of the coupled cladding mode and on the type of the fiber. This makes possible the discrimination of different parameters acting simultaneously on the sensor and offers the possibility to design devices that are particularly sensitive or insensitive to a given stimulus (Bathia, 1999). LPGs written in standard optical fibers offer a temperature sensitivity up to one order of magnitude larger than FBGs and strain sensitivity to almost double by appropriate choice of observed cladding mode. Altering the fiber composition to increase the difference in the thermo-optic coefficients of core and cladding can be a valuable means to achieve higher sensitivities up to 2.75nm/°C (Shu et al., 2001). Among all, one of the most appealing features of LPGs is their intrinsic sensitivity to changes of the SRI because it can serve as a basis for achieving biomolecular and chemical sensors. The first applications of this feature, however, were more like solution concentration sensors since the bare LPG alone does not possess any chemical selectivity (Falciai et al., 2001; Falate et al., 2005). The deposition of thin overlay materials that can change their RI as a consequence of a physic-chemical interaction with the surrounding environment has opened a very interesting niche of applications (De Lisa et al., 2000). Moreover, another major pitfall for bare LPGs is their scarce SRI sensitivity in low index ambient (air, water) while they show maximum sensitivity for SRIs close to the cladding RI, typically around 1.45. In this context, a paradigm shift has been represented by the integration of nano-scale polymer overlays with HRI than the cladding and by the discovery of the modal transition phenomenon (Rees et al., 2002; Del Villar et al., 2005; Z. Wang et al. 2005; Cusano et al., 2005). It is by now very well known that the SRI sensitivity of LPGs can be optimized for the specific measurement environment through the deposition of a HRI thin film by acting on its thickness (ranging in hundreds of nanometres). Sensitivities as high as thousands of nanometres for a unitary change of SRI can be easily obtained and therefore LPGs coated by HRI functional layers have been successfully exploited for chemical and biomolecular sensing (Cusano et al., 2006b; Pilla et al., 2009). Humidity sensing is a fairly investigated application, that was performed with LPGs coated by thin films of different hygrosensitive materials (Tan et al., 2005; Kostantaki et al., 2006; Liu et al., 2007; Venugopalan et al., 2008). A zeolite overlay was used in combination with LPGs to detect the presence of few ppm of toluene and isopropanol vapours (Zhang et al., 2008). A sol-gel derived coating of tin dioxide with optimized thickness for high sensitivity (≈200 nm) was used to detect ethanol vapours claiming a resolution of 1ppm (Gu et al., 2006). A partially etched LPG with cladding substituted by a polymer coating of finely tuned RI and able to perform solid-phase microextraction of organic solvents such as xylene, cyclohexane and gasoline was demonstrated. The extra peculiarity of this study being the interrogation system potentially highly miniaturizable and based on the concept of the cavity ring down spectroscopy (Barnes et. al, 2010). A very sensitive probe for pH was manufactured by means of electrostatically self assembled multilayers without the use of colorants. The transduction principle was the swelling of the overlay as a consequence of increased concentration of hydrogen ions (Corres et al., 2007).

In the never ending quest for increased SRI sensitivity, a growing interest, both theoretical and experimental, has been recently shown also for the possibility to excite surface plasma waves by means of cladding modes (Tang et al., 2006, He et al., 2006). A Pd-coated LPG was used as hydrogen sensor (Wei et al., 2008). A particular dispersion behaviour of one of the cladding modes obtained for a specific grating period at a certain wavelength, the so-called turn around point (TAP), together with functional coatings of synthetic or biological nature, was exploited to obtain ultra-sensitivity for volatile organic compounds or biomolecules detection (Chen et al., 2007; Z. Wang et al., 2009; Topliss et al., 2010). It should be noted that the number of biosensing applications with LPGs is rapidly growing and it can be foreseen that it will represent an area of major interest in coming years (Eggen et al., 2010; Smietana et al., 2011). In this context LPGs realized in Photonic Crystal Fibers are extremely attractive for the possibility to achieve very intense light matter interactions with a unique optofluidic design and with nano-liter sample consumption (Rindorf et al.,2006; He et al., 2011). However, the attractiveness of the SRI sensitivity of LPGs is not limited to the field of sensing applications and it is extended to the optical communication domain for the possibility to develop tuneable filters and optical modulators (Yin et al., 2001; Chung et al., 2004; J. Lee et al., 2007). An interesting and relatively new trend in LPGs made in classical telecom fibers is the fabrication of compound structures characterized by spectral details of finer scale for higher resolution in the measurements of environmental parameters or to obtain a compensation against cross-sensitivities (D. Kim et al, 2006; Pilla et al., 2008; Jiang et al.,2009; Mosquera et al., 2010) .

## 3. Long period gratings in D-shaped fibers

The main advantage of LPGs over short-period FBGs is their intrinsic SRI sensitivity. Nevertheless, the fiber section geometry strongly influences the sensitivity characteristics of the considered LPG, in terms of SRI as well as in terms of the other external parameters able to induce changes in the grating spectrum: temperature, strain, bending, etc. In general, one of the most obvious manner to increase the interaction of the light propagating within an optical fiber and the surroundings is represented by the reduction of their distance from the core layer. The fiber structure that better satisfies such a need avoiding the micro-structuring of the fiber itself is the D-shaped optical fiber. In such a fiber the core can be very close to the flat side of the "D" shape maintaining a certain robustness of the fiber structure, especially if compared with a SMF uniformly thinned - preserving its azimuthal symmetry - to reach the same distance of the core from the surroundings. Obviously, a first type of D-shaped fiber can be readily obtained from a standard SMF (see Fig. 1(a)) by side-polishing (Tseng & Chen, 1992). On the other hand, a special D-shaped fiber is commercially available from KVH Industries, Inc.: it is a polarization maintaining SMF. Such a structure has been successfully exploited in the past for applications in both telecommunications and sensing (Smith et al., 2004, 2006; Smith, 2005; Gibson et al., 2007; Gordon et al., 2007). Note that slightly different geometrical features have been reported for this D-fiber by the different research groups involved with it in the past. However, Fig. 1 tries to compare the transversal geometrical features of a standard SMF with those of the D-fiber supplied by KVH. Differently from the standard SMF, the D-fiber is a three-layer structure. In particular, it presents an elliptical Ge-doped core (major and minor axis of ~5 and ~2.5 μm, respectively, and RI of 1.4756) with the major axis parallel to the flat side, an elliptical inner fluorine-doped depressed cladding (~22×18 $\mu m^2$, RI of 1.441), and a D-shaped undoped silica supercladding (RI of 1.444) The distance of the core layer from the flat surface of the D-fiber is of ~13.5 μm. Note that the maximum transversal dimension of the D-fiber is exactly the same of the standard SMF: 125 μm. Evidently, the possibility to combine LPGs with D-shaped optical fibers have represented and still represent an open challenge for the scientific community. In this section, the scientific efforts already carried out in this field are resumed. The subject is treated as follows: first the attention is focused on the fabrication of LPGs into D-fibers, dividing the category in gratings involving physical modification of the core layer (intra-core LPGs) and gratings obtained by evanescent-wave mechanism; successively the different applications proposed for such structures are discussed.

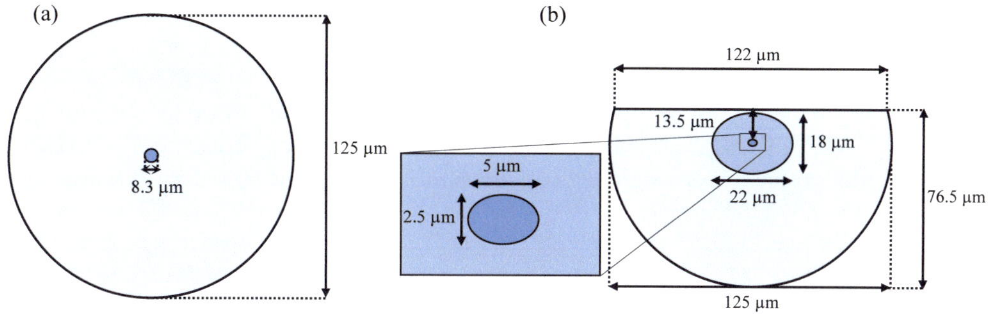


Fig. 1. Schematic diagram of the cross-section of (a) a standard SMF and (b) a D-fiber by KVH Industries, Inc. (not in scale).

### 3.1 Intra-core LPGs

Classically, a LPG is an axially periodic RI variation inscribed in the core of a photosensitive SMF by UV irradiation, which couples light from the fiber core into the cladding modes at discrete wavelengths. The index modulation produces a set of attenuation bands seen in the transmission spectrum of the optical fiber. The first paper presenting a LPG written in D-shaped SMF was dated 2004 (Allsop et al., 2004). Few data were given about the fiber structure: the D-fiber was originally designed for coupler fabrication. The core's radius was 4 µm, and the distance between its center and the flat of the "D" was 9 µm with a cladding radius of 62.5 µm. The core was a composition of $GeO_2/SiO_2$ and the cladding was assumed to be $SiO_2$. The D-fiber was not specifically designed to be photosensitive and so its photosensitivity was increased by hydrogenation at a pressure of 120 Bar for two weeks. The LPGs were fabricated using a frequency doubled argon ion laser at a wavelength of 244 nm with a point-by-point writing technique. Several grating periods were used from 140 to 400 µm with a grating length of 5 cm. Scrutinizing the transmission spectrum during fabrication, it was noticed that the attenuation bands grew in strength with a red shift but this strengthening and red shifting continued post-fabrication with shifts well in excess of 150 nm followed by a roughly comparable blue shift. An example of part of post-fabrication spectral evolution is shown in Fig. 2. The authors hypothesized this behaviour was due to different $H_2$ diffusion rates from core and cladding.

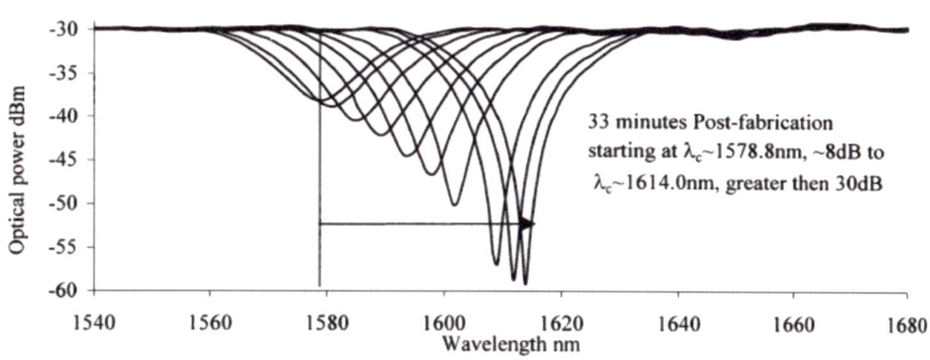


Fig. 2. Post-fabrication spectral evolution of an attenuation band from a D-fiber based LPG (period of 400 µm). Reproduced with permission from (Allsop et al., 2004).

In the same year, LPGs were also fabricated within the D-shaped fiber supplied by KVH (Chen et al., 2004). To provide for comparison, LPGs were UV inscribed in both the D-fiber and standard SMF employing the point-by-point fabrication technique and a continuous-wave frequency-doubled Ar laser of 100 mW power. Prior to UV exposure, the fibers were photosensitized by a standard $H_2$-loading treatment. Following inscription, the gratings were stabilized by thermal annealing at 80°C for 48 hours. Figs. 3(a) and (b) show typical spectra for two 4 cm long LPGs with periods of 490 and 380 µm in standard SMF and D-fiber, respectively. In the D-fiber case, the birefringence results in the presence of two sets of broad loss peaks corresponding, respectively, to the two orthogonal polarization states.

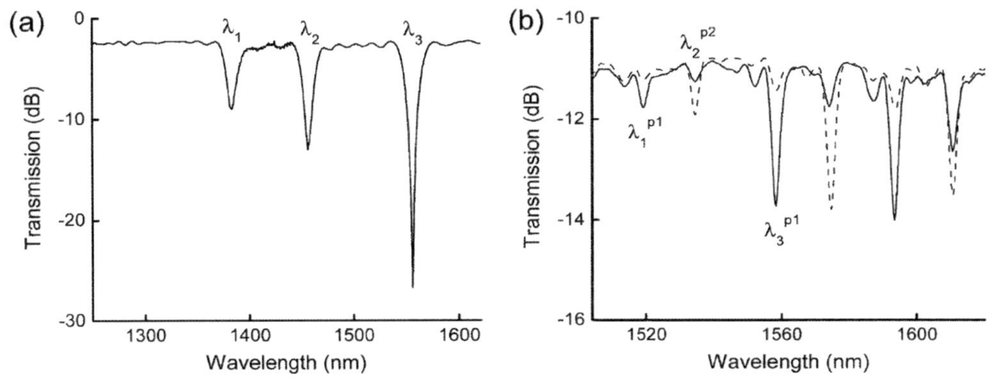


Fig. 3. Typical transmission spectra of LPGs in (a) standard SMF, and (b) D-fiber. There are two sets of resonances in D-fiber grating spectrum, corresponding to the two orthogonal polarization states, $P_1$ and $P_2$. Reproduced with permission from (Chen et al., 2004).

Obviously, D-fiber based LPGs can be also obtained by grating writing in standard SMF and successive side-polishing of the fiber section containing the grating (Tien et al., 2009a). In that case, a 2 cm long LPG with period of 380 µm was written within a $H_2$-loaded standard SMF using a KrF excimer laser with a wavelength of 248 nm and an amplitude mask. After grating writing, the fiber section containing the LPG was double-sided-polished: during the process, the polishing depth was monitored by checking the transmitted light power levels. Fig. 4 shows the original LPG transmission spectrum and that obtained after polishing: a red shift in the range of several nanometres is observable.

Finally, special attention has to be dedicated to the air-gap LPG (AG-LPG) first proposed in 2009 (Fu et al., 2009). Differently from standard UV written LPGs, in fact, here the grating is an axially periodic structural modification of the core layer (periodic AGs). The fabrication steps of such a D-fiber based LPG are resumed in Fig. 5. In particular, starting from a standard SMF, the first step is to side-polish the fiber to yield a flat polished surface on the cladding layer (see Fig. 5(a)). The distance between the core and the flat polished surface is ~12-15 µm. The second step is to coat a negative photoresist with a thickness of ~10 µm on the flat polished surface of the fiber. After it is exposed and developed under UV light, the fiber is coated with a periodic (410 µm) resist (see Fig. 5(b)). Finally, the fiber is HF etched to yield a 3 cm long AG-LPG (see Fig. 5(c)). Note that, owing to the isotropic nature of the HF based etching, experimentally the sections of AGs appear to be ladder-shaped.

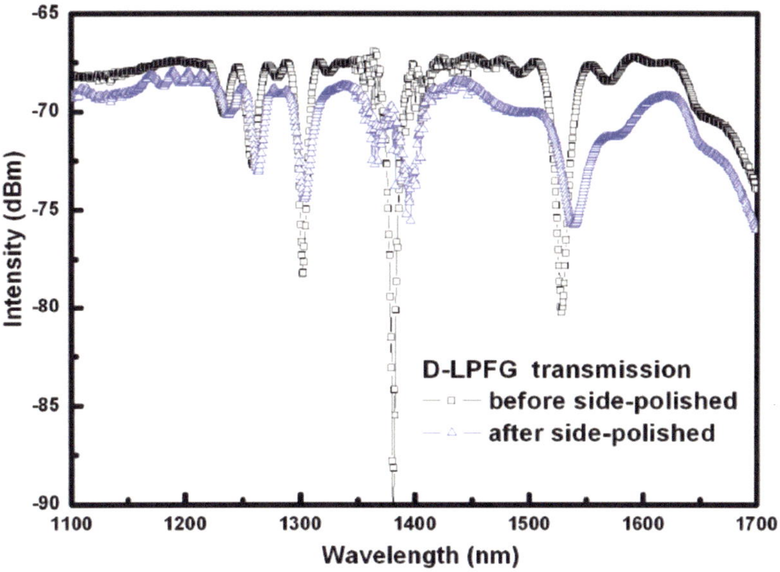


Fig. 4. Transmission spectra before and after side-polishing of the LPG. Reproduced with permission from (Tien et al., 2009a).

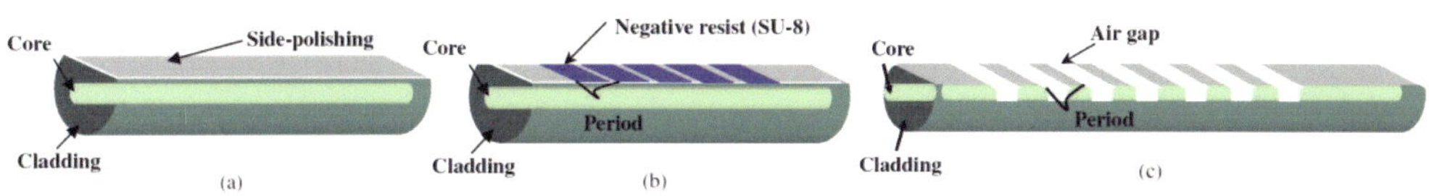


Fig. 5. Schematic diagram of the fabrication steps for the realization of AG-LPGs (not in scale). Reproduced with permission from (Fu et al., 2011).

### 3.2 Evanescent-wave LPGs

Along a LPG, the core periodic modification is substantially necessary to induce an effective RI modulation with the same periodicity on the core mode of the SMF: this is the real perturbation able to force the modal coupling between the core and cladding modes. On the other hand, the direct modification of the interested fiber layer is not the only way to force an effective RI modulation of the modes propagating within that layer. Uniform HRI nano-coatings, for example, have been proved to induce strong changes on the cladding modes effective RIs via evanescent-wave (Cusano et al., 2009b). By exploiting the same principle, HRI coatings should be able to induce changes on the core mode effective RI if the diameter of the cladding layer is opportunely reduced (Cusano et al., 2007). The first evanescent-wave D-fiber based LPG has been demonstrated in 2009 (Jang et al., 2009). The cladding layer of a SMF was substantially reduced using the side-polishing method to enhance the interaction between the core mode and the external medium via evanescent-wave. In particular, the unjacketed fiber was placed in a bent groove (curvature radius of 250 mm) in a quartz block and was held by a UV epoxy. The block was polished until the cladding of the fiber was nearly removed. Successively, the LPG pattern was formed on the side-polished surface using a photolithography process: i) photoresist was spread on the polished surface by spin-coating (thickness of 2.1 µm) and ii) the LPG pattern was formed by UV exposure through a shadow long-period mask followed by a development process. The evanescent-wave LPG was 25 mm long, with a period of 600 µm. Fig. 6(a) and (b) show a schematic diagram of the LPG and a microscope image of a section of the LPG, respectively. In this case, the HRI photoresist increases the effective RI of the core mode along the coated regions, whereas it is left unperturbed elsewhere.

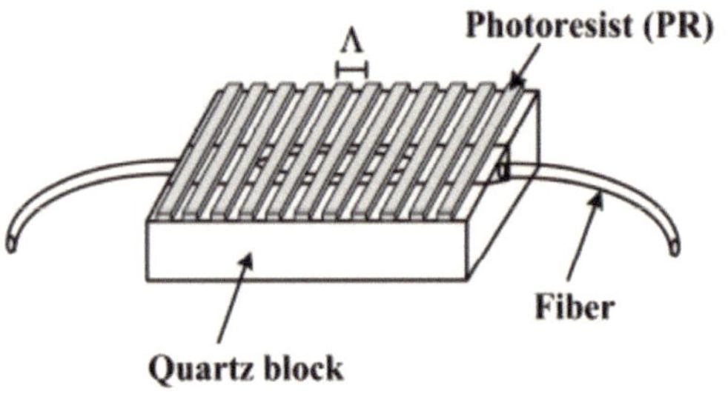

(a)

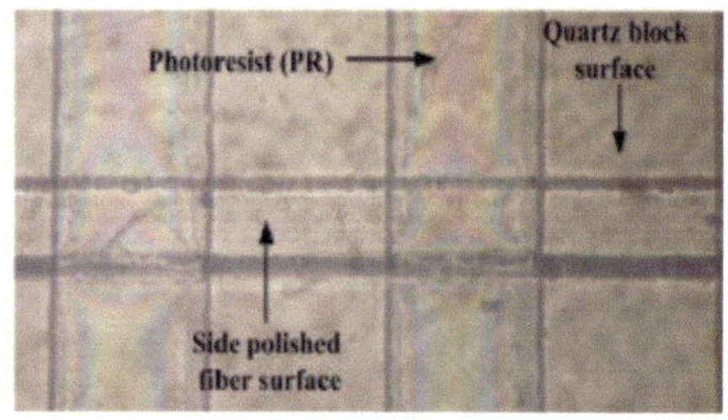

(b)

Fig. 6. Side-polished fiber with a LPG pattern: (a) schematic diagram (not in scale); (b) microscope image. Reproduced with permission from (Jang et al., 2009).

The same principle of operation has been exploited to develop an evanescent-wave LPG also along a D-shaped PCF (H. Kim et al., 2010). Figs. 7(a) and (b) schematically show the device evolution along the longitudinal direction of the fiber at different steps of its fabrication; Fig. 7(c), instead, shows the transversal section of one of the coated regions along the final device. The utilized PCF had pitch of 5.1 µm, air hole size of 1.3 µm, and core and cladding diameter of 10 and 130 µm, respectively. To fabricate the D-shaped PCF, two polishing processes were performed. In the first one, the PCF was placed on a V-groove along a quartz block (curvature radius of 90 cm) and fixed by using a UV-curable epoxy. The structure was ground down on a brass plate with $Al_2O_3$ powder. Then, the slurry on the flat surface was washed by ultrasonic cleaning with de-ionized water and successively dried at 100°C for 10 min. The air holes on the polished surface were covered by using a UV-curable epoxy (RI of 1.56) in order to remove external materials infiltrated into air holes: the transmission loss caused by the epoxy was measured to be less than 0.2 dB. For the second polishing process, the ground PCF was positioned on a polyurethane plate and polished with $CeO_2$ powder to diminish the surface roughness of the D-shaped PCF. Also after this step, the polished surface was washed and dried. The measured residual cladding thickness was ~0.1 µm. The previously described photolithography processes were used for the deposition of a uniform photoresist overlay and for its patterning. Two D-shaped PCF samples with different surface structures were fabricated: one presenting a uniform thin film of resist (thickness d of 3.5 µm) and the other one with the same resist layer patterned with a period of 400 µm to produce the LPG. Figs. 7(d) and (e) show the transmission spectrum of the two D-shaped PCF samples. In both cases, dips associated to modal coupling are present. In particular, the PCF-based thin layer present two different dips at ~800 and ~1320 nm: they are probably due to coupling of the core mode with overlay modes. On the other side, for the surface LPG these two dips are narrower and reduced in depth; in addition also a dip slightly beyond 1000 nm is present, probably the only one effectively associated to the grating.

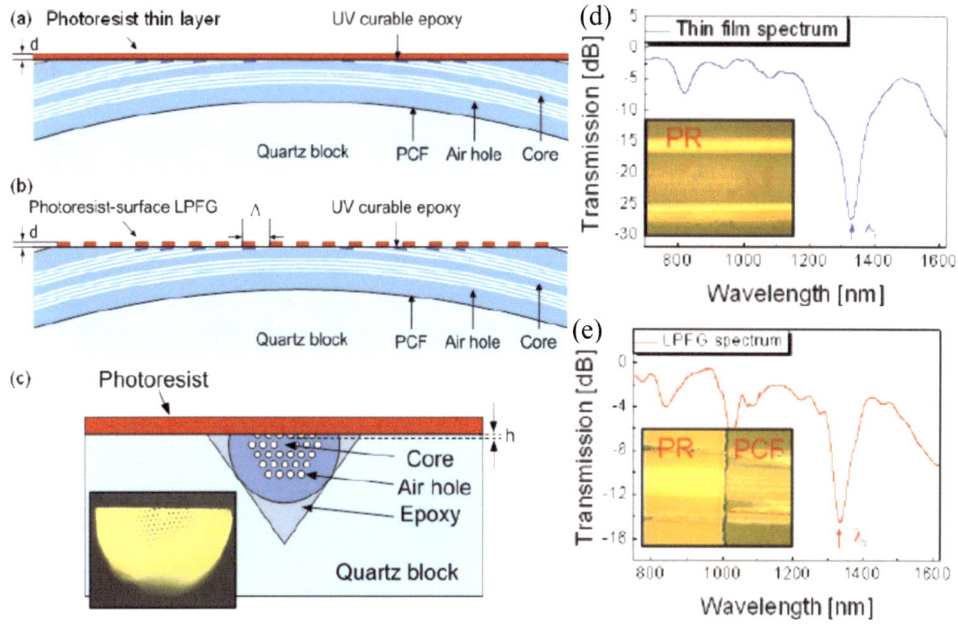


Fig. 7. Schematic diagram (not in scale) of (a) the uniform thin layer along the D-shaped PCF, (b) the PCF-based surface LPG, and (c) the cross section of the D-shaped PCF with the photoresist overlay; transmission spectrum of (d) the PCF-based thin layer and (e) the PCF-based surface LPG. Reproduced with permission from (H. Kim et al., 2010).

The idea at the basis of the previous evanescent-wave LPG configuration is very interesting, but its fabrication procedure shows certain limitations: i) the fiber device has to be integrated in a bulk material for side-polishing, loosing its typical compactness and ii) the adoptable overlay types are limited to photoresist. To overcome this limits, we recently proposed a different approach to realize evanescent-wave D-fiber based LPGs (Quero et al., 2011). First of all, the D-fiber supplied by KVH (see Fig. 1(b)) was adopted: it simply needs superficial etching in correspondence of the flat surface to allow evanescent-wave interaction of the core mode (Fig. 8(a)). It also provides, at this stage, the possibility to tailor the SRI sensitivity of the device by a proper choice of the etching depth. Successively, as proof of concept, a basic polymeric overlay of atactic polystyrene (PS) was uniformly deposited along the fiber by dip-coating technique (Fig. 8(b)). Finally, the overlay was properly confined in correspondence of the core layer on the flat surface of the fiber (Fig. 8(c)) and periodically patterned (Fig. 8(d)) by laser micromachining techniques. The main advantage of this approach relies on the flexibility: PS was used only for validation, several HRI material can be adopted depending on the specific application. During the HF based etching procedure, a particular point to be taken into account is the different etching rates of the three layer constituting the D-fiber structure. In particular, it is necessary to etch the fluorine-doped inner cladding to obtain evanescent-wave interaction of the core mode and such a layer etches ~1.4 times faster than the silica super-cladding. However, the etching depth can be controlled by monitoring the transmitted power: a 2.5 cm etched sample presenting 5% optical power losses was selected. The correspondent transmitted spectrum is reported in Fig. 8(e) (black curve). During the second step, a uniform PS overlay (RI of

~1.59) was deposited along the etched D-fiber by dip-coating (thickness of approximately 1.4 μm): the correspondent transmitted spectrum is shown in red in Fig. 8(e). Evidently, the HRI overlay induces further optical losses, but, above all, forces selective spectral features probably due to a coupling mechanism of the guided light with overlay modes. To avoid that the spectral features of the LPG to be realized would be compromised by such selective spectral features, an excimer laser micromachining system (KrF, λ=248 nm) was used to confine the overlay on the flat surface of the fiber: a strip 2 cm long and 30 μm wide was realized in correspondence of the core. As observable from the spectral response of the device after confinement (green curve in Fig. 8(e)), the selective spectral features were no more observable and the transmission spectrum results quite flat. Finally, the evanescent-wave LPG was realized via laser micromachining by periodically patterning the PS strip with a period of 500 μm (250 μm alternatively coated and uncoated): as shown in Fig. 8(f), the LPG transmission spectrum present three different attenuation bands located at ~1360, ~1440, and ~1530 nm.

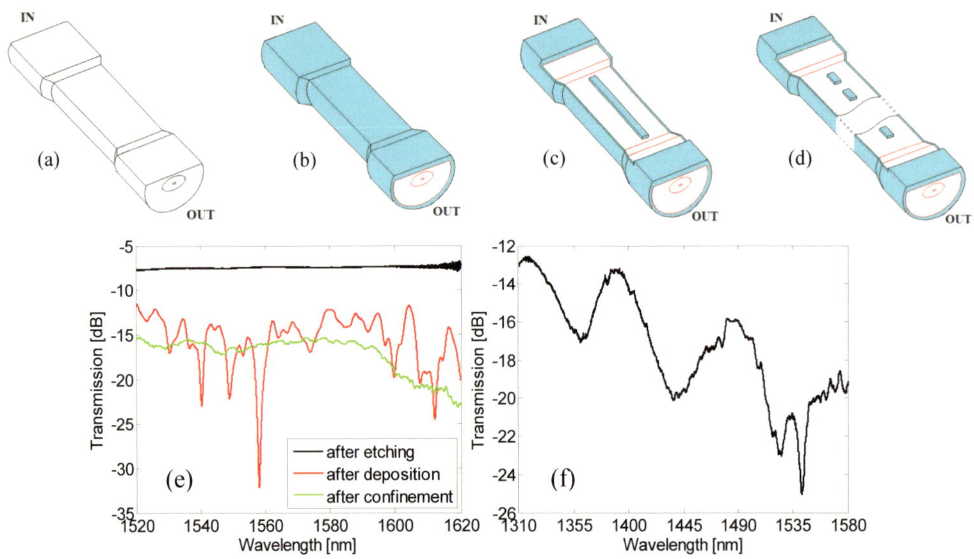


Fig. 8. Evanescent-wave LPFG: (a)-(d) Schematic diagram (not in scale) of the fiber structure step-by-step; transmission spectrum at (e) steps (a)-(c) and (f) step (d). (Quero et al., 2011).

### 3.3 Applications

D-fiber based LPGs have been demonstrated to be very useful devices, especially in the sensing field. In this section, the applications of such devices are briefly resumed: according to the previous sections, first intra-core and then evanescent-wave LPGs will be considered.

The bending and orientational characteristics were the main aim of the first study about D-fiber based LPGs (Allsop et al., 2004). Typically increasing curvature causes splitting of LPG attenuation bands in standard SMF. Differently, the spectral evolution versus bending of one of the first D-fiber based LPGs is shown in Fig. 9. As observable, new bands appear under the influence of bending. Such transmission features are due to mode coupling coefficients that increase with curvature. The most sensitive bend-induced attenuation bands was characterized by a bend sensitivity of $12.55 \pm 2\times10^{-2}$ nm•m, whereas the normal attenuation band had a sensitivity of –1.735 ± $8\times10^{-3}$ nm•m. Also, it was found that the bend-induced bands were sensitive to the orientation of the bend with respect to the flat of the "D", suggesting, in principle, a possible application as directional bend sensor for the D-fiber LPG. The grating was tested also versus temperature, revealing the possibility to use the LPG sensor also to discriminate between temperature and bending effects because it yields a reasonably well-conditioned sensitivity matrix.

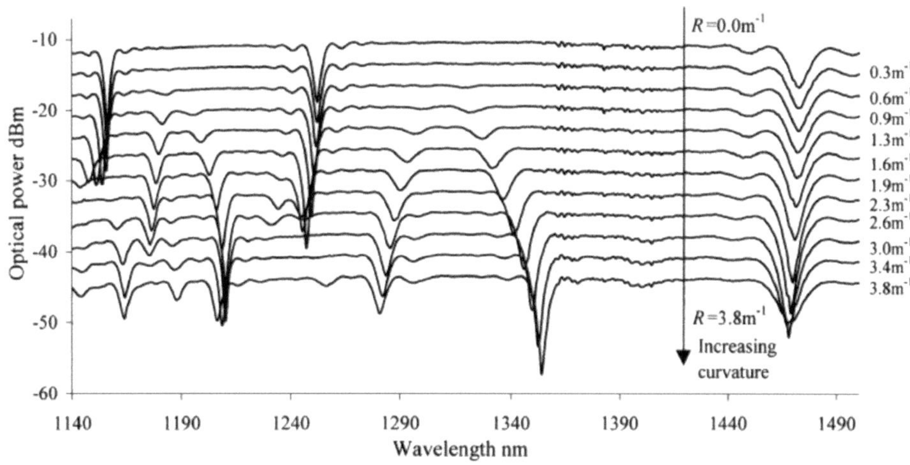


Fig. 9. Evolution of the transmission spectrum of a D-fiber based LPG (period of 380 μm) subjected to bending, showing the appearance of bend induced stop-bands. (Reproduced with permission from (Allsop et al., 2004)).

The application proposed for the first LPG inscribed within the D-fiber by KVH, instead, was as optical chemsensor with cladding etching to enhance sensitivity (Chen et al., 2004). A comparative investigation on the SRI sensitivity was conducted on LPGs in both D-fiber and the standard SMF by measuring the grating response to the aqueous sugar solutions, with sugar concentrations varying from 0% to 60%. The LPGs presented in Fig. 3 were utilized and both of them were tested in standard configuration and after an etching time of 40 min in HF bath at 10% concentration (the etching rate was found to be the same for the two kinds of optical fiber). Evidently, the D-fiber device possessed an intrinsically higher SRI sensitivity and the etching significantly enhances the SRI sensitivity, especially for the D-fiber LPG. However, it is important to note that different grating periods were considered (shorter for the D-fiber) and this surely influenced the SRI sensitivity. The same authors also proposed a dual-parameter in-fiber sensor based on a hybrid LPG-FBG structure (Chen et al., 2005). The simultaneous measurement of temperature (FBG) and SRI (LPG) was proved. Always in 2004, the bending sensitivity characteristics of such D-fiber based LPGs were studied (D. Zhao et al., 2004a; D. Zhao et al., 2004b): their spectral response depends strongly not only on the curvature amplitude but also on the fiber orientation. Potential applications as directional shape sensor, bend-insensitive sensor, and two-axis curvature sensor (here a couple of LPGs is necessary) were hypothesized. Finally, the spectral characteristics of LPGs UV-written within the D-fiber supplied by KVH were also studied by Allsop *et al.* in 2006 (Allsop et al., 2006). The authors were able to fabricate LPGs with overlapping orthogonal polarization state attenuation bands: the use of such bands can considerably simplify the sensor interrogation. However, the spectral sensitivity of both orthogonal polarization states was measured with respect to temperature, rotation and bending. The temperature sensitivity was low compared to LPGs in standard SMF. Moreover, such LPGs devices produced blue and red shifts depending upon the orientation of the bend with measured maximum sensitivities of -3.56 and 6.51 nm•m. The use of neighbouring bands to the overlapping orthogonally polarized attenuation bands to perform simultaneous measurement of temperature and bending was also demonstrated, which yielded a maximum polarization dependence curvature error of ±0.08 $m^{-1}$ and a temperature error of ±0.05°C. Since also the rotation of the bent LPG produced wavelength shifts, this type of LPG may be useful as a shape sensor and the polarization dependence can be reduced by using the overlapping orthogonal polarization state attenuation bands.

The effective scientific interest in D-fiber based LPGs is demonstrated also by the theoretical works regarding SRI measurements based on surface plasmon polariton (Tripathi et al., 2008; Tripathi et al., 2009), opening the way to the design of high performance chemical/biological sensors. The 2008 work studies the SRI sensing characteristics of metal-coated side-polished standard SMF gratings: both FBGs and LPGs. The authors used a simple approach for modelling side-polished SMF (Sharma et al., 1990) and demonstrated that the LPG-based sensor requires shorter grating lengths and higher metal thickness for a given sensitivity, making it more practical to realize. The 2009 work, instead, is focused on the SRI sensitivity characteristics of metal-coated LPGs operating in the power coupling regime corresponding to dual spectral resonance within D-fiber by KVH. The authors used a simple and sufficiently accurate first-order perturbation model (Kumar & Varshney, 1984) and demonstrated that, by an optimum combination of metal thickness and core to flat surface separation, SRI sensitivity as high as 5971 nm/RIU (RIU - RI Unit) can be reached.

The double-sided polished D-fiber LPG proposed by Tien et al. in 2009 was first proposed as magnetic field sensor (Tien et al., 2009a). The magnetic sensing material was a Fe thin film with a thickness of 80 nm, deposited by evaporation coating technique onto the double-sided-polished surface. The maximum blue shift experienced by the 1310 nm attenuation band (see Fig. 4) was 36 nm when the magnetic field was 153 kA/m, corresponding to a sensitivity of about 0.24 nm/(kA/m). The double-sided-polished D-fiber LPG was also proposed for liquid RI measurements in the bare configuration (Tien et al., 2009b), revealing a maximum SRI sensitivity of 143.396 nm/RIU.

To complete the intra-core LPGs category, the applications of AG-LPGs needs to be mentioned: first, they were tested for SRI measurements (Fu et al., 2009). Fig. 10(a) shows the transmission spectrum of a AG-LPG with period of 410 μm for several SRIs (sugar solutions at different concentrations): differently from standard LPGs, a red shift as the SRI grows up is observable. In particular, in the 1.33-1.42 SRI range a linear behaviour has been pointed out for the attenuation band at ~1550 nm with SRI sensitivity of ~620 nm/RIU. This peculiar spectral feature can be explained by considering that the resonant wavelengths $\lambda_{res,m}$ of a LPG with period $\Lambda$ are determined by the following phase-matching condition (Vengsarkar et al., 1996; Shu et al., 2002):

$$\lambda_{res,m} = \left(n_{core}^{eff} - n_{cladding,m}^{eff}\right) \cdot \Lambda \quad (1)$$

where $n^{eff}_{core}$ and $n^{eff}_{cladding,m}$ are the effective RIs of the fundamental core mode and the m-th cladding mode, respectively. For a standard LPG, SRI changes are able to modify only the cladding mode RIs: the higher is the SRI, the higher are the cladding mode RIs, leading to a blue shift of the attenuation bands. In the AG-LPG case, instead, SRI changes influence both the cladding and the core mode effective RIs. In particular, the SRI sensitivity is higher for the core mode and consequently a red shift as the SRI grows up is observable in Fig. 10(a). Successively, the AG-LPG has been proposed as humidity sensor (Fu et al., 2011). In this case, the polished surface of a 500 μm period AG-LPG was coated with a calcium chloride ($CaCl_2$) thin film of ~3 μm thickness. Since $CaCl_2$ is strongly hygroscopic, it is known as a drying agent or desiccant: in practice, the higher the relative humidity (RH) increases, the lower the RI of the $CaCl_2$ thin film decreases. The device was tested in the RH range from 55% to 95% and a linear blue shift was observed for the grating attenuation band, revealing a sensitivity of about 1.36 nm/1%RH. However, when the RH was increased from 85% to 95%, no wavelength shift was observed because of the saturation of the chemical interaction between $CaCl_2$ and $H_2O$ molecules. Finally, the thermal crosstalk was smaller than that of conventional LPGs with consequent less thermal compensation requirements.

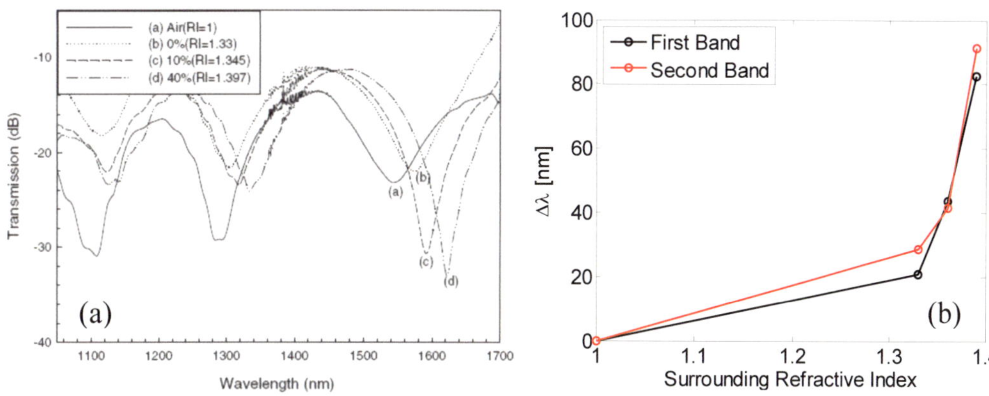


Fig. 10. (a) Transmission spectrum of an AG-LPG with period of 410 μm for several SRIs. (Reproduced with permission from (Fu et al., 2009)). (b) Wavelength shift versus the SRI of the first two attenuation bands of the evanescent-wave LPG. (Quero et al., 2011)

As regards evanescent-wave D-fiber based LPGs, they represent a more recent research conquest. However, their spectral evolution versus the SRI is very similar to that of AG-LPGs (see Fig. 10(a)): a red shift of the attenuation bands is observable for increasing SRIs. The first evanescent-wave LPG was proposed as sensitive DNA biosensor (Jang et al., 2009). The grating was used to detect the hybridization of single strand DNA (ssDNA). The wavelength shift were measured after the binding of the Poly-L-lysine, probe ssDNA and target ssDNA to the surface of the sensor. The overall shift induced by the DNA hybridization was 1.82 nm and the majority of it (0.94 nm) occurred in the first 9 min due to the rapid reaction with DNA hybridization. Recently, the same kind of evanescent-wave LPG - but PCF-based - has been proposed for SRI and temperature measurements demonstrating higher sensitivities as compared with conventional LPGs (H. Kim et al., 2011). As regards the more flexible evanescent-wave D-fiber based LPG recently proposed by us (Quero et al., 2011), it was characterized versus the SRI: Fig. 10(b) shows the wavelength shift of the attenuation bands located at ~1360 and ~1440 nm (see Fig. 8(f)). Without optimization of the device parameters, SRI sensitivity around the water RI of ~700 and ~625 nm/RIU for the first and the second dip, respectively, has been pointed out. As consequence, this configuration represents an extremely attractive technological platform for chemical/biological sensing. In addition, the possibility to use HRI materials with different natures (electro-optical, magneto-optical, etc.) to fabricate such devices would open the way to self-functionalized evanescent-wave LPGs suitable for specific applications.

## 4. Long period gratings in photonic crystal fibers

Photonic Crystal Fibers, thanks to the new ways provided to control and guide light, not obtainable with conventional optical fibers, are driving an exciting and irrepressible research activity all over the World, starting in the telecommunication field and then touching metrology, spectroscopy, microscopy, astronomy, micromachining, biology and sensing. A PCF consists of regularly spaced air holes along the fiber cladding (Russell, 2003). The core of the PCF is formed by the introduction of a defect or a missed hole at the center of the fiber. According to the distinct mechanisms of light propagation in the core region, as shown in Fig. 11, PCFs fall into two general categories: (1) microstructure fiber or holey fiber in which the light is trapped by total internal reflection (TIR) in a solid core, which has a larger refractive index than the cladding region (i.e. index guided IG-PCF - see Fig. 11(a)), and (2) photonic bandgap fiber (PBG) in which the core of the fiber is hollow, and the light is trapped in the central lower-index region by a two-dimensional photonic bandgap created by the periodic cladding (i.e. hollow core HC-PCF - see Fig. 11(b)) (Frazao et al., 2008).

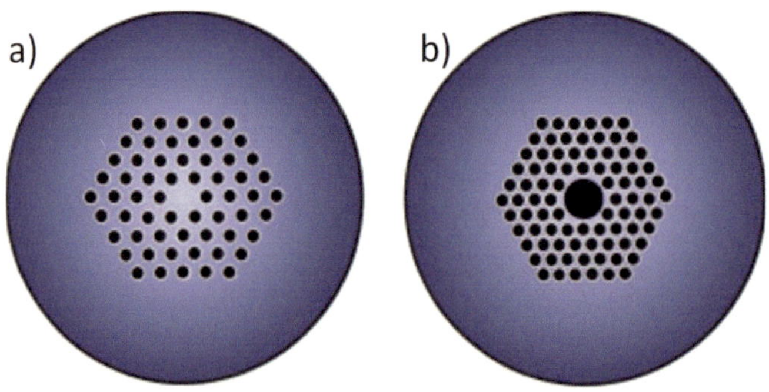


Fig. 11. Two main classes of photonic crystal fibers (PCF): index-guiding PBF (a) and photonic bandgap PBF (b). Reprinted with permission from Frazao, O., Santos, J.L., Araujo, F. M., & Ferreira, L.A., (2008). Optical sensing with photonic crystal fibers. Laser & Photonics Reviews, Vol. 2, No. 6, November 2008, pp.(449–459), ISSN 1863-8899.

In contrast to conventional fibers, most of PCFs are made by use of just a single material, typically fused silica, using the stack and draw technique, and their dispersion, mode-field confinement, single-mode range, and polarization dependence can be greatly controlled by size, shape, and pitch of the air holes. The structure of the PCF enables to have different types of fibers such as endless single mode, double clad, germanium or rare earth doped, high birefringence, and many others with peculiar features due to its manufacturing flexibility. This variety of choices permits the use of PCF in numerous applications spanning from communication components to sensors which measure physical parameters (temperature, pressure, force, etc.), chemical compounds in gas and liquids, and even biosensors. In particular, since 1999 (Eggleton et al. 1999, Espindola et a., 1999, Diez et al., 2000, Kakarantzas et al., 2002, Lim et al., 2004), the writing of LPGs into different types of

PCFs with or without photosensitivity by the use of advanced laser processing techniques along with mechanical and chemical methods was quickly embraced expecting many new applications as sensors (Frazao et al., 2008) and fiber components (Y. Wang et al., 2007). For example, the resonant wavelength of an LPG written in a PCF is blue-shifted with the increase of grating periodicity, contrary to the usual case in a conventional single mode fiber (SMF). The next sections will present an overview of the long-period-grating written in IG-PCF and in HC-PCF.

### 4.1 LPGs in IG-PCFs

The first photochemical grating written in PCF was achieved in 1999 (Eggleton et al. 1999). In this work a special PCF with a photosensitive Ge-doped core was used and the induced refractive index changes originated from the linear absorption of Ge oxygen-deficient centers with the maximum at 242 nm (common single-quantum inscription mechanism). UV-laser exposure is a common technique for writing gratings in Ge-doped fiber, but, however, typical PCFs have no photosensitivity because they are composed of pure silica, which is fully transparent in the UV spectral region. Therefore new non photochemical inscription techniques to fabricate LPGs in PCFs have been explored, that modify the refractive index in the fiber cladding either by heating (using $CO_2$ laser light or an electric arc discharge) or by applying mechanical pressure. Kakarantzas et al. reported the first example of structural LPGs written in pure-silica solid core PCFs (Kakarantzas et al., 2002). As shown in Fig. 12, the gratings are realized by periodic collapse of air holes in the PCF via heat treatment with a $CO_2$ laser. The resulting periodic hole-size perturbation produces core-to cladding- mode conversion, thus creating LPG in the PCF. In contrast with the LPGs written by UV light, which become unstable over time, $CO_2$ laser-induced LPGs are temperature insensitive because of their structural perturbation along the fiber. This property can be utilized to obtain temperature-insensitive PCF-based devices, as demonstrated in (C. Zhao et al., 2008). The $CO_2$ laser irradiation is a flexible, highly efficient, point-by-point, low cost technique for writing very compact (few mm), deep notched (>20dB) LPGs in a pure-silica PCFs without photosensitivity and the writing process can be computer-programmed to produce complicated grating profiles.
Similar to the $CO_2$ laser is the arc induced technique. The LPGs are imprinted in PCFs with the electric arc discharge of a fusion splicer by using a point-by-point technique, which is extremely low cost since it eliminates the need for expensive laser systems or the need for pre-hydrogenation of the fiber (in comparison with UV gratings) and consequent post-thermal annealing to stabilize the gratings (Humbert et al., 2003, Dobb et al., 2006). Repeatability was assured by always maintaining a constant arc current, arc duration and fiber tension. However, the reported number of the grating periods that is needed to achieve a comparable attenuation band is much larger than that of $CO_2$ laser irradiation technique (Ju & Jin, 2010). Practically all the LPGs written in PCFs by local heating rely on glass structure change and fiber deformation. Because of the high fictive temperature of pure silica, a high heating temperature (achieved with intense $CO_2$-laser radiation or an electric arc) is needed to cause significant glass structure change. Thus, one of the serious disadvantages of these methods is the collapsing of fragile PCF holes, especially in the case of PCFs with relatively large holes, which results in a high insertion loss. Another disadvantage originates from the irregularity of period deformation. A recent study shows that by applying tension to the fiber during the writing process, through the mechanism of

frozen-in viscoelasticity, it is possible to write strong gratings in PCFs with a dosage of $CO_2$ laser radiation low enough not to cause any signification fiber structure deformation (H. Lee & Chiang, 2009).

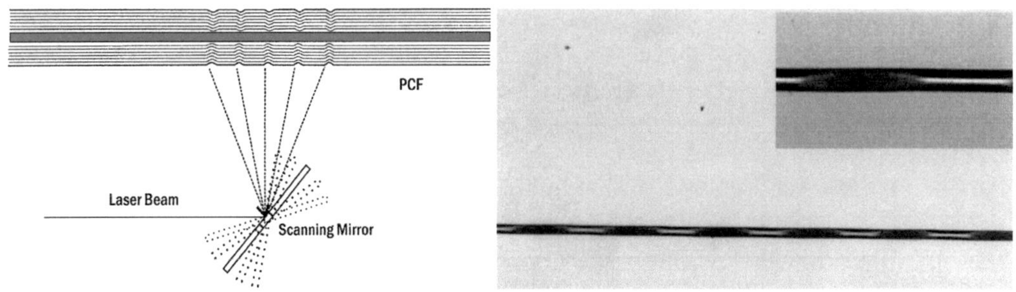


Fig. 12. Fabrication of a LPG based on microcollapsing the PCF holes with a CO2 laser beam (left) and optical micrographs of a section of the LPG after the holes at the parts heated by the $CO_2$ laser have completely collapsed (right). The inset shows a magnification of one period. Reprinted from (Kakarantzas et al., 2002) with permission of Optical Society of America.

More recently the fabrication of LPG in PCFS was successfully demonstrated by use of a high-intensity femtosecond laser via a multiphoton absorption process. In order to avoid some drawbacks of the femtosecond laser technique, such as low efficiency (Allsop et al., 2008) or a $H_2$ -loading pre-procedure (Fotiadi e al., 2007), S. Liu et al. propose to tightly focus femtosecond infrared beam onto the holey inner-cladding region of a PCF (Liu et al., 2010). As shown in Fig. 13, the high intensity femtosecond-laser irradiation results in the filling of air-holes through possibly a laser-induced micro-explosion and redeposition process, which modifies the waveguide structure and forms an LPG. Fig. 13 shows the evolution of transmission spectrum with an increasing number of grating periods. Two resonant dips (A: 1540 nm, 26 dB; B: 1370 nm, 5 dB) were observed for the LPG with 13 periods and the insertion loss is 2 dB. Furthermore, several LPGs were fabricated with similar inscription parameters but with pitch varying from 340 to 390 μm. Fig. 13 shows the measured relation between the resonant wavelength (the resonant dip A) and the grating pitch.
Mechanical pressure has provided another direct and flexible means for LPG inscription in PCF (Lim et al., 2004, D. Lee et al., 2006, Parka et al., 2006). Pressure on the fiber surface with a periodic grooved plate induces periodic index changes in the fiber. With this method the strength and the resonant wavelength of the mechanically induced LPG can be easily tuned simply by adjusting the grating period and the pressure applied on the PCF. The efficiency of the mode coupling between the core mode and a cladding mode varies with pressure. LPGs with proper lengths and periods have to be selected such that mode coupling occurs at the predetermined wavelength. However, the coupling was found to be highly polarization-dependent and dependent on the angular position where stress was applied (D. Lee et al., 2006). The highly polarization-dependent broadband coupling was observed due to the unique beat-length dispersion between the core-mode and cladding-mode, which could find potential applications in wide-band polarization dependent loss (PDL) compensation. Acousto-optical interaction can also cause mode coupling between different modes if phase matching conditions are satisfied. The acoustic LPG has great advantages in terms of tuning

range and speed. Therefore, it has been extensively studied and is enabling numerous practical applications in optical communication. The acoustic LPG built on a PCF can make a function with the tuning range over 1000 nm as a single optical element (Hong et al., 2008). In conclusion, as synthetically illustrated, the integration of LPGs in PCFs can be achieved with different, complementary techniques. This provides new promising platforms for developing novel devices, for application in telecommunication and in sensing, combining the unique properties of the PCFs with the peculiarities of the PLGs.

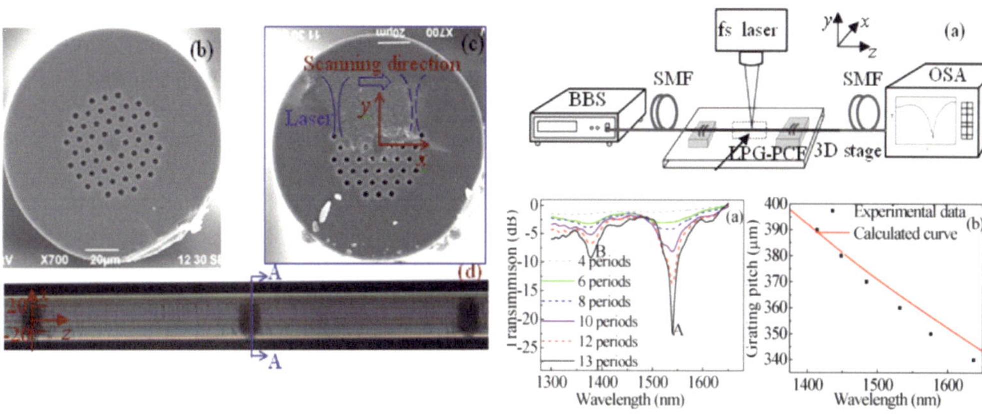


Fig. 13. (Top right a) Experimental setup for LPG fabrication. Scanning electron micrographs of LMA-10 PCF cross section (left b) before and (left c) after femtosecond-laser treatment. (left d) Top view of femtosecond-laser fabricated LPG. (Bottom Right a) Evolution of the transmission spectrum of LPG with increasing number of grating periods. (Bottom Right b) Relation between grating pitch and resonant wavelength. Reprinted with permission from Liu, S., Jin, L., Jin, W., Wang, Y., & Wang, D. N. (2010). Fabrication of Long-Period Gratings by Femtosecond Laser-Induced Filling of Air-Holes in Photonic Crystal Fibers. IEEE Photonics Technology Letters, Vol. 22, No. 22, November 2010, pp. (1635-1637), ISSN 1041-1135 © 2010 IEEE.

#### 4.2 LPGs in HC-PCFs

As discussed above, a large number of gratings have been demonstrated in different types of IG-PCFs by the use of various fabrication techniques while the development of efficient techniques useful to write LPG structure in HC-PCFs (Smith, 2003; Frazao et al., 2008) is a hard challenge for research community. Since more than 95 % of the light propagates in the core-air of an HC-PCF and not in the glass (see Fig. 11(b)), such fibers offer a number of unique features including lower Rayleigh scattering, reduced nonlinearity, novel dispersion characteristics, and potentially lower loss compared to conventional optical fibers (Smith et al., 2003; West et al., 2004). In addition, the hollow core characteristic also enables enhanced light/material interaction, thus providing a valuable technological platform for ultra-sensitive and distributed biochemical sensors. However, periodic index modulations usually required to realize mode coupling in LPG devices are not easy to achieve in HC-PCFs (Ozcana & Demircib, 2004; Frazao et al., 2008). The main issue relies on the difficulty in introducing UV-induced refractive index modulation since up to 95% of the light energy is confined within the air-core. Additionally, the alternative use of localized fiber tapers could

induces holes collapsing in the cladding region preventing the low loss propagation of light in the hollow core.
To bypass these technological difficulties and at same time to keep the concept of the PBG fiber, few years ago some LPG structures were demonstrated in a new kind of bandgap-guiding fibers such as fluid-filled PBG fibers. Such fiber was composed of two different materials and can be fabricated by taking a solid core PCF with air-holes in the cladding and filling the holes with a high index fluid (Kuhlmey, 2009). The first LPG in fluid-filled PBG fibers was formed by inducing periodic mechanical stresses on the fiber in 2006 (Steinvurzel, 2006a, 2006b). The periodic stress-induced deformations of the fiber force light coupling between core mode and higher order modes. On the same kind of fiber also an electric-arc induced LPG was demonstrated (Iredale, 2006). One year later, it was proposed a rewritable self-assembled LPG in air-core PBG fibers (Ozcana & Demircib, 2007). The LPGs were written by filling the air-core region of the fiber with a solution containing polystyrene microspheres. The microspheres are self-assembled into a periodic structure as the liquid inside the fiber evaporates, forming the long-period grating.
In 2008 instead, Wang et al. demonstrated the first example (as authors declared) of gratings written in an HC-PCF (Y. Wang et al, 2008). They proposed the use of a focused $CO_2$ laser beam to periodically deform/perturb air holes along the fiber axis (used HC-PCF: Crystal-Fiber HC-1500-02). Figs 14(a) and 14(b) compare the cross-section of the unperturbed and $CO_2$ treated fiber. The focused $CO_2$ beam scans periodically the HC-PCF causing the ablation of glass on the fiber surface and the partial or complete collapse of air holes in the cladding. The outer rings of air holes in the cladding, facing to the $CO_2$ laser irradiation, were largely deformed; however, little or no deformation were observed in the innermost ring of air holes and in the air core. As a result, periodic index modulations are achieved along the fiber axis due to the periodic perturbation (see Fig. 14(c)). Compared with the fabrication parameters for writing a grating in a solid-core PCF (Ju et al. 2004) a lower average laser power and shorter total time of laser irradiation are typically used to write a LPG in an HC-PCF. A proper choice of the fabrication parameters is critical for the fabrication of such LPG. High energy pulses with a long irradiation time may cause large deformation or collapsing of the holes and thus a higher insertion loss, while low energy pulses with short irradiation time may be insufficient to inscribe an LPG. Fig. 14(d) shows the measured transmitted spectrum of a 40-period LPG. It was retrieved the 3dB-bandwidth is about 5.6nm, which is much narrower than that of the LPGs with same number of grating periods in conventional SMFs (Bhatia et al., 1999) and in IG-PCFs (Morishita & Miyake, 2004). Besides, Wang et al. believe that for the LPG written in HC-PCF, periodic perturbations of the waveguide (geometric) structure could be the dominant factor that causes resonant mode coupling, although the stress relaxation-induced index variation may also contribute a little (Y. Wang et al., 2008, 2010). Recently, the same group theoretically investigated the LPGs fabricated in HC-PCFs with a pulsed $CO_2$ laser (Jin et al., 2011). By the use of the coupled local-mode theory, they numerically modelled the transmission and the polarization properties of the LPGs. They found that resonant couplings are resulted from the periodic modification of the fundamental and the higher order mode fields. As a result, two highly polarization dependent resonant dips are observed. Finally, they also investigated the spectral response versus the grating pitch (see Fig. 14(e)). Accordingly with LPG in IG-PCFs (Morishita & Miyake, 2004; Petrovic, 2008) and in disagreement with LPG in SMFs (Vengsarkar, 1996), the LPGs written in the HC-PCFs have negative relationship between resonant wavelength and grating pitch.

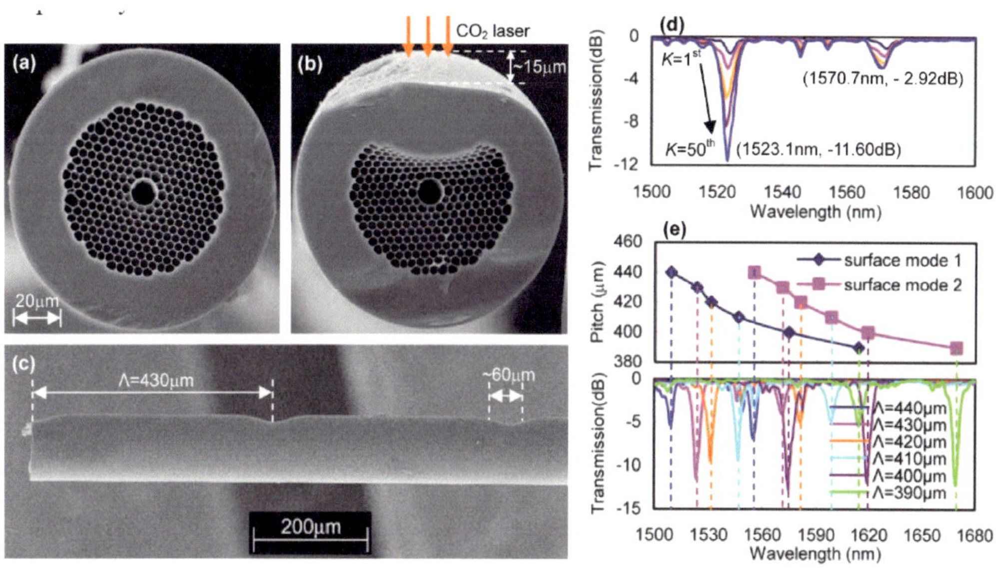


Fig. 14. Scanning electron micrographs of HC-PCF cross-sections (a) before and (b) after $CO_2$ laser irradiation; (c) LPG on HC-PCF with 50 scanning cycles; (d) Evolution of the transmitted spectrum of LPG with 40 periods and a grating pitch of 430µm with increasing number of scanning cycles; (e) Variation of LPFG resonant wavelengths with grating pitch. Reprinted from (Y. Wang et al., 2008) with permission of Optical Society of America.

Recently, the authors of the present work have investigated the possibility to use a modified Electric Arc Discharge (EAD) technique (such as a pressure assisted EAD technique) to fabricate LPGs in HC-PCFs (Iadicicco et al, 2011a, 2011b). The fabrication procedure relies on the combined use of EAD step, to locally heat the HC fiber, and of a static pressure slightly higher than the external one inside the fiber holes, to modify the holes. This procedure permits to preserve the holey structure of the host fiber avoiding any hole collapsing and it enables a local effective refractive index change due to the size and shape modifications of core and cladding holes. EAD procedure has been carried out by a commercial fusion splicer unit (Sumitomo Type-39). To achieve an arc-discharge that would locally heat the fiber (avoiding any permanent distortion) fusion current and arc duration were manually selected to approximately 13 mA and 300ms, respectively. Besides, to force a static pressure inside the fiber holes, one end of the hollow-core fiber was EAD treated in order to force hole collapsing in both core and cladding region while the other end was connected to the needle of a 1ml syringe. Before any EAD step, static pressure inside core and cladding holes was imposed by decreasing the syringe volume of 20%. A microscope image of the HC fiber cross-section before and after the pressure assisted EAD procedure is show in Figs. 15(a) and 15(b), respectively. By image analysis it is possible to retrieve that the external diameter of the fiber is reduced from 120±1 µm to 117±1 µm as well as the inner diameter of the external solid silica region from approximately 70±1 µm to 65±1 µm. Moreover, the core size resulted enlarged passing from 11±1 µm to 13±1 µm whereas none air-hole rings was found collapsed even if changes in shape and size are evident. It is important to remark that further optimization margins exist by controlling the EAD procedure through fusion current, arc duration and syringe volume decreasing while the parameters used for first prototyping allowed to preserve the fiber band-gap since negligible insertion losses were observed. Finally, the LPG writing was possible by spatially repeating the EAD procedure along the host fiber by the use of a micro-controlled translation stage. Fig. 15(c) shows the side view of an LPG with 20 periods and pitch of 400μm. In comparison with grating notches provided by $CO_2$ laser approach in Fig. 14, the pressure assisted EAD procedure permits to avoid the strong impairment of final device and reduce the polarization dependent due to the asymmetric perturbations.

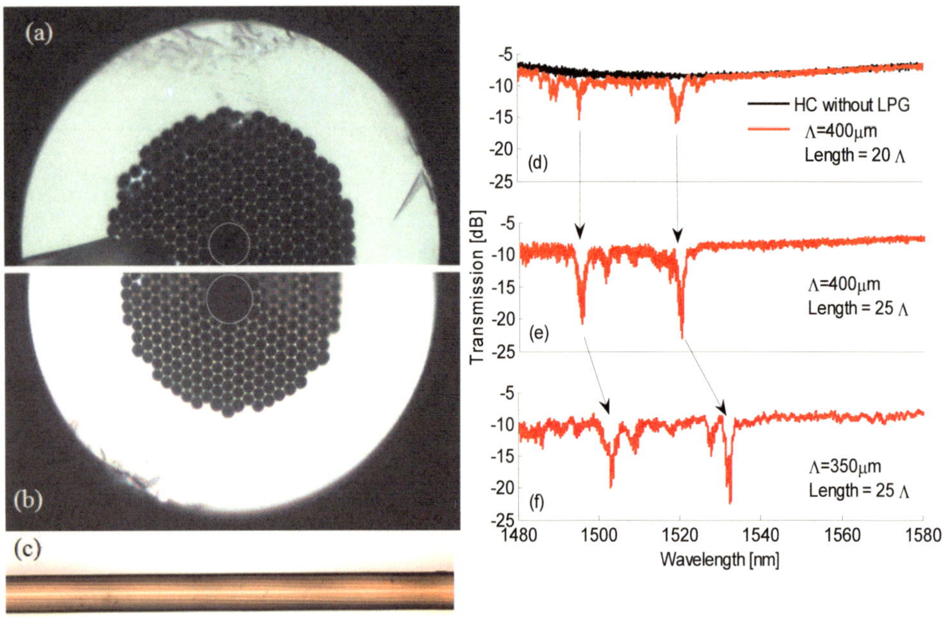


Fig. 15. Optical microscopy images of HC-PCF cross-section (a) before and (b) after EAD procedure; (c) Optical microscopy image of LPG with Λ=400μm (20 periods); transmitted spectra of LPG in HC-PCF with (d) 20 periods and Λ=400μm, (e) 25 periods and Λ=400μm, and (f) 25 periods and Λ=350μm. (Iadicicco et al., 2011).

Fig. 15(d) plots the transmitted spectrum of the LPG realized with 20 periods and a pitch of 400 μm. Despite the limited grating length (approx. 8mm), the transmitted spectrum clearly exhibits attenuation bands due to resonant coupling of the fundamental core mode to leaky higher order modes (core or surface-like). Resonant wavelengths are approximately at 1495.3nm and 1519.8nm with attenuation depth of about 5 dB and 6 dB respectively. The 3dB bandwidth for both resonances was found to be 1.2 nm and 2.0 nm, respectively, and thus narrower than the common LPGs written in SMFs (Bhatia , 1999). Besides the geometric variation, it is worth highlighting that EAD also induces silica refractive index change due to stress relaxation induced by local hot spots. However, accordingly with the LPG based on $CO_2$ laser we believe that the effective refractive index modulation is principally related to geometrical perturbation since the confinement of the fundamental mode (>95%) within air core. Moreover, from Fig. 15(d), it is also possible to observe background oscillations or ripples (in spectra with or without LPG) attributable to different effects: i) Fabry-Perot effect due to HCF-SMF splicing ii) accordingly to [13], HOM (higher order modes) weakly excited in the HC fiber. Besides, Fig. 15(e) plots the spectrum of a LPG with 25 periods and Λ=

400μm. As expected, two attenuation bands at 1495.4 nm and 1520.3 nm (very close to attenuation bands shown in Fig. 15(d) ) with depth of 9 dB and 12 dB and bandwidth of 1.4nm and 1.5 nm, respectively, can be observed. Fig. 15(f), instead, plots the transmitted spectrum of a LPG with 25 periods and Λ= 350μm. It seems that the resonant wavelengths increase with the decrease in the grating pitch, which is opposite to the behavior of LPFGs written in the conventional SMFs (Bhatia, 1999), but agrees with the above presented results (Y. Wang, 2008; Morishita & Miyake, 2004). Here, two attenuation bands at 1503.0 nm and 1532.2 nm with depth of 9 dB and 11 dB and bandwidth of 0.9nm and 1.2 nm, respectively, are evident. Finally, it is worth noting that all the spectra in Figs 15(d)-15(f) exhibit side lobes around the main peaks. These lobes could be attributed to the coupling of the core mode into asymmetric cladding modes probably due to not-perfect symmetry of arc-induced perturbation. Ivanov and Rego demonstrated that the asymmetry in arc perturbation is principally caused by the temperature gradient during arc discharge (Ivanov & Rego, 2007).

### 4.3 Applications of LPGs in PCFs

As LPGs in standard fiber (see section 2) also LPGs in PCFs have found many promising applications for temperature, strain, bend, torsion, pressure, and biochemical sensors as well as they are becoming appealing for communications applications. In this section a brief review of the main applications of LPGs in PCFs is presented.

Concerning the temperature dependence of LPGs in IG-PCF, most of the papers report that temperature sensitivity is of the order of few picometers (up to 10) per degrees centigrade (Dobb et al., 2004; Petrovic et al., 2007; Zhu et al. 2005; Humbert et al., 2004) while the sensitivity of gratings in SMF is reported to be in the range 30-200 pm/°C or more (James & Tatam, 2003; Rego et al. 2005b). The shifts of LPG bands due to thermal changes is principally attributed to two factors such as the thermooptic effect and the thermal expansion of the fiber that force wavelength shifts of opposite signs (Petrovic et al., 2007). As in SMFs, the impact of the former is dominant and determines the sign of the shift. The unique property of all silica–air fibers is that the light propagates mainly through the silica, and thus, the variations in the effective indices of the core and cladding modes with the refractive index of silica nearly cancel each other in the second term. The overall effect is a red wavelength shift of only a few picometers. This explains the very weak sensitivity of the LPG to temperature. The relatively low temperature sensitivity of the LPG in PCFs despite the sensitivity to other external parameters such as strain, bending and refractive index permitted to proposed several sensing configurations. Just to name a few of them, Dobb et al. (Dobb et al. 2004, 2006) demonstrated a temperature-insensitive LPG sensor to measure strain or curvature. The LPG structure with a period of 500 μm was written by EAD. The LPG was characterized for temperature, strain and curvature and showed sensitivities of 0±10 pm/°C, −2.04 ± 0.12 pm/με, and 3.7 nm/m, respectively. Similar study has been conducted from on LPG fabricated by the use of focused $CO_2$ laser beam (Y. Wang et al., 2006; C. Zhao et al., 2008). These structures written with $CO_2$ laser were proposed as strain-insensitive high-temperature PCF sensors (Zhu et al., 2005). For high-temperature applications, LPGs written in PCF with the electric arc technique have also shown an adequate performance, as was reported in (Humbert et al. 2004). Wang et al. also proposed an dipper characterization of the strain and temperature sensitivity as function of the external grooves created by focused $CO_2$ laser (Y. Wang, 2006). They realized two LPGs with same length and pitch but with different notches on the fiber surface: the first one presents evident grooves on the fiber surface achieved with setup provided in (Y. Wang, 2006), the second grating was fabricated by use of the same laser setup but a lower dosage of irradiation and presents no observable grooves on the fiber surface. The responses of the resonant wavelength to tensile strain and temperature of LPGs with and without visible notches is plotted in Fig. 16. The strain sensitivity of the LPG with grooves (−7.6 pm/ με) is about 25 times higher than the LPG without physical deformation (−0.31 pm/ με) (Fig. 16(a)) whereas the temperature sensitivities of the two LPGs are approximately the same (Fig. 16(b)). The asymmetrical structure caused by the periodic grooves introduces microbend when the LPG is axially stretched, which effectively enhanced the effective refractive index change of the LPGs with grooves. However the LPG with visible grooves also demonstrated very strong polarization dependent loss and can be used as in-fiber polarizers with good temperature stability (Y. Wang, 2007).

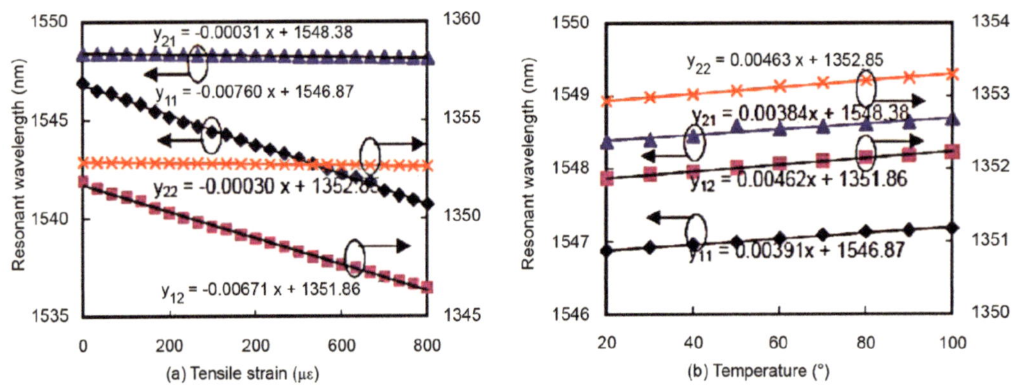


Fig. 16. Resonant wavelength of LPFG via (a) the tensile strain and (b) the temperature: ♦ and ■ refer to LPG with period grooves and ▲ and X refer to LPG without period grooves. Reprinted from (Y. Wang et al. 2006) with permission of Optical Society of America.

Han et al. demonstrated simultaneous measurement of strain and temperature using LPGs written in PCFs with different air-hole sizes (Han et al., 2007). They fabricated LPGs in IG-PCFs with Ge-doped core and different air-hole size by the UV exposure (at 244 nm) through an amplitude metal mask. They experimentally proved that the strain sensitivity of LPFGs depend on the air-hole size. Since all fibers have the same material composition, the LPGs exhibited similar temperature sensitivities regardless of the air-hole size. However, the strain sensitivities of the LPGs are different because of the different cross-sectional areas of the fibers depending on the different air-hole sizes. Similar functionality was reported by Sun et al. with processing based on an artificial neural network (Sun et al, 2007). Pressure sensing using an LPG fabricated in a PCF was presented in (Lim et al. 2004). Later, a hydrostatic pressure sensor using a tapered LPG written in PCF by the electric arc technique was also reported (Bock et al., 2007). The pressure sensitivity was found to be 11.2 pm/bar, a factor of two higher than the value found in standard single-mode fibers.

Concerning to the sensing for chemical application, LPGs in PCF seem an appealing technological platform. Spectral sensitivity to refractive index of the surround medium is possible only if part of guided light comes into contact with surrounding materials (Petrovic et al., 2007). On this topic Dobb et al. experimentally investigated the SRI sensitivity of LPGs written in different fibers by EAD (Dobb et al, 2006). They demonstrated that LPGs written in large mode area (LMA) PCFs could be used for sensing SRI whereas the no-LMA PCFs exhibit minimal changes to SRI. They declare the spectral response of former gratings is similar to the behavior of LPGs fabricated in SMFs. The sensitivity increases as the SRI approaches that of the cladding. Besides, when the SRI and cladding indices are matched, the cladding appears to be infinite and thus, no cladding modes are supported. However the most appealing characteristic of the PCFs is to provide strong light/material interaction inside the fiber-hole, which offers a new features for developing ultra-sensitive and distributed gas and liquid sensors. Rindorf et al. presented LPG in photonic crystal fibers as sensitive biochemical sensor (Rindorf et al., 2006). A layer of biomolecules was immobilized on the sides of the holes of the PCF and by observing the shift in the resonant wavelength of the LPG it was possible to measure the thickness of the layer. The thicknesses of a monolayer of poly-L-lysine and double-stranded DNA was measured with s sensitivity of approximately 1.4nm/1nm. Later, LPG in large-mode-area PCF was presented as highly sensitive refractometers exhibiting 1500 nm/RIU at a refractive index of 1.33 (Rindorf and Bong, 2008). The high sensitivity is obtained by infiltrating the sample into the holes of the PCF to give a strong interaction between the sample and the probing field.

Furthermore, as for the same device in SMF, LPGs in PCF have found useful applications in optical communications systems. For example, a tuneable long-period grating filter in a hybrid polymer-silica PCF was reported in 2002 (Kerbage et al., 2002). The polymer is infused into the fiber holes and it is possible to change the LPG resonance wavelength in a range of 200 nm, with a temperature variation of 10 °C. Recently, by utilizing coupled-mode theory, a design of band stop filter based on optimal LPG parameters (in terms of full-width half-maximum and grating length) in PCF is presented (Seraji et al., 2011). The analysis is presented for optimization of LPG length and number of gratings with respect to air-hole spacing $\Lambda$, hole diameter d, and air filling factor $d/\Lambda$ of the PCF in which LPG is inscribed.

It is worth noting that also the response of LPGs in HC-PCF as function of physical parameters changes in terms of temperature stain and bending was investigated. The responses of the first LPG in the HC-PCF realized by focused $CO_2$ laser to temperature, strain and bend in reported in Fig. 17 (Y. Wang at al., 2008). The temperature sensitivity of the resonant wavelength is about 2.9pm/°C. Accordingly to LPG in IG-PCF, the wavelength sensitivity is one to two orders of magnitude less than those of the LPGs in SMFs (Bhatia & Vengsarkar, 1996). Also, the same LPG (in HC-PCF) was immerged into liquids with different refractive index and the resonant wavelength hardly changed, whereas the LPGs in SMFs are very sensitive to SRI. Differently, with the increase of applied tensile strain, the resonant wavelength of $CO_2$ LPG shifted linearly toward shorter wavelength with a strain sensitivity of -0.83nm/mε and thus with sensitivity higher than that typical of LPGs in SMFs (Bhatia & Vengsarkar, 1996). Finally, when the curvature of LPG was increased to 13.3m-1, the resonant wavelength changed by only ±8pm. From these results Wang et al. declare that LPG in HC-PCF achieved by $CO_2$ laser may be used as a strain sensor without cross-sensitivity to temperature, curvature, and external refractive index.

Finally, Fig. 18 reports the resonant wavelengths of the LPG in HC-PCF realized by EAD as function of the temperate. The grating take into consideration for the thermal characterization is composed by 25 perturbation with pitch of $\Lambda$=400µm. The spectrum of this grating is presented in Fig. 15(b) showing two attenuation bands at 1495.4 nm and 1520.3 nm, respectively (Iadicicco et al., 2011). Thermal characterization in range 30-80 °C

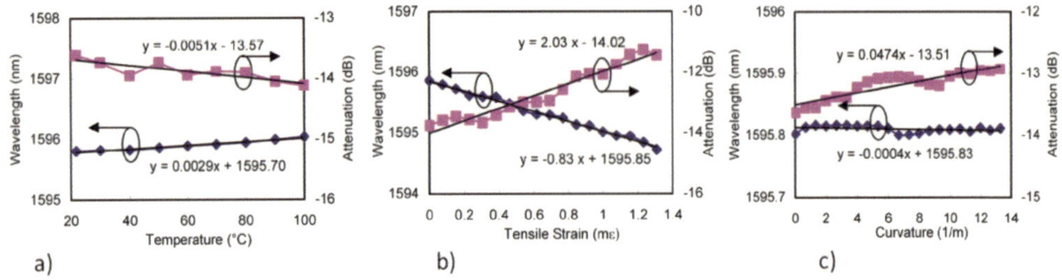


Fig. 17. Measured resonant wavelength and peak transmitted attenuation of the LPG in HC-PCF as functions of a) temperature, b) tensile strain and c) curvature. Reprinted from (Y. Wang et al., 2008) with permission of Optical Society of America.

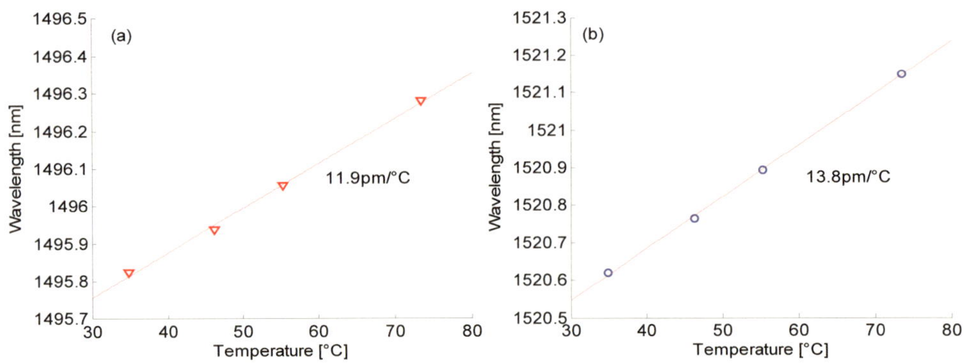


Fig. 18. Measured resonant wavelengths of the LPG in HC-PCF achieved by EAD as functions of temperature for left (a) and right (b) band.

are obtained using a furnace and a commercial FBG-based temperature sensor as reference. Accordingly with other LPG prototype above presented, both attenuation bands exhibit red shift with temperature increasing. Sensitivity of 11.9pm/°C (Fig. 18(a)) and 13.8pm/°C (Fig. 18(b)) are measured for the left and right band respectively. In addition, the same LPG was characterized to different SRIs by using different liquids with RI ranging from 1.33 to 1.47. As expected from the HC-PCFs mode fields distribution, no spectral changes were measured. Compared with the LPG in HC-PCF fabricated by focused $CO_2$ laser (Y. Wang et al., 2008), it exhibits higher sensitivity versus thermal changes even if it is kept significantly lower than LPGs in SMFs (Bhatia et al, 1996). Besides, accordingly with $CO_2$ laser, EAD based LPG exhibits trivial sensitivity to SRI.

In conclusion the LPGs in PCF (both IG and principally HC) act as novel platform to develop future devices with high performance principally for sensing applications. However at the state of the art, they still represent a very young technology requiring more work before of the industrialization of full sensing systems based on LPGs in PCF.

## 5. Conclusions

In this work, the fabrication techniques and application fields of LPGs in new generation optical fibers involving D-shaped fiber and solid- and hollow-core PCFs are reported. Despite the mature status of LPG technology in SMFs, in fact, LPGs transfer into new

hosting fiber with novel features in terms of tuning and sensitivity of the guided field to environmental parameters is not easy. However, to address this issue, a lot of work has been made in the last decade and further work is currently in progress, since LPGs in new generation optical fibers seem to be extremely promising technological platforms.
D-shaped fiber seems the easiest solution to enhance the interaction between the guided mode and the surrounding medium. Among the different configurations, highly attractive is the evanescent-wave LPG made by patterning HRI layer on etched D-fiber surface and exhibiting (in a not optimized prototype) SRI sensitivity around the water RI of ~700 nm/RIU. From these results, evanescent-wave LPGs represent an extremely attractive technological platform for chemical and/or biological sensing. In addition, the possibility to use patterned coating with different properties (electro-optical, magneto-optical, etc.) to fabricate such devices would open the way to self-functionalized evanescent-wave LPGs suitable for specific applications.
Furthermore also LPGs in PCFs act as a novel technological platform for the development of optical fiber devices. It is important to distinguish between different PCFs. LPGs in IG-PCFs are in our opinion a quite-mature technology. Several approaches have been adopted to fabricate grating in IG-PCFs and many works report on their sensing capabilities, ranging from chemical/biological applications to physical parameters detections. On the other side, only few prototypes of LPGs in HC-PCFs have been reported. However we believe that the last configuration represents the most attractive one for chemical/biological applications in light of the strong light/material interaction that HC fibers allow inside the fiber hollow core. On this topic more work is required and it is reasonable to believe that good results will be published in next years

## 6. References


Allsop, T., Gillooly, A., Mezentsev, V., Earthgrowl-Gould, T., Neal, R., Webb, D. J. & Bennion, I. (2004). Bending and orientational characteristics of long period gratings written in D-shaped optical fiber. *IEEE Transactions on Instrumentation and Measurement,* Vol.53, No.1, (February 2004), pp. 130-135, ISSN 0018-9456

Allsop, T., Dobb, H., Mezentsev, V., Earthgrowl, T., Gillooly, A., Webb, D. J. & Bennion, I. (2006). The spectral sensitivity of long-period gratings fabricated in elliptical core D-shaped optical fibre. *Optics Communications,* Vol.259, No.2, (March 2006), pp. 537-544, ISSN 0030-4018

Allsop, T., Kalli, K., Zhou, K., Laia, Y., Smith, G., Dubov, M., Webb, D. J., & Bennion, I. (2008). Long period gratings written into a photonic crystal fiber by a femtosecond laser as directional bend sensors. *Optics Communications,* Vol. 281, No. 20, October 2008, pp. (5092–5096), ISSN 0030-4018

Barnes, J. A., Brown, R. S., Cheung, A. H., Dreher, M. A., Mackey, G., & Loock, H.-P. (2010). Chemical sensing using a polymer coated long-period fiber grating interrogated by ring-down spectroscopy. *Sensors and Actuators B: Chemical,* Vol. 148, No. 1, pp. 221-226. ISSN 0925-4005

Bhatia V. (1999). Applications of long-period gratings to single and multi-parameter sensing. *Opt. Express,* Vol. 4, pp. 457–66, ISSN 1094-4087.

Bhatia V. & Vengsarkar A.M. (1996). Optical fiber long-period grating sensors. *Optics Letters,* Vol. 21, pp. 692-694, ISSN 1539-4794.

Birks, T. A., Roberts, P. J., Russell, P. S. J., Atkin, D. M., & Shepherd, T. J. (1995). Full 2-D photonic bandgaps in silica/air structures. *Electronics Letters*, Vol. 31, No. 22, October 1995, pp. (1941-1943), ISSN 0013-5194

Block, U. L., Dangui, V., Digonnet, M. J. F., & Fejer, M. M. (2006). Origin of Apparent Resonance Mode Splitting in Bent Long-Period Fiber Gratings. *Journal of Lightwave Technology*, Vol. 24, No. 2, pp. 1027-. ISSN 0733-8724

Bock W., Chen J., Mikulic P., Eftimov T., Korwin-Pawlowski M. (2007). Pressure sensing using periodically tapered long-period gratings written in photonic crystal fibers. *Meas. Sci. Technol.*, Vol. 18, pp. 3098–3102, ISSN 1361-6501

Canning J. (2008). Fibre gratings and devices for sensors and lasers. *Laser & Photon. Rev.*, Vol. 2, No. 4, pp. 275-289. ISSN 1863-8899.

Chen, X., Zhou, K., Zhang, L. & Bennion, I. (2004). Optical chemsensors utilizing long-period fiber gratings UV-inscribed in D-fiber with enhanced sensitivity through cladding etching. *IEEE Photonics Technology Letters*, Vol.16, No.5, (May 2004), pp. 1352-1354, ISSN 1041-1135

Chen, X., Zhou, K., Zhang, L. & Bennion, I. (2005). Simultaneous measurement of temperature and external refractive index by use of a hybrid grating in D fiber with enhanced sensitivity by HF etching. *Applied Optics*, Vol.44, No.2, (January 2005), pp. 178-182, ISSN 0003-6935

Chen, X., Zhou, K., Zhang, L., & Bennion, I. (2007). Dual-peak long-period fiber gratings with enhanced refractive index sensitivity by finely tailored mode dispersion that uses the light cladding etching technique. *Applied Optics*, Vol. 46, No. 4, pp. 451-455. ISSN 1559-128X

Chiang, K. S., & Liu, Q. (2006) Long-period gratings for application in optical communications, in *Proc. 5th International Conference on Optical Communications and Networks and 2nd International Symposium on Advances and Trends in Fiber Optics and Applications* (ICOCN/ATFO 2006) (Chengdu, China, Sept. 2006), pp.128-133

Choi, H. Y., Park, K. S., & Lee, B. H. (2008). Photonic crystal fiber interferometer composed of a long period fiber grating and one point collapsing of air holes. *Optics Letters*, Vol. 33, No. 8, 2008, pp. (812–814), ISSN 1539-4794

Chung, K.-W., & Yin, S. (2004). Analysis of a widely tunable long-period grating by use of an ultrathin cladding layer and higher-order cladding mode coupling. *Optics Letters*, Vol. 29, No. 8, pp. 812-814. ISSN 0146-9592

Corres, J. M., del Villar, I., Matias, I. R., & Arregui, F. J. (2007). Fiber-optic pH-sensors in long-period fiber gratings using electrostatic self-assembly. *Optics Letters*, Vol. 32, No. 1, pp. 29-31. ISSN 0146-9592

Cusano, A., Iadicicco, A., Pilla, P., Contessa, L., Campopiano, S., Cutolo, A., & Giordano, M. (2005). Cladding mode reorganization in high-refractive-index-coated long-period gratings: effects on the refractive-index sensitivity. *Optics Letters*, Vol. 30, No. 19, pp. 2536-2538. ISSN 0146-9592

Cusano A., Iadicicco A., Pilla P., Contessa L., Campopiano S., Cutolo A., Giordano M. (2006a). Mode Transition in High Refractive Index Coated Long Period Gratings. *Optics Express*, Vol. 14, No. 1, pp. 19-34, ISSN 1094-4087.

Cusano, A., Iadicicco, A., Pilla, P., Contessa, L., Campopiano, S., Cutolo, A., Giordano, M., & Guerra, G. (2006b). Coated Long-Period Fiber Gratings as High-Sensitivity

Optochemical Sensors. *Journal of Lightwave Technology*, Vol. 24, No. 4, pp. 1776-. ISSN 0733-8724

Cusano, A., Iadicicco, A., Paladino, D., Campopiano, S., Cutolo, A. & Giordano, M. (2007). Micro-structured fiber Bragg gratings. PartII: towards advanced photonic devices. *Optical Fiber Technology*, Vol.13, No.4, (October 2007), pp. 291-301, ISSN 1068-5200

Cusano, A.; Paladino, D.; Iadicicco, A. (2009a). Microstructured Fiber Bragg Gratings. *Journal of Lightwave Technology*, Vol. 27, No. 11, pp. 1663 - 1697. ISSN 0733-8724

Cusano, A., Pilla, P., Giordano, M. & Cutolo, A. (2009b). Modal transition in nano-coated long period fiber gratings: principle and applications to chemical sensing, In: *Advanced Photonic Structures for Biological and Chemical Detection*, X. Fan, (Ed.), 35-76, Springer, ISBN 978-0-387-98060-7, Dordrecht, Heidelberg, London, New York

Del Villar, I., Matias, I., Arregui, F., & Lalanne, P. (2005). Optimization of sensitivity in Long Period Fiber Gratings with overlay deposition. *Optics Express*, Vol. 13, No. 1, pp. 56-69. ISSN 1094-4087

DeLisa, M. P., Zhang, Z., Shiloach, M., Pilevar, S., Davis, C. C., Sirkis, J. S., & Bentley, W. E. (2000). Evanescent Wave Long-Period Fiber Bragg Grating as an Immobilized Antibody Biosensor. *Analytical Chemistry*, Vol. 72, No. 13, pp. 2895-2900. ISSN 0003-2700

Diez, A., Birks, T.A., Reeves, W.H., Mangan, B.J., & Russell, P.St.J. (2000). Excitation of cladding modes in photonic crystal fibers by flexural acoustic waves, *Optics Letters*, Vol. 25 ,No. 20, 2000, pp. (1499–1501), ISSN 1539-4794

Dobb H., Kalli K., & Webb D. J. (2004). Temperature-insensitive long-period grating sensors in photonic crystal fiber. *Electron. Lett.*, Vol. 40, pp. 657–658, ISSN 0013-5194.

Dobb, H., Kalli, K., & Webb, D. J., (2006). Measured sensitivity of arc-induced long-period grating sensors in photonic crystal fibre. *Optics Communications*, Vol. 260, No 1, April 2006,pp. (184-191), ISSN 0030-4018

Eggen, C. L., Lin, Y. S., Wei, T., & Xiao, H. (2010). Detection of lipid bilayer membranes formed on silica fibers by double-long period fiber grating laser refractometry. *Sensors and Actuators B: Chemical*, Vol. 150, No. 2, pp. 734-741. ISSN 0925-4005

Eggleton, B.J., Westbrook, P.S., Windeler, R.S., Spalter, S., & Strasser, T.A. (1999). Grating resonances in air silica microstructured optical fiber. *Optics Letters*, Vol. 24 , No. 21, 1999, pp. (1460–1462), ISSN 1539-4794

Erdogan T. (1997). Fiber Grating Spectra. *J. of Lightwave Technology*, Vol. 15, No. 8, pp. 1277-1295, ISSN 0733-8724

Espindola, R.P., Windeler, R.S., Abramov, A.A., Eggleton, B.J., Strasser, T.A. & Di Giovanni, D.J. (1999). External refractive index insensitive air-clad long-period fiber grating. *Electronics Letters*, Vol. 35, No. 4, February 1999, pp. (327–328), ISSN 0013-5194

Falate, R., Kamikawachi, R. C., Müller, M., Kalinowski, H. J., & Fabris, J. L. (2005). Fiber optic sensors for hydrocarbon detection. *Sensors and Actuators B: Chemical*, Vol. 105, No. 2, pp. 430-436. ISSN 0925-4005

Falciai, R., Mignani, A. G., & Vannini, A. (2001). Long period gratings as solution concentration sensors. *Sensors and Actuators B: Chemical*, Vol. 74, No. (1-3), pp. 74-77. ISSN 0925-4005

Fotiadi, A. A., Brambilla, G., Ernst, T., Slattery, S. A., & Nikogosyan, D. N., (2007). TPA-induced long-period gratings in a photonic crystal fiber: Inscription and temperature sensing properties. *Journal of Optical Society of America B*, Vol. 24, No. 7, Jul. 2007, pp. (1475–1481), ISSN 1520-8540

Frazao, O., Santos, J.L., Araujo, F. M., & Ferreira, L.A., (2008). Optical sensing with photonic crystal fibers. *Laser & Photonics Reviews*, Vol. 2, No. 6, November 2008, pp.(449–459), ISSN 1863-8899

Fu, M.-Y., Lin, G.-R., Liu, W.-F., Sheng, H.-J., Su, P.-C. & Tien, C.-L. (2009). Optical fiber sensor based on air-gap long-period fiber gratings. *Japanese Journal of Applied Physics*, Vol.48, No.12, (December 2009), 120211 (3 pages), ISSN 0021-4922

Fu, M.-Y., Lin, G.-R., Liu, W.-F. & Wu, C.W. (2011). Fiber-optic humidity sensor based on an air-gap long period fiber grating. *Optical Review*, Vol.18, No.1, (January 2011), pp. 93-95, ISSN 1340-6000

Gibson, R., Kvavle, J., Selfridge, R. & Schultz, S. (2007). Improved sensing performance of D-fiber/planar waveguide couplers. *Optics Express*, Vol.15, No.5, (March 2007), pp. 2139-2144, ISSN 1094-4087

Gordon, J. D., Lowder, T. L., Selfridge, R. H. & Schultz, S. M. (2007). Optical D-fiber-based volatile organic compound sensor. *Applied Optics*, Vol.46, No.32, (November 2007), pp. 7805-7810, ISSN 0003-6935

Gu, Z., Xu, Y., & Gao, K. (2006). Optical fiber long-period grating with solgel coating for gas sensor. *Optics Letters*, Vol. 31, No. 16, pp. 2405-2407. ISSN 0146-9592

Han Y-G, Song S., Kim G.H., Lee K., Lee S., Lee J.H., Jeong C.H., Oh C.H., & Kang H.J., (2007). Simultaneous independent measurement of strain and temperature based on long-period fiber gratings inscribed in holey fibers depending on air-hole size *Optics Letters*, Vol. 32, pp. 2245-2247, ISSN 1539-4794.

He, Y.-J., Lo, Y.-L., & Huang, J.-F. (2006). Optical-fiber surface-plasmon-resonance sensor employing long-period fiber gratings in multiplexing. *Journal of the Optical Society of America B*, Vol. 23, No. 5, pp. 801-811. ISSN 0740-3224

He, Z., Tian, F., Zhu, Y., Lavlinskaia, N., & Du, H. (w.d.). Long-period gratings in photonic crystal fiber as an optofluidic label-free biosensor. *Biosensors and Bioelectronics*, In Press, Corrected Proof. ISSN 0956-5663

Hong, K. S., Park, H. C., Kim, B. Y., Hwang, I. K., Jin, W. Ju, J., & Yeom, D. I. (2008). 1000 nm tunable acousto-optic filter based on photonic crystal fiber. *Applied Physics Letters*, Vol. 92, No. 3, pp. 031110-1–031110-3, ISSN 1077-3118

Humbert, G., Malki, A., Fevrier, S., Roy, P., & Pagnoux, D. (2003), Electric arc-induced long-period gratings in Ge-free air-silica microstructure fibres. *Electronics Letters*, Vol. 39, No. 4, February, pp. (349-350), ISSN 0013-5194

Humbert, G., Malki, A., Fevrier, S., Roy, P., & Pagnoux, D. (2004). Characterizations at high temperatures of long-period gratings written in germanium-free air–silica microstructure fiber. *Opt. Lett.*, Vol. 29, pp. 38–40, ISSN 1539-4794.

Iadicicco A., Campopiano S., Giordano M., Cusano A. (2007). Spectral behavior in thinned long period gratings: effects of fiber diameter on refractive index sensitivity. *Applied Optics*, Vol. 46, p. 6945-6952, ISSN: 0003-6935.

Iadicicco A., Campopiano S., Cutolo A., Korwin-Pawlowski M.L., Bock W.J., CusanoA. (2008). Refractive Index Sensitivity in Thinned UV and Arc Induced Long-Period Gratings: A Comparative Study. *International Journal on Smart Sensing and Intelligent Systems*, Vol. 1, p. 354-369, 2008, ISSN: 1178-5608.

Iadicicco A., Campopiano S., & Cusano A. (2011a). Long Period Gratings in Hollow Core Fibers by Pressure Assisted Arc Discharge Technique. *Photonic Technology Letters*, Vol. pp, No 99 (in press - online available), ISSN 1041–1135.

Iadicicco A., Campopiano S., Cutolo A., & Cusano A. (2011b). Long Period Grating in Hollow Core Fibers: Fabrication and Characterization. *IEEE Sensors Conference*, (in press).

Iredale, T.B.; Steinvurzel, P.; Eggleton, B.J., (2006). Electric-arc-induced long-period gratings in fluid-filled photonic bandgap fibre, *Electronics Letters*, Vol 42, No 13, pp 739 - 740, ISSN: 0013-5194

Ivanov O.V.& RegoG. (2007). Origin of coupling to antisymmetric modes in arc-induced long-period fiber gratings. *Opt. Express*, Vol. 15, pp. 13936-13941, ISSN 1094-4087.

James, S. W., & Tatam, R. P. (2003). Optical fibre long-period grating sensors: characteristics and application. *Measurement Science and Technology*, Vol. 14, No. 5, pp. R49-R61. ISSN 0957-0233

Jang, H. S., Park, K. N., Kim, J. P., Sim, S. J., Kwon, O. J., Han, Y.-G. & Lee, K. S. (2009). Sensitive DNA biosensor based on a long-period grating formed on the side-polished fiber surface. *Optics Express*, Vol.17, No.5, (March 2009), pp. 3855-3860, ISSN 1094-4087

Jiang, M., Zhang, A. P., Wang, Y.-C., Tam, H.-Y., & He, S. (2009). Fabrication of a compact reflective long-period grating sensor with a cladding-mode-selective fiber end-face mirror. *Optics Express*, Vol. 17, No. 20, pp. 17976-17982. ISSN 1094-4087

Jin L., Jin W., Ju J., & Wang Y. (2011). Investigation of Long-Period Grating Resonances in Hollow-Core Photonic Bandgap Fibers. *J. Lightwave Technol.*, Vol. 29, pp. 1707-1713, ISSN 0733-8724.

Ju J., Jin W., & Demokan M. S. (2004). Two-mode operation in highly birefringent photonic crystal fiber, *IEEE Photon. Technol. Lett.*, Vol. 16, pp. 2472-2474, ISSN 1041–1135.

Ju, J., Jin, W., & Ho, H. L. (2008). Compact in-fiber interferometer formed by long-period gratings in photonic crystal fiber. *IEEE Photonics Technology Letters*, Vol. 20, No. 23, December 2008, pp. (1899- 1901), ISSN 1041–1135

Ju, J., & Jin, W. (2011). Long Period Gratings in Photonic Crystal Fibers. *Photonic Sensors*, DOI: 10.1007/s13320-011-0020-9, ISSN 2190-7439

Kaiser P.V., Astle H.W. (1974). Low-loss single material fibers made from pure fused silica. *Bell System Technical Journal*, Vol. 53, No. 6, pp. 1021-1039. ISSN 1538-7305.

Kakarantzas, G., Birks, T.A., & Russell, P.St.J. (2002). Structured long-period gratings in photonic crystal fiber. *Optics Letters*, Vol. 27, No. 12, June 2002, pp. (1013–1015), ISSN 1539-4794

Kalachev, A. I., Nikogosyan, D. N., & Brambilla, G. (2005). Long-period fiber grating fabrication by high-intensity femtosecond pulses at 211 nm. *Journal of Lightwave Technology*, Vol. 23, No. 8, pp. 2568- 2578. ISSN 0733-8724

Kashyap R. (1999). Fiber Bragg Gratings, San Diego: Academic Press. ISBN 0124005608, 9780124005600

Kerbage C, Steinvurzel P., Hale A., Windeler R.S., Eggleton B.J. (2002). Microstructured optical fiber with tunable birefringence. *Electron. Lett.*, Vol. 38, pp. 310–312, ISSN 0013-5194.

Kersey A.D., Davis M.A., Patrick H.J., LeBlanc M., Koo K.P.,. Askins C.G, Putnam M.A., & Friebele E.J.. (1997). Fiber Grating Sensors. *Journal of Lightwave Technology*, Vol. 15, No. 8, pp. 1442- 1463. ISSN 0733-8724.

Kim, D. W., Zhang, Y., Cooper, K. L., & Wang, A. (2006). Fibre-optic interferometric immuno-sensor using long period grating. *Electronics Letters*, Vol. 42, No. 6, pp. 324-325. ISSN 0013-5194

Kim, H.-J., Kwon, O.-J., Han, Y.-G., Lee, M. K. & Lee, S. B. (2010). Surface long-period fiber gratings inscribed in photonic crystal fibers. *Journal of the Korean Physical Society*, Vol.57, No.6, (December 2010), pp. 1956-1959, ISSN 0374-4884

Kim, H.-J., Kwon, O.-J., Lee, S. B. & Han, Y.-G. (2011). Measurement of temperature and refractive index based on surface long-period gratings deposited onto a D-shaped photonic crystal fiber. *Applied Physics B: Lasers and Optics*, Vol.102, No.1, (January 2011), pp. 81-85, ISSN 0946-2171

Knight, J. C., Birks, T. A., Russell, P. S. J. , & Atkin, D. M.. 1996. All-silica single-mode optical fiber with photonic crystal cladding. *Optics Letters*, Vol. 21, No. 19, 1996, pp. 1547–1549. ISSN 539-4794

Konstantaki, M., Pissadakis, S., Pispas, S., Madamopoulos, N., & Vainos, N. A. (2006). Optical fiber long-period grating humidity sensor with poly(ethylene oxide)/cobalt chloride coating. *Applied Optics*, Vol. 45, No. 19, pp. 4567-4571. ISSN 1559-128X

Kuhlmey B.T., Eggleton B.J., & Wu D.K.C. (2009). Fluid-Filled Solid-Core Photonic Bandgap Fibers. *IEEE Journal of Lightwave Techonology*, Vol. 27, pp. 1617-1630, ISSN 0733-8724.

Kumar, A. & Varshney, R. K. (1984). Propagation characteristics of highly elliptical core optical waveguides: a perturbation approach. *Optical and Quantum Electronics*, Vol.16, No.4, (July 1984), pp. 349-354, ISSN 0306-8919

Lee, D., Jung, Y., Jeong, Y. S., Oh, K., Kobelke, J., Schuster, K., & Kirchhof, J. (2006). Highly polarization-dependent periodic coupling in mechanically induced long period grating over air-silica fibers. Optics Letters, Vol. 31, No. 3, 2006, pp. (296–298), ISSN 1539-4794

Lee, H. W., & Chiang, K. S. (2009). $CO_2$ laser writing of long-period fiber grating in photonic crystal fiber under tension. *Optics Express*, Vol. 17, Issue 6, March 2009, pp. (4533-4539), ISSN 1094-4087

Lee, J., Chen, Q., Zhang, Q., Reichard, K., Ditto, D., Mazurowski, J., Hackert, M., & Shizhuo, Y. (2007). Enhancing the tuning range of a single resonant band long period grating while maintaining the resonant peak depth by using an optimized high index indium tin oxide overlay. *Applied Optics*, Vol. 46, No. 28, pp. 6984-6989. ISSN 1559-128X

Lim, J.H., Lee, K.S., Kim, J.C., & Lee, B.H. (2004). Tunable fiber gratings fabricated in photonic crystal fiber by use of mechanical pressure, *Optics Letters*, Vol. 29, No. 4, February 2004, pp. (331–333), ISSN 1539-4794

Liu Y., Wang L., Zhang M., Tu D., Mao X., & Liao. Y. (2007). Long-Period Grating Relative Humidity Sensor With Hydrogel Coating. *IEEE Photonics Technology Letters*, Vol. 19, No. 12, pp. 880-882. ISSN 1041-1135

Liu, S., Jin, L., Jin, W., Wang, Y., & Wang, D. N. (2010). Fabrication of Long-Period Gratings by Femtosecond Laser-Induced Filling of Air-Holes in Photonic Crystal Fibers. *IEEE Photonics Technology Letters*, Vol. 22, No. 22, November 2010, pp. (1635-1637), ISSN 1041–1135

Morishita, K. & Miyake, Y. (2004). Fabrication and resonance wavelengths of long-period gratings written in a pure-silica photonic crystal fiber by the glass structure change. *Journal of Lightwave Technology*, Vol. 22, No. 1, 625-630, ISSN 0733-8724.

Mosquera, L., Sàez-Rodriguez, D., Cruz, J. L., & Andrès, M. V. (2010). In-fiber Fabry-Perot refractometer assisted by a long-period grating. *Optics Letters*, Vol. 35, No. 4, pp. 613-615. ISSN 0146-9592

Ozcana A. & Demircib U. (2007). Rewritable self-assembled long-period gratings in photonic bandgap fibers using microparticles. *Optics Communications*, Vol 270, No 2, pp 225-228, ISSN 0030-4018

Parka, K. N., Erdogan, T., & Lee, K. S., (2006). Cladding mode coupling in long-period gratings formed in photonic crystal fibers. *Optics Communications*, Vol. 266, No. 2, 2006, pp. (541–545), ISSN 0030-4018

Patrick, H. J., Kersey, A. D., & Bucholtz, F. (1998). Analysis of the Response of Long Period Fiber Gratings to External Index of Refraction. *Journal of Lightwave Technology*, Vol. 16, No. 9, pp. 1606-. ISSN 0733-8724

Petrovic J., Dobb H., Mezentsev V. K., Kalli K., Webb D. J., & BennionI. (2007). Sensitivity of LPGS in PCFs fabricated by an electric arc to temperature, strain and external refractive index. *Journal of Lightwave, Technol.*, Vol. 25, pp. 1306–1312. ISSN 0733-8724

Petrovic J. (2008). Modelling of Long Period Gratings in Photonic Crystal Fibres and Sensors Based on Them. InTech *Book Modelling and Simulation*, Chapter 22, 417, ISBN 978-3-902613-25-7.

Pilla, P., Foglia Manzillo, P., Giordano, M., Korwin-Pawlowski, M. L., Bock, W. J., & Cusano, A. (2008). Spectral behavior of thin film coated cascaded tapered long period gratings in multiple configurations. *Optics Express*, Vol. 16, No. 13, pp. 9765-9780. ISSN 1094-4087

Pilla, P., Manzillo, P. F., Malachovska, V., Buosciolo, A., Campopiano, S., Cutolo, A., Ambrosio, L., Giordano, M., & Cusano, A. (2009). Long period grating working in transition mode as promising technological platform for label-free biosensing. *Optics Express*, Vol. 17, No. 22, pp. 20039-20050. ISSN 1094-4087

Quero, G., Crescitelli, A., Paladino, D., Consales, M., Buosciolo, A., Giordano, M., Cutolo, A. & Cusano, A. (2011). Evanescent wave long-period fiber grating within D-shaped optical fibers for high sensitivity refractive index detection. *Sensors and Actuators B: Chemical*, Vol.152, No.2, (March 2011), pp. 196-205, ISSN 0925-4005

Rees, N. D., James, S. W., Tatam, R. P., & Ashwell, G. J. (2002). Optical fiber long-period gratings with Langmuir?Blodgett thin-film overlays. *Optics Letters*, Vol. 27, No. 9, pp. 686-688. ISSN 0146-9592

Rego, G., Marques, P. V. S., Santos, J. L., & Salgado, H. M. (2005a). Arc-Induced Long-Period Gratings. *Fiber and Integrated Optics*, Vol. 24, No. 3, pp. 245-259. ISSN 0146-8030

Rego G., Marques P.V.S., Salgado H.M., & Santos J.L. (2005b). Simultaneous measurement of temperature and strain based on arc-induced long-period fibre gratings. *Electron. Lett.*, Vol. 41, No. 2, pp. 60–62, Jan, ISSN 0013-5194.

Rego, G., Ivanov, O. V., & Marques, P. V. S. (2006). Demonstration of coupling to symmetric and antisymmetric cladding modes in arc-induced long-period fiber gratings. *Optics Express*, Vol. 14, No. 21, pp. 9594-9599. ISSN 1094-4087

Rindorf, L., Jensen, J. B., Dufva, M., Pedersen, L. H., Hoiby, P. E., & Bang, O. (2006). Photonic crystal fiber long-period gratings for biochemical sensing. *Optics Express*, Vol. 14, No. 18, September 2008, pp.(8224-8231), ISSN 1094-4087

Rindorf L. & Bang O. (2008). Highly sensitive refractometer with a photonic crystal-fiber long-period grating. *Optics Letters*, Vol. 33, pp. 563–564, , ISSN 1539-4794.

Russell R. (2003). Photonic crystal fibers. *Science*, Vol. 229, pp. 258–362. ISSN 1095-9203.

Seraji F.E., Farsinezhad S., & Anzabi L.C. (2011). Optimization of long-period grating inscribed in large mode area photonic crystal fiber for design of band stop filter. *Optik - International Journal for Light and Electron Optics*, Vol. 122, N. 2011, pp. 58–62, ISSN 0030-4026.

Sharma, A., Kompella, J. & Mishra, P. K. (1990). Analysis of fiber directional coupler half blocks using a new sinple model for single mode fiber. *Journal of Lightwave Technology*, Vol.8, No.2, (February 1990), pp. 143-151, ISSN 0733-8724

Shu, X., Zhang, L., & Bennion, I. (2002). Sensitivity Characteristics of Long-Period Fiber Gratings. *Journal of Lightwave Technology*, Vol. 20, No. 2, pp. 255-. ISSN 0733-8724

Smietana, M., Bock, W. J., Mikulic, P., Ng, A., Chinnappan, R., & Zourob, M. (2011). Detection of bacteria using bacteriophages as recognition elements immobilized on long-period fiber gratings. *Optics Express*, Vol. 19, No. 9, pp. 7971-7978. ISSN 1094-4087

Smith C.M., Venkataraman N., Gallagher M.T., Muller D., West J.A., Borrelli N.F.,. Allan D.C, & Koch K.W. (2003). Low-loss hollow-core silica/air photonic bandgap fibre. *Nature* Vol. 424, pp. 657-659. ISSN: 0028-0836

Smith, K. H., Markos, D. J., Ipson, B. L., Schultz, S. M., Selfridge, R. H., Barber, J. P., Campbell, K. J., Monte, T. D. & Dyott, R. B. (2004). Fabrication and analysis of a low-loss in-fiber active polymer waveguide. *Applied Optics*, Vol.43, No.4, (February 2004), pp. 933-939, ISSN 0003-6935

Smith, K. H. (2005). In-fiber optical devices based on D-fiber, *Ph.D. Dissertation at Brigham Young University*, April 2005, Available from http://contentdm.lib.byu.edu/ETD/image/etd738.pdf

Smith, K. H., Ipson, B. L., Lowder, T. L., Hawkins, A. R., Selfridge, R. H. & Schultz, S. M. (2006). Surface-relief fiber Bragg gratings for sensing applications. *Applied Optics*, Vol.45, No.8, (March 2006), pp. 1669-1675, ISSN 0003-6935

Sun J., Chan C.C., Dong X.Y., &. Shum. (2007). Application of an artificial neural network for simultaneous measurement of temperature and strain by using a photonic crystal fiber long-period grating. *Meas. Sci. Technol.* Vol. 18, pp. 2943–2948, ISSN 1361-6501.

Smith C.M., & Venkataraman N, Gallagher M. T., Muller D., West J. A., Borrelli N.F., Allan D.C., & Koch K.W. (2003). Low-loss hollow-core silica/air photonic bandgapfibre. Nature Vol. 424, pp. 657-659, 2003, ISSN 1476- 4687.

Steinvurzel P., Moore E.D., Mägi E.C., Kuhlmey B.T., & Eggleton B.J. (2006a). Long period grating resonances in photonic bandgap fiber. Opt. Express, Vol. 14, No. 7, pp. 3007-3014, ISSN 1094-4087

Steinvurzel P., Moore E.D., Mägi E.C., & Eggleton B.J. (2006b). Tuning properties of long period gratings in photonic bandgap fibers, *Optics Letters.*, Vol. 31, No. 14, pp. 2103-2105, ISSN 1539-4794

Tan, K. M., Tay, C. M., Tjin, S. C., Chan, C. C., & Rahardjo, H. (2005). High relative humidity measurements using gelatin coated long-period grating sensors. *Sensors and Actuators B: Chemical*, Vol. 110, No. 2, pp. 335-341. ISSN 0925-4005

Tang, J.-L., Cheng, S.-F., Hsu, W.-T., Chiang, T.-Y., & Chau, L.-K. (2006). Fiber-optic biochemical sensing with a colloidal gold-modified long period fiber grating. *Sensors and Actuators B: Chemical*, Vol. 119, No. 1, pp. 105-109. ISSN 0925-4005

Tien, C.-L., Hwang, C.-C., Liu, W.-F. & Lin, T.-W. (2009a). Magnetic field sensor based on D-shaped long period fiber gratings, *Proceedings of SPIE 7508 (2009 International Conference on Optical Instruments and Technology: Advanced Sesnor Technologies and Applications)*, 750811, ISBN 978-0-819-47894-8, Shanghai, China, October 19-21, 2009

Tien, C.-L., Lin, T.-W., Hsu, H.-Y. & Liu, W.-F. (2009b). Double-sided polishing long period fiber grating sensors for measuring liquid refractive index, *Proceedings of 2009 Asia Communications and Photonics (ACP) Conference and Exhibition*, ISBN 978-1-55752-877-3, Shanghai, China, November 2-6, 2009

Topliss, S. M., James, S. W., Davis, F., Higson, S. P. J., & Tatam, R. P. (2010). Optical fibre long period grating based selective vapour sensing of volatile organic compounds. *Sensors and Actuators B: Chemical*, Vol. 143, No. 2, pp. 629-634. ISSN 0925-4005

Tripathi, S. M., Kumar, A., Marin, E. & Meunier, J.-P. (2008). Side-polished optical fiber grating-based refractive index sensors utilizing the pure surface plasmon polariton. *Journal of Lightwave Technology*, Vol.26, No.13, (July 2008), pp. 1980-1985, ISSN 0733-8724

Tripathi, S. M., Marin, E., Kumar, A. & Meunier, J.-P. (2009). Refractive index sensing characteristics of dual resonance long period gratings in bare and metal-coated D-shaped fibers. *Applied Optics*, Vol.48, No.31, (November 2009), pp. G53-G58, ISSN 1539-4522

Tseng, S.-M. & Chen, C.-L. (1992). Side-polished fibers. *Applied Optics*, Vol.31, No.18, (June 1992), pp. 3438-3447, ISSN 0003-6935

Vasil'ev, S. A., Medvedkov, O. I., Korolev, I. G., Bozhkov, A. S., Kurkov, A. S., & Dianov, E. M. (2005). Fibre gratings and their applications. *Quantum Electronics*, Vol. 35, No. 12, pp. 1085-1103. ISSN 1063-7818

Vengsarkar A.M., Lemaire P.J., Judkins J.B., Bhatia V., Erdogan T., & Sipe J.E., (1996). Long-period fiber gratings as band-rejection filters. *Journal of Lightwave Technology*, Vol. 14, pp. 58–65, Jan, , ISSN 0733-8724

Venugopalan, T., Sun, T., & Grattan, K. T. V. (2008). Long period grating-based humidity sensor for potential structural health monitoring. *Sensors and Actuators A: Physical*, Vol. 148, No. 1, pp. 57-62. ISSN 0924-4247

Wang Y., Xiao L., Wang D., & Jin W. (2006). Highly sensitive long-period fiber-grating strain sensor with low temperature sensitivity. *Optics Letters*, Vol. 31, pp. 3414-3416, ISSN 1539-4794.

Wang, Y., Xiao, L., Wang, D. N., & Jin, W. (2007). In-fiber polarizer based on a long-period fiber grating written on photonic crystal fiber. *Optics Letters*, Vol. 32, No. 9, May 2007, pp. (1035-1037), ISSN 1539-4794

Wang Y., Jin W., Ju J., Xuan H., Ho H.L., Xiao L., & Wang D., (2008). Long period gratings in air-core photonic bandgap fibers, *Opt. Exp.*, Vol. 16, No. 4, pp. 2784–2790, Feb, ISSN 1094-4087

Wang Y. (2010). Review of long period fiber gratings written by CO2 laser. *Journal of Applied Physics* Vol. 108, No. 081101, pp. 081101-1 081101-18, ISSN 1089-7550

Wang, Z., Heflin, J. R., Stolen, R. H., & Ramachandran, S. (2005). Highly sensitive optical response of optical fiber long period gratings to nanometer-thick ionic self-assembled multilayers. *Applied Physics Letters*, Vol. 86, No. 22, pp. 223104. ISSN 0003-6951

Wang, Z., Heflin, J. R., Van Cott, K., Stolen, R. H., Ramachandran, S., & Ghalmi, S. (2009). Biosensors employing ionic self-assembled multilayers adsorbed on long-period fiber gratings. *Sensors and Actuators B: Chemical*, Vol. 139, No. 2, pp. 618-623. ISSN 0925-4005

Wei, X., Wei, T., Xiao, H., & Lin, Y. S. (2008). Nano-structured Pd-long period fiber gratings integrated optical sensor for hydrogen detection. *Sensors and Actuators B: Chemical*, Vol. 134, No. 2, pp. 687-693. ISSN 0925-4005

West J. A., Smith C. M., Borrelli N. F., Allan D. C., & Koch K. W. (2004). Surface modes in air-core photonic band-gap fibers. *Opt. Express*, Vol. 12, pp. 1485-1496, ISSN 1094-4087.

Xuewen S., Allsop, T., Gwandu, B., Lin Zhang, & Bennion, I. (2001). High-temperature sensitivity of long-period gratings in B-Ge codoped fiber. *IEEE Photonics Technology Letters*, Vol. 13, No. 8, pp. 818-820. ISSN 1041-1135.

Yin, S., Chung, K.-W., & Zhu, X. (2001). A novel all-optic tunable long-period grating using a unique double-cladding layer. *Optics Communications*, Vol. 196, No. 1-6, pp. 181-186. ISSN 0030-4018

Zhao, C.-L., Xiao, L., Ju, J., Demokan M.S., & Jin,W. (2008). Strain and temperature characteristics of a long-period grating written in a photonic crystal fiber and its application as a temperature-insensitive strain sensor. *IEEE Journal of Lightwave Techonology*, Vol. 26, No. 2, January 2008, pp. (220-227), ISSN 0733-8724.

Zhao, D., Zhou, K., Chen, X., Zhang, L., Bennion, I., Flockhart, G., MacPherson, W. N., Barton, J. S. & Jones, J. D. C. (2004a). Implementation of vectorial bend sensors using long-period gratings UV-inscribed in special shape fibres. *Measurement Science and Technology*, Vol.15, No.8, (August 2004), pp. 1647-1650, ISSN 0957-0233

Zhao, D., Chen, X., Zhou, K., Zhang, L., Bennion, I., MacPherson, W. N., Barton, J. S. & Jones, J. D. C. (2004b). Bend sensors with direction recognition based on long-period gratings written in D-shaped fiber. *Applied Optics*, Vol.43, No.29, (October 2004), pp. 5425-5428, ISSN 0003-6935

Zhang, J., Tang, X., Dong, J., Wei, T., & Xiao, H. (2008). Zeolite thin film-coated long period fiber grating sensor for measuring trace chemical. *Optics Express*, Vol. 16, No. 11, pp. 8317-8323. ISSN 1094-4087

Zhu Y., ShumP., Bay H.-W., Yan M., Yu X., Hu J., Hao J., & Lu C. (2005). Strain-insensitive and high-temperature longperiod gratings inscribed in photonic crystal fiber. *Opt. Lett.*, Vol. 30, pp. 367–369, ISSN 1539-4794.

15

# Optical Vortices in a Fiber: Mode Division Multiplexing and Multimode Self-Imaging

S.N. Khonina, N.L. Kazanskiy and V.A. Soifer
*Image Processing Systems Institute of the Russian Academy of Sciences,*
*S.P. Korolyov Samara State Aerospace University,*
*Russia*

## 1. Introduction

The optical vortices (Dennis et al., 2009; Desyatnikov et al., 2005; Soskin & Vasnetsov, 2001) or angular harmonics exp(imφ) describe a wavefront peculiarity, or helical dislocation, when in passing around the origin of coordinates the light field phase acquires a phase shift of 2πm, where m is the optical vortex's order. The generation and propagation of the laser vortices in free space has been studied fairly well, meanwhile, the excitation of individual vortex modes and obtaining desired superpositions thereof in optical fibers present a greater challenge (Berdague & Facq, 1982; Bolshtyansky et al., 1999; Dubois et al., 1994; Karpeev & Khonina, 2007; Mikaelian, 1990; Soifer & Golub, 1994; Thornburg et al., 1994; Volyar & Fadeeva, 2002).

Note that the most interesting is the excitation and propagation of pure optical vortices that are not step- or graded-index fiber modes. However decomposition of the light fields in terms of angular harmonics has a number of advantages over other bases, including modal ones, when dealing with problems of laser beam generation and analysis and mode division multiplexing. As distinct from the classical LP-modes, the angular harmonics are scale-invariant when coupled into the fiber and selected at the fiber's output using diffractive optical elements (DOEs) (Dubois et al., 1994; Karpeev & Khonina, 2007; Soifer & Golub, 1994; Thornburg et al., 1994). This gives much freedom in choosing parameters of an optical scheme, allowing one to effectively counteract noises, as it will be demonstrated below.

A term "mode division multiplexing" (MDM) is used for multimodal optical fibers when describing methods for data transmission channel multiplexing, with each spatial fiber mode being treated as a separate channel that carries its own signal (Berdague & Facq, 1982; Soifer & Golub, 1994). The essence of mode division multiplexing is as follows: laser beams as a linear superposition of fiber modes can be used to generate signals that will effectively transmit data in a physical carrier - a multimodal fiber. The data transmitted can be contained both in the modal composition and in the energy portion associated with each laser mode.

The MDM concept has not yet been turned to practical use because a definite mode superposition with desired between-mode energy distribution is difficult to excite. Another reason is that there is energy redistribution between modes when transmitting data in real fibers over long distances. However, for optical fibers 1-2 m long - for example, used in endoscopy - the modes do not mix at small bendings (when the curvature radius is much greater than fiber's core radius), acquiring only a radius-related phase delay.

A major problem with the MDM is exciting a definite modal superposition with a desired energy distribution between modes. Lower-order modes (e.g., LP11) can be excited by applying a periodic fiber deformation (squeezing or bending) and with a tilted grating written in a photosensitive fiber by two interfering laser beams. Higher-order modes LPm1 can be generated through the off-axis coupling of laser light into the fiber's end at a definite angle or by DOE in which the complex amplitude of mode distribution was encoded. Using diffractive optical elements any set of modes with designed weights can be effectively generated and selected (Soifer & Golub, 1994; Karpeev & Khonina, 2007).

We discuss linear superpositions of LP-modes of a stepped-index fiber in the first section. As an alternative to the superposition of classical LP-modes used to carry signal in a light fiber we propose a superposition of angular harmonics that can be derived as a special combination of LP-modes also featuring modal properties in an optical fiber.

Imposing the certain conditions on mode's compound it is possible to form laser beams with the definite self-reproduction (Kotlyar et al., 1998), while mode's weights and phase shift between modes provide approximation of desirable cross-section distribution of laser beam intensity on the certain distances (Almazov & Khonina, 2004).

The light field periodic self-reproduction in the gradient-index media was analytically studied using the ray tracing approach and wave theory in (Mikaelian, 1980). The self-reproduction was treated in the above studies as self-focusing, i.e. a periodic focusing of radiation. The analytical expressions for the refractive index of the medium where the phenomenon occurs have been derived.

In the second section we numerically simulate the behavior of multi-mode light fields in the circular parabolic graded-index fibers which propagate linearly polarized Laguerre-Gaussian (LG) modes in the weak guidance approximation. Analytical formulae describing the propagation of a linear composition of the LG modes in a fiber are rather simple, thus making it possible to simulate the propagation of a certain light field (image) along a definite fiber via decomposing it into the LG modes (Snyder & Love, 1987). Note that the accuracy of image representation is essential. The more modes are found in the linear combination, the more adequate is the image approximation. Also, with regard to the aforementioned application, it would be interesting to determine the self-reproduction periods of the chosen mode superposition.

Unfortunately, the use of an arbitrary number of modes satisfying definite criteria is impossible for the following two reasons: (1) a multi-mode fiber is able to transmit only a limited number of modes determined by its radius and the core's refractive index and (2) the more modes participate in the approximation, the greater is their general period, i.e. the image self-reproduction will occur more rarely. Besides, the image is disintegrated even under a minor change in the fiber length (of about 0.1.mm) as a result of temperature variations, mechanical deformation, etc. Thus, the "direct" image recognition from the intensity pattern becomes difficult if possible at all. However, the images can be fairly accurately recognized from the distribution pattern of the squared modules of the mode coefficients (Bolshtyansky & Zel'dovich, 1996), which are preserved at any distance in the ideal fiber.

The propagation of the electromagnetic wave in the medium can be modeled in several ways. The most common technique is to describe the propagation using Maxwell's

equations, from which vectorial wave equations defining the electric and magnetic field components can be deduced. If the relative change of the medium refractive index per wavelength is significantly smaller than unity, the Helmhotz equation can be written for each scalar component of the vector field (Agrawal, 2002).
For weakly nonuniform media, the approximation based on a periodic array of identical optical elements placed in a uniform medium is also valid. In particular, for the parabolic-index medium, this array comprises circular converging lenses. For the limiting case of an infinitely large number of lenses with an infinitesimally small separation we derive an integral operator to describe the propagation of light in a medium with parabolic refractive index in the scalar theory. This integral operator is analogous to the Fresnel transform that describes, with the same accuracy, the propagation of light in a uniform medium.

## 2. Vortical laser beams in a weakly guiding stepped-index fibers

Let us consider a circular stepped-index optical fiber, in which the core of radius $a$ has the refractive index of $n_1$, and the cladding of radius $b$ has the refractive index of $n_2$. For most popular commercial fibers, the core-cladding index contrast, $\Delta n = n_1 - n_2$, is less than 1%. For such fibers, termed weakly guiding, assuming $n_1 \cong n_2$, we can consider in place of hybrid modes of the propagating electromagnetic field their linearly polarized superpositions (Cherin, 1987; Gloge, 1971; Marcuse, 1972; Yeh, 1990).
Considering that for the LP-mode the transverse field is essentially linearly polarized, a complete set of modes takes place when only one electric and one magnetic component are predominant. In this case, it is possible, for example, to consider the electric vector **E** directed along the $x$-axis, and a perpendicular magnetic vector **H**, directed along the $y$-axis. Note, also, that the time-averaged power flux (Re[**E**×**H**]/2) appears to be proportional to the electric vector intensity (Gloge, 1971; Yeh, 1990). All above considered, we will consider the LP-modes in the scalar form:

$$\Psi_{pq}(r,\phi,z) = \exp\left(-i\beta_{pq}z\right)T_p(\phi)R_{pq}(r) = \exp\left(-i\beta_{pq}z\right)\begin{Bmatrix}\cos(p\phi)\\ \sin(p\phi)\end{Bmatrix}\begin{cases}\dfrac{J_p(u_{pq}r/a)}{J_p(u_{pq})}, & 0 \le r \le a\\[2ex] \dfrac{K_p(w_{pq}r/a)}{K_p(w_{pq})}, & a \le r \le b\end{cases} \tag{1}$$

where the parameters

$$u = a\sqrt{(kn_1)^2 - \beta^2}\,, \tag{2}$$

$$w = a\sqrt{\beta^2 - (kn_2)^2}\,, \tag{3}$$

$$u^2 + w^2 = V^2 \equiv \left(\frac{2\pi}{\lambda}\right)^2 a^2\left(n_1^2 - n_2^2\right), \tag{4}$$

are derived from an equation for eigen-values:

$$\frac{uJ_{p-1}(u)}{J_p(u)} = -\frac{wK_{p-1}(w)}{K_p(w)}\,. \tag{5}$$

In Eq. (4), $\lambda$ is the wavelength of laser light in air. In Eq. (1), the first-kind Bessel functions $J_m(x)$ describe the field in the fiber core,whereas the modified Bessel functions $K_m(x)$ are for the cladding.

We consider the propagation of a linear superposition of LP-modes in an ideal stepped-index optical fiber:

$$U_0(r,\phi) = \sum_{p,q\in\Omega} C_{pq}\Psi_{pq}(r,\phi) \tag{6}$$

where $C_{pq}$ are the complex coefficients, $\Psi_{pq}(r,\varphi)$ are the modes of Eq. (1) at $z$=0, whose angular component is represented in a different way without loss of generality:

$$\Psi_{pq}(r,\phi,z) = \exp\left(-i\beta_{pq}z\right)\exp\left(ip\phi\right)\begin{cases} \dfrac{J_p(u_{pq}r/a)}{J_p(u_{pq})}, & 0 \le r \le a \\ \dfrac{K_p(w_{pq}r/a)}{K_p(w_{pq})}, & a \le r \le b \end{cases} \tag{7}$$

Although Eqs. (1) and (7) are connected via a simple relation, they describe modes with somewhat different properties. By way of illustration, the modes in Eq. (1) are real at $z$=0, but they do not have an orbital angular momentum. Thus, for each mode in Eq. (7), the linear density of z-projection of the orbital angular momentum is proportional to the first index $p$ (Allen et al., 2003).

For the field in Eq. (6) with the modes of Eq. (7), the z-projection of the orbital momentum (Kotlyar et al., 2002):

$$J_{z0} = -\frac{\sum_{p,q\in\Omega} p\left|C_{pq}\right|^2}{\omega\sum_{p,q\in\Omega}\left|C_{pq}\right|^2} \tag{8}$$

The modes show a key property of invariance to the propagation operator in a given medium, implying that in propagation the mode structure remains unchanged, acquiring only a phase shift. In particular, the cross-section of the field in Eq. (7) will remain unchanged at any distance, being equal to its value at $z$=0:

$$\left|\Psi_{pq}(r,\phi,z)\right|^2 = \left|R_{pq}(r)\exp(ip\phi)\exp(-i\beta_{pq}z)\right|^2 = R_{pq}^2(r) = \left|\Psi_{pq}(r,\phi)\right|^2. \tag{9}$$

Because the expression in Eq. (9) is $z$-independent in a perfect fiber, it can be used as an additional parameter to characterize individual modes in Eq. (7) or modal groups with an identical first index.

Figure 1 shows cross-section distributions for some modes of Eq. (7) for a stepped-index fiber with cut-off number $V$=8.4398. These modal characteristics remain unchanged upon propagation in a perfect fiber, with only phase changes taking place. For comparison, shown in Figs. 1d and 1e are phases at z=0 and z=100 μm, respectively.

The numerical simulation parameters are as follows: core radius is $a$=5 μm, cladding radius is $b$=62.5 μm, the respective refractive indices of the core and cladding are $n_1$=1.45

and $n_2$=1.44. Optical fibers with the above-specified parameters are normally used in a unimodal regime for wavelengths λ=1.31 μm and λ=1.55 μm. However, for the wavelength of λ=0.633 μm of a He-Ne laser a few-mode regime occurs (Khonina et al., 2003), meaning that several modes are propagated. For used parameters there are 11 propagating modes with $|p|\leq 5$.

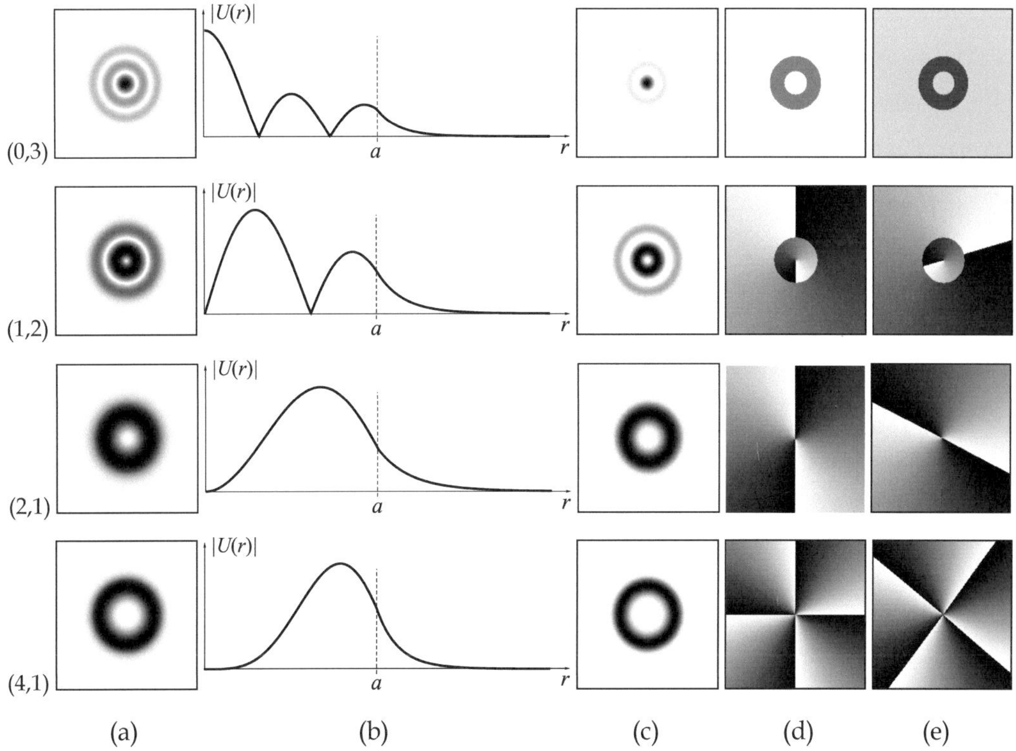


Fig. 1. The $(p,q)$ modes: (0,3), (1,2), (2,1), (4,1): (a) transverse amplitude distribution (negative), (b) radial amplitude cross-section, and (c) transverse intensity distribution (negative) in the plane $z$=0; transverse phase distribution (white: zero phase, black: 2π) in the planes (d) $z$=0 and (e) $z$=100 μm.

### 2.1 Multimode laser beam self-imaging in a weakly guiding stepped-index fibers

In the general case, the field in Eq. (6) does not show invariance regarding an individual mode in Eq. (9). However, by fitting a modal composition in Eq. (6) it is possible to find a modal superposition showing some self-reproduction properties.

In a perfect fiber at distance $z$, the superposition in Eq. (6) has the complex distribution

$$U(r,\phi,z)=\sum_{p,q\in\Omega} C_{pq}\Psi_{pq}(r,\phi,z), \tag{10}$$

where $\Psi_{pq}(r,\phi,z)=\Psi_{pq}(r,\phi)\cdot\exp(-i\beta_{pq}z)$, $\beta_{pq}$ are the propagation constants.

For any pair of modes, the intensity at distance $z$

$$\left|C_{p_iq_i}\Psi_{p_iq_i}(r,\phi,z)+C_{p_jq_j}\Psi_{p_jq_j}(r,\phi,z)\right|^2=\left|C_{p_iq_i}\right|^2R^2_{|p_i|q_i}(r)+\left|C_{p_jq_j}\right|^2R^2_{|p_j|q_j}+$$
$$+2\left|C_{p_iq_i}\right|\left|C_{p_jq_j}\right|R_{|p_i|q_i}(r)R_{|p_j|q_j}(r)\cos\left[(\arg C_{p_iq_i}-\arg C_{p_jq_j})+(p_i-p_j)\phi+(\beta_{|p_i|q_i}-\beta_{|p_j|q_j})z\right] \quad (11)$$

is other than the intensity at $z$=0:

$$\left|C_{p_iq_i}\Psi_{p_iq_i}(r,\phi)+C_{p_jq_j}\Psi_{p_jq_j}(r,\phi)\right|^2=\left|C_{p_iq_i}\right|^2R^2_{|p_i|q_i}(r)+\left|C_{p_jq_j}\right|^2R^2_{|p_j|q_j}+$$
$$+2\left|C_{p_iq_i}\right|\left|C_{p_jq_j}\right|R_{|p_i|q_i}(r)R_{|p_j|q_j}(r)\cos\left[(\arg C_{p_iq_i}-\arg C_{p_jq_j})+(p_i-p_j)\phi\right] \quad (12)$$

because the former has a cosine term.
By imposing definite conditions on all pairs of constituent modes in the superposition in Eq. (6), it is possible to obtain fields featuring special properties of intensity distribution self-reproduction.

*Invariance in the entire region of propagation*

In propagation, a change in the transverse field distribution is due to intermode dispersion caused by a difference between mode propagation constants $\beta_{pq}$. For the function of the form (7) only modes with identical indices ($|p|$,$q$) will have the same propagation velocities. Thus, at any interval (in a perfect fiber) the invariance is shown only by a mode pair superposition given by

$$C_{|p|q}\Psi_{|p|q}(r,\phi)+C_{-|p|q}\Psi_{-|p|q}(r,\phi)\,. \quad (13)$$

In this case, in Eq. (11) we have

$$\cos\left[(\arg C_{|p|q}-\arg C_{-|p|q})+(|p|+|p|)\phi+(\beta_{|p|q}-\beta_{-|p|q})z\right]=\cos\left[(\arg C_{|p|q}-\arg C_{-|p|q})+2|p|\phi\right]$$

and the cross-section intensity ceases to depend on $z$, remaining unchanged. The form of the intensity distribution is entirely determined by the coefficients $C_{pq}$ (see Fig. 2).
In a particular case, when $\left|C_{|p|q}\right|=\pm\left|C_{-|p|q}\right|$ we get classical LP-modes in the form of Eq. (1) (first row in Fig. 2). It is noteworthy that the complex coefficient arguments have no effect on the value of the orbital angular momentum for the superposition of Eq. (6). Thus, with the coefficient amplitudes remaining unchanged, we obtain a rotated classical LP-mode whose orbital angular momentum is also zero (second row in Fig. 2).
Changes in the coefficient amplitude cause both the cross-section structure and the orbital angular momentum to be changed. For the third and bottom rows in Fig. 2, the respective values of the orbital angular momentum in Eq. (8) are different and equal to 0.6 and 0.923.

*Invariance on the interval [0,z].*

Besides, superpositions that approximately (to some accuracy) preserve the cross-section intensity distribution may be of interest. In this case, for all constituent mode pairs the following condition should be met:

$$\left|\cos\left[(p_i-p_j)\phi+(\beta_{p_iq_i}-\beta_{p_jq_j})z\right]-\cos\left[(p_i-p_j)\phi\right]\right|<\varepsilon\,, \quad (14)$$

where $\varepsilon$ is small and defines the "recession" of different modes on the $z$-axis. Such between-mode "delay" can be defined as a small phase shift $\phi_\varepsilon$:

$$\left|\beta_{p_i q_i} - \beta_{p_j q_j}\right| z \le \varphi_\varepsilon \,. \tag{15}$$

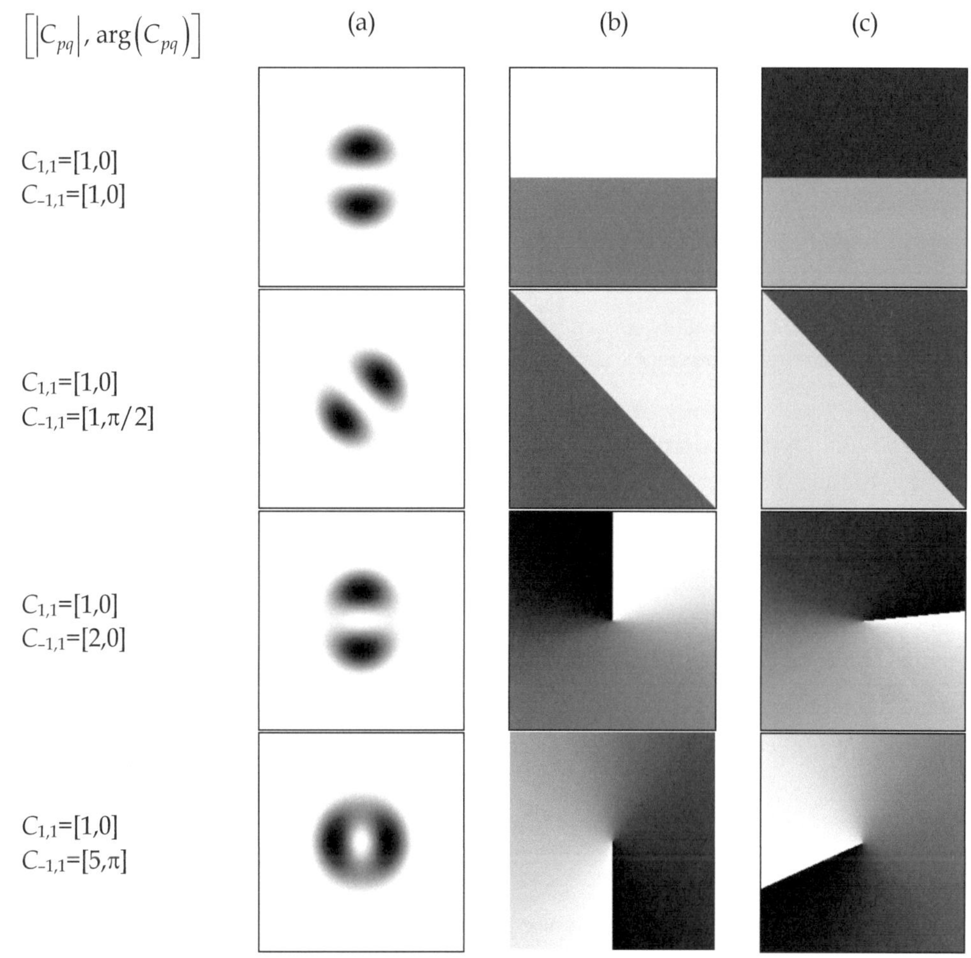


Fig. 2. Superposition of the $(p,q)$ modes: (1,1)+(−1,1) with different complex coefficients: transverse distribution of (a) intensity, and (b) phase in the plane $z$=0, and (c) phase distribution at distance $z$=200 m.

Formalizing the condition of the interval-specific invariance to a desired accuracy makes possible an automated procedure for selecting admissible superpositions from the entire set of fiber modes. The algorithm can be realized as an exhaustive search of modes with selection of superpositions satisfying the condition formulated.

For instance, putting on the 10 μm interval the admissible phase shift equal to $\phi_\varepsilon=\pi/18$, the algorithm allowed us to select 59 superpositions (containing 2-5 modes, regarding the index $p$ sign) from the set of 11 propagating modes for used parameters.

Figure 3 shows the transverse distributions of amplitude, intensity, and phase at different distances for a single superposition, namely, $(p,q)$: (1,2)+(−1,2)+(3,1)+(−3,1), with the $C_{pq}$ coefficients chosen to be the same.

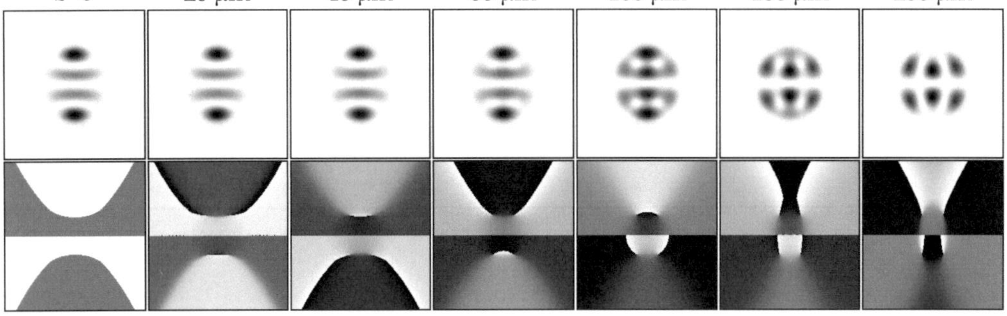


Fig. 3. Propagation of the superposition $(p,q)$: (1,2)+(−1,2)+(3,1)+(−3,1): transverse distributions of intensity (top row), and phase (bottom row) at different distances $z$.

As seen from Fig. 3, intensity of the multimode beam remains practically constant up to distance 40 μm. Unfortunately, since intermode dispersion in a stepped-index fiber being very high, putting the phase shift $\phi_\varepsilon \leq \pi/18$ makes possible only 17 superpositions on the 20 μm interval and 8 superpositions on the interval 40 μm. Note that there are just 8 superpositions of Eq. (10) which are admissible on any interval.
The number of superpositions that preserve their form on any interval can be essentially extended if considering a rotation-accurate invariance or "rotating" fields.

*All-region, rotation-accurate propagation invariance*

Assuming rotation-accurate invariance, mode pairs in the superposition must obey the following condition:

$$\cos\left[(p_i - p_j)\phi + (\beta_{|p_i|q_i} - \beta_{|p_j|q_j})z\right] = \cos\left[(p_i - p_j)(\phi + \phi_0)\right], \tag{16}$$

where $\varphi_0$ is some angle.
From Eq. (16), the rotation condition for any pair in the superposition is

$$\frac{\beta_{|p_i|q_i} - \beta_{|p_j|q_j}}{p_i - p_j} z = \phi_0 , \tag{17}$$

The exact condition in Eq. (17) complies with any two-mode superpositions, given $|p_i| \neq |p_j|$, since at $|p_i| = |p_j|$ there will occur the rotation by angle $\varphi_0=0$, i.e. the total invariance dealt with in the previous section. Thus, exciting various mode pairs enables obtaining fields that preserve their structure (except for rotation) at any interval. There may be 154 such superpositions, which exceeds 8 purely invariant syperpositions. By way of illustration, Fig. 4 shows the propagation at distance 150 m of invariant, rotating mode pairs.

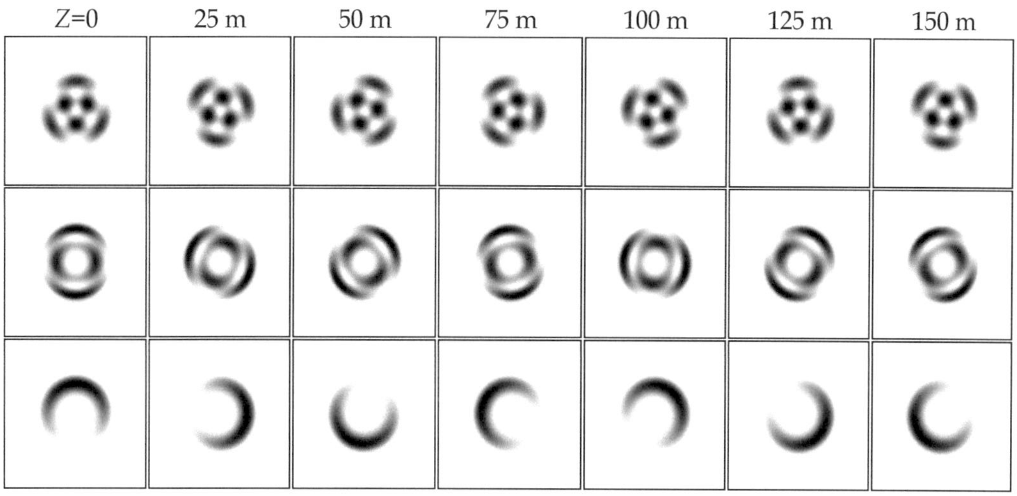


Fig. 4. Propagation of rotating modal pairs, $(p,q)$: (1,2)+(−2,1) – top row, (3,2)+(5,1) – middle row, (4,1)+(5,1) – bottom row; the intensity distribution is shown at different distances $z$.

From Fig. 4, the superpositions are seen to have symmetry of order

$$s = |p_1 - p_2| . \tag{18}$$

Note that due to symmetry, the transverse intensity distribution is self-reproduced $s$ times at a distance of one full revolution.

For such a pair, the rotation rate is given by

$$\vartheta = \frac{\beta_{|p_1|q_1} - \beta_{|p_2|q_2}}{p_1 - p_2}, \tag{19}$$

with the rotation direction corresponding to the sign of Eq. (18).

It is noteworthy that the rotation rate of interference pattern for the constituent modes in Eq. (19) is not related to the orbital angular momentum, depending on the propagation constants rather than the mode coefficients. In particular, for the mode pairs in Fig. 4, considering equal coefficients, Eqs. (9) and (24) take the values: for (1,2)+(−2,1), $\omega J_{z0}=0.5$, $\vartheta=0.54$; for (3,2)+(5,1), $\omega J_{z0}=-4$, $\vartheta=-0.35$; and for (4,1)+(5,1), $\omega J_{z0}=-4.5$, $\vartheta=1.02$.

Note that the transverse energy distribution of a beam composed of two modes can be varied by varying the mode coefficients. The intensity distribution itself will be preserved in propagation in a perfect fiber.

*Rotation-accurate invariance on the [0,z] interval*

Rotating superpositions containing more than two modes become possible by assuming a small error in the self-reproduction of the transverse intensity distribution. In this case, the following condition should be met for any two modal pairs in the superposition, $(p_i,q_i)+(p_j,q_j)$ and $(p_k,q_k)+(p_l,q_l)$:

$$\max \Delta_{ij}^{kl} - \min \Delta_{ij}^{kl} \le \phi_\varepsilon , \tag{20}$$

where $\Delta_{ij}^{kl} = \phi_{ij} - \phi_{kl}$, $\phi_{ij} = \left| \dfrac{\beta_{p_i q_i} - \beta_{p_j q_j}}{p_i - p_j} \right| z$, $\varphi_\varepsilon$ is the admissible rotation mismatch angle on the entire interval $[0,z]$.

Figure 5 shows the propagation of a superposition of three modes $(p,q)$: (2,2)+(–4,1)+(5,1) with identical coefficients at distance 1500 μm. This superposition obeys the condition in Eq. (25) with admissible mismatch angle of $\varphi_\varepsilon \le \pi/36$ on the interval up to 150 μm. The (-2,2)+(4,1)+(–5,1) superposition, with symmetric index signs, shows a similar property. No other more-than-two modal combinations were found.

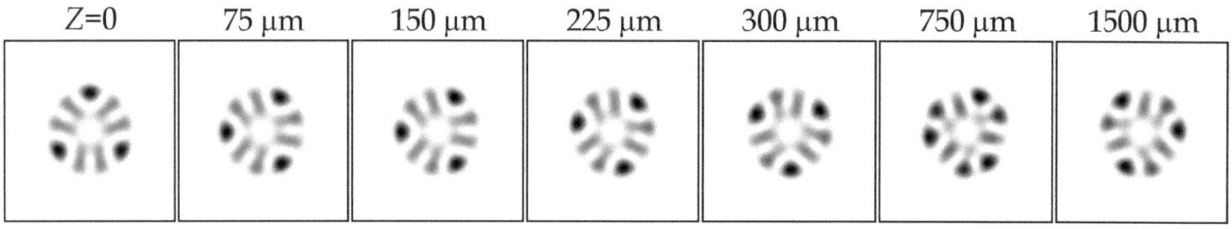


Fig. 5. Propagation of a superposition of $(p,q)$: (2,2)+(–4,1)+(5,1): transverse distributions of intensity at different distances $z$.

It is seen from the above that the more-than-two-mode superpositions preserve their structure at a short interval of about one hundred microns, followed by the structure's disintegration. After a while (period), however, the beam cross-section is self-reproduced.

*Periodic self-reproduction*

For a two-mode superposition there is always a self-reproduction period $z_0$ defined as

$$\cos\left[(p_1 - p_2)\phi + (\beta_{p_1 q_1} - \beta_{p_2 q_2}) z_0\right] = \cos\left[(p_1 - p_2)\phi\right] \Rightarrow \left|\beta_{p_i q_i} - \beta_{p_j q_j}\right| z_0 = 2\pi m, \tag{21}$$

where $m$ is integer.

However, once the distance $z_L$ is set, it would be of greater interest to identify possible modal superpositions that will be self-reproduced at this distance to a certain admissible accuracy. Such superpositions can be formed as mode pairs satisfying the condition:

$$\left| \left[ (\beta_{p_i q_i} - \beta_{p_j q_j}) z_L \right]_\pi \right| \le \varphi_\varepsilon, \tag{22}$$

where $\phi_\varepsilon$ is the admissible, reduced phase shift and $[\ldots]_\pi$ denotes reduction to the interval $[-\pi,\pi]$.

For example, putting $z_L$=1089.4 um (which is close to the self-reproduction period for two modes, (0,3)+(5,1)), specifying the admissible reduced phase shift equal to $\phi_\varepsilon \le \pi/12$ allows a set of 41 admissible superpositions, each containing 2-5 modes, to exist. In particular, Fig. 6 shows how a five-mode superposition, $(p,q)$: (0,3)+(3,2)+(–3,2)+(5,1)+(–5,1), propagates at the interval from $z$=0 to $z_L$. In the $z_L$-plane the superposition is self-reproduced with an error of $\delta$=0.48% for intensity. Note that the complex field correlation at $z$=0 and $z_L$ is close to unity: $\eta$=0.989.

It is noteworthy that at a half-period distance, $z_L/2$ = 544.7 μm, the transverse intensity distribution equals the original one rotated by 180 degrees (see fig. 6). Thus, it is possible to increase the number of points where a field is self-reproduced if considering the rotation-accurate self-reproduction.

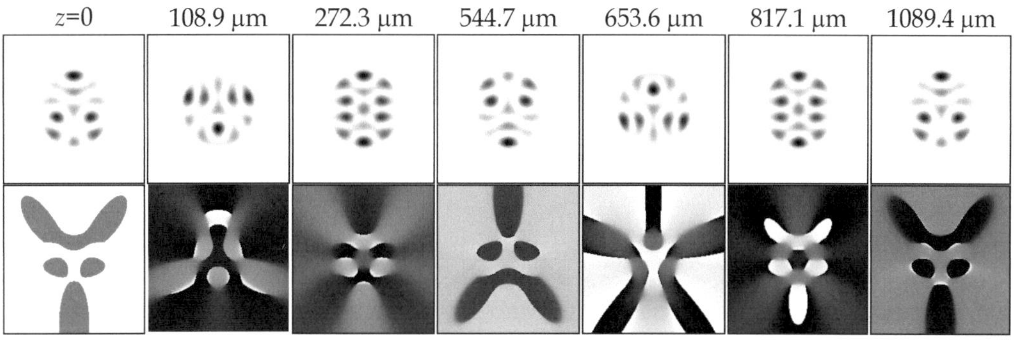


Fig. 6. In propagation, the superposition of $(p,q)$ modes: (0,3)+(3,2)+(−3,2)+(5,1)+(−5,1) is nearly self-reproduced at distance $z_L$=1089 μm (intensity and phase distributions are depicted at various distances $z$).

*Periodic rotation-accurate self-reproduction*

Similar to the previous section, putting the distance $z_L$ (e.g. fiber's length), we consider mode superpositions self-reproduced at this distance (period) up to a rotation-angle, with an admissible mismatch (otherwise, the set will only contain two-mode superpositions). In this case, the mode pairs in superposition should obey the condition:

$$\max \Delta_{ij}^{kl} - \min \Delta_{ij}^{kl} \leq \phi_{\varepsilon} , \tag{23}$$

where $\Delta_{ij}^{kl} = \left| \left[ \phi_{ij} - \phi_{kl} \right]_{\pi} \right|$, $\phi_{ij} = \dfrac{(\beta_{p_i q_i} - \beta_{p_j q_j}) z_L}{p_i - p_j}$, $\varphi_\varepsilon$ is the admissible mismatch angle in the $z_L$-plane.

Putting $z_L = 1$ m and the admissible mismatch angle equal to $\varphi_\varepsilon \leq \pi/9$, it is possible to obtain a set of 173 allowed superposition, each containing from 2 to 3 modes. Figure 7 shows the propagation of a three-mode superposition of $(p,q)$: (2,1)+(3,1)+(4,1) on the interval from $z$=0 to $z_L$ (at point $z_L$ = 1 m, the mismatch angle being $\varphi_\varepsilon \leq \pi/30$).

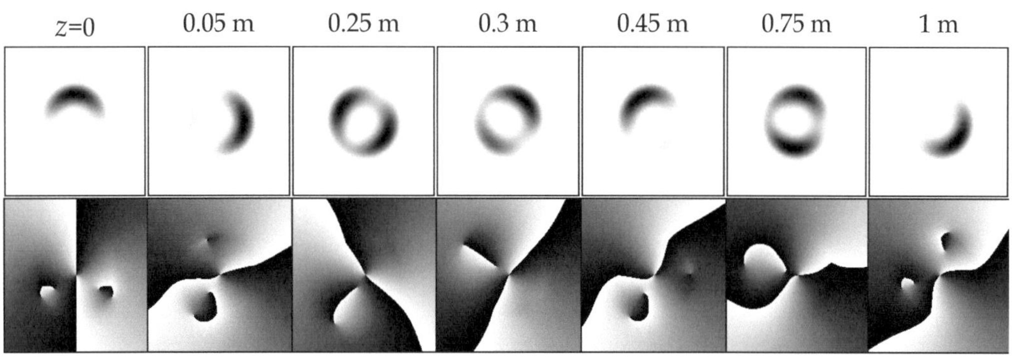


Fig. 7. In propagation, the superposition of $(p,q)$ modes: (2,1)+(3,1)+(4,1), is nearly self-reproduced at distance $z_L$=1 m (intensity and phase distributions are depicted at various distances $z$).

From Fig. 7, the superposition's intensity is also seen to be self-reproduced (to some accuracy) in other planes. However, this work was not aimed at identifying all points of self-reproduction for a definite superposition. The problem addressed was as follows: based on given physical characteristics of a stepped-index optical fiber (thickness, length, and parameters of material) it was required to identify the entire possible set of propagating modes and modal superpositions that show various self-reproduction properties to a designed accuracy.

## 2.2 Experimental excitation and detection of angular harmonics in a stepped-index optical fiber

When angular harmonics (optical vortices) are coupled into a fiber or selected at output using DOEs they show the scale invariance that provides much freedom in choosing optical scheme parameters. As shown below, this provides effective means for preventing system noise.

We describe natural experiments on selective excitation of both separate angular harmonics and their superposition. We used a DOE that was able to form beams with phase singularity $\exp(im\varphi)$ of order $m$=–1 and $m$=–2 and a superposition $\exp(im_1\varphi)+\exp(im_2\varphi)$, $m_1$=–1, $m_2$=2 (see Fig. 8). The multi-level DOEs were fabricated using e-beam lithography at the University of Joensuu (Finland). The DOEs parameters are: 32 quantization levels, diameter is 2.5 mm, and discretization step is 5 μm. Spiral DOEs were fabricated for wavelength λ=633 nm.

Selection was performed using multi-order DOEs (Khonina et al., 2003) matched to angular harmonics, which were also fabricated at Joensuu University. Shown in Fig. 9 is the 8-order binary DOE to detect spiral singularities with different numbers.

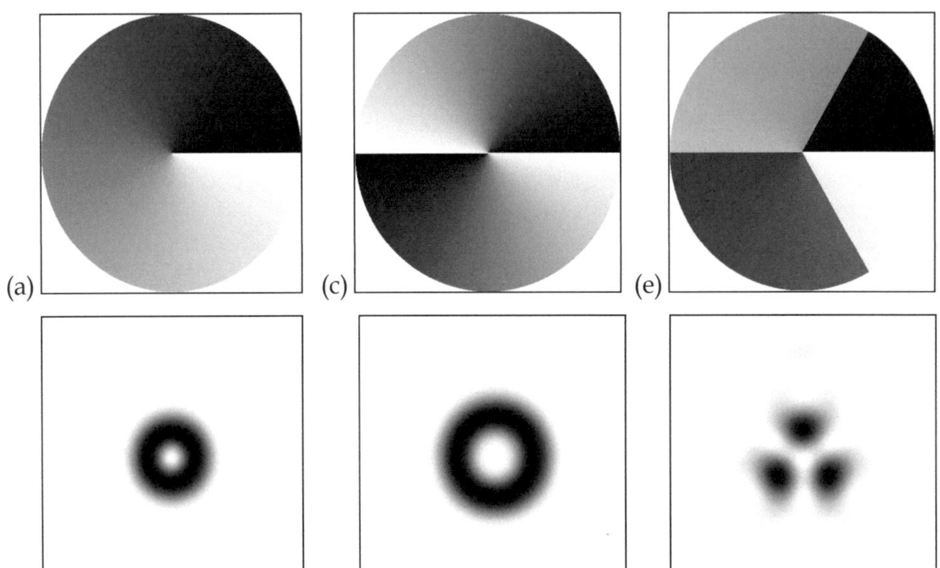


Fig. 8. Generation of light fields with phase singularity exp(inφ): DOE phase for (a) m=–1, (c) m=–2 and (e) a superposition of $m_1$=–1 and $m_2$=2, and (b), (d), (f) corresponding far-field intensity distributions.

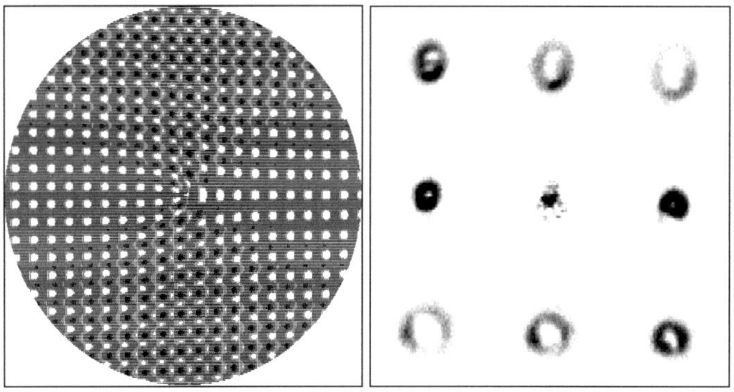

| | | |
|---|---|---|
| 2 | 3 | 4 |
| 1 | | -1 |
| -4 | -3 | -2 |

(a) (b) (c)

Fig. 9. Binary DOE matched to 8 different number angular harmonics: (a) phase, (b) corresponding patterns for the diffraction orders for plane wave and (c) the accordance scheme of angular harmonics' numbers and diffraction orders.

Diffraction order patterns are also put in correspondence with the numbers of the angular harmonics. The DOE parameters are: diameter is 10 mm, discretization step is 5 μm, and microrelief height for wavelength $\lambda$=633 nm.

First, following the procedure described in (Karpeev et al., 2005), the system was adjusted for coupling the principal mode. At this stage, the mode-generating DOE's substrate, being already put into the beam, is displaced to prevent the phase microrelief region from getting into the beam path. At the output, the Gaussian beam of the principal mode is collimated and then passed through a DOE matched to the angular harmonics and a Fourier stage. The scale at the Fourier stage output plane is related to both the output beam's diameter and the Fourier stage focal length. For angular harmonics, these parameters can be independently changed, as distinct from the classical modes where the beam size is rigidly connected with the DOE parameters. Besides, increasing focal length and correspondingly increasing scale help reduce noise. This is due to the high-frequency nature of noises resulting from high-frequency discretization of the phase DOEs, with noise level becoming lower closer to the optical axis. Thus, with the optical system's overall size allowing a decrease, for the noise impact to be reduced, lower carrier frequencies need to be chosen (on the assumption that there is no order overlapping).

The experiments were conducted with three beam-generation DOEs, which, accordingly, generated the first- and second-order angular harmonics as well as their superposition are coupled into a fiber. Figure 10 shows intensity distributions in the output plane when the corresponding beam is excited in a fiber.

When a first-order optical vortex is excited the intensity peak appears at the center of the corresponding order with the near-noise intensity (no more than 10% of peak intensity) found at the other orders (fig. 10a). Next, a second-order optical vortex was excited (fig. 10b). It was found that depending on the position of the beam-generation DOE in the illuminating beam, the intensity peak can emerge in diffraction orders corresponding to the second-order harmonics of both signs. It is possible to excite both any separate mode and their combination featuring about the same intensity. Note, however, that in this case the excitation selectivity is lower, compared with the first-order harmonic.

A third experiment was on excitation of a superposition of the opposite-sign, first- and second-order angular harmonics (fig. 10c). The emergence of the first- and second-order

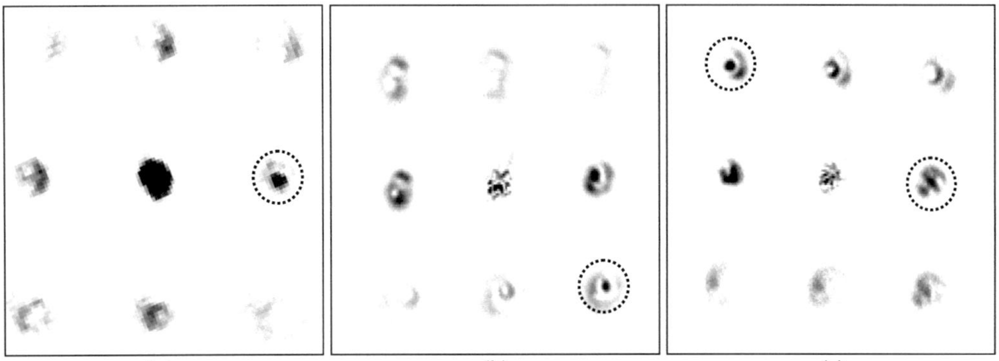

(a) (b) (c)

Fig. 10. Output intensity distributions for different angular harmonic coupled into a fiber: (a) m=–1, (b) m=–2 and (c) a superposition of $m_1$=–1 and $m_2$=2.

singularity, with central intensity maximum seen at the corresponding orders and near-noise intensity at the other orders (no more than 15% of maximum). It should be noted that the maximum corresponding to the second-order angular harmonic is 10% weaker than the maximum of the first-order harmonic. It may be due to inadequate resolution, because space resolution requirements for different-order angular harmonics are different.

## 3. Multimode self-imaging in a weakly guiding parabolic fiber

In a gradient parabolic fiber, the refractive index is given by

$$n^2(r) = n_0^2\left(1 - 2\Delta\frac{r^2}{r_0^2}\right) = n_0^2\left(1 - \alpha^2 r^2\right), \tag{24}$$

where $r$ is the radius of the cylindrical coordinate system; $n_0$ is the refractive index on the fiber's optical axis; $r_0$ is a characteristic fiber radius; $\Delta$ is the dispersion parameter of the medium refractive index; and $\alpha = \sqrt{2\Delta}/r_0$ is a constant that defines the curvature of the refractive index profile.

It has been known (Snyder & Love, 1987; Soifer & Golub, 1994) that the solution of the Helmholtz equation in the cylindrical coordinates is given by the superposition of the Laguerre-Gaussian (GL) modes

$$\Psi_{nm}(r,\varphi,z) = \frac{1}{\sigma_0}\sqrt{\frac{n!}{\pi(n+|m|)!}}\cdot\left(\frac{r}{\sigma_0}\right)^{|m|} L_n^{|m|}\left(\frac{r^2}{\sigma_0^2}\right)\exp\left(-\frac{r^2}{2\sigma_0^2}\right)\exp(im\varphi)\exp(\pm i\beta_{nm}z), \tag{25}$$

where $L_n^m(\xi) = \frac{1}{n!}e^{\xi}\xi^{-m}\frac{d^n}{d\xi^n}\left\{e^{-\xi}\xi^{n+m}\right\}$ are the Laguerre polynomials, $\sigma_0 = \left(\lambda r_0/\pi n_0\right)^{1/2}(2\Delta)^{-1/4}$ is the effective radius of the LG modes, $\beta_{nm} = \left[k^2 n_0^2 - 4\left(2n+|m|+1\right)/\sigma_0^2\right]^{1/2}$ is a parameter proportional to the mode phase speed, $n$ is a non-negative integer number, $m$ is integer.

Propagation of an image in an ideal weakly guiding graded-index fiber can be described through a superposition of LG modes (Almazov & Khonina, 2004; Kotlyar et al., 1998). The approximation of an arbitrary image by the LG mode superposition is given by

$$F(r,\varphi)=\sum_{n,m\in\Omega} C_{nm}\Psi_{nm}(r,\varphi)\,, \tag{26}$$

where coefficients $C_{nm}$ can be derived from

$$C_{nm}=\int_0^{\infty}\int_0^{2\pi} F(r,\varphi)\Psi^*_{nm}(r,\varphi)r\,\mathrm{d}r\,\mathrm{d}\varphi\,. \tag{27}$$

Then the beam (26) propagated distance z will have the following appearance

$$F(r,\varphi,z)=\sum_{n,m\in\Omega} C_{nm}\Psi_{nm}(r,\varphi,z)\,. \tag{28}$$

The cut-off condition is taken from

$$kn_0 \le \beta_{nm} \le kn_0\sqrt{1-2\Delta}\,. \tag{29}$$

Modelling the propagation of different test images (a cross, a triangle, a line-segment) through a fiber produces similar results: the image is disintegrated at a distance of about 0.1 mm, whereas the coefficient distribution is preserved at any distance to a 0.2% accuracy, which is close to the computation error (see Figs. 11-13). Hence, we can infer that the image recognition from the distribution of squared modules of the expansion coefficients $\boldsymbol{C_{nm}}$ has advantages over intensity-based recognition, on the understanding that the fiber has no considerable nonhomogeneities and bending resulting in changed coefficients and energy redistribution between the modes.

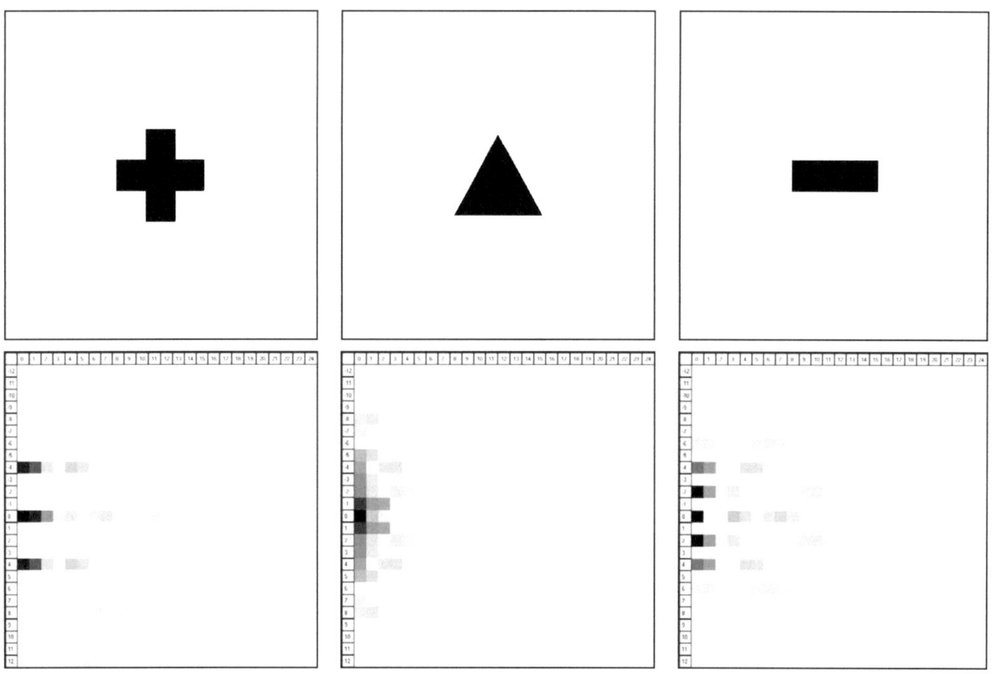

Fig. 11. Distribution of the squared modules of the coefficients of image expansion into the LG modes.

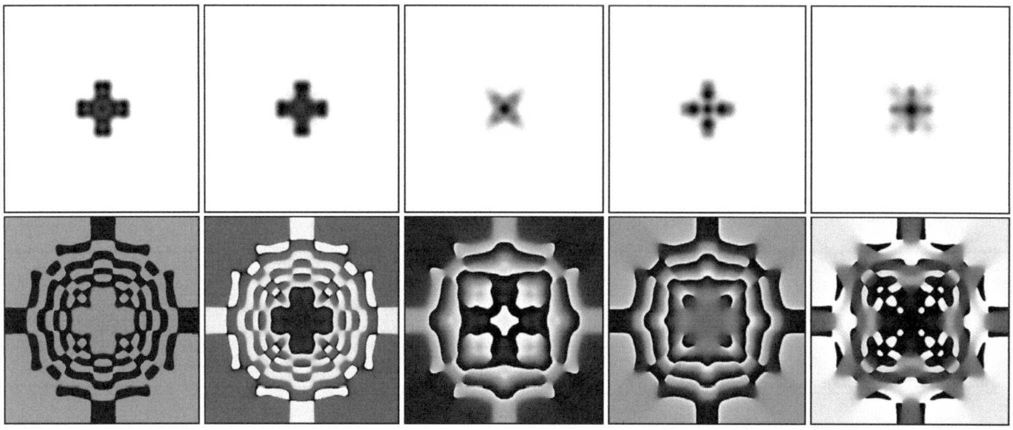


Fig. 12. Intensity and phase distribution for the cross image decomposition at different distances.

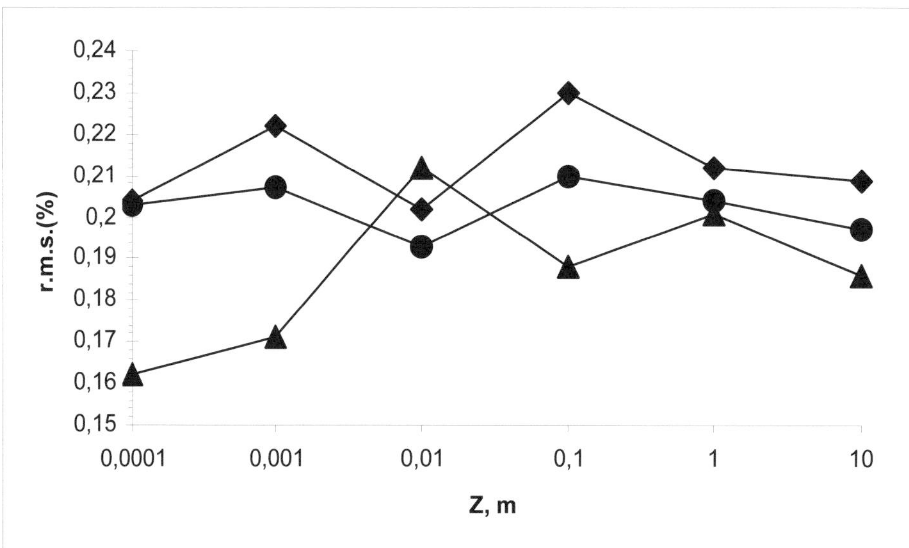


Fig. 13. The r.m.s. of the squared modules of amplitude coefficients $\mathbf{C}_{\mathbf{nm}}$ (%) vs distance (♦ - cross, • - triangle and ▲ - horizontal line-segment).

### 3.1 Self-imaging in a weakly guiding parabolic fiber

From expression (25) it is possible to define the period of self-reproduction $z_{nm}$ for each single mode in the superposition (28). The image will be periodically reproduced at a distance $Z$, such that $Z/z_{nm}$ is an integer for any $n$, $m$ of the constituent modes found in the composite image. Since the $z_{nm}$ are irrational in the general case, there is no a general period even for a two-mode composition. However, we are able to obtain local self-reproduction periods where the image is reproduced to a sufficient accuracy. After the image has

propagated through several such distances the phase mismatch error will increase until it reaches a margin of visible image disintegration. However, having passed some distance the image will again enter a certain local stability zone with approximate self-reproduction points found at close intervals. It stands to reason that the greater number of modes are included into the image approximation, the greater is the self-reproduction period. For an ideal image composed of infinite number of modes the period is equal to infinity, i.e. there are no self-reproduction points. Thus, to be able to visually recognize the images we must impose an additional strict limitation on the approximation quality. Figures 14 and 15 show the patterns of the intensity and phase for the cross image at different distances in a circular graded-index fiber with parabolic refractive index distribution and the following parameters: $r_0$=25 μm; $n_0$=1.5; Δ=0.01; λ=0.63 μm.
It is seen from Figs. 14 and 15 that the cross image within the fiber has a local period of about 1.105 mm (the first group of self-reproduction points) and a large period of about 2.9 m (the second group of self-reproduction points). From Fig. 15 it is seen that after 2.9 m the reproduced image very closely matches the initial image (Fig. 14, $z$=0). Obviously, there are larger self-reproduction periods at which the superposition is reproduced to a greater accuracy.

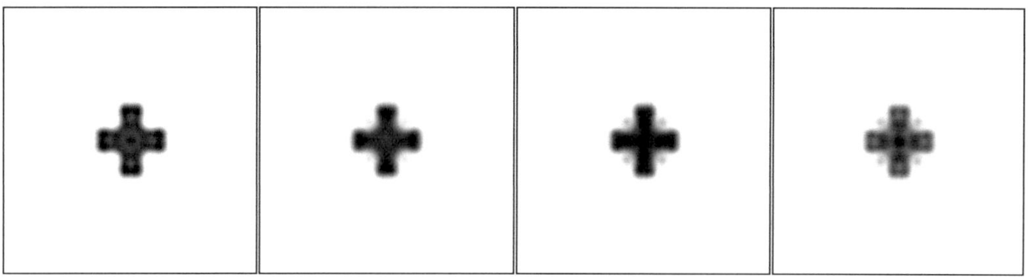

Z=0 Z= 1.094547 mm Z= 4.411702 mm Z= 7.733058 mm

Fig. 14. Image self-reproduction - the "cross" image decomposition at different distances **z** from the first group of image self-reproduction points.

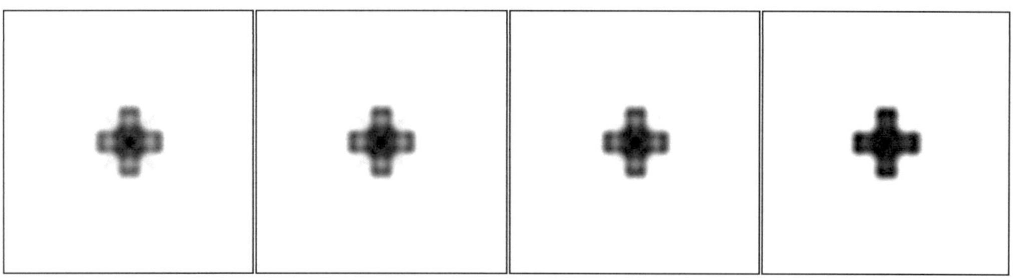

Z= 2890.746699 mm Z= 2891.853399 mm Z= 2892.958399 mm Z= 2897.381399 mm

Fig. 15. Image self-reproduction - the "cross" image decomposition at different distances **z** from the second group of image self-reproduction points.

Figure 16 shows similar patterns of the decomposition of a triangle and a horizontal line-segment. Figure 17 shows examples of test superpositions composed of a few number of modes.

It should be noted that arbitrary mode superpositions propagated in a fiber appear to have the same local self-reproduction periods. This fact can be due to existence of general local periods for the entire set of modes propagated in the fiber. Because the self-reproduction periods for separate LG modes in a given fiber are similar and found in the range from 420 nm ((0,0) mode) to 426 nm (higher-order modes), the value of a local general period is much greater than an individual mode period.

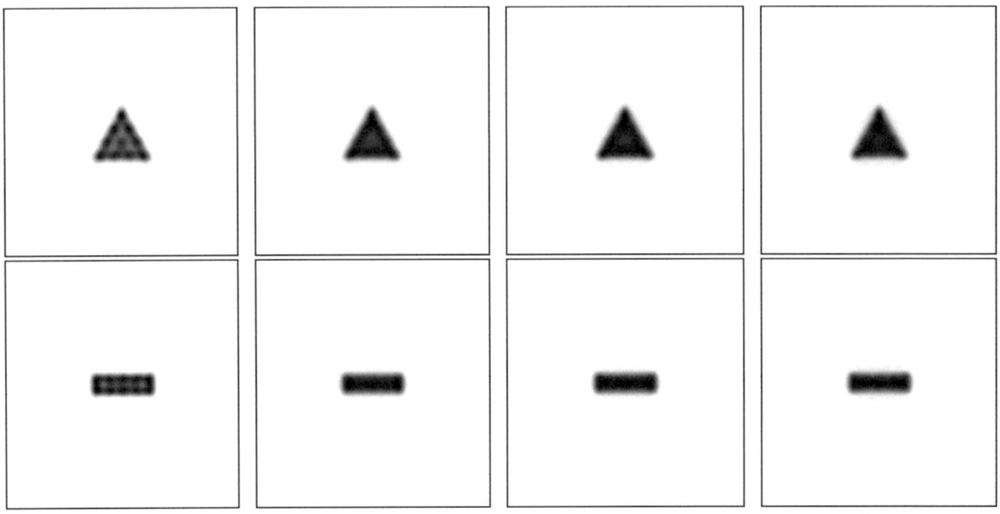

Z=0 Z= 1.094547 mm Z= 2.199985 mm Z= 3.305423 mm

Fig. 16. Image self-reproduction – the "triangle" and "horizontal line" images decomposition at different distances **z** from the first group of image self-reproduction points.

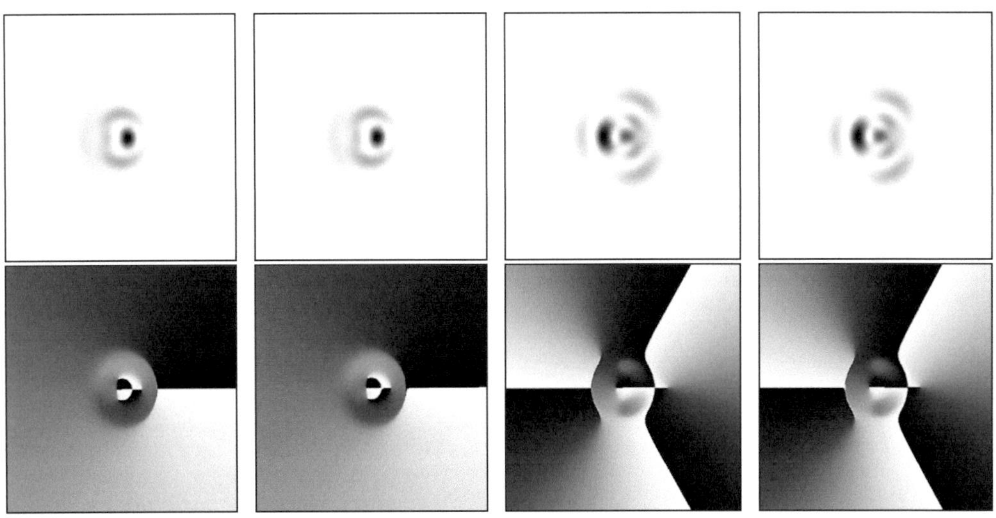

Z=0 Z= 3.305423 mm Z=0 Z= 1.094547 mm

Fig. 17. Self-reproduction of the test mode decomposition (n,m): (0,-1)+(1,0)+(2,1) (the first and second columns), and (n,m): (0,1)+(1,3)+(2,0) (the third and fourth) at different distances **z** from the first group of image self-reproduction points.

It remains to note that finding the points of approximate image self-reproduction is a real computational challenge even for a comparatively small number of modes propagated in a 25 μm fiber. The problem is solved via successive search of the distance **Z** and finding a value at which all the quotients are integer. The method has other disadvantages. For example, it can provide only the boundaries of the intervals over which the desired mode composition is reproduced with a certain phase delay error, but is unable to identify an optimal within-interval point. This makes topical the development of a new, more efficient method of searching for the image self-reproduction points

### 3.2 Propagation of laser vortex beams in a parabolic optical fiber

The propagation of the electromagnetic wave in the medium can be modeled in several ways. The most common technique is to describe the propagation using Maxwell's equations, from which vectorial wave equations defining the electric and magnetic field components can be deduced. If the relative change of the medium refractive index per wavelength is significantly smaller than unity, the Helmhotz equation can be written for each scalar component of the vector field.

We have looked into the propagation of monochromatic light beams with helical phase singularity in a nonuniform medium, including a parabolic-index waveguide. We have proposed an approximation of the differential operator of propagation in a weakly nonuniform medium, which allows the propagation of light beams in the nonuniform medium to be treated as the propagation in a uniform medium through an array of thin optical elements. Using the limiting passage to an infinitely large number of lenses put at an infinitesimally small distance, a paraxial integral operator to describe the light field propagation in a parabolic medium has been derived (Khonina et al., 2010):

$$\begin{aligned} E(x,y,z) \approx & -\frac{ik\alpha}{2\pi\sin(\alpha z)}\exp\{ikz\}\exp\left\{\frac{ik\alpha}{2\tan(\alpha z)}\left[x^2+y^2\right]\right\}\times \\ & \times\int_{-\infty}^{\infty}\int_{-\infty}^{\infty} E_0(\xi,\eta)\exp\left\{\frac{ik\alpha}{2\tan(\alpha z)}\left[\xi^2+\eta^2\right]\right\}\exp\left\{-\frac{ik\alpha}{\sin(\alpha z)}\left[\xi x+\eta y\right]\right\}d\xi d\eta. \end{aligned} \tag{30}$$

This integral operator makes it possible to simulate the propagation of arbitrarily shaped light beams, being indefinite in a general sense at distances multiple to the half-period. At $\alpha \to 0$ this integral operator is reduced to the Fresnel transform that describes, with the same accuracy, the propagation of light in a uniform medium. The integral in Eq. (30) has a period of $z_T = 2\pi/\alpha$.

At distances multiple to a quarter of period, the distribution $F(x,y,z) = E(x,y,z)\exp\{-ikz\}$ has the following specific features:

- at distance $z = z_T / 4$, the distribution $F(x, y, z)$ is defined by the Fourier transform of the initial distribution;
- at distance $z = z_T / 2$, the inverted distribution is formed: $-E_0(-x, -y)$;
- at distance $z = 3z_T / 4$, the distribution $F(x, y, z)$ is the inverse Fourier transform of the initial distribution;
- at distance $z = z_T$, the initial distribution $E_0(x, y)$ is formed.

We performed the numerical simulation of the paraxial integral operator in Eq. (30) by the sequential integration method based on the quadrature Simpson formulae in a bounded square region. In a general sense, the paraxial integral is not defined at distances multiple to $z_T / 2$, where the inverted and equi-initial intensity distributions are to be formed, so that the numerical simulation based on the quadrature formulae produces a completely erroneous result. At these distances, the integral in Eq. (30) needs to be treated in a general sense.

Figure 18 depicts the numerically simulated propagation of the LG mode $\Psi_{0,1}$ partially shielded with an opaque screen. The simulation is based on the paraxial integral operator in Eq. (30).

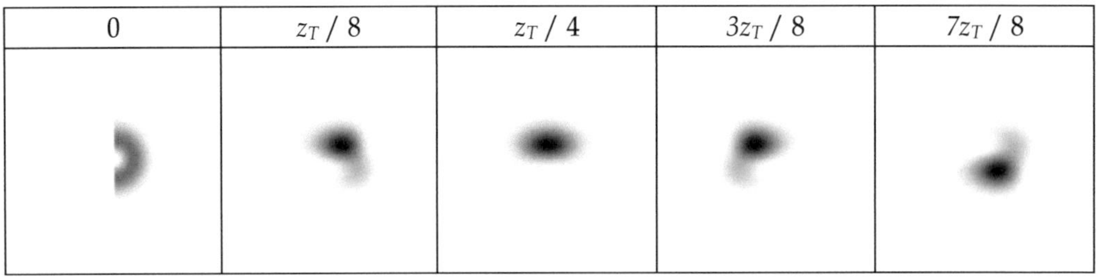


Fig. 18. Numerically simulated propagation of the mode $\Psi_{0,1}$ shielded by an aperture on the left ($\sigma = \sigma_0$).

The above results suggest that although the use of the integral operator makes it possible to model the propagation of arbitrarily shaped beams, computational challenges arise at definite distances on the optical axis.

An alternative method for modeling the propagation of light based on the decomposition of the input light beam into the medium eigenmodes has also been discussed. The effect of the operator in Eq. (30) on the LG modes can be found in ( Striletz & Khonina, 2008). Here, we only give the final relation

$$\Psi_{nm}(r,\varphi,z)=\frac{1}{\sigma_0}\sqrt{\frac{n!}{\pi(n+|m|)!}}\frac{\sigma}{\sigma(z)}\left(\frac{r}{\sigma(z)}\right)^{|m|}L_n^{|m|}\left(\frac{r^2}{\sigma^2(z)}\right)\exp\left\{i\beta_{nm}(r,z)-\frac{r^2}{2\sigma^2(z)}+im\varphi\right\}, \quad (31)$$

where

$$\sigma(z)=\sigma\left[\cos^2(\alpha z)+\sigma_0^4\sin^2(\alpha z)/\sigma^4\right]^{1/2} \quad (32)$$

is the effective radius of the LG mode;

$$\beta_{nm}(r,z)=kz+(2n+|m|+1)\left[\arctan\left\{\frac{\sigma^2}{\sigma_0^2}\frac{1}{\tan(\alpha z)}\right\}-\frac{\pi}{2}\right]+\left(1-\frac{\sigma^2}{\sigma^2(z)}\right)\frac{1}{\tan(\alpha z)}\frac{r^2}{2\sigma_0^2} \quad (33)$$

is the function that defines the phase velocity.

In particular, when the waveguide is illuminated by the LG eigenmode ($\sigma = \sigma_0$), Eq. (31) takes the form of Eq. (25).

If the initial radius $\sigma$ is smaller than the effective radius $\sigma_0$of the fiber eigenmode, the beam radius $\sigma(z)$ at first increases, attaining a maximum of $\sigma_{max} = \sigma_0^2/\sigma$ at points

$z_s = \pi(s - 1/2)/\alpha$, $s \in \mathbf{N}$, where the Fourier image of the initial beam is formed. Then, the radius decreases, attaining a minimum of $\sigma_{\min} = \sigma_0$ at points $\sigma(z)$, $z_s = \pi s/\alpha$, $s \in \mathbf{N}$. However, if $\sigma$ is larger than $\sigma_0$, then, $\sigma(z)$ at first decreases till $\sigma_{\min} = \sigma_0^2/\sigma$ and then increases up to the initial value.

Figure 19 shows the intensity distributions for a LG mode superposition, whose propagation is defined by Eq. (31).

The expansion coefficients for the Gaussian vortex beam have been deduced (Khonina et al., 2010) in the analytical form and can be used for the non-paraxial modeling.

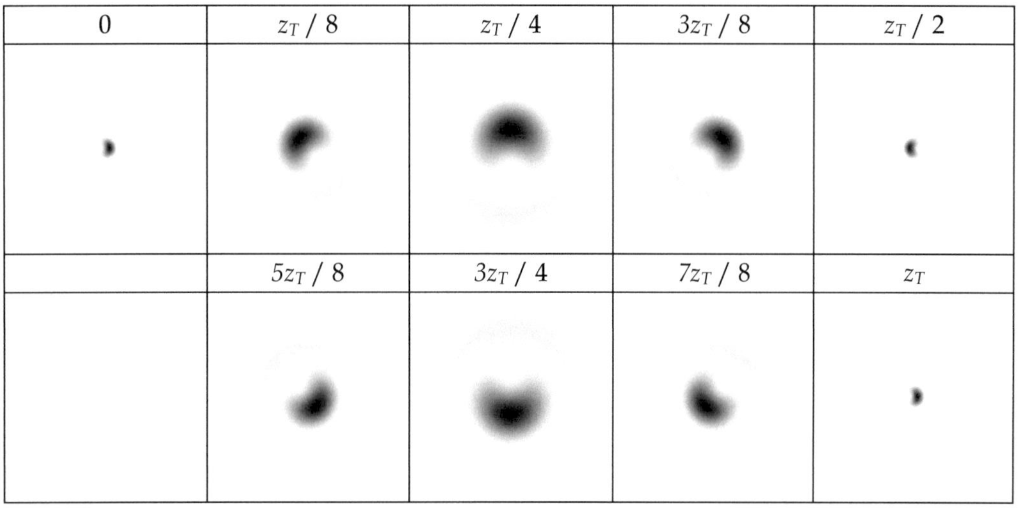


Fig. 19. Intensity distribution of the mode superposition $\Psi_{0,0} + \Psi_{1,-1}$ ($\sigma = \sigma_0/2$).

A Gaussian vortex beam with an arbitrary initial radius $\sigma$ is given by

$$E_0(r,\varphi) = \frac{1}{\sigma\sqrt{\pi}} \exp\left\{-\frac{r^2}{2\sigma^2}\right\} \exp\{i\mu\varphi\}, \tag{34}$$

where $\mu$ is an arbitrary constant.

The result of application of the integral operator in Eq. (30) to the input vortex beam in Eq. (34) is most easily represented by a superposition of the LG modes in Eq. (31) with z=0:

$$E_0(r,\varphi) = \sum_{n,m} C_{nm} \Psi_{nm}(r,\varphi,0) \tag{35}$$

Considering the normalization properties of the LG modes, the coefficients $C_{nm}$ are defined as

$$C_{nm} = A_{n,m} \int_0^{\infty} \left(\frac{r}{\sigma}\right)^{|m|} L_n^{|m|}\left(\frac{r^2}{\sigma^2}\right) \exp\left\{-\frac{r^2}{\sigma^2}\right\} r dr \int_0^{2\pi} \exp\{i(\mu - m)\varphi\} d\varphi. \tag{36}$$

Using the replacement $\xi = r^2/\sigma^2$ and taking the integral with respect to the variable $\varphi$, we obtain

$$C_{nm} = \sqrt{\frac{n!}{(n+|m|)!}} \frac{\exp\{2\pi i(\mu - m)\} - 1}{2\pi i(\mu - m)} \int_0^\infty \xi^{\frac{|m|}{2}} L_n^{|m|}(\xi) \exp\{-\xi\} d\xi. \quad (37)$$

Using the Laguerre polynomials in the form of Eq. (25) and integrating Eq. (37) $n$ times by parts, we obtain

$$\int_0^\infty \xi^{\frac{|m|}{2}} L_n^{|m|}(\xi) \exp\{-\xi\} d\xi = \frac{1}{n!} \left(\frac{|m|}{2}\right)_n \int_0^\infty \xi^{\frac{|m|}{2}} \exp\{-\xi\} d\xi = \frac{1}{n!} \left(\frac{|m|}{2}\right)_n \Gamma\left(\frac{|m|}{2} + 1\right), \quad (38)$$

where $(x)_n = \begin{cases} 1, & n = 0 \\ x(x+1) \cdot \ldots \cdot (x+n-1), & n \neq 0 \end{cases}$; and $\Gamma(x)$ is Gamma function.

Finally, we obtain

$$C_{nm} = \frac{\left(\frac{|m|}{2}\right)_n \Gamma\left(\frac{|m|}{2} + 1\right)}{\sqrt{(n+|m|)!\,n!}} \frac{\exp\{2\pi i(\mu - m)\} - 1}{2\pi i(\mu - m)}. \quad (39)$$

If $\mu$ is integer, the expression in Eq. (39) is not equal to zero only at $m = \mu$. In this case, the propagation of the vortex beam in the parabolic fiber is described by a superposition of the functions in Eq. (31):

$$E(r, \varphi, z) = \sum_{n=0}^\infty \frac{\left(\frac{|\mu|}{2}\right)_n \Gamma\left(\frac{|\mu|}{2} + 1\right)}{\sqrt{(n+|\mu|)!\,n!}} \Psi_{n,\mu}(r, \varphi, z). \quad (40)$$

Notice that the above relation also holds in a non-paraxial region because the modes in Eqs. (25) and (31) only differ by the propagation constant that has no effect on the decomposition coefficients.

Figure 20 respectively give the intensity, $|E(r,\varphi,z)|^2$, and phase, $\arg\{E(r,\varphi,z)\}$, distributions of the Gaussian beam in Eq. (34) for $\mu = 1$. The computations have been conducted using Eq. (40) for a finite number of terms ($n_{\max} = 50$) and by the numerical integration in Eq. (30). The figures suggest that there is a good qualitative agreement between the two methods. However, the numerical integration is seen to result in a minor asymmetry.

Propagation of vortex laser beam in a parabolic fiber has also been numerically simulated by the well known Beam Propagation Method (BPM) with use of BeamPROP simulation tool (RSoft Design, USA). The calculations were conducted for the wavelength of $\lambda = 633$ nm. The waveguide parameter $\alpha = 17.88$ mm$^{-1}$, $\alpha = 26.82$ mm$^{-1}$ and $\alpha = 35.76$ mm$^{-1}$, the waveguide width 30 μm, index on the waveguide axis $n_0 = 1.5$. Sampling step was 0.1 μm along $x$- and $y$- axes and 0.05 μm along $z$-axis. Simulation area has the sizes 90 μm along $x$- and $y$- axes and 300 μm along $z$-axis.

If the light field in initial plane $E_0(x, y)$ has the form of $A(r)\exp(in\varphi)$, it is obvious that intensity in transverse planes will be repeated with the period $\pi/\alpha$ instead of $2\pi/\alpha$. It can be seen in Fig. 21. For mentioned values of parameter $\alpha$ periods will be the following: $T = 175$ μm, $T = 120$ μm and $T = 88$ μm.

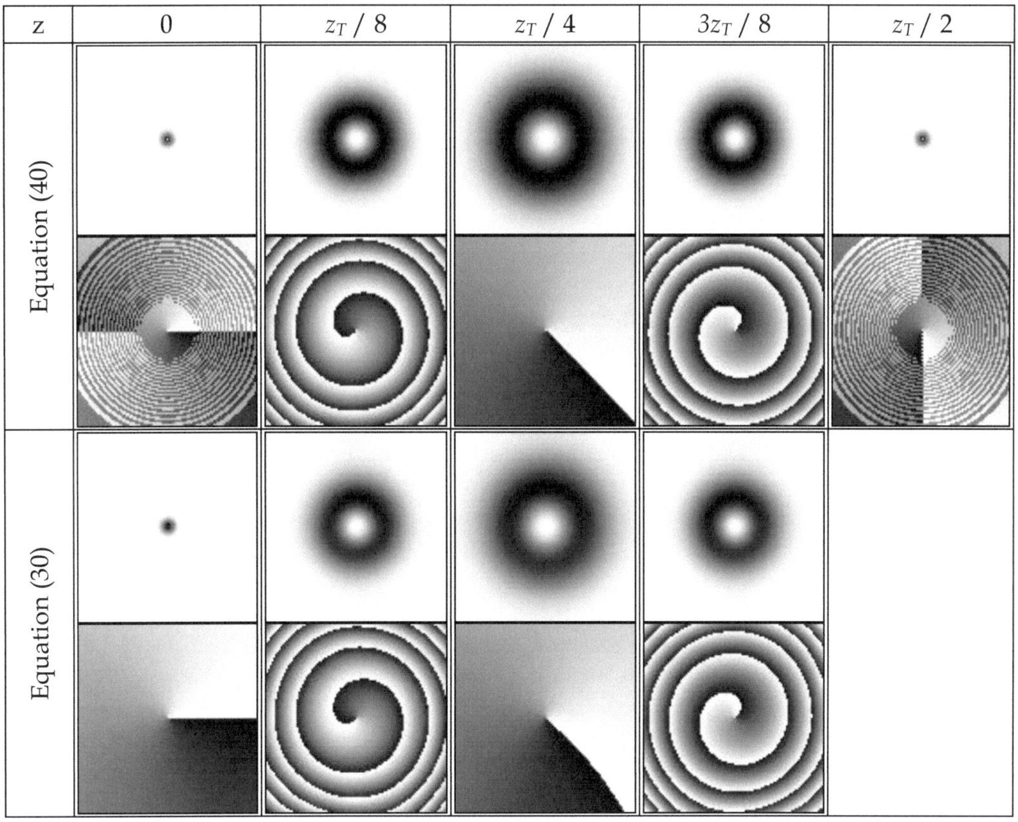


Fig. 20. Propagation Gaussian optical vortex with $\mu = 1$ ($\sigma = \sigma_0/2$).

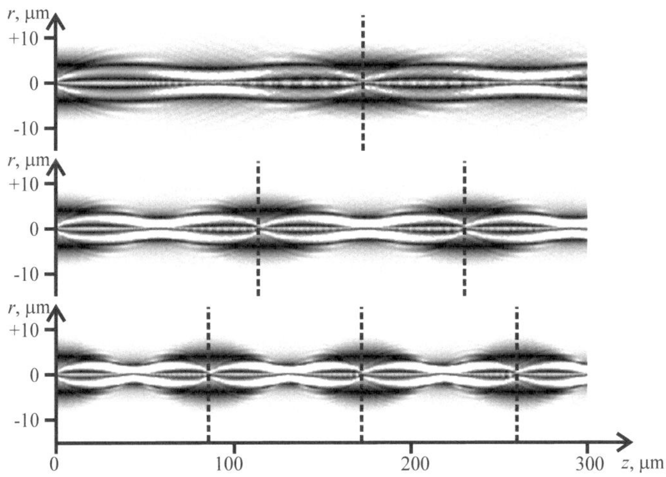


Fig. 21. Propagation of Gaussian optical vortex in parabolic waveguide for various values of parameter $\alpha$ : $\alpha$ = 17.88 mm$^{-1}$ (top part), $\alpha$ = 26.82 mm$^{-1}$ (central part) and $\alpha$ = 35.76 mm$^{-1}$ (bottom part). Dashed lines mean periods of diffraction patterns (i.e. planes $z = p\pi / \alpha$ , where p are integer numbers).

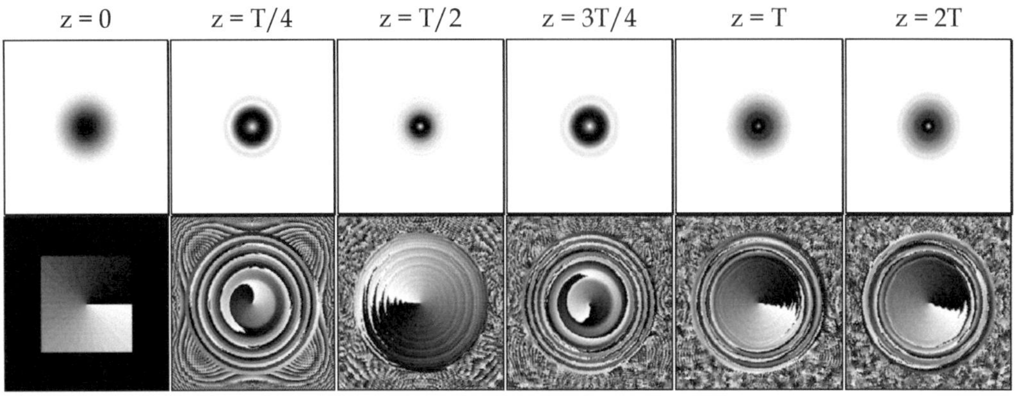


Fig. 22. Propagation Gaussian optical vortex with $\mu = 1$ using BPM.

So, variations in the transverse distribution of the light beam have been shown to be periodic for all beams other than fiber eigenmodes.

## 4. Conclusion

In this work:

- Linearly polarized modes of a weakly guiding fiber with a non-zero orbital angular momentum have been discussed. Conditions (expressed through the mode indices) for various self-reproduction types of multi-mode laser fields (invariance, rotation, periodic self-reproduction of the field transverse intensity distribution) have been deduced;
- An algorithm for generating a set of modal superpositions showing various self-reproduction properties to a designed accuracy has been developed;
- Experiments on excitation of lower-order angular harmonics and their superpositions in a stepped-index few-mode optical fiber have been conducted;
- An algorithm for finding the self-reproduction periods of a linear superposition of the Laguerre-Gauss modes in a circular graded-index fiber is developed. In terms of self-reproduction accuracy, various types of periods (local and general) have been identified. It has been found that arbitrary mode superpositions in a specific fiber have the same local self-reproduction periods, which is owing to the existence of general local periods of the entire set of the fiber modes.
- We have looked into the propagation of monochromatic light beams with helical phase singularity in a nonuniform medium, including a parabolic-index waveguide. Variations in the transverse distribution of the light beam have been shown to be periodic for all beams other than fiber eigenmodes.
- An alternative method for modeling the propagation of light based on the decomposition of the input light beam into the medium eigenmodes has also been discussed. The result of application of the integral operator to the non-paraxial Laguerre-Gauss modes with an arbitrary initial effective radius has been analytically derived.

The revealed features of vortex beams propagation in an optical fiber expand opportunities of fiber optics in various applications, including, additional compression of information channels and new degrees of freedom in coding and protection of the information.

## 5. References


Agrawal, G.P. (2002). *Fiber-Optic Communication Systems* (3rd Edition), John Wiley

Allen, L., Barnett, S.M., and Padget, M. J. (2003). *Optical Angular Momentum*, Institute of Physics, Bristol

Almazov, A.A., Khonina, S.N. (2004). Periodic self-reproduction of multi-mode laser beams in graded-index optical fibers. *Optical Memory and Neural Networks (Information Optics)*, Vol. 13, No. 1, pp. 63-70

Berdague, S., Facq, P. (1982). Mode division multiplexing in optical fibers. *Appl. Optics*, Vol. 21, pp. 1950-1955,

Bolshtyansky, M. A., Zel'dovich, B. Ya. (1996). Transmission of the image signal with the use of multimode fiber. *Optics Comm.*, Vol. 123, pp. 629-636

Bolshtyansky, M. A., Savchenko, A. Yu., Zel'dovich, B. Ya. (1999). Use of skew rays in multimode fibers to generate speckle field with nonzero vorticity. *Opt. Lett.*, Vol. 24, No. 7, pp. 433-435

Cherin, A.H. (1987). *An introduction to optical fibers*, McGraw-Hill book Co., Singapore

Dennis, M. R., O'Holleran, K., Padgett, M. J. (2009). Singular Optics: Optical Vortices and Polarization Singularities. *Progress in Optics*, Vol. 53, pp. 293-363

Desyatnikov, A. S., Kivshar, Yu. S., and Torner, L. (2005) Optical vortices and vortex solitons. *Progress in Optics*, ed. E. Wolf, Vol. 47, pp. 291-391, Elsevier, Amsterdam

Dubois, F., Emplit, Ph., and Hugon, O. (1994). Selective mode excitation in graded-index multimode fiber by a computer-generated optical mask. *Opt. Lett.*, Vol. 19, No. 7, pp. 433-435

Gloge, D. (1971). Weakly guided fibers. *Appl. Opt.*, Vol. 10, pp. 2252-2258

Karpeev, S. V., Pavelyev, V. S., Soifer, V. A., Khonina, S.N., Duparre, M., Luedge, B., Turunen, J. (2005). Transverse mode multiplexing by diffractive optical elements. *Proc. SPIE*, Vol. 5854, pp. 1-12

Karpeev, S.V., Khonina, S.N. (2007). Experimental excitation and detection of angular harmonics in a step-index optical fiber. *Optical Memory & Neural Networks (Information Optics)*, Vol. 16, No. 4, pp. 295-300

Khonina, S. N., Skidanov, R. V., Kotlyar, V. V., Jefimovs, K., Turunen, J. (2003). Phase diffractive filter to analyze an output step-index fiber beam. *Optical Memory & Neural Networks (Information Optics)*, Vol. 12, No. 4, pp. 317-324

Khonina, S. N., Striletz, A. S., Kovalev, A. A., Kotlyar, V. V. (2010). Propagation of laser vortex beams in a parabolic optical fiber. *Proceedings SPIE*, Vol. 7523, pp. 75230B-1-12

Kotlyar, V.V., Soifer, V.A., Khonina, S.N. (1998). Rotation of multimodal Gauss-Laguerre light beams in free space and in a fiber. *Optics and Lasers in Engineering*, Vol. 29, No. 4-5, pp. 343-350

Kotlyar, V. V., Khonina, S. N., Soifer, V. A., Wang, Y. (2002). Measuring the orbital angular momentum of the light field using a diffractive optical element. *Avtometriya*, Vol. 38, No. 3, pp. 33-44

Marcuse, D. (1972). *Light transmission optics*, Van Nostrand Reinhold Co., New York

Mikaelian, A.L. (1980). Self-focusing media with variable index of refraction. *Progress in Optics*, Vol. 27, pp. 279-345

Mikaelian, A. L. (1990). *Optical Methods in Informatics*, Moscow, Nauka (Science) Publishers

Snyder, A. W., Love, J. D. (1987). *Optical Waveguide Theory*, Moscow Radio & Sviaz Publishers

Soifer, V.A., Golub, M.A. (1994). *Laser beam mode selection by computer-generated holograms*, CRC Press, Boca Raton

Soskin, M. S., Vasnetsov, M. V. (2001). Singular optics. *Progress in Optics*, Vol. 42, pp. 219-76

Striletz, A. S., Khonina, S. N. (2008). Matching and analysis of methods based on use of the differential and integral operators to describe the laser light propagation in a weakly nonuniform medium. *Computer Optics*, Vol. 32, No. 1, pp. 33-38 - in Russian

Thornburg, W. Q., Corrado, B. J., and Zhu, X. D. (1994). Selective launching of higher-order modes into an optical fiber with an optical phase shifter. *Opt. Lett.*, Vol. 19, No. 7, pp. 454-456

Volyar, A. V., Fadeeva, T. A. (2002). Dynamics of topological multipoles: II. Creation, annihilation, and evolution of nonparaxial optical vortices. *Optics and Spectr.*, Vol. 92, No. 2, pp. 253-262

Yeh, C. (1990). *Handbook of fiber optics. Theory and applications*, Academic Press Inc., New York

# Use and Limitations of Single- and Multi-Mode Optical Fibers for Exoplanet Detection

Julien F.P. Spronck, Debra A. Fischer
and Zachary A. Kaplan
*Yale University*
*USA*

## 1. Introduction

Optical fibers are of a great importance in diverse areas of modern observational astronomy. Particularly, in the field of exoplanet detection, they have become an essential part of most current and future instruments because of their filtering and stabilizing capability.
In this chapter, we will discuss the use of optical fibers and some limitations in two exoplanet detection methods: nulling interferometry (Section 2) and the radial velocity method (Section 3). We will present simulations, experiments and observations that demonstrate improvements of the instrument performances in the field of exoplanet detection due to the use of optical fibers, as well as some of their limitations.

## 2. Single-mode fibers in nulling interferometry

Nulling interferometry is a direct exoplanet detection method, aimed at the detection of an Earth-like planet around a Sun-like star (Bracewell, 1978; Colavita et al., 2010; Mennesson et al., 2011). It consists in combining light from several telescopes in such a way that a quasi-perfect destructive interference occurs for the star light. In such an instrument, the light coming from a potential planet orbiting the star would experience a (partially) constructive interference because of the optical path differences between the arms of the interferometer for an off-axis point source (i.e. the planet).
Single-mode fibers are used in all state-of-the-art wide-band nulling interferometers because they provide natural wavefront filters, essential for a quasi-perfect destructive interference (Mennesson et al., 2002; Wallner et al., 2003).
In addition to canceling the light from the star and thus making possible direct detection of planets, nulling interferometry should also offer the possibility to obtain spectral information from the planet if destructive interference can be achieved simultaneously for all wavelengths in a wide spectral band (typically from 5-18 $\mu$m would be the optimal wavelength range because it is where the brightness ratio between the star and the planet is minimal)(Angel et al., 1986; Angel & Woolf, 1997). To realize that, very stringent requirements must be fulfilled in terms of amplitude, phase and polarization of the beams to be combined for all wavelengths. Most nulling interferometers use achromatic phase shifters (Rabbia et al., 2003) to create an on-axis destructive interference independent of the wavelength and must

also use an achromatic amplitude-matching device. The use of single-mode fibers in a nulling interferometer can affect this achromaticity condition because the coupling of light into a waveguide is wavelength-dependent. This coupling can therefore chromatically affect both the amplitude and the phase of the beam.

In particular, two beams with slightly different wavefronts will have different wavelength-dependent coupling efficiencies. This results in different wavelength-dependent amplitudes and phases, which will limit the performance of the interferometer. A measure for this performance is called the rejection ratio: it is the ratio between the intensities corresponding to constructive and destructive interferences.

In this section, we will calculate the wavelength-dependent coupling efficiencies of aberrated beams into a single-mode fiber and analyze the influence of aberrations on the rejection ratio and therefore on the performance of the nulling interferometer. From these results, we will quantitatively derive the wavefront quality required to allow the detection of Earth-like planets. We will then show that amplitude, optical path difference and dispersion corrections can be used to reduce the effect of induced wavelength-dependent coupling efficiencies and relax the tolerances on optical quality.

### 2.1 Definitions

Let us consider the case of a two-beam nulling interferometer. We will assume that a perfectly achromatic $\pi$-phase shift has been introduced between the beams in order to get destructive interference for all wavelengths.

Each of the beams $i$ ($i = 1$ or 2) has a distorted wavefront $W_i$, which can be described in terms of normalized Zernike polynomials (Noll, 1976),

$$W_i(x,y) = \sum_j a_j^{(i)} Z_j(x,y). \tag{1}$$

In this representation, each polynomial represents an aberration and the coefficient $a_j^{(i)}$ gives the RMS contribution of the corresponding aberration to the total wavefront.

As explained in Section 2, the beams are then focused onto a single-mode fiber that acts as a wavefront filter. Indeed, the field at the output of the fiber is given (all losses neglected) by the fundamental mode of the fiber, multiplied by a complex factor $\xi_i$ called the complex coupling efficiency that represents the part of the field that is coupled in the fiber (Mennesson et al., 2002; Wallner & Leeb, 2002). This holds for any incoming field and therefore, all wavefront distortions are taken care of by the optical fiber. However, different wavefronts will induce different (wavelength-dependent) coupling efficiencies and this will limit the rejection ratio.

The field in the focal plane is given by the Fourier transform of the field in the entrance pupil,

$$E_i(X,Y,\lambda) = \iint \exp\left[j\frac{2\pi}{\lambda}W_i(x,y)\right] \exp\left[-j\frac{2\pi}{\lambda f}(xX+yY)\right] dxdy, \tag{2}$$

where $\lambda$ is the wavelength, $f$ the focal length of the focusing optics, $(X,Y)$ and $(x,y)$ are respectively the coordinates in the focal plane and in the entrance pupil plane.

The complex coupling efficiency $\xi_i$ of beam $i$ is then given by the overlap integral between the incident field $E_i$ and the fundamental mode of the fiber $F_0$,

$$\xi_i(\lambda) = \frac{\iint E_i(X,Y,\lambda)F_0^*(X,Y,\lambda)dXdY}{\iint |F_0(X,Y,\lambda)|^2\, dXdY}, \tag{3}$$

where $^*$ denotes the complex conjugate.
The rejection ratio R is the ratio between intensities corresponding to constructive and destructive interferences. Therefore, we have

$$R = \frac{\int |\xi_1(\lambda) + \xi_2(\lambda)|^2\, d\lambda}{\int |\xi_1(\lambda) - \xi_2(\lambda)|^2\, d\lambda}. \tag{4}$$

**2.2 Influence of each aberration on the rejection ratio**

We consider a spectral band going from 500 to 650 nm. This spectral band was chosen to match an existing experimental set-up. We will first assume that one of the beams has a perfect plane wavefront ($a_j^{(1)} = 0$ for all j), while the second wavefront is distorted. In this first simulation, we will study the influence of each aberration separately by setting the coefficient $a_j^{(2)} = 30$ nm (wavefront at roughly $\lambda/20$ RMS) and calculate the rejection ratio as a function of the Zernike index $j$ (each index represents a different type of aberration). The results are depicted in Figure 1 (black squares).

We see a "wave" pattern in the rejection ratio as a function of Zernike index. Each of these waves corresponds to a different radial order of the Zernike polynomials. For each radial order, the rejection ratio is minimal for zeroth azimuthal order (radially symmetric) and increases with azimuthal order (towards higher spatial frequencies). The rejection ratio also increases with radial order, since the fiber is less sensitive to high spatial frequencies.

The aberrated wavefront introduced amplitude and phase mismatches between the two beams. There are therefore a few corrections that we can apply to improve the rejection ratio. We can first use an achromatic intensity-matching device, e.g. a knife-edge (which is achromatic at first order) to match the global intensities of the two beams (see Figure 1, blue diamonds). Then, we can use an optical delay line to match the optical path differences (OPD) between the beams (see Figure 1, red stars). Finally, we can compensate for dispersion differences by adding glass plates with variable thicknesses (see Figure 1, magenta crosses) (Spronck et al., 2008; Spronck et al., 2009).

We see that OPD and dispersion compensation only improves the rejection ratio for the fourth (defocus), the twelfth (spherical aberration) and the twenty-fourth (6th order spherical aberration) Zernike polynomials. Indeed these polynomials have a zero azimuthal frequency (radial symmetry) and we can show that the coupling efficiencies corresponding to non-zero azimuthal frequencies are real. Therefore, for these aberrations, no phase corrections can increase the rejection ratio, only the amplitude correction can. Note that the limitation of the rejection ratio is due to a wavelength-dependent amplitude mismatching, for which we cannot easily compensate. Therefore, these results will strongly depend on the width of the spectral band.

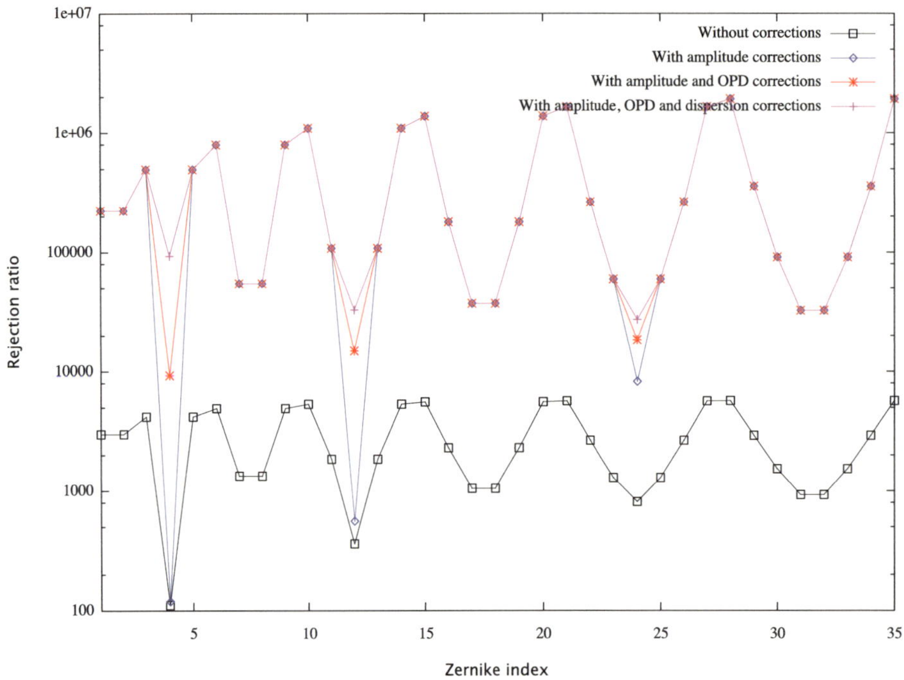


Fig. 1. Rejection ratio as a function of Zernike index when $a_j^{(1)} = 0$ for all $j$ and $a_j^{(2)} = 30$ nm (black squares). The blue diamonds correspond to the rejection ratio after an achromatic intensity matching. For the red stars, the OPD have been matched (additionally to the intensity matching). The magenta crosses corresponds to rejection ratio with intensity, OPD and dispersion correction.

### 2.3 Rejection ratio with randomly chosen wavefronts

In this other simulation, we randomly chose the coefficients $a_j^{(1)}$ and $a_j^{(2)}$ for both wavefronts in such a way that these wavefronts have a standard deviation of 30 nm RMS ($\lambda/20$) (see Figure 2). We found the average rejection ratio with such wavefronts after 35 simulations is of the order of $10^3$ without corrections and $10^6$ with amplitude, OPD and dispersion corrections. We then repeated this simulation with different wavefront standard deviations and plotted the average rejection ratio as a function of RMS wavefront quality (see Figure 3). From this, we derive that the necessary RMS wavefront quality to obtain a $10^6$-rejection ratio is 40 nm RMS ($\lambda/15$). This means that the surface figure of the optics (we only considered here the case of reflective optics) should be better than $\lambda/30/\sqrt{N_{opt}}$ where $N_{opt}$ is the total number of surfaces encountered by the beams.

It is important to realize that these results highly depend on the desired spectral band and cannot directly be translated in a general requirement. However, this is meant to indicate the limitations of single-mode fibers in nulling interferometry. Note also that it will be easier to meet the requirements in the IR where nulling interferometers mainly perform.

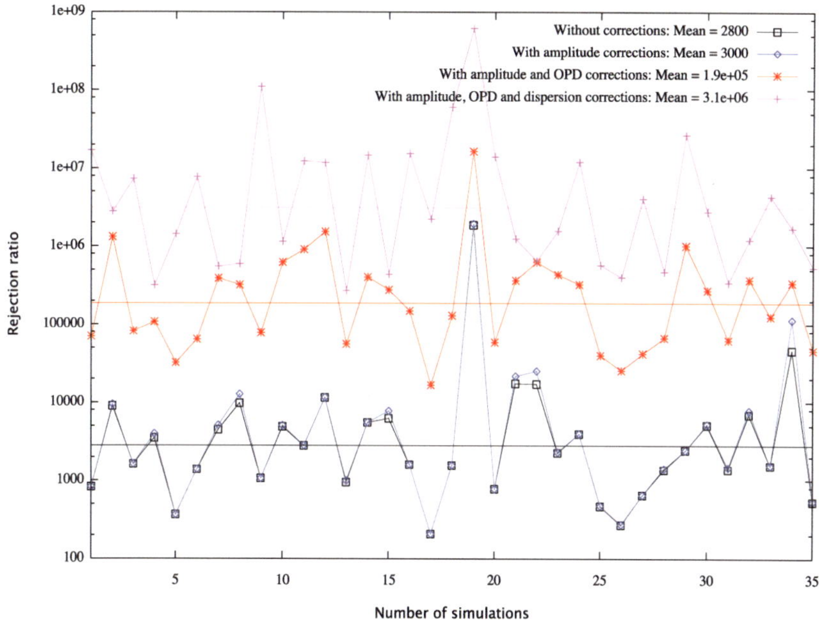


Fig. 2. Rejection ratio for two beams with randomly chosen wavefronts that have a standard deviation of 30 nm RMS.

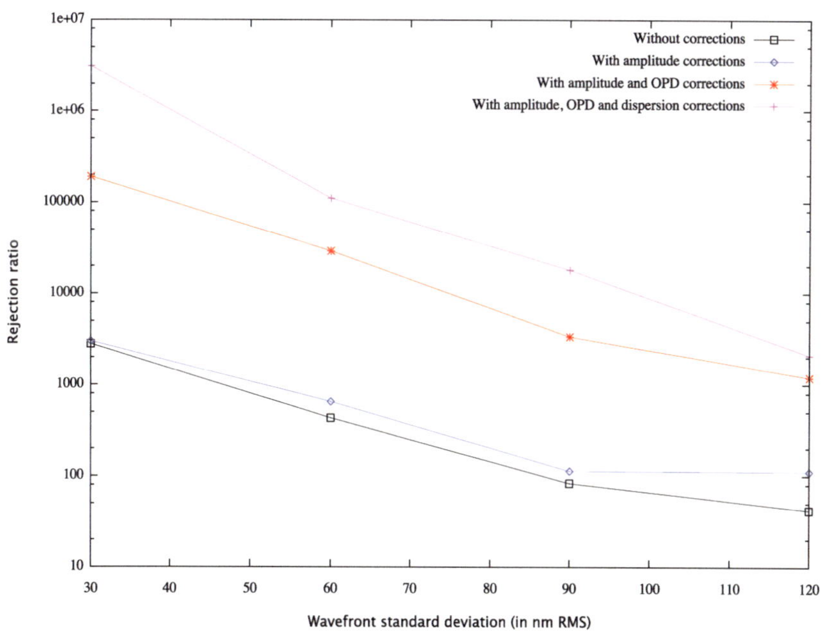


Fig. 3. Rejection ratio as a function of RMS wavefront quality.

## 3. Multi-mode fibers for high-precision radial velocities

Since the discovery of the first exoplanet by Mayor & Queloz (1995), more than 500 planets have been found using the radial velocity method. Currently, the state-of-the-art spectrometers, such as HARPS (Mayor et al., 2003) on the 3.6-m telescope in La Silla and HIRES on Keck I (Vogt et al., 1994), typically achieve precisions of 1-3 m $s^{-1}$ (Howard et al., 2010; Mayor & Udry, 2008). This only permits the detection of planets with amplitudes larger than the measurement errors, typically Super Earth or Neptune-mass planets in relatively short period orbits, or more massive Jupiter-like planets out to several AU. The detection of true Earth analogs requires Doppler precisions on the order of 10 cm $s^{-1}$, corresponding to spectral line shifts across one ten-thousandth of a pixel. Further complicating the analysis, the periodicity of this shift occurs over time scales of months or years for the most interesting planets in the so-called habitable zone. This top level requirement for a measurement precision of 10 cm s-1 leads to the demand for an instrument that exceeds the stability of current instruments.

In order to reach the desired precision, we must reduce errors in the model of the instrumental profile, which cross-talk with our measurement of the Doppler shift. In older spectrographs, the starlight is coupled from the telescope to the instrument using a narrow slit. However, the slit illumination is rapidly varying because of changes in seeing, focus and guiding errors. Changes in slit illumination affect the spectrum in two ways. Since the spectral lines are direct images of the slit, changes in slit illumination yield changes in the shape of the spectral lines. Additionally, variations in slit illumination can result in changes in the illumination of the spectrograph optics. This will in turn introduce different aberrations, which will change the instrumental response. Mathematically, these two effects are modeled simultaneously by convolving the spectrum with the instrumental profile (IP), in such a way that any variability impedes our ability to recover Doppler shifts with the desired precision. If the instrumental profile were unchanging, variations in the final extracted spectrum would be dramatically reduced. Thus, instrumental profile stability has become a focus of current instrumentation work.

Optical fibers provide an excellent way to reduce variability in the illumination of the spectrograph. Fibers have been used since the 1980's to couple telescopes to high-precision spectrographs (Heacox & Connes, 1992). The throughput of fibers was initially low, however, they offered unprecedented convenience in mechanical design. The attribute of fibers that is particularly important today for high-precision Doppler measurements is the natural ability of optical fibers to scramble light (Barden et al., 1981; Heacox, 1980; 1986; 1988) and produce a more uniform and constant output beam. Because light from the telescope must be efficiently coupled into the fiber, the fiber diameters must match the typical image size (generally 100 microns or more), so multi-mode fibers are required.

Other sources of errors come from environmental changes within the spectrograph. Temperature, pressure or mechanical variations cause the spectrum to shift and to change. These errors will not be solved by replacing the slit by a fiber.

### 3.1 Laboratory characterization

We have carried out laboratory measurements to better understand scrambling properties of fibers with different geometries (circular, square, octagonal), different lengths, and different fiber diameters. While testing these fibers, we have noticed that the optical properties vary

widely from fiber to fiber. This is even true for supposedly identical commercial fibers from the same manufacturer and same production batch.
At an observatory, the illumination of the fiber will vary due to guiding, focusing errors and seeing changes. To characterize the scrambling properties of the fiber under similar conditions, we scan the incoming beam across the fiber and examine the output beam.
As described by Hunter & Ramsey (1992), two characteristics are of importance when it comes to the output beam: the far-field and the near-field patterns. The far-field is the cross-sectional intensity distribution of the diverging beam. The far-field will be projected onto the collimator, the grating and the rest of the spectrograph optics. Variations in the far-field will therefore cause different parts of the grating and the optics to be illuminated. This will in turn introduce different aberrations, which will change the instrumental profile. The near-field pattern is the intensity distribution across the output face of the fiber. The spectral lines are direct images of the fiber output face, so variations in the near-field pattern are also important in the stability of the final spectrum. Commonly (but erroneously), the term near-field is used to describe the image of the output face of the fiber by an optical system. We will adopt this definition throughout this chapter.

### 3.1.1 Experimental set-up

The set-up used for the fiber characterization measurements is depicted in Figure 4.
We focus the light from either a green He-Ne laser or a LED onto a single-mode fiber that is used to create a star-like point source. Light coming from the single-mode fiber is then collimated (by lens $L_2$) and re-focused (by lens $L_3$) onto the test multi-mode fiber. Light reflected from the fiber front surface is re-directed using a beam-splitter and re-imaged onto a CCD ($CCD_1$) to check the alignment of the beam with respect to the fiber front surface. A translation stage allows us to move the fiber with respect to the incoming beam and therefore simulate guiding errors. Light coming out of the test fiber is then collimated (by lens $L_7$) and re-focused (by lens $L_5$) onto a CCD ($CCD_2$). Lens $L_6$ moves in and out of the light path to enable measurements of the near-field (out) and the far-field (in) patterns.

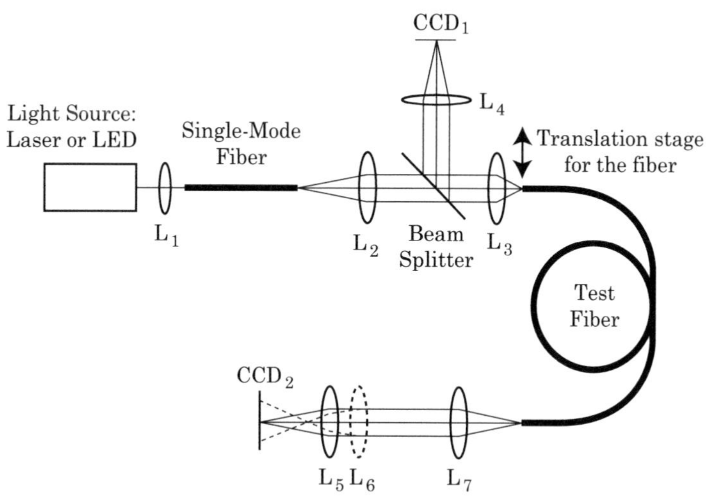


Fig. 4. Schematic drawing of our set-up.

#### 3.1.2 Scrambling versus fiber length

We first measured the amount of scrambling as a function of fiber length. All fibers used in this test were 50-micron fibers from Polymicro (FBP050070085) with lengths of 5, 20 or 40 m. We used a similar set-up as described in Section 3.1.1. A green He-Ne laser was coupled into the single-mode fiber. A couple of lenses were used to re-image the single-mode fiber onto the test fiber. The used imaging system was not of very high quality, so that the spot size was a significant fraction of the fiber core.

For this experiment, we inserted a mirror on a kinematic mount between the two lenses and tilted it to scan the image across the fiber face and simulate guiding errors. The far-field pattern was recorded as a function of mirror tilt (or equivalently of beam position on the fiber). To eliminate the speckle pattern caused by modal interference, we agitated the test fiber.

Figure 5 depicts the far-field patterns as a function of beam position on the fiber for three different fiber lengths (5, 20 and 40m). The left columns corresponds to input that was well centered on the fiber. In the right columns, the beam is increasingly displaced from the fiber center. For the 5-m fiber, we clearly see rings when moving away from the center, which become dominant rather quickly. The 5-m fiber quickly develops a ring pattern. Rings are also seen for the 20-m fiber, but not until the image is much further displaced from the center. No ring pattern appears for the 40-m fiber; the far-field distribution seems almost independent of the spot position.

Any type of variation in the far-field pattern is undesirable since it will induce variations in the illumination of the grating and spectrograph optics that will cause varying instrumental profile.

The rings occur because light is propagating through the cladding: they only appear when the spot was large enough to overlap with the cladding (i.e., when the spot was slightly off-center). Because light does not propagate very well in the cladding, there is a dependence on fiber length and the 40-m fiber is long enough that this pattern is not seen in the output beam.

These measurements were confirmed by measuring three fibers of each length. They all exhibited the same behavior.

There are two important consequences of these measurements. First, longer fiber will be better for scrambling. On the other hand, longer fiber will have a lower throughput. For example, the Polymicro FBP fibers have 15 dB/km losses due to absorption at 500 nm. For a 5-m fiber, that gives a throughput of 99 % for a 5-m fiber and 90 % for a 40-m fiber. A trade-off between scrambling and throughput is needed given a specific application.

The second consequence is that if good scrambling is desired, the cladding should never be illuminated. Cladding illumination can be avoided by appropriate masking (in the fiber input plane or more easily, in an intermediate focal plane). The mask alignment can be critical.

#### 3.1.3 Scrambling with circular fibers

In this test, we used the set-up depicted in Figure 4. A green LED (50-nm FWHM) was used as light source. In terms of coherence length, a standard He-Ne laser would be more appropriate than the LED for very high-resolution spectrographs. However, because of its low coherence, the LED makes it possible to measure reproducible and precise fiber outputs without agitating the fiber.

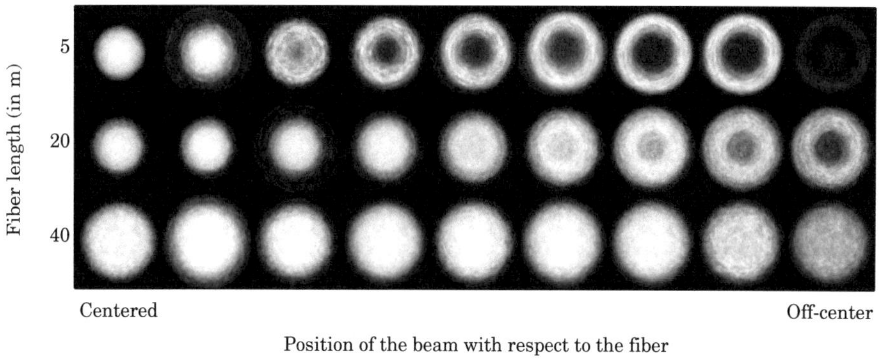


Fig. 5. Scrambling of 50-micron fibers of various lengths

The single-mode fiber was re-imaged onto the test multi-mode fiber using a pair of diffraction-limited aspheric lenses. This time, the resulting spot onto the test fiber was much smaller than the fiber. Using a commercial camera, we imaged the input face of the test fiber and the spot. This way we could carefully position the spot with respect to the fiber and we could also make sure that the test fiber was exactly in the image plane (and thus the spot was well in focus when entering the fiber).

We scanned the spot with respect to the fiber by moving the fiber (which was on a differential screw stage with a precision of 1 $\mu$m). For every fiber position, we checked the spot position with the camera. We then recorded both far-field and near-field patterns for every fiber position.

Figure 6 shows the far-field (top row) and near-field (bottom row) as a function of fiber position (from cladding to cladding) for a 15-m long 100-micron Polymicro fiber (FBP100120140). The far-field pattern shows strange non-radially symmetric structures but both far-field and near-field distributions are nearly independent of fiber position. However, looking closer at the near-field (see Figure 7), systematic variations can be seen for different fiber positions. This position memory is evidence of non-perfect scrambling by the fiber and will limit the instrumental profile stability of a high-resolution spectrograph, since guiding errors will directly translate into variations in near-field patterns.

### 3.1.4 Scrambling with octagonal fibers

It has been suggested that fibers with different geometries (square, hexagonal, octagonal) were better scramblers and were therefore more suitable for high-precision radial velocities (Avila et al., 2010; Chazelas et al., 2010). We purchased 20-m octagonal fibers from CeramOptec with a 200-micron octagonal core and a 672-micron round cladding.

We repeated the measurements presented in Section 3.1.3. Figure 8 summarizes the results. The top row is the far-field, the middle row is the near-field and the bottom row shows the spot position across the input fiber face.

The far-field is better behaved in terms of symmetry than it was for the circular fiber but is not as position independent for the octagonal fiber. On the other hand, the near-field (see

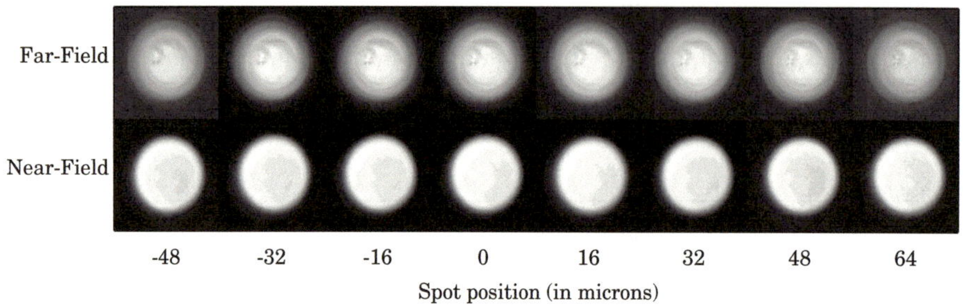


Fig. 6. Scrambling of a circular fiber.

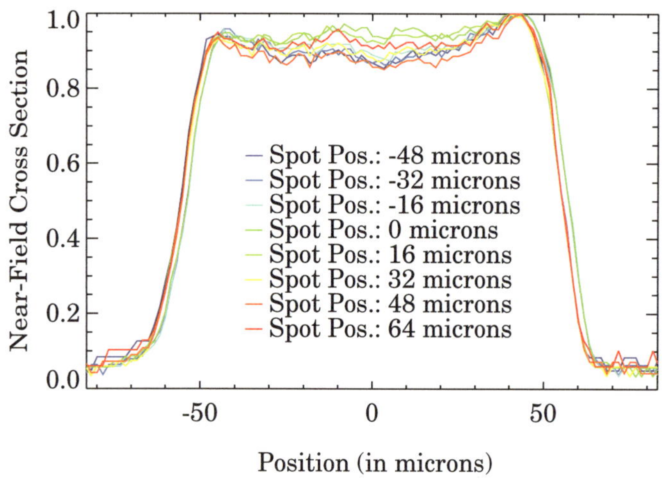


Fig. 7. Scrambling of a circular fiber.

Figure 9) shows no systematic variations and seem independent of fiber position (within the measurement precision).

The fact that the near-field is so independent on beam position is very encouraging for use in high-resolution spectrographs, as it probably yields a very stable instrumental profile. In contrast, the far-field is not as good and depending on the local quality of the spectrograph optical components, can contribute to some variations in instrumental profile.

### 3.2 Results at Lick observatory

In 2009, we have installed a fiber feed for the Hamilton spectrograph on the 3-m telescope at Lick Observatory (Spronck et al., 2010). The key results are presented in this section.

#### 3.2.1 Comparison between slit and fiber using the Hamilton spectrograph

In August 2010, extensive tests were carried out to quantify the improvement in instrumental profile stability brought by the fiber scrambler and to identify the remaining sources of error.

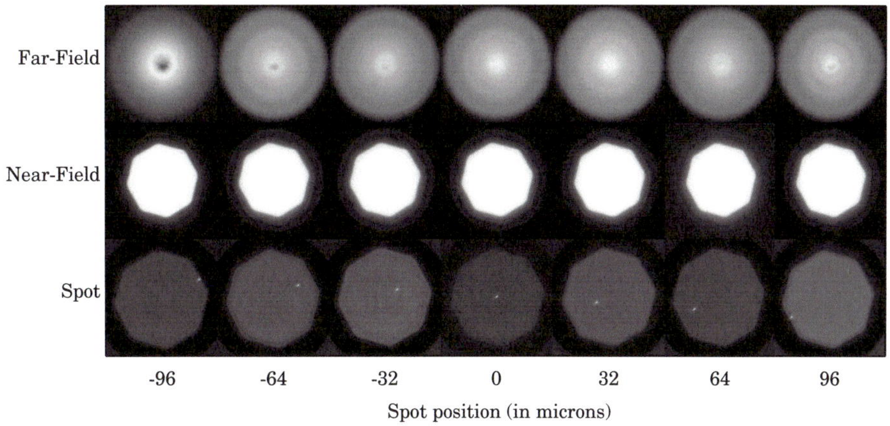


Fig. 8. Scrambling of an octagonal fiber.

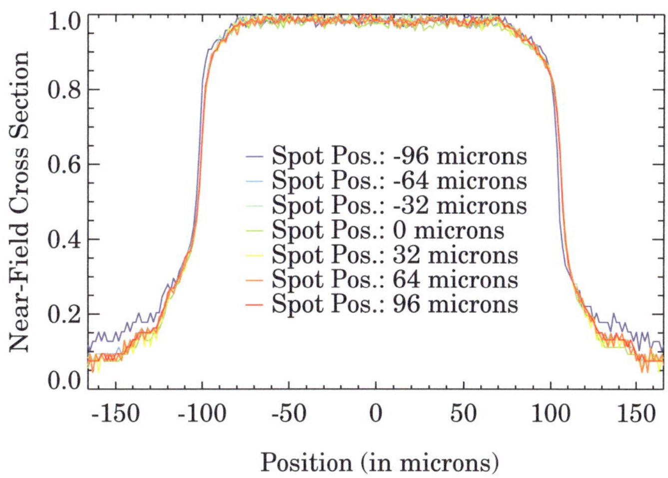


Fig. 9. Scrambling of an octagonal fiber.

Observations of stars with known constant radial velocity were made on two consecutive nights. The weather and seeing conditions were nearly identical for both nights. The fiber scrambler was installed for the first night, and the regular observing slit (640 $\mu$m wide) was used on the second night.

On both nights, an iodine cell was used. As starlight passes through the cell, the molecular iodine imposes thousands of absorption lines in the stellar spectrum. We use an extremely high resolution ($R \approx 1{,}000{,}000$), high SNR Fourier Transform Spectrum (FTS) of the iodine cell to model the instrumental profile, which when convolved with the product of the stellar spectrum and the iodine FTS spectrum reproduces the observed spectrum. The instrumental profile must be modeled for small wavelength segments of the echelle spectrum to account for 2-D spatial variations. Although there are some asymmetries in the wings of the IP, a single Gaussian gives, for our purpose, a reasonable fit to the composite IP. We fitted a Gaussian to the instrumental profile for each of the spatial regions on the CCD and calculated the average full-width half maximum (FWHM) of the Gaussian across the entire detector (iodine region). Figure 10 depicts the evolution of the average FWHM for the slit observations (blue squares) and for the fiber observations (red filled circles) through time. The abscissa in this plot is the sequential observation number through the night. The time-dependence variation of the IP for the slit observations (blue squares) is quite dramatic. For both nights, the same sequence of observations were taken: a set of B stars, 50 observations of the velocity standard star HD 161797, a second set of B stars, 50 observations of the velocity standard star HD 188512 and a third set of B stars.

The smooth functional dependence on time for slit observations strongly suggests that the dominant factor in the instrumental profile variation is the changing illumination of the slit due to monotonic changes in seeing or tracking through different hour angles (which might result in different input angles into the fiber). The peak-to-valley (PTV) amplitude of the variation is about 8% throughout the night.

Figure 10 also shows significant improvement in instrumental profile stability due to the fiber scrambler (red solid dots). However, there is still a slight linear (upward) trend in the fiber data (1%-2% PTV), indicative of incomplete scrambling with the fiber. After removing the linear trend, the residual fluctuation is of the order of 1% PTV.

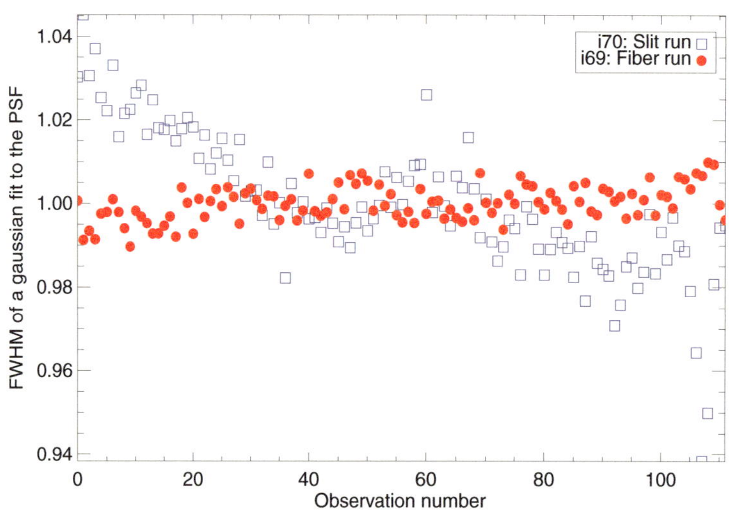


Fig. 10. Average FWHM of a Gaussian fit to the instrumental profile for all observations during Night 1 using the fiber (red filled circles) and Night 2 using the slit (blue squares).

#### 3.2.2 Results with a double scrambler using the Hamilton spectrograph

In August 2010, a double scrambler (Avila, 1998; Hunter & Ramsey, 1992) was designed and built. In this double scrambler, a ball-lens transforms the image of the fiber end in a pupil that is then injected into a second fiber. The light from the second fiber is then sent to the spectrograph. Because of time constraints, the double scrambler was not optimized and as a consequence, the throughput when used in the Hamilton spectrograph was rather low (15% as opposed to 55-60% with one fiber only).

The double scrambler test consisted in taking alternative sets of five B-star observations with the regular fiber scrambler (one fiber only) and with the double scrambler throughout the same night. For each observation, we calculated the instrumental profile for each region of the CCD and fitted it with a Gaussian. We then calculated the average FWHM of the fit across the entire detector.

Figure 11 depicts the evolution of the average FWHM for the single fiber observations (blue) and for the double scrambler observations (red) through the night. Different symbols correspond to different sets of B stars. Even though the scale is different from Figure 10 (with the slit observations), we can still see a linear trend in the fiber data in Figure 11, indicating imperfect fiber scrambling. In this case, the amplitude of the variation is about 3%.

The IP obtained with the double scrambler is significantly more stable throughout the night, with no significant (above errors) systematic trend.

Instrumental noise can be broken down into two main components: errors due to coupling of the light to the instrument (varying fiber illumination due to guiding, tracking, seeing and focusing) and environmental instability (mechanical, temperature or pressure). The double scrambler results prove that coupling errors are the dominant source of instrumental noise.

Residual fluctuations from observation to observation have an amplitude of 1%, which is large for precise radial velocities. The source for these fluctuations has not yet been identified but possible culprits include modal noise in the fiber, photon noise and modeling errors. We do not expect the environmental instability to be responsible for residual fluctuations because of the short time scale of the variability.

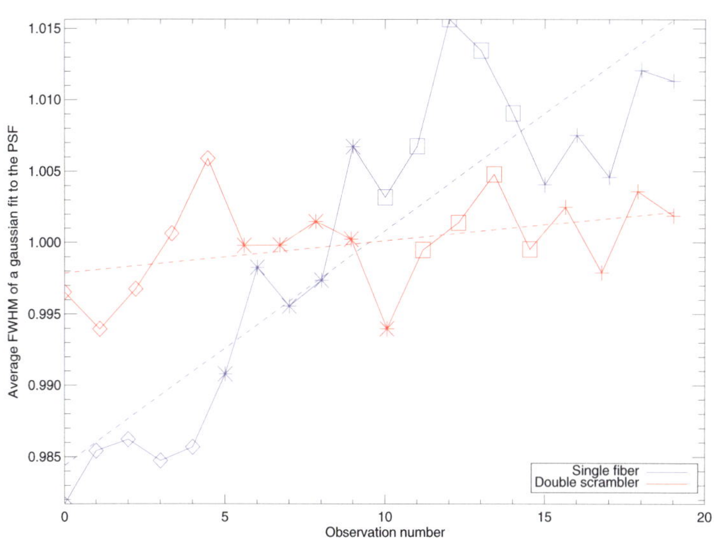


Fig. 11. Average FWHM of a Gaussian fit to the instrumental profile for B-star observations taken with the fiber (blue) and with the double scrambler (red). All observations were taken during the same night alternating with the fiber and with the double scrambler.

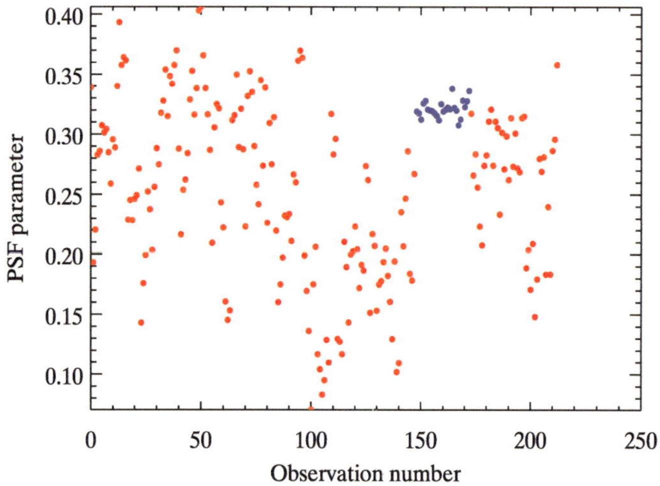


(a)

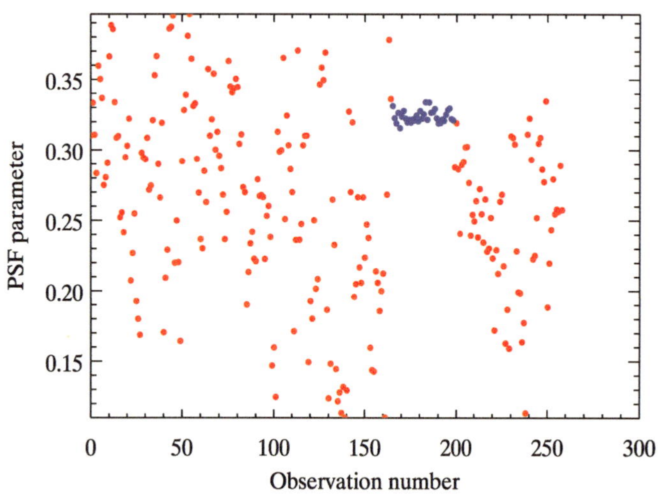


(b)

Fig. 12. Instrumental profile parameters for all HIRES observations of (a) HD 26965 and (b) HD 32147. Red dots correspond to slit observations and blue dots correspond to fiber observations.

### 3.3 Results at Keck observatory

During the last week of September 2010, we repeated the tests performed at Lick Observatory at the Keck telescope using the HIRES spectrograph. The larger aperture telescope at Keck helped to keep the exposure times short so that a large data set could be acquired and so that barycentric errors were minimized. We designed and built a prototype fiber scrambler with a 200-micron 20-m Polymicro fiber

We collected data on two nights; on 30 September 2010, we used the fiber scrambler and on 1 October 2010, we obtained a similar set of data with the usual slit. We observed sets of 25 observations for the standard stars HD 26965 and HD 32147. Figure 12(a) and (b) depict one of the parameters used to model the instrumental profile for all existing observations of (a) HD 26965 and (b) HD 32147. The red filled circles correspond to slit observations and exhibit an RMS scatter of 0.066 for HD 32147, while the fiber observations for the same star (blue filled circles) exhibit a dramatically reduced RMS scatter of 0.0044, demonstrating a factor of 15 improvement in the IP stability.

## 4. Conclusion

In the fist part of this chapter, we have studied the performances of a two-beam nulling interferometer with distorted wavefronts. We have studied the influence of each individual aberration and we have seen that aberrations will induce phase and amplitude mismatches between the beams that can partially be compensated. Unfortunately the wavelength-dependence of these mismatches will limit the rejection ratio. We have seen that the interferometer will be more sensitive to lower order aberrations (both radial and azimuthal orders). In particular, aberrations that will mostly limit the rejection ratio are radially symmetric aberrations (such as defocus, spherical aberration and sixth order spherical aberration). For the considered spectral band (500-650 nm), we quantified the wavefront and surface quality needed to have a rejection ratio of $10^6$. The quality of the wavefront should be better than $\lambda/15$ RMS. This result depends on the width of the spectral band. Even though single-mode fibers are essential parts of nulling interferometers, they are not perfect modal filters and will eventually limit the performances of the instrument if care is not taken in the optical design.

In the second part of this chapter, we have characterized the scrambling properties of multi-mode fibers as a function of length and cross-sectional geometry. We conclude that longer fibers perform better in terms of scrambling (but have lower throughput) because light in the cladding will not propagate efficiently. We also conclude that the best scrambling is achieved when the cladding is not illuminated. We see evidence of non-perfect scrambling in the near-field of the circular fiber, while the octagonal fiber has a very well-behaved near-field. This implies that the octagonal fiber should therefore yield a very stable instrumental profile. However, we find that the far-field of the octagonal fiber is not as good and therefore octagonal fibers will only be helpful if the grating and other spectrograph optics have excellent optical quality.

To summarize, in order to measure spectral line shifts smaller than one ten-thousandth of a pixel and stable for many months, we must reduce errors in our instrumental profile, which cross-talks with our measurement of the Doppler shift. The instrumental noise can be broken down in coupling errors (slit or fiber illumination) and environmental instability. These results show that coupling errors are the dominant source of instrumental noise. We show that double scrambler observations have a more stable IP than fiber observations, which have a more stable IP than slit observations. The double scrambler data still has residual RMS scatter. The source of this has not yet been identified but is likely to be modal noise, photon noise or modeling errors. We do not expect that the residual scatter can be caused by environmental effects due to the random nature of the variability.

While some fibers are clearly better than others at scrambling light, modal noise will always limit the ability of multi-mode fibers to perfectly scramble light and therefore multi-mode fibers will produce some variability in the instrumental profile. Whether this is important or not depends on the level of precision needed in terms of radial velocities. When looking for Earth analogs, a Doppler precision of 10 cm $s^{-1}$ or better will be required. At this level of precision, everything becomes relevant.

## 5. Acknowledgments

We acknowledge the support of the Planetary Society, who made possible the development and installation of the fiber feeds at Lick and Keck Observatory. We also would like to thank the National Science Foundation for their support.

## 6. References

Angel, J. R., Cheng, A. Y. S. & Woolf, N. J. (1986). A space telescope for ir spectroscopy of earthlike planets, *Nature* 232: 341–343.

Angel, J. & Woolf, N. J. (1997). An imaging nulling interferometer to study extrasolar planets, *The Astrophysical Journal* 475(1): 373–379.

Avila, G. (1998). Results on Fiber Characterization at ESO, *in* S. Arribas, E. Mediavilla, & F. Watson (ed.), *Fiber Optics in Astronomy III*, Vol. 152 of *Astronomical Society of the Pacific Conference Series*, p. 44.

Avila, G., Singh, P. & Chazelas, B. (2010). Results on fibre scrambling for high accuracy radial velocity measurements, *Society of Photo-Optical Instrumentation Engineers (SPIE) Conference Series*, Vol. 7735 of *Society of Photo-Optical Instrumentation Engineers (SPIE) Conference Series*.

Barden, S. C., Ramsey, L. W. & Truax, R. J. (1981). Evaluation of some fiber optical waveguides for astronomical instrumentation, *PASP* 93: 154–162.

Bracewell, R. N. (1978). Detecting nonsolar planets by spinning infrared interferometer, *Nature* 274(5673): 780–781.

Chazelas, B., Pepe, F., Wildi, F., Bouchy, F., Perruchot, S. & Avila, G. (2010). New scramblers for precision radial velocity: square and octagonal fibers, *Society of Photo-Optical Instrumentation Engineers (SPIE) Conference Series*, Vol. 7739 of *Society of Photo-Optical Instrumentation Engineers (SPIE) Conference Series*.

Colavita, M. M., Serabyn, E., Ragland, S., Millan-Gabet, R. & Akeson, R. L. (2010). Keck Interferometer nuller instrument performance, *Society of Photo-Optical Instrumentation Engineers (SPIE) Conference Series*, Vol. 7734 of *Society of Photo-Optical Instrumentation Engineers (SPIE) Conference Series*.

Heacox, W. (1980). A Optical Fiber Spectrograph Coupler, *in* A. Hewitt (ed.), *Optical and Infrared Telescopes for the 1990's*, p. 702.

Heacox, W. D. (1986). On the application of optical-fiber image scramblers to astronomical spectroscopy, *AJ* 92: 219–229.

Heacox, W. D. (1988). Wavelength-precise slit spectroscopy with optical fiber image scramblers, *in* S. C. Barden (ed.), *Fiber Optics in Astronomy*, Vol. 3 of *Astronomical Society of the Pacific Conference Series*, pp. 204–235.

Heacox, W. D. & Connes, P. (1992). Optical fibers in astronomical instruments, *A&A Rev.* 3: 169–199.

Howard, A. W., Marcy, G. W., Johnson, J. A., Fischer, D. A., Wright, J. T., Isaacson, H., Valenti, J. A., Anderson, J., Lin, D. N. C. & Ida, S. (2010). The Occurrence and Mass Distribution of Close-in Super-Earths, Neptunes, and Jupiters, *Science* 330: 653.

Hunter, T. R. & Ramsey, L. W. (1992). Scrambling properties of optical fibers and the performance of a double scrambler, *PASP* 104: 1244–1251.

Mayor, M., Pepe, F., Queloz, D., Bouchy, F., Rupprecht, G., Lo Curto, G., Avila, G., Benz, W., Bertaux, J.-L., Bonfils, X., Dall, T., Dekker, H., Delabre, B., Eckert, W., Fleury, M., Gilliotte, A., Gojak, D., Guzman, J. C., Kohler, D., Lizon, J.-L., Longinotti, A., Lovis, C., Megevand, D., Pasquini, L., Reyes, J., Sivan, J.-P., Sosnowska, D., Soto, R., Udry, S., van Kesteren, A., Weber, L. & Weilenmann, U. (2003). Setting New Standards with HARPS, *The Messenger* 114: 20–24.

Mayor, M. & Queloz, D. (1995). A jupiter-mass companion to a solar-type star, *Nature* 378: 355–359.

Mayor, M. & Udry, S. (2008). The quest for very low-mass planets, *Physica Scripta Volume T* 130(1): 014010.

Mennesson, B., Ollivier, M. & Ruilier, C. (2002). Use of single-mode waveguides to correct the optical defects of a nulling interferometer, *J. Opt. Soc. Am. A* 19(3): 596–602.

Mennesson, B., Serabyn, E., Hanot, C., Martin, S. R., Liewer, K. & Mawet, D. (2011). New Constraints on Companions and Dust within a Few AU of Vega, *ApJ* 736: 14.

Noll, R. (1976). Zernike polynomials and atmospheric turbulence, *Journal of the Optical Society of America* 66(3): 207–211.

Rabbia, Y., Gay, J., Rivet, J.-P. & Schneider, J.-L. (2003). Review of Concepts and Constraints for Achromatic Phase Shifters, *GENIE - DARWIN Workshop - Hunting for Planets*, Vol. 522 of *ESA Special Publication*.

Spronck, J. F. P., Los, J. W. N. & Pereira, S. F. (2008). Compensation and optimization of dispersion in nulling interferometry, *Society of Photo-Optical Instrumentation Engineers (SPIE) Conference Series*, Vol. 7013 of *Society of Photo-Optical Instrumentation Engineers (SPIE) Conference Series*.

Spronck, J. F. P., Los, J. W. N. & Pereira, S. F. (2009). Dispersion in nulling interferometry for exoplanet detection: experimental validation, *Journal of Optics A: Pure and Applied Optics* 11(1): 015510.

Spronck, J. F. P., Schwab, C. & Fischer, D. A. (2010). Fiber-stabilized PSF for sub-m/s Doppler precision at Lick Observatory, *Society of Photo-Optical Instrumentation Engineers (SPIE) Conference Series*, Vol. 7735 of *Society of Photo-Optical Instrumentation Engineers (SPIE) Conference Series*.

Vogt, S. S., Allen, S. L., Bigelow, B. C., Bresee, L., Brown, B., Cantrall, T., Conrad, A., Couture, M., Delaney, C., Epps, H. W., Hilyard, D., Hilyard, D. F., Horn, E., Jern, N., Kanto, D., Keane, M. J., Kibrick, R. I., Lewis, J. W., Osborne, J., Pardeilhan, G. H., Pfister, T.,

Ricketts, T., Robinson, L. B., Stover, R. J., Tucker, D., Ward, J. & Wei, M. Z. (1994). HIRES: the high-resolution echelle spectrometer on the Keck 10-m Telescope, *in* D. L. Crawford & E. R. Craine (ed.), *Society of Photo-Optical Instrumentation Engineers (SPIE) Conference Series*, Vol. 2198 of *Society of Photo-Optical Instrumentation Engineers (SPIE) Conference Series*, p. 362.

Wallner, O. & Leeb, W. (2002). Minimum length of a single-mode fiber spatial filter, *J. Opt. Soc. Am. A* 19: 2445–2448.

Wallner, O., Leeb, W. & Flatscher, R. (2003). Design of spatial and modal filters for nulling interferometers, Vol. 4838 of *Proc. SPIE*, pp. 668–679.

17

# Time and Frequency Transfer in Optical Fibers

Per Olof Hedekvist and Sven-Christian Ebenhag
*SP Technical Research Institute of Sweden, Sweden*

## 1. Introduction

The development towards more services in the digital domain, based on computers and server logs at different locations and in different networks, increases the need for high precision time indication. Even though GPS can support this with sufficient precision, many users do not have access to outdoor antennas. Furthermore, there is vulnerability in the weak radio-transmission from the satellites (NSTAC) as well as the dependence on the continuous replacement of old and outdated satellites (Chaplain). Therefore, alternative systems to support precise time are needed. Standardization of time transfer of a master clock is done for example in the IRIG system, but this one-way time transfer system do not take variations in transfer time into account, mainly because it is supposed to work on short distances (IRIG). In additional efforts to meet this request, several time and frequency transfer methods using optical fibers have been developed or are under development, using dedicated fibers (Kihara; Jefferts; Ebenhag2008; Kéfélian), dedicated capacity in existing fiber networks (Calhoun) or already existing synchronization in active fiber networks (Emardson, Ebenhag2010a). A similarity of all these techniques is the need for two-way communication to compensate for the inevitable variations of propagation time, such as variation of temperature and mechanical stress along the transmission path. A two-way connection may however be undesirable when many users are connected in one network, or when user privacy is requested. As an alternative, a one-way transmission over fiber optic wavelength division multiplexing network with detection of variation in propagation time has been presented (Ebenhag2010b, Hanssen).

The general conception of fiber optic communication is the transmission of digital data from one user to another, and through recovery of the phase variation of the bit-slots after reception, the exact time it has taken to transfer the data is of low importance. The individual packets of the data may even follow different paths with different propagation time, and still be interpreted correctly at the user end. Physical effects such as noise, dispersion and polarization dependence are important, but as long as each bit can be detected correctly, slow variations in propagation time do not affect the communication. When the fiber is used to transmit time or frequency however, the physical properties of the transmission link become very important. Even though time and frequency may appear as two faces of the same parameter, there are differences in the requirement of a transmission link. For time transfer, any variations in the delay through the link must be compensated for, either in a real time compensator or through post processing. For frequency transfer, the frequency shift caused by the rapidity of a change in the fiber delay must be handled.

During the last years of the 20th century, the development and installation of optical fiber communication systems increased rapidly, and after a few slow years, the deployment has gained new speed. All continents are connected with submarine fiber networks, and all major cities have installed fibers at least for their long distance communication. In regular optical communication however, the propagation time through the fiber is of no major concern. Slow variations are handled through clock recovery at the receiver end. Therefore, little or no efforts have been made to develop transmission links with stable net transmission time. The development of synchronous networks, e.g. following the first version of Synchronous Digital Hierarchy (SDH), was left as soon as the control system could handle asynchronous routing between different links. With the increase of the need for precise time and frequency transfer over optical fibers, the time and time variations is however of outmost importance. This chapter will be a review of the published work, covering both the transfer of low frequency and time, and the necessary techniques for accurate optical frequency transmission. Even though the similarities are apparent, the transmission of frequency and the transmission of time require completely different properties.

### 1.1 Definition of time

When the definition of time was changed in 1972 from Greenwich Mean Time (GMT) to Universal Coordinated Time (UTC) (OICM), the need to compare clocks became more imminent. While GMT is determined from observations of the sun, UTC is the addition of seconds from Cesium oscillators around the world. These devices are to be compared constantly, and since there are more than 300 oscillators on almost 60 different locations around the world (BIPM), the preferred technique has been over radio transmission, and presently utilizing satellites. As the society moves into an ever increasing request for connectivity, with the subsequent needs for verification, identification, encryption etc. many systems rely on the time signal given. To ensure the quality of time information, and to make it robust towards radio based disturbances, there have been several suggestions on how to communicate between the participating clock laboratories using alternative techniques, and with the long distances at hand, the choice of optical fibers is obvious.

One second is presently defined as the duration of 9 192 631 770 periods of the radiation corresponding to the transition between the two hyperfine levels of the ground state of the Cesium 133 atom (OICM). This definition has been official since 1967, and it does also correspond to the realization of a second. To increase the accuracy further, is there ongoing research on optical clocks. Optical clocks are defined by the output of an optical frequency standard and can offer an extremely high frequency precision and stability, exceeding the performance of the best Cesium atomic clocks. A challenge in the early years of optical clocks was to relate the stable optical frequency to a microwave frequency standard such as a Cesium atomic clock. This was solved with the realization of frequency combs from femtosecond mode-locked lasers (Paschotta). Optical clocks compared to microwave standards such as Cesium atomic clocks have some key advantages:

- There are certain atoms and ions with extremely well-defined clock transitions that promise higher accuracy and stability than the best microwave atomic clocks. The anticipated (but not yet demonstrated) relative frequency uncertainty of atomic optical clocks for long enough averaging times (possibly a few days) is of the order of $10^{-18}$ (Paschotta).

- The high optical frequencies themselves are of high importance because these allow precise clock comparisons within much shorter times. For example, a $10^{-15}$ precision can be achieved in a few seconds if the compared frequencies are in the optical range, whereas a full day would be required for microwave clocks.
- Optical signals can easily be transported over long distances using fibers whereas microwave cables are more expensive and have much higher losses.

Therefore, it is to be expected that in the near future the Cesium clock as the fundamental timing reference will be replaced with an optical clock, although it is at the moment not clear which type of optical clock would be used as such a standard. The definition of the second will then be changed to refer to an optical frequency rather than to a microwave frequency. However, even after that profound change, Cesium clocks (and other non-optical atomic clocks, such as Rubidium clocks) will continue to play an important role in technological applications as they can be simpler and more compact than optical clocks (Paschotta). In the purpose to be able to compare two optical clocks, the optical wave must be compared. To manage this, the optical link must be stable when it comes to frequency.

### 1.2 Temperature of trunk fiber

In all utilization of optical fiber, the influence of the environment must be handled, even though the solutions depend on the application, knowledge about which properties to take into account, and their magnitudes, is of equal importance. In time and frequency transfer, the surrounding temperature is the main source for variations and to estimate the size, some data is analyzed.

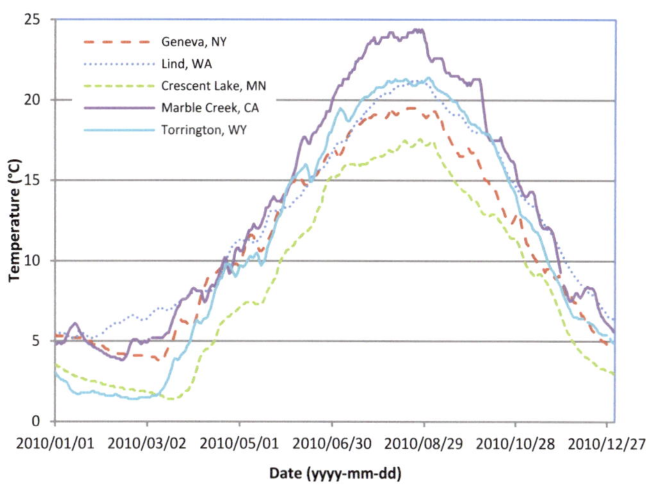


Fig. 1. Soil temperature at 40″ depth, at five US locations.

Most fiber in the terrestrial networks is buried in the ground, at a depth of about 1 - 2m. A common misconception is that this would be a stable environment with respect to temperature. Figure 1 shows the measured soil temperature at 40″ depth (approx. 1 m) at five different US locations, measured daily during 2010 (NRCS). The locations are all in the northern part of the country, with warm summers and cold winters, and represents examples of the worst conditions within the dataset with respect to temperature variations.

### 1.3 Temperature of fiber in amplifier stations

The temperature of the fiber when it is installed into a repeater station, for amplification, routing, or any other process, cannot be presumed stable unless verified. While many end nodes are in rooms with controlled temperature, most inline amplifiers reside in small buildings with less stringent environment control. Figure 2 shows the temperature detected at 9 of the power supply cards of the amplifiers along one of the routes between Borås and Stockholm, Sweden. The actual temperature is high since the sensor is located close to a heat emitter, but the variations are caused by a variation in room temperature. In these stations, the affected fiber length is short, but the variations are fast. Furthermore, if the link is equipped with dispersion compensated fiber, these spools will be affected by the local indoor temperature variation and may cause a difference in propagation time for signals in opposite directions (Ebenhag2007).

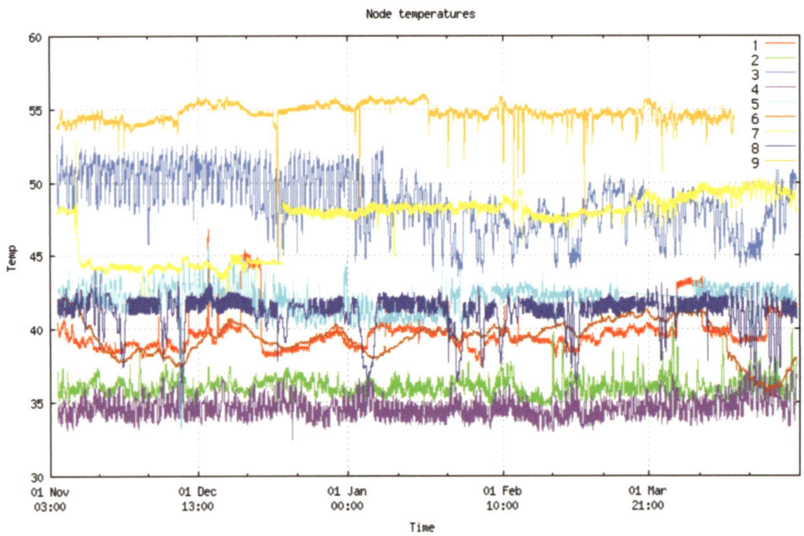


Fig. 2. Temperature measured in power supply in 9 telecom amplifier stations.

## 2. Time transfer

The unique characteristic of time, which also complicates the transmission, is that it is ever-changing, and the required information is both the actual time-of-day, (TOD) and the time that has passed since this information was created. It can for many applications be sufficient to estimate an approximate delay, and accept the variations, but for better accuracy than µs, the transmission time must be constantly estimated or measured and taken into account.
The output time $t_{out}(t)$ from an uncompensated fiber can be described by eq.(1)

$$t_{out}(t) = t_{in}(t) + \tau_{fiber}(t) \tag{1}$$

Where $t_{in}$ is the time information from the transmitter clock and $\tau_{fiber}(t)$ is the varying delay through the fiber. For increased accuracy, the equation can be elaborated to:

$$t_{out}(t) = t_{in}(t) + \tau_{fiber,0} + \tau_{fiber,det}(t) + \tau_{fiber,rnd}(t) \tag{2}$$

Where $\tau_{fiber,0}$ is the delay through the fiber at t=0, $\tau_{fiber,det}(t)$ includes any delay variations that can be determined, and $\tau_{fiber,rnd}(t)$ are the remaining, random variations of transfer delay. The main effort of any time transfer is to minimize the undetermined variations of the delay, through complementary measurements to the actual signal transfer.

### 2.1 Two-way time transfer

Two-way time transfer presumes that the system is bidirectional, and that the propagation time is equal in both directions (or at least with a deterministic and measurable difference). It can be schematically described through figure 3.

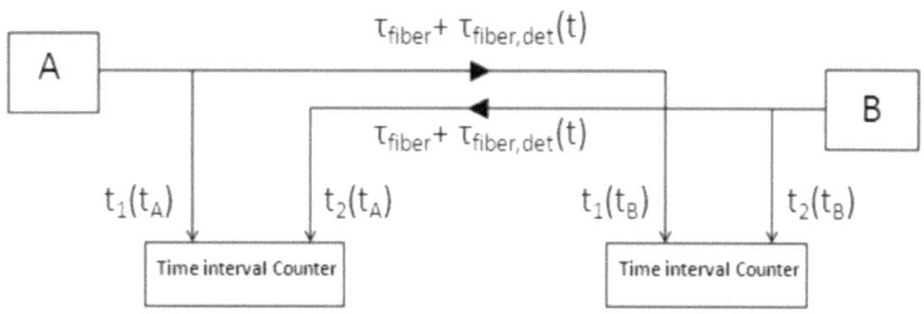


Fig. 3. Schematic system for two-way time transfer.

A well-defined signal is transmitted from point A, and the time it leaves the sender is measured with respect to the master clock A; $t_1(t_A)$. When it arrives at point B, the arrival time is measured with respect to the local clock B; $t_1(t_B)$. In addition, another well-defined signal is transmitted back from B to A, resulting in the time stamps $t_2(t_B)$ and $t_2(t_A)$. Assuming that the delay through the fiber, in both directions, is $\tau_{fiber}+ \tau_{fiber,det}(t)$, equations (3) - (5) is derived

$$t_1(t_B) = t_1(t_A) + \tau_{fiber} + \tau_{fiber,det}(t) \tag{3}$$

$$t_2(t_A) = t_2(t_B) + \tau_{fiber} + \tau_{fiber,det}(t) \tag{4}$$

$$t_1(t_B) = t_1(t_A) + t_2(t_A) - t_2(t_B) \tag{5}$$

Thus, the relationship between signal emitters A and B can be determined from measured data, and the calculations can be made at either end of the link.

Time transfer over optical fibers includes two-way transfer based on transmission on a dedicated fiber, a dedicated channel, and the piggy-back technique on existing traffic. Even though most of these techniques is based on measurements of delay, and corrections afterwards, some short distance transfer is achieved in real-time, were the output signal is corrected as the transmission characteristics change (Ebenhag2008). One-way time transfer based on two-wavelength transmission is also described in detail in this chapter.

#### 2.1.1 Time transfer over dedicated capacity

Any transmission of a signal over a dedicated capacity requires that the network owner allocate bandwidth for the connection. It could be a channel space in a wavelength division multiplexed (WDM) system, or a whole fiber. Transmitting a signal over a dedicated fiber is to some extent the simplest technology, since there are no interference from adjacent channels that has to be taken into account, and the modulation format can be chosen arbitrarily.(Smotlacha; Amemiya). There are no major differences to transmit over a dedicated channel, i.e. using one wavelength in the vicinity of others, with the exception of any constraints induced by interchannel interference.

### 2.1.2 Time transfer over shared capacity

To minimize any unnecessary bandwidth allocation, it is advantageous to operate on an active channel, where data-communication uses all, or at least most of, the available capacity. An early approached used the data transmission of SONET OC-3 at 155,52 Mbit/s and locked this repetition rate to the master 5 MHz. Furthermore, a synchronization signal was generated in the data-stream at 1 pps (Calhoun). Thus, it would be possible to share the time and frequency transfer capacity with active communication, where time transfer only need a well defined sequence once per second.

An even less bandwidth consuming technique uses an existing well defined sequence of a digital communication protocol for time transfer (Emardson, Ebenhag2010a). It can thereby be called a 'piggy-back' technique. Time transfer using this technique relies on an existing, continuous transmission of digital data. In this case, a sync sequence is detected in all locations of the two-way transmission, and the time stamp defining of the occasions is transfer separately, as a low bandwidth signal. The piggy back technique has been presented at 10 Gbit/s on the SONET and SDH protocol, where data is transmitted in 125 μs long frames and every frame start with a sequence of 192 A1 bytes, followed by 192 A2 bytes[1]. If every occurrence of a frame start sequence is detected at both transmitters and both receivers of a fiber link, and all data is sent to a computational node, the necessary timing information can be calculated for accurate time transfer. The repetitive structure of the transmission enables a simplification, where it is sufficient to detect one sequence/s, and with the knowledge of 125 μs interval between sequences, time transfer can be extracted even though the four measurements correspond two four different sequences.

## 2.2 One-way time transfer

When the surrounding temperatures of the fiber vary, it affects both the transfer time and the dispersion, which can be measured at the receiving end of the fiber. Since there is an unambiguous relationship between these two parameters, the correlation between them can be used to estimate one from the other. The measurement technique for fiber dispersion is well known (Vella) and the variation with respect to temperature has been studied previously (Hatton; Walter). This property is utilized in the one-way time transfer, and the scale coefficient for a specific fiber link must be individually characterized.

In a fully operational solution, the time from the Master clock is distributed to a Slave clock, with a precision better than what it would be in the case of a single signal was transmitted. The system is described schematically in figure 4.

At the transmitting end, a Master clock controls two lasers, and at the receiving end a slave clock makes an interpretation of the two signals, received after transmission over two wavelengths, to enhance its precision. The thin and thick lines are electrical cables and optical fibers, respectively, and the open line on top symbolizes the outdoor transmission fiber of arbitrary length while the dashed regions indicates indoor environment.

---

[1] A1 = [11110110], A2 = [01101000]

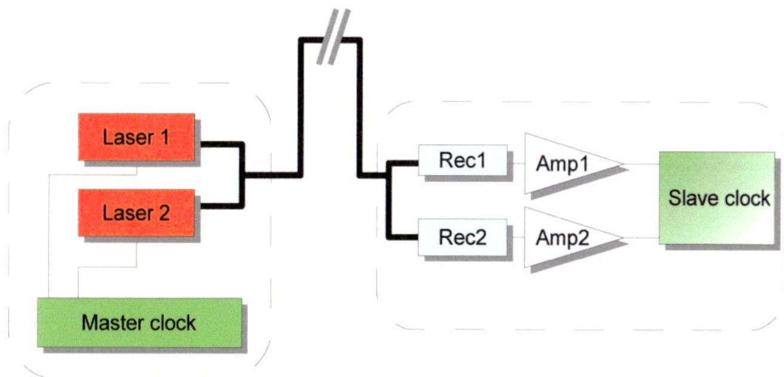


Fig. 4. Schematic system for one-way fiber based time transfer.

### 2.2.1 Theory

The theory for one-way dual wavelength optical fiber time and frequency transfer is based on the transit time τ for propagation of a single mode in a fiber (Cochrane) expressed as the group velocity for a certain distance L and the wavelength λ.

$$\tau = \frac{L}{c}\left(n - \lambda\frac{dn}{d\lambda}\right) \tag{6}$$

where $n$ is the refractive index and $c$ is the speed of light in vacuum. The transit time τ, sometimes known as the group delay time, in a fiber is thus dependent on the refractive index and the wavelength. This means that two different wavelengths will propagate at different velocity in the same fiber. A standard single mode fiber is temperature dependent, to an extent shown in previous studies (Walter), and the most important factor to include in the calculations. By calculating the derivative of the transit time with respect to temperature, both wavelength and refractive index will be taken into account as follows:

$$\left.\frac{d\tau}{dT}\right|_{\lambda_N} = \frac{1}{c}\left(\frac{dL}{dT}\left(n - \lambda\frac{dn}{d\lambda}\right) + L\left(\frac{dn}{dT} - \lambda\frac{d^2n}{dTd\lambda}\right)\right) \qquad \text{N=1,2} \tag{7}$$

The variation in transit time as a function of temperature can thus be calculated where $\lambda_N$ N=1,2; represents the two wavelengths. The equations for the two wavelengths are subtracted from each other, resulting in:

$$\left.\frac{d\tau}{dT}\right|_{\lambda_1-\lambda_2} = \frac{1}{c}\left(\frac{dL}{dT}\left(\left(n_{\lambda_1} - n_{\lambda_2}\right) + \lambda_2\frac{dn_{\lambda_2}}{d\lambda_2} - \lambda_1\frac{dn_{\lambda_1}}{d\lambda_1}\right) + L\frac{d}{dT}\left(\left(n_{\lambda_1} - n_{\lambda_2}\right) + \lambda_2\frac{dn_{\lambda_2}}{d\lambda_2} - \lambda_1\frac{dn_{\lambda_1}}{d\lambda_1}\right)\right) \tag{8}$$

This expression shows how the refractive indices of the two wavelengths are influenced by temperature, and based on this the variations in propagation time can be calculated. The time transfer technique uses the property that the variations are different, but correlated, which also is supported by experimental results later on.

### 2.2.2 Numerical simulations

The difference in transit time through the fiber will, as shown in eq (8) depend on the variation of length, L, and the variation in refractive index, n. Both these effects will affect the chromatic dispersion of the fiber, but through different properties.

#### 2.2.2.1 Variations in refractive index

The refractive index of the fiber can be described by eq. (9), called the Sellmeier equation (Sellmeier; Ghosh)

$$n^2 = A + \frac{B}{1-C/\lambda^2} + \frac{D}{1-E/\lambda^2} \tag{9}$$

Where λ is the wavelength in μm and the Sellmeier coefficients A, B, C, D and E have been empirically fitted with respect to temperature, T, for different glasses. Using the data for fused Silica (Ghosh), results in:

| Sellmeier coefficient | Fitted constants ($SiO_2$) |
|---|---|
| A | 6,90754*10$^{-6}$T + 1,31552 |
| B | 2,35835*10$^{-5}$T + 0,788404 |
| C | 5,84758*10$^{-7}$T + 1,10199*10$^{-2}$ |
| D | 5,48368*10$^{-7}$T + 0,91326 |
| E | 100 |

Table 1. Empirically fitted values for Sellmeier coefficients

From these equations, the material dispersion can be calculated as[2]:

$$D_M(\lambda) = \frac{1}{cn}\left[-\frac{4}{\lambda^5}\left\{\frac{BC^2}{(1-C/\lambda^2)^3} + \frac{DE^2}{(1-E/\lambda^2)^3}\right\} + \lambda\left(\frac{dn}{d\lambda}\right)^2 + 3n\frac{dn}{d\lambda}\right] \tag{10}$$

where

$$\frac{dn}{d\lambda} = -\frac{1}{n\lambda^3}\left(\frac{BC}{(1-C/\lambda^2)^2} + \frac{DE}{(1-E/\lambda^2)^2}\right) \tag{11}$$

Using these parameters, the material dispersion of $SiO_2$ is calculated and shown in figure 5. It may vary slightly in communication fibers where the silica is doped with small amount of other substances. Nevertheless the overall behavior is comparable.

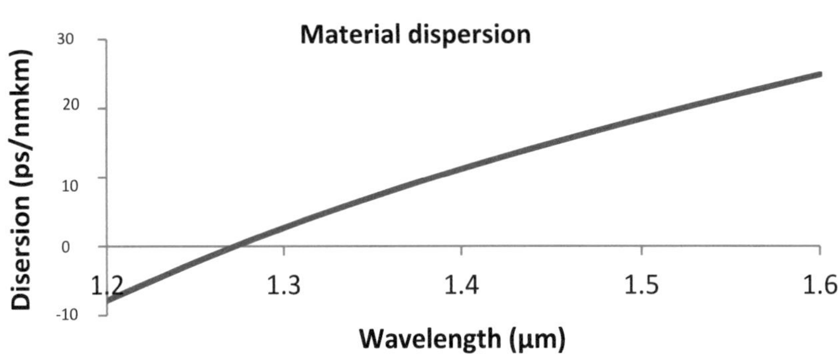


Fig. 5. Calculation of material dispersion in Fused Silica at 20°C.

[2] This equation is corrected with respect to the reference, where the left side of the equation begins with a "-".

From the equations (6)-(11), it is possible to estimate the amount of propagation time variations with respect to temperature. Assuming a fiber where material dispersion is dominant (as is the case in standard single mode fiber), at a length of 20 km and measurement at 1530 nm and 1560 nm, the result is shown in figure 6. The slope of the calculated dispersion is -0,0016 ps/nmkm°C, which is comparable to previously reported results -0,0025 ps/nmkm°C for NZDSF (non-zero dispersion shifted fiber) and -0,0038 ps/nmkm°C for large core fiber (Walter).

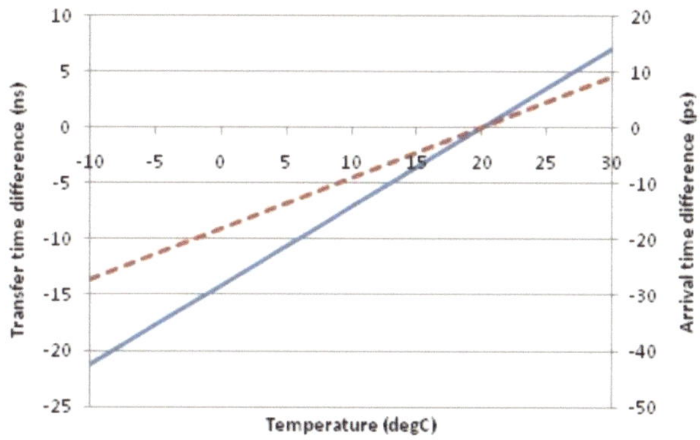


Fig. 6. Temperature dependence of transfer time (solid blue, left axis) and arrival time difference (dashed red, right axis).

The solid curve (left axis) shows the transfer time for a signal at 1530 nm, and the dashed curve shows the arrival time difference for two signals at 1530 nm and 1560 nm. Both curves are normalized with respect to the value at 20°C, and it is apparent that the propagation time within a single, 20 km long fiber varies with almost 30 ns when affected by 40°C temperature difference. The calculations also suggests that this variation can be detected and compensated for, using transmission at two wavelengths and a measurement system that can measure time variations on ps level with sufficient precision.

**2.2.2.2 Variations of length**

This evaluation assumes that the cabling or mounting will stretch the fiber at increasing temperature, however leaving the volume intact. The variations in dimensions of the glass are assumed to be negligible. If the core of the fiber is modelled as a glass cylinder, of length *L* and diameter *d*, a geometrical approach gives that the variation in temperature will change the length with $\Delta L(T\text{-}T_0)$ and the diameter with $\Delta d(T\text{-}T_0)$, such that

$$\frac{\Delta d(T-T_0)}{d} = -\frac{\Delta L(T-T_0)}{2L} \tag{12}$$

where T is the temperature and $T_0$ is the reference temperature.
This change in diameter will change the dispersion according to the variation in waveguide dispersion (Gloge; Keiser):

$$D_W(\lambda) = -\frac{n_2\Delta}{c\lambda} V \frac{d^2(Vb)}{dV^2} \tag{13}$$

where $n_2$ is the refractive of the cladding and $\Delta$ is the relative difference of refractive index in the core and in the cladding. V and b are the normalized frequency and the normalized propagation constant, respectively, and can be found through:

$$V = ka\sqrt{n_1^2 - n_2^2} \cong kan_2\sqrt{2\Delta} \quad (14)$$

$$b = \frac{(\beta/k)^2 - n^2}{n_1^2 - n_2^2} \quad (15)$$

where $k$ is the free-space propagation constant, $\beta$ is the propagation constant and $a = d/2$ is the fiber core radius. From these equations, it is apparent that fibers with notable waveguide dispersion, e.g. dispersion shifted fibers, dispersion compensating fibers etc, will have different response to a change in diameter $d$, than standard fibers where material dispersion is dominant. However, this response must be evaluated for each fiber design, since the term $V(d^2(Vb)/dV^2)$ is between 0 and 1,2 with a maximum at $V \approx 1{,}2$. These equations show nevertheless that the system of detecting a variation in propagation time through a fiber with substantial waveguide dispersion is possible, but must be optimized for the actual fiber parameters.

### 2.2.3 Experimental setup

The experimental setup for the verification of the proposed time and frequency transfer technique is shown in figure 7. Two lasers at wavelengths 1530 nm and 1560 nm are directly modulated by a 10MHz reference oscillator and the light is launched into the SMF through a 50/50 power combiner. The reference oscillator is a frequency stabilized H-maser used as Master clock. In the experiment, the oscillator is also used as reference to the measurement equipment, connected as indicated by the lower line, in order to evaluate the technique. Furthermore, to increase sensitivity, the signal from the oscillator is connected to the LO-ports of the two double balanced mixers at the output of the transmitted signal paths. The equipment within the dashed frame is held within a controlled environment, and the spools of SMF are placed outdoors together with a temperature sensor for monitoring and comparison with transfer time variations. The total sum of fiber length is 12,761.5 m, including 187.6 m of transfer fiber between the lab and the outdoor fiber spools. At the receiving end, the two wavelengths are separated in a 50/50 power splitter, filtered in optical band-pass filters and

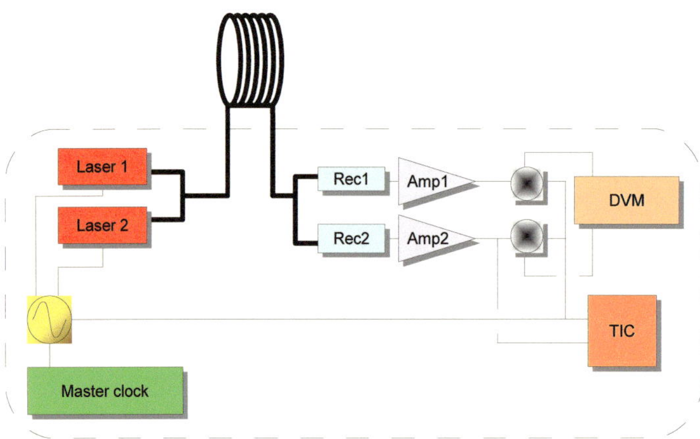


Fig. 7. Experimental setup. Rec1 and Rec2 include optical pre-amplification, optical band pass filter, photodiode and electrical trans-impedance amplifier. Amp1 and Amp2 are electrical amplifiers, DVM digital voltmeter and TIC is time-interval counter. Thin lines symbolize electrical wires and thick lines optical fibers.

detected in two 10 Gb/s p-i-n receivers. The signals are amplified and connected to the RF ports of two double balanced mixers. One of the signals is also divided and connected to the reference time interval counter (TIC), which measures the total propagation time between the transmitter and the receiver. The output of the TIC is interpreted as the precision of an uncompensated one-way time and frequency transmission. By measuring the voltages of the two output ports of the mixers in a digital voltmeter (DVM), a correction signal is achieved and can be used for a real-time delay control of the uncompensated signal.

#### 2.2.4 Experimental results

In Figure 8, the result from six days of measurement is plotted over time with the one-way method (blue, left scale), and the estimated delay from the two-wavelength time difference (red, left scale). The estimated transfer time $T_{est}$ is made through empirical fitting, and follows the equation:

$$T_{est} = F_1 \arccos(I_1 - I_2) + F_2 \qquad (16)$$

where $I_1$ and $I_2$ are the output voltages from the two mixers, normalized with the maximum level of each output. The numerical values of the fitting parameters, $F_1$ and $F_2$ resulting in the lowest residual error (rms) are shown in table 2.

| Compensation parameter | Fitted constant |
|---|---|
| $F_1$ | $1{,}58 \cdot 10^{-8}$ s |
| $F_2$ | $1{,}71 \cdot 10^{-9}$ s |

Table 2. Empirically fitted compensation constants.

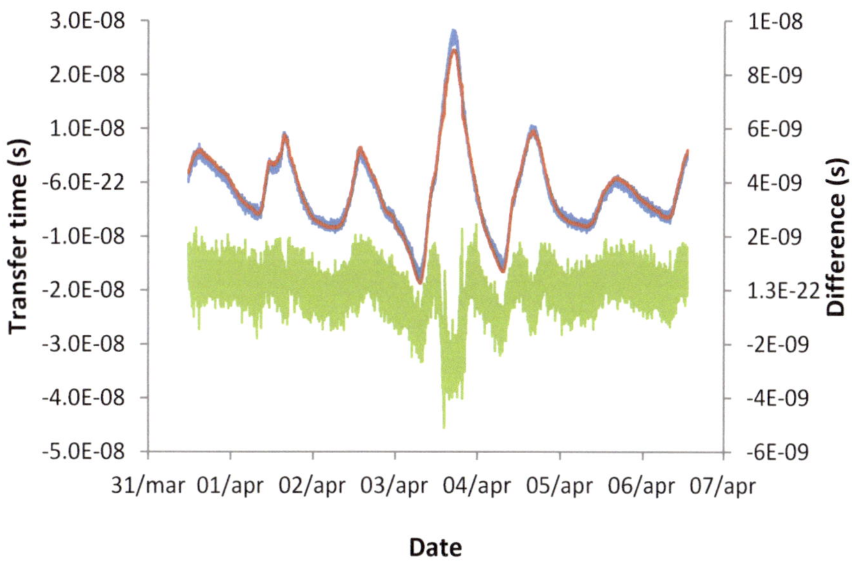


Fig. 8. Measured variations during six days. The uncompensated one-way transfer time (blue, left axis) is compared with the compensation signal from two-wavelength difference measurement (red, left axis). The residual error (green, right axis) is an order of magnitude lower.

The difference between the measured time delay and the compensation signal is shown in the final curve (green, right axis). The stability of the output signal is thereby enhanced from 7,7 ns rms to 0,9 ns rms.

## 3. Frequency transfer

While time transfer stability is compensated for the actual difference in optical path length, the frequency transfer is sensitive for how fast the delay changes. In comparison to equation (1), the output frequency of an uncompensated fiber is described by:

$$f_{out}(t) = f_{in}(t) + \frac{d\varphi(t)}{dt} \tag{17}$$

where $f_{in}(t)$ and $f_{out}(t)$ are the momentaneous input and output frequencies, respectively, and $\tau(t)$ is the time varying delay through the fiber. The derivative $d\varphi(t)/dt$ arises from the change in $\tau(t)$ with respect to the period of the microwave frequency, such that

$$\frac{d\varphi(t)}{dt} = 2\pi f_{in}(t)\frac{d\tau(t)}{dt} \tag{18}$$

### 3.1 Optical transfer of microwave frequency

When the fiber link is used to transfer a microwave frequency modulated on top of an optical carrier, this variation will only be notable over long distances, or if the fiber is installed in harsh environment (open air, sunlit roofs etc.). A two-way frequency transfer will then schematically be implemented as shown in figure 9. The control equipment adjusts the input signal to the phase modulator of the transmitted and returned signal, such that the total phase variation after a round-trip in the fiber link is cancelled out.

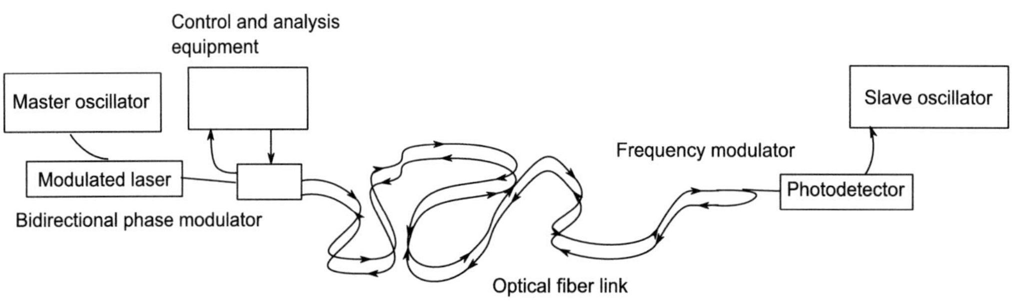


Fig. 9. Schematic frequency transfer in microwave domain.

### 3.2 Optical comb

One key invention for optical frequency transfer, as well as for other techniques, is the optical comb (KVA). By generating short optical pulses with a constant repetition rate, the corresponding spectrum will consist of a comb of equidistant peaks. T. Hänsch and J. Hall managed in to broaden this spectrum to exceed one octave of optical tones, which enabled new measurements (Hall; Holzwarth).

Figure 10 illustrates the comb structure of the optical spectrum. If one of the lowest frequencies in the spectrum, $\nu_1$, is doubled, it will create a new frequency, $2\nu_1$, close to one of the highest in the comb, $\nu_d$. Since the difference between the two frequencies is known,

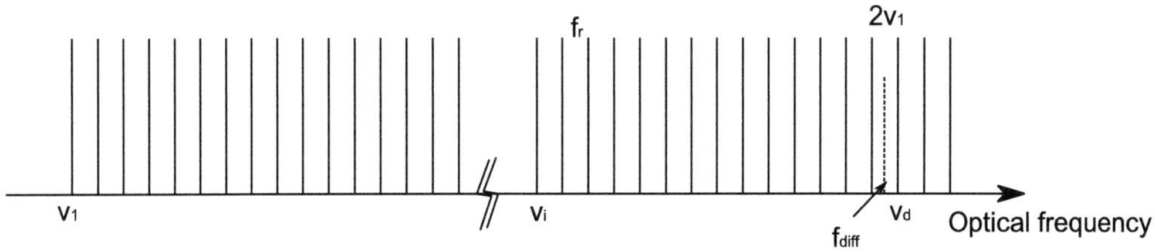


Fig. 10. Schematic spectrum of optical comb spanning one octave.

every optical frequency in the comb can be determined at comparable accuracy of a microwave frequency. With the parameters $f_r$ and $f_{diff}$ describing the repetition frequency of the pulses creating the comb, and the measured difference frequency between $2\nu_1$ and $\nu_d$, respectively, equations (19) and (20) results in the determination of an arbitrary optical frequency $\nu_i$.

$$\vartheta_i = \vartheta_1 + N_i f_r \tag{19}$$

$$\vartheta_1 = N_d f_r + f_{diff} \tag{20}$$

### 3.3 Optical frequency transfer

To be able to compare two optical clocks at different locations, optical frequency transfer over fiber is the only option. Figure 11 shows the basic technique, but does not cover all details. It can be described as follows. The optical clock A emits a wavelength corresponding to the atom or ion in use, usually not within the telecommunication bands. Therefore, an ultra-stable wavelength at approximately 1550 nm is also created in lab A. Through an optical comb, the frequency relation between these two wavelengths can be determined.

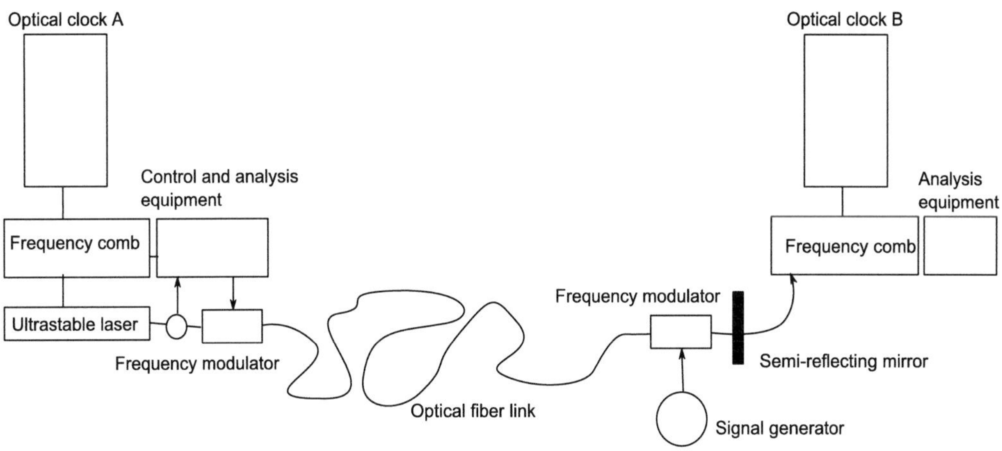


Fig. 11. Schematic setup for optical frequency transfer.

The light from the ultra-stable laser is launched through an optical frequency modulator (usually an acousto-optical modulator) and transferred through the fiber to lab B, where another frequency modulator is passed. A semi-reflecting mirror (often the Fresnel-reflection of the glass-air interface is sufficient) lets the light return along the same path. After the return to lab A, the received signal is compared with the transmitted, and the

modulation is adjusted to counteract any phase variations induced through the fiber. The modulator in lab B is used to offset the return signal, whereas scattering effects in the fiber will deteriorate the signal when sent at the same wavelength in both directions.
Finally, the light entering lab B is stable with respect to variations in the fiber, and can be compared with the light emitted from Optical clock B, through another optical comb. Since all this comparison must be performed through analog signal interference in the optical domain, the ultra-stable frequency transfer must be performed in real-time, where any perturbation in the fiber must be corrected on the fly. It is also significant that where a microwave frequency can be transferred between two labs through a fiber pair, with the addition of an increased uncertainty, optical frequency transfer must be performed through a bi-directional two-way transfer in a single fiber.
Successful experiments with optical frequency transfer has been reported from several groups, bridging distances up to 480 km and connecting labs with optical clocks. (Jiang; Foreman; Terra).

## 4. Conclusion

In conclusion, fiber optics is shown to be an advantageous channel for precise time and frequency transfer, both for comparing next generation optical clocks and to support the emerging users of network time with high precision. For long baseline comparisons, there may however be a need for new components and connection schemes, and the development towards better and more precise links is in its beginning. The ultimate target to reach trans-Atlantic and trans-Pacific distances will require much future effort, however definitely achievable.

## 5. References

Amemiya, M.; Imae, M.; Fujii, Y.; Suzuyama, T. and Ohshima, S. (2005) "Time and Frequency Transfer and Dissemination Methods Using Optical Fiber Network", *Topical Meeting on Precise Time and Time Interval, PTTI'05*, paper 99 2005

BIPM (2010), "BIPM Annual Report on Time Activities", Bureau International Des Poids et Mesures, Vol 5, 2010. *http://www.bipm.org/en/scientific/tai/time_ar2010.html.*

Calhoun, M.; Kuhnle, P.; Sydnor, R.; Stein, S. & Gifford, A. (1996) "Precision Time and Frequency Transfer Utilizing SONET OC-3", Proceedings of the 28$^{th}$ Topical Meeting on Precise Time and Time Interval, pp 339 - 348, Dec 3-5, Reston, Va, 1996.

Chaplain, C.T. "Global Positioning System, Significant Challenges in Sustaining and Upgrading Widely Used Capabilities", United States Government Accountability Office report GAO-09-670T, 2009

Cochrane, K.; Bailey, J. E.; Lake P. & Carlson, A. (2001) "Wavelength-dependent measurements of optical-fiber transit time, material dispersion, and attenuation" *Applied Optics*, Vol 40, No 1, January 2001

Ebenhag, S.-C.; Jarlemark, P.; Hedekvist, P.O. & Emardson, R. (2007), " Time Transfer Using an Asynchronous Computer Network: an Analysis of Error Sources"*European Frequency and Time Forum*, EFTF'07, 2007.

Ebenhag, S.C.; Hedekvist, P.O.; Rieck, C.; Skoogh, H.; Jarlemark, P. & Jaldehag, K. (2008) "Evaluation of Output Phase Stability in a Fiber Optic Two-Way Frequency

Distribution System", Proceedings of the Precise Time and Time Interval Meeting 2008. Paper 11 (2008).

Ebenhag, S.C.; Hedekvist, P.O.; Jarlemark, P.; Emardson, R.; Jaldehag, K.; Rieck C. and Löthberg P. (2010a) "Measurements and Error Sources in Time Transfer Using Asynchronous Fiber Network", *IEEE Trans. Instr. Meas.*, vol. 59, pp. 1918 - 1924, (2010)

Ebenhag, S.C. and Hedekvist, P.O. (2010b) "Fiber Based One-way Time Transfer with Enhanced Accuracy", *European Frequency and Time Forum EFTF'10*, 2010

Emardson, R.; Hedekvist, P.O.; Nilsson, M.; Ebenhag, S.C.; Jaldehag, K.; Jarlemark, P.; Rieck, C.; Johansson, J.; Pendrill, L.; Löthberg P. and Nilsson, H. (2008) "Time Transfer by Passive Listening over a 10 Gb/s Optical Fiber", IEEE Trans. Instr. Meas.,vol. 57, pp. 2495 - 2501 (2008)

Foreman, S.M.; Ludlow, A.D.; de Miranda, M.H.G.; Stalnaker, J.E.; Diddams, S.A. & Ye, J.. (2007), "Coherent optical phase transfer over a 32-km fiber with 1 s instability <10(-17). *Physical Review Letters*. 2007 Oct 12;99(15):153601. Epub 2007 Oct 9

Ghosh, G.; Endo M. and Iwasaki, T.(1994) "Temperature-Dependent Sellmeier Coefficients and Chromatic Dispersions for Some Optical Fiber Glasses", *J. Lightwave Technol.* Vol. 12, pp 1338 - 1342, 1994.

Gloge, D. (1971) "Dispersion in weakly guided fibers", Appl. Opt., vol 10, pp 2442 - 2445, Nov. 1971.

Hall, J.L.; Ye, J.; Diddams, S.A.; Ma, L.-S.; Cundiff S.T. & Jones, D.J.(2001) "Ultrasensitive spectroscopy, the ultrastable lasers, the ultrafast lasers, and the seriously nonlinear fiber: a new alliance for physics and metrology", *IEEE. J. Quant. Electr.* 37, 1482, 2001

Hanssen, J.; Crane, S. & Ekstrom, C. (2011) "One-way Temperature Compensated Fiber Link", *European Frequency and Time Forum EFTF'11*, 2011.

Hatton, W.H. & Nishimura, M,, "Temperature dependence of chromatic dispersion in single mode fiber", *Journal of Lightwave Technology*, Vol LT-4, No10, October 1

Holzwarth, R.; Zimmermann, M.; Udem, Th. & Hänsch, T.W. (2001)"Optical clockworks and the measurement of laser frequencies with a mode-locked frequency comb" *IEEE J. Quant. Electr.* 37, 1493, 2001

IRIG (2004), "IRIG Serial Time Code Formats", Timing Committee, Telecommunications and Timing Group, Range Commanders Council, IRIG Standard 200-04.

Jefferts, S.R.; Weiss, M.; Levine, J.; Dilla, S. & Parker, T. E. (1996) "Two-Way Time Transfer through SDH and Sonet Systems", *European Frequency and Time Forum EFTF'96*, 1996

Jiang, H.; Kéfélian, F.; Crane, S.; Lopez, O.; Lours, M.; Millo, J.; Holleville, D.; Lemonde, P.; Chardonnet, Ch.; Amy-Klein, A. & G. Santarelli, G. (2008)" Long-distance frequency transfer over an urban fiber link using optical phase stabilization" *J. OSA B*, Vol. 25, Issue 12, pp. 2029-2035 (2008)

Kéfélian, F.; Jiang, H.; Lemonde P. & Santarelli, G. (2009): "Ultralow-frequency-noise stabilization of a laser by locking to an optical fiber-delay line", *Optics Lett.*, Vol 34, No 7, 2009

Keiser, G. (1991), "Optical Fiber Communication, 2nd ed", McGraw-Hill, 1991

Kihara, M.; Imaoka, A.; Imae, M. & Imamura, K. (2001) "Two-Way Time Transfer through 2.4 Gb/s Optical SDH Systems", *IEEE Trans. Instr. Meas.*, vol. 50, pp. 709-715, 2001

KVA, (2005), "What limits the measurable?" and "Quantum-mechanical theory of optical coherence - Laser-based precision spectroscopy and optical frequency comb techniques", Royal Swedish Academy of Sciences (Kungliga Vetenskapsakademin), Supplementary information on the Nobel Prize in Physics 2005, *http://nobelprize.org/nobel_prizes/physics/laureates/2005/info.pdf* and *http://nobelprize.org/nobel_prizes/physics/laureates/2005/phyadv05.pdf*

NCRS, National Resources Conservation Service, Data available at *http://www.wcc.nrcs.usda.gov/scan/*

NSTAC, "NSTAC Report to the President on Commercial Communications Reliance on the Global Positioning System (GPS)", National Security Telecommunications Advisory Committee (NSTAC) Publications, Feb. 28, 2008

OICM "The International System of Units (SI)", Report from Organisation Intergouvernementale de la Convention de Metre, 8th ed. 2006

Paschotta, R.; (2008), "Encyclopedia of Laser Physics and Technology Volume 1 and 2", *Wiley- VCH Verlag GmbH &Co, Weinheim*

Sellmeier, W. (1871) "Zur Erklärung der abnormen Farbenfolge im Spectrum einiger Substanzen", *Annalen der Physik und Chemie* 219, 272-282, 1871.

Smotlacha, V.; Kuna, A. & Mache, W. (2010) "Time Transfer Using Fiber Links", *European Frequency and Time Forum, EFTF'10,* 2010

Terra, O.; Grosche, G. & Schnatz, H.: "Brillouin amplification in phase coherent transfer of optical frequencies over 480 km fiber".*Opt. Express* 18, 16102-16111 (2010).

Vella, P.J.; Garrel-Jones, P.M. & Lowe, R.S. (1985), "Measuring Chromatic Dispersion of Fibers", US Patent 4551019, Nov. 5, 1985.

Walter, A. & Schaefer, G. (2002), "Chromatic Dispersion Variations in Ultra-Long-haul Transmission Systems Arising from Seasonal Soil Temperature Variations", *Conference on Optical Fiber Communication*, OFC'02, 2002.

# 18

# Applications of the Planar Fiber Optic Chip

Brooke M. Beam[1], Jennifer L. Burnett[2],
Nathan A. Webster[2] and Sergio B. Mendes[2]
[1]*University of Arizona*
[2]*University of Louisville*
*USA*

## 1. Introduction

The planar fiber-optic chip (FOC) technology combines the sensitivity of an attenuated total reflection (ATR) element with the ease of use of fiber-optic based spectrometers and light sources to create an improved platform for spectroscopic analysis of molecular adsorbates. A multi-mode optical fiber mounted in a V-groove block was side-polished to create a planar platform that allows access to the evanescent field escaping from the fiber core and has been previously applied to absorbance and spectroelectrochemical measurements of molecular thin-films. Light generated in a surface-confined thin molecular film can be back-coupled into the FOC platform when the conditions for light propagation within the waveguide are met. In this chapter the current applications of the FOC platform will be presented including spectroelectrochemical measurements, fluorescence detection of a bioassay, a broadband fiber optic light source, and Raman interrogation of molecular adsorbates.

In recent years, both planar waveguide-based and fiber-optic-based chemical sensors and biosensors have been developed in an attempt to meet the need for miniature, multifunctional, and sensitive sensor platforms. (Bradshaw *et al.*, 2005; Kuswandi *et al.*, 2001; Monk & Walt, 2004; Plowman *et al.*, 1998; Potyrailo *et al.*, 1998; Reichert, 1989; Tien, 1971; Wolfbeis, 2006) The benefits of fiber optic platforms have led several manufactures of analytical instrumentation to develop inexpensive fiber compatible equipment such as readily available fiber-coupled light sources and spectrometers with standard distal end fiber coupling schemes. Fiber coupled sensing architectures, utilizing the fiber as the optical signal transduction platform, have been developed for various geometries including distal end, tapered, de-clad cylindrical core, U-shape de-clad cylindrical core, and biconical tapered optical fibers. (Leung *et al.*, 2007; McDonagh *et al.*, 2008) Simple distal end fiber optic sensors are commercially available where the exposed core on a cleaved and polished end of a fiber is used as the sensing platform. However, the distal end geometry is limited by low sensitivity due to the small interaction area, analogous to the single-pass transmission absorbance measurement. A second more fragile distal end sensor geometry uses a tapered fiber where the fiber core is etched with HF into a point. The tapered fiber increases the evanescent field amplitude and penetration depth, thus increasing the sensitivity of the platform. Tapered fiber optic sensors are primarily used as fluorescence detection platforms in biochemical and clinical applications. (Anderson *et al.*, 1993; Anderson *et al.*, 1994; Anderson *et al.*, 1994; Golden *et al.*, 1992; Grant & Glass, 1997; Maragos & Thompson, 1999; Thompson & Maragos, 1996; Wiejata *et al.*, 2003; Zhou *et al.*, 1997)

Previous studies, which have taken advantage of the convenience of fiber coupled instrumentation and the increased sensitivity of the total internal reflection geometry, have used a fiber optic with the cladding removed to create a sensing element around the cylindrical fiber core. The exposed core region serves as an ATR element that can be used for absorbance measurements to detect volatile organic compounds (Blair *et al.*, 1997), probe dye solutions (Ruddy *et al.*, 1990), monitor methane gas (Tai *et al.*, 1987) and ammonium ion (Malins *et al.*, 1998) concentrations, and determine solution pH using indicator doped sol-gel coatings (Gupta & Sharma, 1997; MacCraith, 1993), or an indicator doped polymer film. (Egami *et al.*, 1996) Several investigators have worked to further increase the sensitivity of the de-clad cylindrical core fiber optic sensors by tapering the fiber optic (Guo & Albin, 2003; Gupta *et al.*, 1994; Mackenzie & Payne, 1990; Mignani *et al.*, 1998) or bending the sensing region (i.e. into a U-shape). (Khijwania & Gupta, 1998; Khijwania & Gupta, 2000) Fiber optic sensors using the tapered fiber geometry include a humidity (Bariain *et al.*, 2000), temperature (Diaz-Herrera *et al.*, 2004), hydrogen gas (Villatoro *et al.*, 2005), and bovine serum albumin sensors. (Leung *et al.*, 2007; Preejith *et al.*, 2003) U-shaped fiber optic sensors have been used to detect humidity (Gupta & Ratnanjali, 2001), pH (Gupta & Sharma, 2002), and ammonium ion concentrations. (Potyrailo & Hieftje, 1998) Such fiber optic sensor architectures employ signal transduction through a fragile cylindrical probing interface, which can be problematic for several applications where a robust platform or planar deposition technologies are required. Clearly, a supported planar interface would be advantageous for using standard planar deposition technologies such as Langmuir-Blodgett (LB)-deposited thin-films (Doherty *et al.*, 2005; Flora *et al.*, 2005) and planar supported lipid bilayers. (McBee *et al.*, 2006) In addition, due to its more robust supported platform, a planar design would be amenable for integration into microfluidic systems and sensor arrays. The FOC platform is schematically shown in Figure 1.

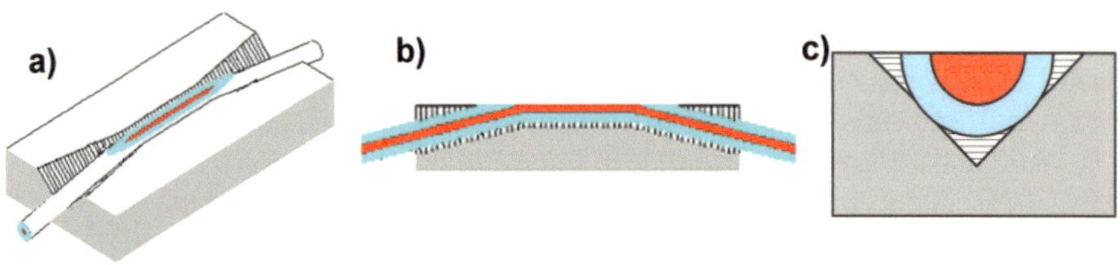


Fig. 1. Fiber Optic Chip (FOC) schematic of a side polished fiber mounted in a V-groove where red represents the exposed fiber core sensing platform. a) Top down view; b) Side view; c) Cross section. Figure modified from Beam *et al.*, 2007 and Beam *et al.*, 2009.

### 1.1 FOC manufacture

The FOC is a D-shaped, side polished fiber optic platform with access to the evanescent field escaping from the fiber core. Fabrication of the FOC begins with stripping the jacket off of a small central section, 2 to 4 cm, of an optical fiber to expose the cladding. The optical fibers used for this work are a 50 μm core/125 μm cladding multimode, step-index optical fiber (Thorlabs AFS50/125Y), with 0.22 numerical aperture (NA). The stripped section of the optical fiber is then mounted in a V-groove substrate using a two-part epoxy; the V-groove acts as a platform for spectroscopic investigation as well as supports the fragile fiber during the polishing proceedure. Prior to mounting the fiber, the edges of the V-groove block must be polished to a 2° taper. Using a custom built assembly jig to keep tension on the fiber during hotplate curing, the fiber is laid into the V-groove and optical grade epoxy (Epotek 301) is applied liberally to ensure permanent immobilization of the fiber. The first generation FOC platforms were produced by side-polishing an optical fiber in a glass V-groove mount, but subsequent improvements on the FOC manufacturing process include replacing the glass V-groove with a customized etched silicon wafer support, improved polishing processing, and finally generating arrays of side polished fibers (Figure 2).

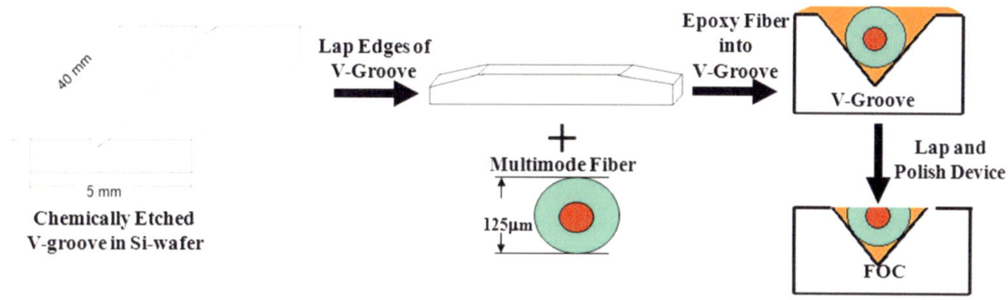


Fig. 2. Schematic for construction of FOC devices.

Initially glass V-groove mounts were purchased from Mindrum with dimensions of 40 mm long, 2 mm wide, and 1.33 mm tall; however, there was a limited supply and the glass V-grooves were not uniform requiring careful charaterization of each piece prior to use. Later V-groove mounts were produced using a chemical etching process (Kendall, 1979) to create a channel in a Si-wafer. Due to the crystalline structure of silicon, the resultant channel has two sloping walls forming a V shape. Creating the V-groove begins with a 500 µm thick Si-wafer with a minimum of a 1 µm oxidized layer. A layer of hexamethyldisilazane (HMDS) primer followed by a layer of photo-resist (Shipley 1813) is spin-coated on the wafer and cured on a hot plate. A slotted mask is placed on top of the wafer using a mask aligner (Süss MicroTec). The slots are of the desired width for the eventual V-groove. The masked wafer is then exposed to UV light for 7 seconds. The wafer is then placed in developer (Microdeposit MF-319) leaving photo-resist in the areas that the mask covered and exposing the wafer surface in the slot formation. The wafer is covered in buffered oxide etchant (BOE) to erode the oxide layer of the exposed wafer, etching the masked pattern into the oxide layer. The BOE will remove 100nm/min of the oxidized layer, so at minimum the wafer should remain in the BOE for a period of 10 minutes. The remaining photo-resist is then removed using a solvent rinse. Finally, the V-grooves are formed through chemical etch in 45% KOH, which is set on a magnetic stirrer and heated to 55° C. It should be noted that the etching rate of the KOH increases with temperature. The KOH etches the silicon at a much faster rate than the $SiO_2$ creating grooves only in the areas without an oxidized layer. The angle between the sloping walls and the face of the substrate, 54.74°, is set by the silicon crystalline structure. Etching will terminate once the (1,1,1) plane is reached; therefore, the depth of the V-groove is pre-determined by the width of the lines in the mask. The resultant V-groove is approximately 240 µm wide and 170 µm deep. Once the chemical etching of the V-grooves is complete, the Si-wafer is diced into approximately 40 mm by 5 mm long strips with a V-groove running longwise through the center of each or the wafer can remain intact to create a FOC array base structure.

Side polishing the fiber to create the D-shaped geometrey of the FOC is achieved using a two part lapping process, where the cladding is slowly polished away exposing the core of the multimode fiber. All lapping and polishing steps are performed on a Lapmaseter model 12 with a cast iron lapping plate covered with a polyurethane pad. The FOC device is mounted onto a custom machined spindle carrier with brass sleeve bearings to hinder parallax motion. If this wobble is not corrected for, the FOC will not be polished evenly leading to the outer edges eroding at a faster rate. First, a coarse grit slurry composed of 1% 1-μm alumina powder (MetMaster SF-RF-1P) is used to lap the device at a rate of about 20 rpm. When the measured width of the exposed cladding is approximately 115 μm, the slurry is changed to a fine grit polishing solution composed of 1% 0.5-μm cerium oxide (Logitech OCON 260). Polishing continues until the center of the fiber is reached, measuring approximately 125 μm across the width of the exposed cladding. Once lapping is complete, FC-PC connectors are attached to the optical fiber ends on both sides of the device.

The sensitivity of an FOC device is intrinsically dependent upon the specific geometry of the side-polished fiber. The fundamental limit of the elliptical flattened area is determined by the structure of the V-groove mount, evenly mounting the fiber in epoxy, and the efficiency of exposure of the fiber core through the polishing process. The depth to which the fiber has been polished is determined by measuring the width of exposed fiber. The width of the fiber is monitored using a standard optical microscope (VWR Vista Vision) and periodically measuring from the boundary of the cladding and epoxy on either side of the fiber. Measurements are taken periodically throughout the lapping process to ensure the fiber is polishing evenly.

### 1.2 Broadband absorbance measurements on the FOC platform

The initial application of the FOC platform was to examine the broadband absorbance characteristics of molecular thin-films. A schematic of the experimental set-up for general absorbance measurements on the FOC is shown in Figure 3a. The thin-film absorbance sensitivity enhancement of the FOC device was evaluated and compared to previously existing technologies. The sensitivity factor ($S$) of a device, defined in Equation 1, is a scaling factor of the device absorbance ($A_{FOC}$) with respect to the conventional absorbance measured ($A_{transmission}$) in direct transmission and used to quantify the sensitivity enhancement of the FOC and ATR platforms.

$$S \equiv \frac{A_{FOC,film}}{A_{transmission}} = \frac{A_{FOC,film}}{\varepsilon_{film}\,\Gamma_{film}} \tag{1}$$

Where $\varepsilon$ is the molar absorptivity and $\Gamma$ is the molecular surface coverage of the film under test. Absorbance of a polyion self-assembled film of poly (diallyldimethylammonium chloride) (P+) and Nickel (II) phthalocyaninetetrasulfonic acid (Ni (TSPc)) on both the FOC and ATR (Doherty *et al.*, 2002) platforms were used to compare the sensitivity perfomance of the two techniques. Figure 3b shows a comparison of the P+/Ni (TSPc) absorbance spectra on the ATR and FOC normalized by interaction length. Currently, the FOC yields thin-film absorbance values comparable with ATR instrumentation; however, the FOC eliminates the complex coupling optics and alignment procedures required to make such measurements using ATR instrumentation. (Beam *et al.*, 2007)

Further refinements in the FOC platform promise to substantially increase its sensitivity. The lower order modes of a fiber (those with optical rays propagating at a small angle from the fiber axis and described by a greater effective refractive index, $N$) do not provide a

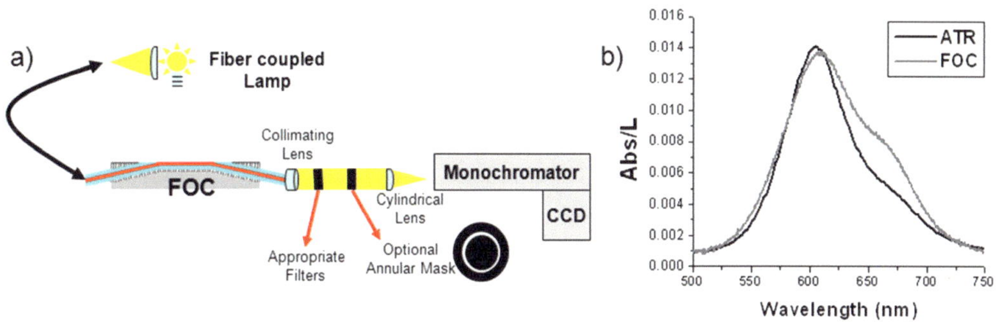


Fig. 3. a) Instrument schematic for FOC absorbance measurements. b) Spectra are of a self-assembled polyion film of P+ and Ni (TSPc), and the spectra are normalized by their interaction length (*L*) of 44 mm (for the ATR spectra) and 17.2 mm (for the FOC spectra). Modified from Beam *et al.*, 2007.

strong interaction with the molecules adsorbed on the active surface of the FOC. Removing these lower order modes from the optical beam prevents collecting average absorbance measurements which are unduly weighted toward the less sensitive traveling waves inside the fiber. (Gloge, 1971; Ruddy *et al.*, 1990) To select the modes allowed to propagate in the FOC an annular mask that only transmits a ring of light of a defined angle has been used. A mask delivering light with a low effective index, therefore working only with the highest order modes that the fiber can support, was shown to double the measured thin-film absorbance on the FOC compared to that measured for the same film without a mask. (Beam *et al.*, 2007)

### 1.3 The electroactive-fiber optic chip (EA-FOC)

Spectroelectrochemical measurements provide complimentary spectroscopic and electrochemical analytical data which have found applications using fiber coupled techniques. UV-Vis (VanDyke & Cheng, 1988), FTIR (Shaw & Geiger, 1996), and Raman (Hartnagel *et al.*, 1995) fiber coupled spectroelectrochemical measurements have been obtained using the distal end of a fiber optic probe as the working electrode. These fiber optic probes, however, suffer from the limited optical pathlength of transmission absorbance spectroelectrochemical measurements. Over the last decade the sensitivity of spectroelectrochemical measurements has been significantly enhanced by using monochromatic and broadband ATR platforms, (Doherty *et al.*, 2002; Winograd & Kuwana, 1969) multi-mode waveguides, and single-mode waveguides. (Bradshaw *et al.*, 2003; Dunphy *et al.*, 1997; Dunphy *et al.*, 1999; Itoh & Fujishima, 1988) A significant hindrance for these ATR and waveguide based spectroelectrochemical technologies has been interfacing the sensor platform with standard, commercially available spectroscopic instrumentation; thus, only one field portable instrument has been developed by Heinemann and coworkers to spectroelectrochemically detect ferrocyanide. (Monk *et al.*, 2002; Stegemiller *et al.*, 2003)

The first application of the FOC as a fully integrated fiber coupled spectroelctrochemical platform, was termed the electroactive-fiber optic chip (EA-FOC). To create the EA-FOC we coat the FOC with a thin-film of indium-tin oxide (ITO) as the working electrode (Figure 4a) and probe electrochemically driven changes in absorbance for surface confined redox species.

(Beam *et al.*, 2009) The sensitivity enhancement of the EA-FOC platform is calculated using the methylene blue (MB) redox couple. Additionally, the properties of the EA-FOC are demonstrated by probing the redox spectroelectrochemistry of an electrodeposited film of the conducting polymer poly (3,4-ethylenedioxythiophene) (PEDOT).

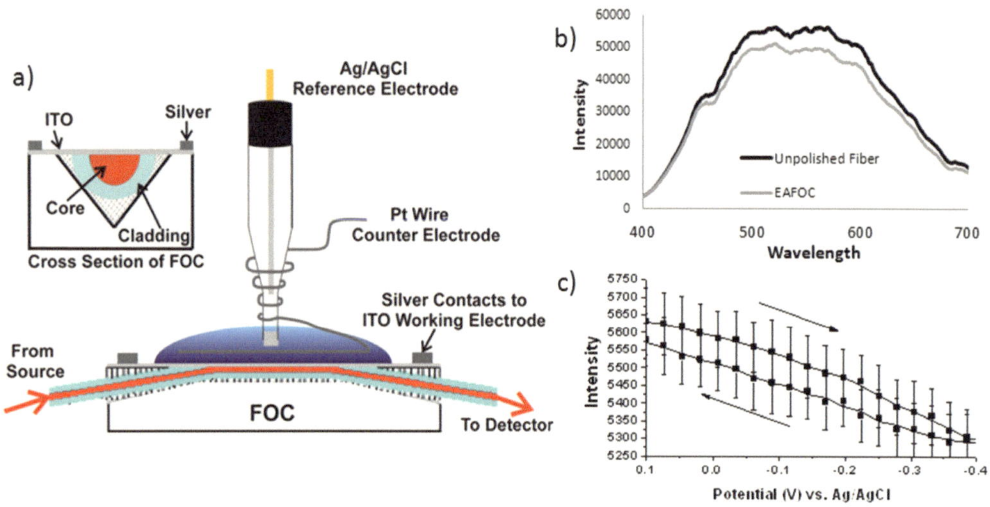


Fig. 4. The EA-FOC a) Schematic of the EA-FOC with labeled electrical contacts and a cross section (inset) b) comparison of the transmission spectra of an unpolished fiber with that of the EA-FOC and c) the EA-FOC out-coupled intensity as a function of potential at 665 nm in a 0.1M $KNO_3$ aqueous solution. Modified from Beam *et al.*, 2009.

### 1.3.1 Optical effects of ITO on the FOC

The optical properties of ITO are dependent on the electrochemical properties of the material. ITO is generally transparent in the visible region where the short wavelength cut off is determined by absorption due to the band gap of the material. The long wavelength cut off is due to scattering of free electrons and is determined by the plasma resonance frequency. As the free carriers within the material increases the plasma resonance wavelength decreases. Therefore, there is a trade off between increasing the free carrier concentration of ITO to improve the electrical properties and decreasing the transmission wavelength window. (Hartnagel *et al.*, 1995) For the ITO sputtered onto the FOC device, the minimum absorptivity coefficient was estimated to be $5 \times 10^{-3}$ at 500 nm (or a propagation loss of ~ 0.5 dB/cm). The transmission of the ITO film on the FOC will affect the optical properties of the device platform, and the broadband transmission of the EAFOC device is slightly decreased by the addition of ITO (Figure 4b).

Before discussing spectroelectrochemical measurements made on the EA-FOC, it is important to evaluate the optical background of the device. Figure 4c plots the out-coupled intensity from the EA-FOC versus potential in an electrolyte solution without electrochemically active analytes. The linear decrease of intensity with potential is attributed to a change in the ITO absorptivity, which is due to the increase in free carrier concentration within the film as the applied potential decreases. To account for the affect of applied potential on the background signal of the EA-FOC, absorbance measurements at each potential were calculated from solvent blanks recorded at a corresponding potential. Additionally, there is a slight hysteresis between the forward and backward potential sweeps due to ion diffusion. Equilibration of the ITO electrode in the electrolyte solution after 10 potential scans stabilized the magnitude of the hysteresis allowing analytical measurements to be collected.

### 1.3.2 Spectroelectrochemical measurements

The spectroelectrochemistry of adsorbed monolayers of methylene blue (MB) has been previously evaluated on both the ATR (Itoh & Fujishima, 1988) and waveguide-based (Dunphy *et al.*, 1997) platforms; therefore, the MB redox couple is used to compare the EA-FOC measurements with these well-known techniques. MB electrostatically adsorbs to the ITO surface in its native oxidized form. The surface adsorbed MB undergoes a chemically reversible 2-electron reduction to the transparent leuco form of the dye at ~ - 0.27 V versus a Ag/AgCl reference electrode. For the micromolar solution concentrations used in this study, the bulk MB absorbance does not contribute appreciably to the EA-FOC spectroelectrochemical response. Potential dependent spectra of MB on the EA-FOC (Figure 5a) shows absorbance maxima for both the monomer (665 nm) and aggregate forms of this dye (605 nm). (Bergmann & O'Konski, 1963) Simultanteous optical and electrochemical detection of the MB redox couple allow for the calculation of the sensitivity of the EA-FOC using the the electrochemically determined surface coverage and the experimentally measured absorbance, using the Beer's Law relationship in equation 1. The sensitivity of the EA-FOC was calculated to be 40 ± 2 or 20.6/cm, which is comparable to sensitivities calculated for the FOC devices. (Beam *et al.*, 2009; Beam *et al.*, 2007).

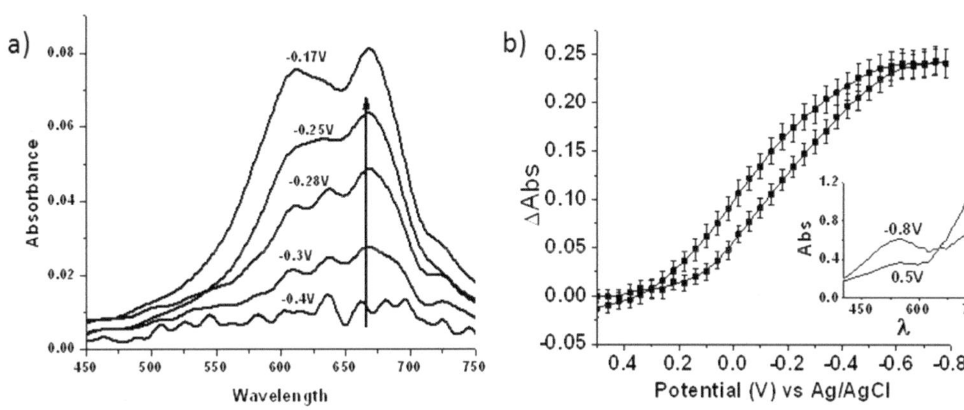


Fig. 5. Spectroelectrochemistry with the EA-FOC a) Potential dependent broadband absorbance spectra of an MB film and b) Absorbance difference (at 550 nm) verus potential for PEDOT film (inset: broadband absorbance spectra for reduced/oxidized polymer on the EA-FOC). Modified from Beam *et al.*, 2009.

The EA-FOC was used to electrochemically polymerize an ultra-thin film, estimated to be 0.3% of a monolayer, of poly (3,4-ethylenedioxythiophene) (PEDOT) and probe its electrochemical properties. The voltammogram of ultra-thin films of PEDOT has broad voltammetric peaks which are poorly distinquishable from the non-faradaic background.

However, PEDOT undergoes a reversible oxidation from the neutral dark blue form of the polymer to the almost transparent single polaron state and upon further oxidation the bipolaron state of the polymer. (Chen & Inganas, 1996) The spectroelectrochemical measurement in Figure 5b of the change in absorbance at the $\lambda_{max}$ (550 nm) versus potential on the EA-FOC illustrates the electrochromic behavior of the polymer. The EA-FOC only monitors the appearance/disappearance of the dark blue neutral form of the polymer and does not indicate the state of the polymer upon oxidation. (Beam *et al.*, 2009) The EA-FOC has the requisite sensitivity to monitor optical redox changes in submonolayer surface coverages of molecular thin-films.

## 2. Fluorescence bioassay

Fluorescence detection architectures are of particular importance for biosensing applications where the fluorescence signal is detected against a zero background enabling low limits of detection, typically in the nano- to femto-molar range. Optical transducers have the advantages of being non-destructive, sensitive, and can be used for real-time and kinetic measurements. Fluorescence signal transduction has widespread applications due to the commercial availability of a variety of fluorescent labels which only require simple modification procedures for incorporation with biomolecules. Several reviews and books have been published which discuss the different fluorescent biosensor designs and applications. (Collings & Caruso, 1997; Cunningham, 1998; Janata *et al.*, 1994; Marazuela & Moreno-Bondi, 2002; Taitt *et al.*, 2005; Thompson, 2006)
Commonly biosensor architectures require immobilization of the biological recognition event onto a surface for which the evanescent field of optical waveguide platforms is specifically suited. Fiber-coupled sensor platforms do not require the bulky free-space optics used for fluorescence microscopy, total internal reflection fluorescence (TIRF), and planar waveguide techniques. Therefore, integrated excitation and emission fiber optic platforms have been constructed using different structures including a de-clad cylindrical core and tapered optical fibers. The FOC is the first demonstration of a multi-mode side polished fiber as a planar integrated excitation and emission platform.

### 2.1 Mechanism of back-coupled fluorescence

According to Snell's law, light traveling in a lower refractive index medium is refracted at a planar interface with angles below the critical angle in a high-index medium, such as a slab waveguide. For light to be guided within a waveguide it must be launched at angles greater than the critical angle; therefore, light from a lower refractive index medium cannot in principle be guided (Figure 6a). However, for surface confined fluorophores, the proximity of the fluorophores to the waveguiding structure allows coupling of the evanescent photons into guided modes of the waveguide termed back-coupled fluorescence. In other words, the electromagnetic near field, created by the oscillation of the excited dipole from the surface confined fluorophore, overlaps with the evanescent tail of the waveguide modes and meets the conditions for light propagation within the waveguide (Figure 6b). (Carniglia *et al.*, 1972) Harrick and Loeb first applied the principle of back-coupled fluorescence using an ATR element to detect a fluorescently labeled self-assembled thin-film of bovine serum albumin. Fiber optic based back-coupled fluorescence biosensors were first presented by Andrade *et al.* in 1985 and theoretically explored by Glass *et al.* and Marcuse.

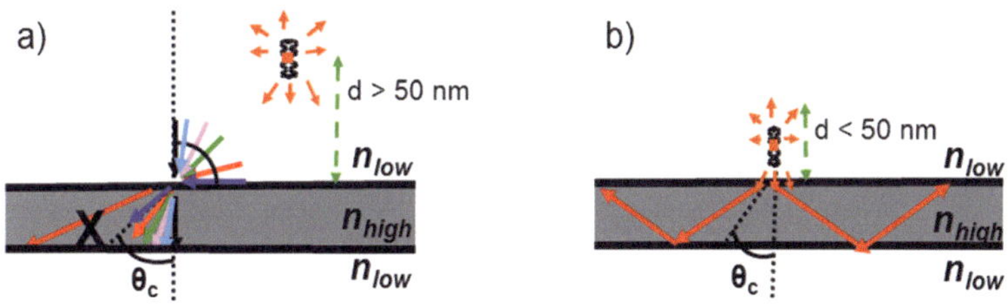


Fig. 6. Back-coupled fluorescence: a) light propagating from fluorophores far away from the waveguiding structure will be refracted at angles less than the critical angle, and therefore will not excite waveguide modes within the structure. b) Light propagating from fluorophores within close proximity of the waveguiding structure will back-couple fluorescence into the waveguide because the evanescent photons, or the near-field, of the fluorophore will overlap with the evanescent tail of propagating modes in the structure.

### 2.2 Fiber optic based fluorescence sensors

The pioneering research utilizing a de-clad quartz fiber to collect back-coupled fluorescence from immobilized biomolecules was presented by Sutherland *et.al*, Andrade *et.al*, and Glass *et.al*. Biosensors based on receptor proteins (Garden *et al.*, 2004; Rogers *et al.*, 1991; Rogers *et al.*, 1989), antibody-antigen interactions (Anis *et al.*, 1993; Bier *et al.*, 1992; Devine *et al.*, 1995; Eenink *et al.*, 1990; McCormack *et al.*, 1997; Oroszlan *et al.*, 1993; Shriver-Lake *et al.*, 1995; Toppozada *et al.*, 1997; Walczak *et al.*, 1992), sandwich immunoassay (Geng *et al.*, 2006; Kapoor *et al.*, 2004), and oligonucleotides (Abel *et al.*, 1996; Graham *et al.*, 1992; Pandey & Weetall, 1995) have been presented employing the de-clad fiber geometry. However, the de-clad fiber architecture is limited by the fragile nature of the fiber platform and inefficient fluorescence back-coupling due to the sharp V-number mismatch.

The V-number, or waveguide parameter, of a waveguide platform can be used to calculate the number of modes the structure will support (Equation 2).

$$V = \frac{2\pi r}{\lambda}\sqrt{\left(n_{core}^2 - n_{clad}^2\right)} \tag{2}$$

Where $\pi$ and $\lambda$ have their usual meanings, $r$ is the radius of the fiber, $n_{core}$ and $n_{clad}$ are the refractive index of the core and cladding respectively. For example, a 50 µm fiber with an $n_{core}$ of 1.460 and $n_{clad}$ of 1.443 will have a V-number of 62 at 560 nm. For the de-clad fiber sensor geometry, the value of $n_{clad}$ should be replaced with the aqueous medium surrounding the sensing platform (1.33); therefore, the V-number is 169 at 560 nm in the sensing region. Thus, approximately 60% of the modes in the sensing region of the fiber will not propagate in the fiber. To complicate the matter further, the back-coupled fluorescence primarily propagates in the higher order modes of the de-clad fiber, which are the non-propagating modes in the clad fiber.

One method researchers have employed to minimize back-coupled fluorescence loss due to V-number mismatch is to increase the value of $n_{clad}$ in the sensing region of a de-clad fiber. Potyrailo and Hieftje have immobilized reagents sensitive to ammonia, humidity, and oxygen in the polymer cladding of optical fibers, thus ensuring no change in the value of $n_{clad}$. (Potyrailo & Hieftje, 1998; Potyrailo & Hieftje, 1999) An alternative strategy to increase $n_{clad}$ in the sensing region of a de-clad fiber is the application of a sol-gel cladding containing an analyte sensitive dye. (Browne *et al.*, 1996; Kao *et al.*, 1998; MacCraith *et al.*, 1993; O'Keeffe

*et al.*, 1995) A second method to match the V-number between the clad and unclad sensing region of fiber optic sensors is to decrease *r* in the sensing region through etching the de-clad fiber. Fluorescent fiber sensors, where the de-clad sensing region has been step- or taper-etched, exhibit a 20 to 50 fold improvement in sensitivity. (Anderson *et al.*, 1994; Anderson *et al.*, 1994) Tapered fiber optic biosensor platforms have been applied to sandwich assays (Golden *et al.*, 1992; Zhou *et al.*, 1997), immunosensors (Anderson *et al.*, 1993; Maragos & Thompson, 1999; Thompson & Maragos, 1996), and measuring pH. (Grant & Glass, 1997) The feasibility of collecting fluorescence using a single-mode biconical tapered fiber has also been explored. (Wiejata *et al.*, 2003)
The FOC provides a planar, robust, supported, side-polished multimode fiber platform for fluorescence biosensing applications. A related platform using a single-mode fiber in a bent configuration to collect the luminescence of a rhodamine 6-G film has been previously reported; however, this device was limited to single wavelength detection and relied on frequency modulated detection. (Poscio *et al.*, 1990) The ability to simply collect broadband fluorescence will enable the FOC device to be used in a broad range of sensor configurations using fluorescence detection systems, including on-chip, fully integrated excitation and sequential optical characterization of luminescent analytes. The first generation FOC device demonstrated back-coupled fluoresence with a drop cast film of CdSe semiconductor nanoparticles (SC-NP) as a luminescent model system. (Beam *et al.*, 2007) Here, the FOC is applied to quantitatively characterize a biotin-Streptavidin binding event as a model bioassay system.

### 2.3 BSA-biotin/streptavidin-CY bioassay on the planar fiber optic chip

A bioassay of surface-adsorbed biotin with fluorescently labeled streptavidin was chosen to quantitatively explore back-coupled fluorescence collection by the FOC. The small molecule biotin interacts non-covalently with the streptavidin protein and is highly specific, with one of the largest known binding constants ($K_a \cong 10^{15}$ $M^{-1}$). Bovine serum albumin (BSA) labeled with biotin (Sigma) adsorbs to the surface of the FOC and is transparent in the visible region. The back-coupled fluorescence is collected from the fluorescently labeled (Cascade Yellow, CY: Invitrogen) streptavidin bound to the surface adsorbed biotin. The Cascade Yellow dye was chosen for labeling due to its large Stokes shift (~150 nm) and short excitation wavelength. A fluorophore with a large Stokes shift is very valuable for back-coupled fluorescence measurements because of the magnified inner-filter effects of the fluorophores on the waveguide platform. Back-coupled fluorescence propagates in the waveguide modes; therefore, the emitted light is available in the evanescent field for re-absorption by the same film. The concentration dependence of the bioassay and the efficiency of the back-coupled fluorescence collected by the FOC are examined.
A representative spectrum of a BSA-biotin/streptavidin-CY film is plotted in Figure 7a along with two control experiments. The first control confirms that there is no contribution to the fluorescence from the BSA-biotin thin-film (blue line). The second control consisted of a BSA film, which was not labeled with biotin, to be incubated with the fluorescently tagged streptavidin to test for non-specific adsorption (green line). The contribution to the back-coupled fluorescence from non-specific adsorption of streptavidin-CY was shown to be below the detection limit of the experimental set-up. Finally, to confirm the back-coupled fluorescence resulted from the Cascade Yellow dye, the corrected back-coupled fluorescence spectrum is compared with the Cascade Yellow spectrum supplied by the manufacturer.

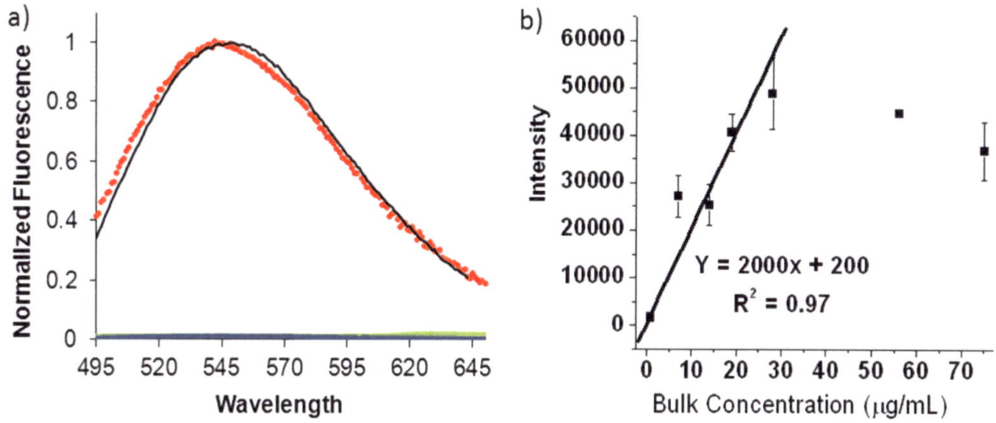


Fig. 7. Back-coupled fluorescence for a biotin/streptavidin bioassay. a) Cascade Yellow flourescence (black) spectra from the manufacturer (Invitrogen). Back-coupled fluorescence spectra for a BSA-biotin/streptavidin-CY film (red), background fluorescence from the buffer and biotin-BSA film was shown to be negligible (blue), and fluorescence due to nonspecific adsorption of the strepavidin was shown to be below the detection limit of the measurement (green) by using an unlabeled film of BSA. b) Bioassay calibration plot of the streptavidin-CY bulk concentration versus the average maximum collected fluorescence intensity on the FOC.

The concentration dependence of the collected back-coupled fluorescence for the bioassay was determined by varying the concentration of streptavidin-CY while maintaining the same adsorption conditions for the BSA-biotin film. The fluorescence for three BSA-biotin/streptavidin-CY films was collected for each bulk solution concentration (1, 7, 14, 19, 28, 56, and 75 μg/mL). A plot of bulk concentration versus the average fluorescence maximum intensity illustrates a linear dependence of fluorescence with bulk concentration and a $R^2$ value close to 0.97 for bulk concentrations less than 28 μM (Figure 7b). A self limiting surface coverage at bulk concentrations ≥ 28 μM was observed. The subsequent slight decline in fluorescence for the higher bulk concentrations is attributed to luminescence quenching and photobleaching of dyes in close proximity to each other. The observed bulk concentration limit of detection for the biotin/streptavidin bioassay is 15 nM, which is on the order of the LOD reported for several de-clad fiber fluorescence sensors. (Anis *et al.*, 1993; Devine *et al.*, 1995; Eenink *et al.*, 1990; Graham *et al.*, 1992; McCormack *et al.*, 1997; Oroszlan *et al.*, 1993; Pandey & Weetall, 1995; Shriver-Lake *et al.*, 1995; Sutherland *et al.*, 1984) However, the measured pathlength of the FOC is only ~ 24 mm for this device compared to the ~ 60 mm pathlength of most cylindrical core de-clad fiber sensor platforms. The de-clad fiber sensor platforms require large fiber cores (~ 600 μm) with low modal surface interaction to increase mechanical strength, and therefore, a long interaction length is necessary. The supported planar substrate of the side polished FOC increases the mechanical robustness of the platform; thus, smaller core (50 μm), more surface sensitive fibers are used. One method which could be employed to decrease the LOD of the FOC based fluorescence sensor, and to surpass the de-clad fiber platforms, is to increase the physical pathlength of the FOC device.

### 2.4 Back-coupled fluorescence collection efficiency

To quantify the fluorescence collection of the FOC using the BSA-biotin/streptavidin-CY bioassay an efficiency calculation was conducted. The backcoupled fluorescence collection efficiency is calculated from the power ratio of the actual fluorescence collected ($P_{FOC}$) and the calculated total power of fluorescence of the film ($P_{calc}$) (Equation 3).

$$Efficiency = \frac{P_{FOC}}{P_{calc}} = \frac{P_{FOC}}{P_o L\left(1-10^{-A_{405}}\right)Q_{cy}} \tag{3}$$

$P_{FOC}$ is the area under a gaussian fit curve to the corrected fluorescence spectrum detected (mW). The values in the denominator are an expression of the calculated fluorescence power propagating in all directions, where $P_o$ is the power launched into the fiber from the 405 nm laser (mW); $L$ is defined as 1-Loss of the FOC; $\left(1-10^{-A_{405}}\right)$ is the percent of power absorbed at 405 nm by the dye and available to be converted to fluorescence; and $Q_{cy}$ is the quantum yield of the Cascade Yellow labeled streptavidin. The calculated back-coupled fluorescence efficiency is 0.02% of the light emitted by the Cascade Yellow dye. For comparison, only 2% of the light emitted from an isotropic emitter is typically collected with a lens. To improve device performance and decrease the detection, the FOC should be engineered to more efficiently collect back-coupled fluorescence. The back-coupled fluorescence efficiency could be improved upon by increasing the numerical aperture/refractive index of the fiber; however, the extent of V-number mismatch of the FOC platform should be evaluated in conjunction with alternative FOC architectures.

## 3. Bright and broadband-guided light source

Field portable optical sensing devices require light sources that are robust, compact, spectrally broad, and bright. The ideal fiber coupled light source will have high power per unit area and unit solid angle; thus, yielding high power per guided mode inside an optical fiber. Using the back-coupled light mechanism of fluorescent thin-films deposited on the FOC platform, a fully integreated broadband, bright guided light source is created. The FOC light source is produced by pumping an aluminum tris-hydroxyquinoline thin-film capped with a silicon dioxide overlayer. A directly fiber coupled broadband FOC source extending from 405 nm to 650 nm is produced with an output of 1.8 mW, which is significantly brighter than a fiber-coupled tungsten source and spectrally borader than a light emitting diode (LED) source.

### 3.1 Fiber optic light sources

Bright and spectrally broad light sources are currently required for several applications including optical coherence tomography, optical spectroscopies, and chemical/biological sensing. Recently, several promising technologies have been developed to fulfill those needs. In particular, superluminescent LEDs (Zhang *et al.*, 2009), supercontinuum-generation light sources (Berge *et al.*, 2007), and amplified spontaneous emission (Samuel & Turnbull, 2007) are becoming increasingly useful for many applications. Despite those developments, a cost-effective light source that is fiber-coupled, spectrally broad, and bright is still in quite demand, especially in the visible and ultra-violet regions of the spectrum.

### 3.2 Planar fiber optic chip broadband light source

The planar geometry of the FOC device is amenable to standard thin-film deposition architectures such as thermal deposition. The back-coupled fluorescence from an organic fluorophore deposited on the polished surface of the FOC platform is used to create a fully integrated fiber optic broadband light source. With the growing interest in organic LEDs and photovoltaic devices, a vast number of inexpensive, easily processable, high quantum yield fluorescent organic materials are commercially available. (Kafafi, 2005; Li *et al.*, 2007) These fluorescent organic compounds garner much interest because of their broad emission spectrum when compared to inorganic compounds such as GaN and Si. Tris-aluminum 8-hydroxyquinoline (Alq3) was chosen as a fluorescent material for the FOC light source due to its broad emission in the visible region. A 45 nm thick film of Alq3 was deposited on the FOC using thermal deposition. Both oxygen and water can cause degradation of the Alq3 thin-film (Burrows *et al.*, 1994; McElvain *et al.*, 1996); therefore, FOC light source devices must be protected from the ambient environment to ensure continued operation. Encapsulation of the FOC device was achieved with a 100 nm film of $SiO_2$ deposited, by electron beam evaporation, over the Alq3 film without breaking vacuum. Emission of the Alq3 film on the FOC device is achieved with pumping the film with a 405 nm GaN laser diode which is fiber coupled into the FOC platform. Figure 8a shows a picture of the active region of the FOC light source, where luminescence in the Alq3 film is back-coupled into guided modes of the fiber, next to the light out-coupled from the FOC fiber.

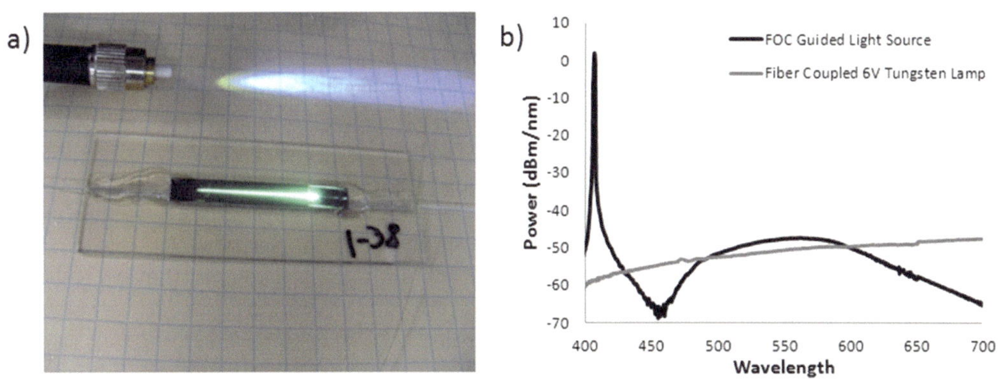


Fig. 8. a) Image of the FOC light source device; b) comparison of the fiber output of the FOC light source with a fiber coupled 6V tungsten-halogen lamp.

A comparison of the out-coupled spectrum from the FOC light source with a fiber coupled 6V tungsten lamp are plotted in Figure 8b. The light from the thermal source (6V tungsten lamp) is coupled into the same patch cable as the FOC using similar high precision aspheric optics. The fluorescence from the FOC light source has an increased out-coupled intensity over a 100 nm range from 480 to 580 nm overlapping with the Alq3 emission peak compared to the fiber coupled thermal source. The spectra resulting from the FOC light source includes a large intensity peak at 405 nm for the laser used to pump the film. The measured power intensity out-coupled from the FOC light source was measured to be 1.8 mW.

While optical pumping was used for these experiments, there is the potential to use an electrically pumped system to directly drive the FOC light source producing broadband spectra with high brightness. Using an ITO film as a transparent electrode, organic layers could be built onto the FOC structure very similar to that of an organic light emitting diode. Combining emission spectra for multiple organic films would allow an even broader output spectrum to be achieved. With the amount of research in organic fluorescent compounds with increased quantum efficiences and variety of wavelength ranges, improving the power output and wavelength range of future FOC light source devices will be developed.

## 4. Raman spectroscopy

Raman spectroscopy is a well-established analytical technique that can identify chemical and physical properties through interactions with the vibrational states of a particular analyte. This section presents investigations of excitation and collection of Raman scattering using the FOC for thin molecular films. Thin-film Raman measurements were achieved with the added signal enhancement of surface enhanced raman spectroscopy (SERS). Gold nanoparticles are deposited on the FOC surface to enhance the Raman signal of a 4-aminothiophenol film and the Raman signal was both excited and collected by the FOC in decoupled instrument schemes. In a similar approach, Raman scattering of carbon nanotubes was demonstrated, setting the stage for the FOC as a platform for interesting chemical analysis.

### 4.1 Raman scattering

Raman spectroscopy relies on the inelastic scattering of incident light with Raman active molecular thin-films. Typically, in an elastic event known as Rayleigh scattering, the excited molecule relaxes back to the initial ground state and light of equal energy to the incident light is reemitted. Raman scattering occurs when interactions between molecular vibrations and rotations with the incident light result in lower frequency, Stokes, or higher frequency, anti-Stokes, shifts from the incident frequency of light. Raman spectra are independent of the initial frequency of the incident light, and the resultant energy spectrum is a signature of the vibrational/rotational states of the probed molecules.

Raman scattering occurs for only one out of every $10^6$-$10^8$ scattering events, making it a very weak signal. (Smith & Dent, 2005) To improve upon this small cross section, researchers have utilized the effects of localized surface plasmon resonance. A localized surface plasmon resonance occurs when small metallic structures are irradiated by light. Similar to a lightning rod, these structures induce an electric-field enhancing corona effect. This effect relies on the size of the metallic structure to be small compared to the wavelength of the incident light, and the electric-field will concentrate in areas of greatest curvature. Surface enhanced Raman spectroscopy (SERS) occurs when Raman active molecules are in the presence of roughened metallic surfaces or nanoparticles. The electric-field amplitude will generate a larger intensity of the incident light as well as amplify Raman scattering. The SERS amplification effect has lead to reported Raman signal enhancements of $10^6$ (Felidj *et al.*, 2003), $10^{11}$ (Gupta & Weimer, 2003), even $10^{14}$ (Kneipp *et al.*, 1997). (Willets & Van Duyne, 2007)

Increasing Raman spectroscopy sensitivity has been sought after in recent years, ultimately reaching single molecule detection. (Kneipp *et al.*, 1997; Xu *et al.*, 1999) Particularly, thin-film characterization is of interest to a growing number of fields yet analysis by conventional commercial Raman microscope instruments is difficult due to the convolution between analyte and substrate Raman activity. The unique geometry of the FOC allows for analyzing a strip of sample rather than a single spot as found in conventional Raman machine. We present here the extended ability of the FOC to both excite and collect Raman scattering of thin-films.

### 4.2 Fiber optic raman probes

The use of distal end fiber probes for Raman scattering has been ongoing for some time. Although mechanisms have been described for propagating the excitation beam and collected Raman scattering through a single fiber (Potyrailo *et al.*, 1998), these devices often utilize separate launching and collection fibers. In some schemes, a single launching fiber is surrounded by a bundle of collection fibers. Raman spectroscopy using distal end fiber probes has been demonstrated in a number of chemical (Khijwania *et al.*, 2007; McCreery *et al.*, 1983; Tiwari *et al.*, 2007) and biomedical (Krafft *et al.*, 2007; Lima *et al.*, 2008; Vo-Dinh *et al.*, 1999) applications. However, the overlap between the illumination cone and the collection cone is often poor for these sensor platforms, weakening the already very faint Raman signal. To improve sensitivity of Raman sensors modifications such as GRIN lens application to the distal end (Mo *et al.*, 2009) or tapered fibers (Stokes et al., 2004) have been implemented. Although these changes have shown some improvement in overall signal to noise ratio, the move to a planar fiber optic chip offers the advantages of simplifying optical alignment and providing a larger surface area for interaction.

Exposing the core of the optical probe allows for the use of evanescent field interactions of the exposed fiber core with immobilized analytes. Here the fiber core may be functionalized to capture the analyte or to enhance the Raman signal as in the case of a SERS substrate. In a D-shaped device similar to the FOC, Zhang *et al.* were able to demonstrate the excitation of the Raman analyte Rhodamine 6G by a SERS functionalized planar probe (Zhang *et al.*, 2005). Near-field interactions do not rely on the optical alignment of the system; therefore, a more streamlined approach would only use a single fiber for both excitation and collection of the Raman signal. Coupled excitation and collection of SERS for an exposed core fiber has been demonstrated for thin-film and aqueous samples (Stokes & Vo-Dinh, 2000); however, the exposed fragile core limits the applications of this sensor architecture.

### 4.3 Raman spectroscopy with the FOC

A single fiber is used to deliver the excitation beam and collect the scattered Raman signal to form a fully integrated system. At the boundary of the exposed core of the FOC, adsorbed analytes interact with the evanescent field of the excitation light. The Raman signal of the analyte may then be launched into the fiber through near-field coupling. To test the FOC for its function in Raman spectroscopy, the excitation and collection of the Raman signal was decoupled. In the excitation scheme, laser light (632.8 nm) was launched into the fiber and the adsorbed analyte was excited. The scattered Raman signal was then collected by external optical components mounted over the FOC and delivered to a spectrometer connected to a CCD for data acquisition. For the collection scheme, external optics were used to deliver the excitation laser beam to the planar surface of the FOC. The Raman signal was then coupled into the FOC and guided by the fiber to the data acquisition set-up. Since, in both schemes, Rayleigh scattering of the laser line was generated, a notch filter was placed in the beam path before the spectrometer to remove as much of the initial laser beam as possible. Both schemes are shown in Figure 9.

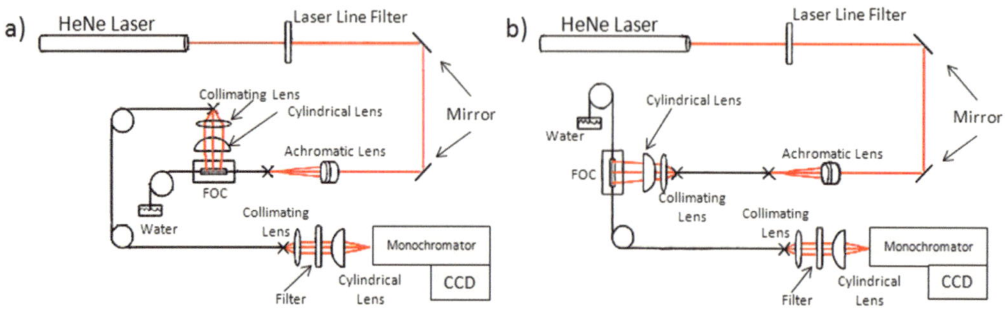


Fig. 9. The experimental set-up for the a) excitation scheme in which the light is delivered to the analyte by evanescent waves, and the b) collection scheme in which the light is directly focused on the FOC surface.

As a proof of concept study, the FOC was tested in the excitation scheme for the generation of Raman scattering of bulk media. A commercial optical gel (Cargille) was chosen for its potential as a higher index capping media for thin-films. The gel was deposited in a thin, even layer across the active area of the FOC. The Raman signal of the gel was induced by the FOC and collected by the external optics as seen in Figure 10. Demonstrating the feasibility of Raman excitation by the FOC sets the stage for the chip to be employed in other bulk media investigations without the need for surface enhancement. However, the more interesting objective was to determine the utility of the FOC in generating Raman scattering of smaller volumes of analyte such as in the case of thin-films.

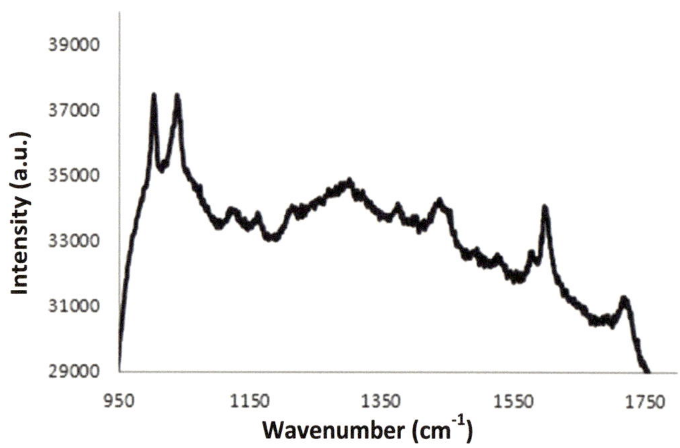


Fig. 10. Raman scattering of commericially available optical gel excited by the FOC.

#### 4.4 Surface enhanced raman of thin films with the FOC

For thin-film analysis, the FOC was first functionalized with gold nanoparticles to create a SERS substrate. A seed based method (Wei *et al.*, 2004; Wei & Zamborini, 2004) was performed directly on the FOC platform. In short, the process required three stages. First, the FOC was functionalized with 3- mercaptopropyltrimethoxysilane (MPTMS). Next, 5 nm gold nanoparticles were adsorbed to the MPTMS functionalized FOC platform. Finally, a gold growth solution catalyzed the growth of the nanoparticles to rods, platelets, and spheroids on the order of 20 nm in diameter.

The first analyte investigated was 4-aminothiophenol (p-ATP). Raman activity of p-ATP in the presence of gold SERS substrates has been well documented. (Baia *et al.*, 2006; Guo *et al.*, 2007; Wang *et al.*, 2008) The analyte was deposited onto the FOC by submersion in an ethanol solution overnight. The FOC was washed and dried under a nitrogen flow before analysis of the thin-film. As shown in Figure 11a, the Raman signal of p-ATP was excited by the FOC and collected by external optics in the excitation scheme. Similar results were achieved by external excitation of the analyte and collection of the Raman signal via the FOC in the Collection Scheme. Both spectra are consistent with published data of p-ATP Raman activity as well as spectra of the p-ATP coated FOC collected by a commercial machine (Renishaw Invia). To attain acceptable signal-to-noise ratios, integration times of 5 to 10 minutes were used. Signal was detected for sub mW incident laser power, suggesting a low power light source could be used in a field version of the FOC Raman sensor platform.

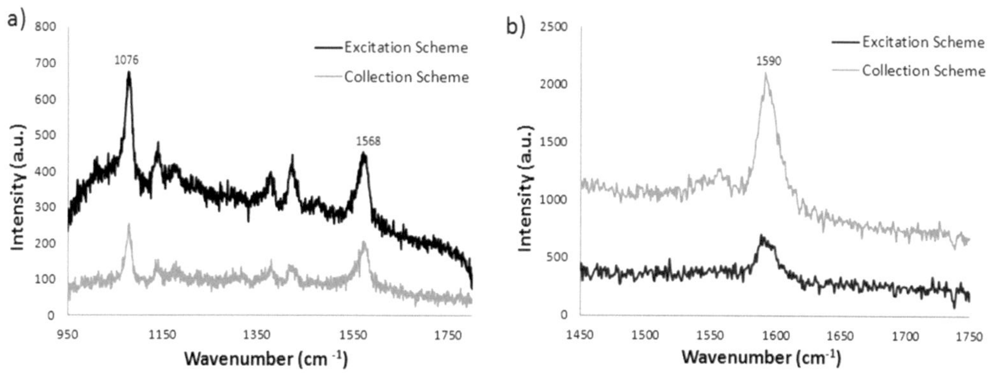


Fig. 11. a) Raman spectroscopy of p-ATP using both the excitation and collection schemes on the FOC. b) Excitation and collection of CNT G-band Raman signal by the FOC.

The second analyte studied with the Raman FOC device coated with gold nanoparticles was carbon nanotubes (CNT) in an isopropanol solution. In this case, the CNT were deposited by a drop cast method and remained in solution during analysis. The G-band of the CNT spectra was examined in both the excitation and collection schemes using the FOC. The data, shown in Figure 11b, was consistent with established Raman data for the CNT G-band. (Dresselhaus *et al.*, 2005; Kneipp *et al.*, 2004) Pairing the FOC device with the specific and nondestructive technique of Raman spectroscopy is a very desirable application.

## 5. Conclusion

In this chapter the current applications of the FOC platform were presented. FOC fabrication consists of an optical fiber mounted in a V-groove block, side-polished to create a planar platform that allows access to the evanescent field escaping from the fiber core. Currently, the FOC yields thin-film abosrbance sensitivity comparable with existing ATR instrumentation; however, it eliminates the complex coupling optics and alignment procedures used with planar waveguide based instrumentation. Spectroelectrochemical measurements on an ITO coated FOC platform have been previously demonstrated for both the potential dependent spectra of a methylene blue film and 0.3% of a monolayer of a conductive polymer film. Additionally, light generated in a surface-confined thin molecular film can be back-coupled into the FOC platform when the conditions for light propagation within the waveguide are met. Back-coupled light into the FOC was used for the first time here to expand the applications of the FOC platform to a fluorescence bioassay, a broadband fiber optic light source, and Raman interrogation of molecular adsorbates. A BSA-biotin/streptavidin-CY fluorescence bioassay was demonstrated on the FOC with a limit-of-detection of 15 nM and calculated back-coupled fluorescence efficiency of 0.02%. A 1.8 mW directly fiber coupled broadband FOC source extending from 405 nm to 650 nm is produced with a $SiO_2$ capped Alq3 film deposited on the FOC, which is significantly brighter than a fiber-coupled tungsten source and spectrally broader than a LED source. Finally the FOC platform was modified with gold nanoparticles to create a surface enhanced Raman substrate for detection of 4-aminothiophenol and carbon nanotubes. Contrary to the un-clad fiber approach, the FOC with a supported planar interface can facilitate the use of conventional planar deposition technologies and provide a robust planar platform that is amenable for integration into various sensor applications.

## 6. Acknowledgments

This work was supported by the National Science Foundation under grants Number DBI-0352449, CTS-0428885, CHE-0517963, and CHE-0515963, the Kentucky Science and Technology Corporation under grant KSEF-1869-RDE-012, and the Science and Technology Center-Materials and Devices for Information Technology Research Grant Number DMR-0120967. B.M.B. acknowledges fellowship support from a TRIF Proposition 301 (Arizona) Graduate Fellowship in Photonics. The authors would like to thank Neal R. Armstrong, Jill Craven, Jinuk Jang, Yevgeniy Merzlyak, Gary Paysnoe, Clayton Shallcross, and Joseph Lynch for their scientific contributions to this work.

## 7. References

Abel, A. P., M. G. Weller, et al. (1996). Fiber-Optic Evanescent Wave Biosensor for the Detection of Oligonucleotides. *Analytical Chemistry*, Vol. 68, pp. 2905-2912

Anderson, G. P., J. P. Golden, et al. (1993). A Fiber Optic Biosensor: Combination Tapered Fibers Designed for Improved Signal Acquisition. *Biosensors and Bioelectronics*, Vol. 8, 5, pp. 249-256

Anderson, G. P., J. P. Golden, et al. (1994). An Evanescent Wave Biosensor-Part 2: Fluorescent Signal Acquisition from Tapered Fiber Optic Probes. *IEEE Transactions on Biomedical Engineering*, Vol. 41, 6, pp. 585-591

Anderson, G. P., J. P. Golden, et al. (1994). An Evanescent Wave Biosensor-Part I: Fluorescent Signal Acquisition from Step-Etched Fiber Optic Probes. *IEEE Transactions on Biomedical Engineering*, Vol. 41, 6, pp. 578-584

Andrade, J. D., R. A. Vanwagenen, et al. (1985). Remote Fiber-Optic Biosensors Based on Evanescent-Excited Fluoro-Immunoassay: Concept and Progress. *IEEE Transactions on Electron Devices*, Vol. ED-32, 7, pp. 1175-1179

Anis, N. A., M. E. Eldefrawi, et al. (1993). Reusable Fiber Optic Immunosensor for Rapid Detection of Imazethapry Herbicide. *Journal of Agricultural and Food Chemistry*, Vol. 41, pp. 843-848

Baia, M., F. Toderas, et al. (2006). Probing the enhancement mechanisms of SERS with p-aminothiophenol molecules adsorbed on self-assembled gold colloidal nanoparticles. *Chemical Physics Letters*, Vol. 422, 1-3, pp. 127-132

Bariain, C., I. R. Matias, et al. (2000). Optical Fiber Humidity Sensor Based on a Tapered Fiber Coated with Agarose Gel. *Sensors and Actuators B*, Vol. 69, pp. 127-131

Beam, B. M., N. R. Armstrong, et al. (2009). An Electroactive Fiber Optic Chip for Spectroelectrochemical Characterization of Ultra-Thin Redox Active Films. *Analyst*, Vol. 134, pp. 454-459

Beam, B. M., R. C. Shallcross, et al. (2007). Planar Fiber-Optic Chips for Broadband Spectroscopic Interrogation of Thin-Films. *Applied Spectroscopy*, Vol. 61, 6, pp. 585-592

Berge, L., S. Skupin, et al. (2007). Ultrashort filaments of Light in Weakly Ionized, Optically Transparent Media. *Reports on Progress in Physics*, Vol. 70, pp. 1633-1713

Bergmann, K. and C. T. O'Konski. (1963). A Spectroscopic Study of Methylene Blue Monomer, Dimer and Complexes with Montmorillonite. *Nature*, Vol. 67, pp. 2169-2177

Bier, F. F., W. Stocklein, et al. (1992). Use of a fibre optic immunosensor for the detection of pesticides. *Sensors and Actuators B*, Vol. 7, pp. 509-512

Blair, D. S., L. W. Burgess, et al. (1997). Evanescent Fiber-Optic Chemical Sensor for Monitoring Volatile Organic Compounds in Water. *Analytical Chemistry*, Vol. 69, pp. 2238-2246

Bradshaw, J. T., S. B. Mendes, et al. (2003). Broadband Coupling into a Single-Mode, Electroactive Integrated Optical Waveguide for Spectroelectrochemical Analysis of Surface-Confined Redox Couples. *Analytical Chemistry*, Vol. 75, 5, pp. 1080-1088

Bradshaw, J. T., S. B. Mendes, et al. (2005). Planar Integrated Optical Waveguide Spectroscopy. *Analytical Chemistry*, Vol. 77, 1, pp. 29A-36A

Browne, C. A., D. H. Tarrant, et al. (1996). Intrinsic Sol-Gel Clad Fiber-Optic Sensors with Time-Resolved Detection. *Analytical Chemistry*, Vol. 68, 14, pp. 2289-2295

Burrows, P. E., V. Bulovic, et al. (1994). Reliability and Degredation of Organic Light Emitting Devices. *Applied Physics Letters*, Vol. 65, pp. 2922-2924

Carniglia, C. K., L. Mandel, et al. (1972). Absorption and Emission of Evanescent Photons. *Journal of the Optical Society of America*, Vol. 62, 4, pp. 479-486

Chen, X. and O. Inganas. (1996). Three-Step Redox in Polythiophenes: Evidence from Electrocheistry at an Ultramicroelectrode. *Journal of Physical Chemistry*, Vol. 100, pp. 15202-15206

Collings, A. F. and F. Caruso. (1997). Biosensors: Recent Advances. *Reports on Progress in Physics*, Vol. 60, pp. 1397-1445

Cunningham, A. J. (1998). *Introduction to Bioanalytical Sensors* John Wiley and Sons, Inc., New York

Devine, P., N. A. Anis, et al. (1995). A Fiber-Optic Cocaine Biosensor. *Analytical Biochemistry*, Vol. 227, pp. 216-224

Diaz-Herrera, N., M. C. Navarrete, et al. (2004). A Fibre-Optic Temperature Sensor Based on the Deposition of a Thermochromic Material on an Adiabatic Taper. *Measurement Science and Technology*, Vol. 15, pp. 353-358

Doherty, W. J., C. L. Donley, et al. (2002). Broadband Spectroelectrochemical Attenuated Total Reflectance Instrument for Molecular Adlayer Studies. *Applied Spectroscopy*, Vol. 56, 7, pp. 920

Doherty, W. J., A. G. Simmonds, et al. (2005). Molecular Ordering in Monolayers of an Alkyl-Substituted Perylene-Bisimide Dye by Attenuated Total Reflectance Ultraviolet-Visible Spectroscopy. *Applied Spectroscopy*, Vol. 59, 10, pp. 1248-1256

Dresselhaus, M. S., G. Dresselhaus, et al. (2005). Raman spectroscopy of carbon nanotubes. *Physics Reports-Review Section of Physics Letters*, Vol. 409, 2, pp. 47-99

Dunphy, D. R., S. B. Mendes, et al. (1997). The Electroactive Integrated Optical Waveguide: Ultrasensitive Spectroelectrochemistry of Submonolayer Adsorbates. *Analytical Chemistry*, Vol. 69, 15, pp. 3086-3094

Dunphy, D. R., S. B. Mendes, et al. (1999). Spectroelectrochemistry of Monolayer and Submonolayer Films Using an Electroactive Integrated Optical Waveguide. In: *Interfacial Electrochemistry: Theory, Experiment and Applications*. A. Wieckowski. pp. (513-525), Marcel Dekker, Inc., New York

Eenink, R. G., H. E. de Bruijn, et al. (1990). Fibre-Fluorescence Immunosensor Based on Evanescent Wave Detection. *Analytica Chimica Acta*, Vol. 238, pp. 317-321

Egami, C., K. Takeda, et al. (1996). Evanescent-Wave Spectroscopic Fiber Optic pH Sensor. *Optics Communications*, Vol. 122, pp. 122-126

Felidj, N., J. Aubard, et al. (2003). Optimized surface-enhanced Raman scattering on gold nanoparticle arrays. *Applied Physics Letters*, Vol. 82, 18, pp. 3095-3097

Flora, W. H., S. B. Mendes, et al. (2005). Determination of Molecular Anisotropy in Thin Films of Discotic Assemblies Using Attenuated Total Reflectance UV -Visible Spectroscopy. *Langmuir*, Vol. 21, 1, pp. 360-368

Garden, S. R., G. J. Doellgast, et al. (2004). A Fluorescent Coagulation Assay for Thrombin Using a Fiber Optic Evanescent Wave Sensor. *Biosensors and Bioelectronics*, Vol. 19, pp. 737-740

Geng, T., J. Uknalis, et al. (2006). Fiber-Optic Biosensor Employing Alexa-Fluor Conjugated Antibody for Detection of *Escherichia coli* O157:H7 from Ground Beef in Four Hours. *Sensors*, Vol. 6, pp. 796-807

Glass, T. R., S. Lackie, et al. (1987). Effect of Numerical Aperture on Signal Level in Cylindrical Waveguide Evanescent Fluorosensors. *Applied Optics*, Vol. 26, 11, pp. 2181-2187

Gloge, D. (1971). Weakly Guiding Fibers. *Applied Optics*, Vol. 10, 10, pp. 2252-2258

Golden, J. P., L. C. Shriver-Lake, et al. (1992). Fluorometer and Tapered Fiber Optic Probes for Sensing in the Evanescent Wave. *Optical Engineering*, Vol. 31, 7, pp. 1458-1462

Graham, C. R., D. Leslie, et al. (1992). Gene Probe Assays on a Fibre-Optic Evanescent Wave Biosensor. *Biosensors and Bioelectronics*, Vol. 7, pp. 487-493

Grant, S. A. and R. S. Glass. (1997). A Sol-Gel Based Fiber Optic Sensor for Local Blood pH Measurements. *Sensors and Actuators B*, Vol. 45, pp. 35-42

Green, N. M. (1975). *Advances in Protein Chemistry*, Vol. 29, pp. 85-133

Guo, S. and S. Albin. (2003). Transmission Properties and Evanescent Wave Absorption of Cladded Multimode Fiber Tapers. *Optics Express*, Vol. 11, 3, pp. 215-223

Guo, S. J., Y. L. Wang, et al. (2007). Large-scale, rapid synthesis and application in surface-enhanced Raman spectroscopy of sub-micrometer polyhedral gold nanocrystals. *Nanotechnology*, Vol. 18, 40,

Gupta, B. D. and Ratnanjali. (2001). A Novel Probe for a Fiber Optic Humidity Sensor. *Sensors and Actuators B*, Vol. 80, pp. 132-135

Gupta, B. D. and D. K. Sharma. (1997). Evanescent Wave Absorption Based Fiber Optic pH Sensor Prepared by Dye Doped Sol-Gel Immobilization Technique. *Optics Communications*, Vol. 140, pp. 32-35

Gupta, B. D. and N. K. Sharma. (2002). Fabrication and Characterization of U-Shaped Fiber-Optic pH Probes. *Sensors and Actuators B*, Vol. 82, pp. 89-93

Gupta, B. D., C. D. Singh, et al. (1994). Fiber Optic Evanescent Field Absorption Sensor: Effect of Launching Condition and the Geometry of the Sensing Region. *Optical Engineering*, Vol. 33, 6, pp. 1864-1868

Gupta, R. and W. A. Weimer. (2003). High enhancement factor gold films for surface enhanced Raman spectroscopy. *Chemical Physics Letters*, Vol. 374, 3-4, pp. 302-306

Harrick, N. J. and G. I. Loeb. (1973). Multiple Internal Reflection Fluroescence Spectrometry. *Analytical Chemistry*, Vol. 45, 4, pp. 687-691

Hartnagel, H. L., A. L. Dawar, et al. (1995). *Semiconducting Transparent Thin Films* Institute of Physics Publishing, Philadelphia

Invitrogen. (2007). Spectra-Cascade Yellow Goat Anti-Mouse IgG Antibody/pH 8.0, Available from: <http://probes.invitrogen.com/servlets/spectra?fileid=10995ph8>

Itoh, K. and A. Fujishima. (1988). An Application of Optical Waveguides to Electrochemistry: Construction of Optical Waveguide Electrodes. *Journal of Physical Chemistry*, Vol. 92, pp. 7043-7045

Janata, J., M. Josowicz, et al. (1994). Chemical Sensors. *Analytical Chemistry*, Vol. 66, pp. 207R-228R

Kafafi, Z. H. (2005). *Organic Electroluminescence* Taylor & Francis Group, Boca Raton, FL

Kao, H. P., N. Yang, et al. (1998). Enhancement of Evanescent Fluorescence from Fiber-Optic Sensors by Thin-Film Sol-Gel Coatings. *Journal of the Optical Society of America A*, Vol. 15, 8, pp. 2163-2171

Kapoor, R., N. Kaur, et al. (2004). Detection of Trophic Factor Activated Signaling Molecules in Cells by a Compact Fiber-Optic Sensor. *Biosensors and Bioelectronics*, Vol. 20, pp. 345-349

Kendall, D. L. (1979). Vertical Etching of Silicon at Very High Aspect Ratios. *Annual Review of Material Science*, Vol. 9, pp. 373-403

Khijwania, S. K. and B. D. Gupta. (1998). Fiber Optic Evanescent Field Absorption Sensor with High Sensitivity and Linear Dynamic Range. *Optics Communications*, Vol. 152, pp. 259-262

Khijwania, S. K. and B. D. Gupta. (2000). Maximum Achievable Sensitivity of the Fiber Optic Evanescent Field Absorption Sensor Based on the U-Shaped Probe. *Optics Communications*, Vol. 175, pp. 135-137

Khijwania, S. K., V. S. Tiwari, et al. (2007). A fiber optic Raman sensor for hydrocarbon detection. *Sensors and Actuators B-Chemical*, Vol. 125, 2, pp. 563-568

Kneipp, K., H. Kneipp, et al. (2004). Surface-enhanced Raman scattering on single-wall carbon nanotubes. *Philosophical Transactions of the Royal Society of London Series a-Mathematical Physical and Engineering Sciences*, Vol. 362, 1824, pp. 2361-2373

Kneipp, K., Y. Wang, et al. (1997). Single molecule detection using surface-enhanced Raman scattering (SERS). *Physical Review Letters*, Vol. 78, 9, pp. 1667-1670

Krafft, C., M. Kirsch, et al. (2007). Methodology for fiber-optic Raman mapping and FTIR imaging of metastases in mouse brains. *Analytical and Bioanalytical Chemistry*, Vol. 389, 4, pp. 1133-1142

Kuswandi, B., R. Andres, et al. (2001). Optical Fibre Biosensors Based on Immobilized Enzymes. *The Analyst*, Vol. 126, pp. 1469-1491

Leung, A., P. M. Shankar, et al. (2007). Real-Time Monitoring of Bovine Serum Albumin at Femtogram/mL Levels on Antibody-Immobilized Tapered Fibers. *Sensors and Actuators B*, Vol. 123, pp. 888-895

Leung, A., P. M. Shankar, et al. (2007). A Review of Fiber-Optic Biosensors. *Sensors and Actuators B*, Vol. 125, pp. 688-703

Li, Z., Z. R. Li, et al. (2007). *Organic Light-Emitting Materials and Devices* CRC Press Taylor & Francis Group, Boca Raton

Lima, C. J., M. T. T. Pacheco, et al. (2008). Catheter with dielectric optical filter deposited upon the fiber optic end for Raman in vivo biospectroscopy applications. *Spectroscopy-an International Journal*, Vol. 22, 6, pp. 459-466

MacCraith, B. D. (1993). Enhanced Evanescent Wave Sensors Based on Sol-Gel Derived Porous Glass Coatings. *Sensors and Actuators B*, Vol. 11, pp. 29-34

MacCraith, B. D., C. M. McDonagh, et al. (1993). Fibre Optic Oxygen Sensor Based on Fluroescence Quenching of Evanescent-Wave Excited Ruthenium Complexes in Sol-Gel Derived Porous Coatings. *Analyst*, Vol. 118, pp. 385-388

Mackenzie, H. S. and F. P. Payne. (1990). Evanescent Field Amplification in a Tapered Single-Mode Optical Fibre. *Electronics Letters*, Vol. 26, 2, pp. 130-132

Malins, C., M. Landl, et al. (1998). Fibre Optic Ammonia Sensor Employing Novel Near Infrared Dyes. *Sensors and Actuators B*, Vol. 51, pp. 359-367

Maragos, C. M. and V. S. Thompson. (1999). Fiber-Optic Immunosensor for Mycotoxins. *Natural Toxins*, Vol. 7, pp. 371-376

Marazuela, M. D. and M. C. Moreno-Bondi. (2002). Fiber-Optic Biosensors: An Overview. *Analytical and Bioanalytical Chemistry*, Vol. 372, pp. 664-682

Marcuse, D. (1988). Launching Light into Fiber Cores from Sources Located in the Cladding. *Journal of Lightwave Technology*, Vol. 6, 8, pp. 1273-1279

McBee, T. W., L.-Y. Wang, et al. (2006). Characterization of Proton Transport Across a Waveguide-Supported Lipid Bilayer. *Journal of the American Chemical Society*, Vol. 128, 7, pp. 2184-2185

McCormack, T., G. O'Keeffe, et al. (1997). Optical Immunosensing of Lactate Dehydrogenase (LDH). *Sensors and Actuators B*, Vol. 41, pp. 89-96

McCreery, R. L., M. Fleischmann, et al. (1983). FIBER OPTIC PROBE FOR REMOTE RAMAN SPECTROMETRY. *Analytical Chemistry*, Vol. 55, 1, pp. 146-148

McDonagh, C., C. S. Burke, et al. (2008). Optical Chemical Sensors. *Chemical Reviews*, Vol. 108, pp. 400-422

McElvain, J., H. Antoniadis, et al. (1996). Formation and Growth of Black Spots in Organic Light-Emitting Diodes. *Journal of Applied Physics*, Vol. 80, pp. 6002-6007

Mignani, A. G., R. Falciai, et al. (1998). Evanescent Wave Absorption Spectroscopy by Means of Bi-Tapered Multimode Optical Fibers. *Applied Spectroscopy*, Vol. 52, 4, pp. 546-551

Mo, J., W. Zheng, et al. (2009). High Wavenumber Raman Spectroscopy for in Vivo Detection of Cervical Dysplasia. *Analytical Chemistry*, Vol. 81, 21, pp. 8908-8915

Monk, D. J., T. H. Ridgway, et al. (2002). Spectroelectrochemical Sensing Based on Multimode Selectivity Simultaneously Achievable in a Single Device. 15. Development of Portable Spectroelectrochemical Instrumentation. *Electroanalysis*, Vol. 15, 14, pp. 1198-1203

Monk, D. J. and D. R. Walt. (2004). Optical Fiber-Based Biosensors. *Analytical and Bioanalytical Chemistry*, Vol. 379, pp. 931-945

O'Keeffe, G., B. D. MacCraith, et al. (1995). Development of a LED-Based Phase Fluorimetric Oxygen Sensor Using Evanescent Wave Excitation of a Sol-Gel Immobilized Dye. *Sensors and Actuators B*, Vol. 29, pp. 226-230

Oroszlan, P., G. L. Duveneck, et al. (1993). Fiber-Optic Atrazine Immunosensor. *Sensors and Actuators B*, Vol. 11, pp. 301-305

Pandey, P. C. and H. H. Weetall. (1995). Detection of Aromatic Compounds Based on DNA Intercalation Using a Evanescent Wave Biosensor. *Analytical Chemistry*, Vol. 67, pp. 787-792

Plowman, T. E., S. S. Saavedra, et al. (1998). Planar Integrated Optical Methods for Examining Thin-Films and Their Surface Adlayers. *Biomaterials*, Vol. 19, pp. 341-355

Poscio, P., C. Depeursinge, et al. (1990). Realization of a Miniaturized Optical Sensor for Biomedical Applications. *Sensors and Actuators A*, Vol. 23, 1-3, pp. 1092-1096

Potyrailo, R. A. and G. M. Hieftje. (1998). Distributed Fiber-Optic Chemical Sensor with Chemically Modified Plastic Cladding. *Applied Spectroscopy*, Vol. 52, 8, pp. 1092-1095

Potyrailo, R. A. and G. M. Hieftje. (1998). Oxygen Detection by Fluorescence Quenching of Tetraphenylporphyrin Immobilized in the Original Cladding of an Optical Fiber. *Analytica Chimica Acta*, Vol. 370, pp. 1-8

Potyrailo, R. A. and G. M. Hieftje. (1999). Use of the Original Silicone Cladding of an Optical Fiber as a Reagent-Immobilization Medium for Intrinsic Chemical Sensors. *Fresenius Journal of Analytical Chemistry*, Vol. 364, pp. 32-40

Potyrailo, R. A., S. E. Hobbs, et al. (1998). Optcal Waveguide Sensors in Analytical Chemistry: Today's Instrumentation, Applications and Trends for Future Development. *Fresenius Journal of Analytical Chemistry*, Vol. 362, pp. 349-373

Potyrailo, R. A., S. E. Hobbs, et al. (1998). Optical waveguide sensors in analytical chemistry: today's instrumentation, applications and trends for future development. *Fresenius Journal of Analytical Chemistry*, Vol. 362, 4, pp. 349-373

Preejith, P. V., C. S. LIm, et al. (2003). Total Protein Measurement Using a Fiber-Optic Evanescnet Wave-Based Biosensor. *Biotechnology Letters*, Vol. 25, pp. 105-110

Reichert, W. M. (1989). Evanescent Detection of Adsorbed Films: Assessment of Optical Considerations for Absorbance and Fluorescence Spectroscopy at the Crystal/Solution and Polymer Solution Interfaces. *Critical Reviews in Biocompatibility*, Vol. 5, 2, pp. 173-205

Rogers, K. R., M. E. Eldefrawi, et al. (1991). Pharmacological Specificity of a Nicotinic Acetylcholine Receptor Optical Sensor. *Biosensors and Bioelectronics*, Vol. 6, pp. 507-516

Rogers, K. R., J. J. Valdes, et al. (1989). Acetylcholine Receptor Fiber-Optic Evanescent Fluorosensor. *Analytical Biochemistry*, Vol. 182, pp. 353-359

Ruddy, V., B. D. MacCraith, et al. (1990). Evanescent Wave Absorption Spectroscopy Using Multimode Fibers. *Journal of Applied Physics*, Vol. 67, 10, pp. 6070-6074

Samuel, I. D. W. and G. A. Turnbull. (2007). Organic Semiconductor Lasers. *Chemical Reviews*, Vol. 107, 4, pp. 1272-1295

Shaw, M. J. and W. E. Geiger. (1996). A New Approach to Infrared Spectroelectrochemistry Using a Fiber-Optic Probe: Application to Organometallic Redox Chemistry. *Organometallics*, Vol. 15, pp. 13-15

Shriver-Lake, L. C., J. P. Golden, et al. (1995). Use of Three Longer-Wavelength Fluorophores with the Fiber-Optic Biosensor. *Sensors and Actuators B*, Vol. 29, pp. 25-30

Smith, E. and G. Dent, Eds. (2005). *Modern Raman Spectroscopy: a practical approach*, John Wiley and Sons Ltd, Chichester

Stegemiller, M. L., W. R. Heineman, et al. (2003). Spectroelectrochemical Sensing Based on Multimode Selectivity Simultaneously Achievable in a Single Device. 11. Design and Evaluation of a Small Portable Sensor for the Determination of Ferrocyanide in Hanford Waste Samples. *Environmental Science and Technology*, Vol. 37, pp. 123-130

Stokes, D. L. and T. Vo-Dinh. (2000). Development of an integrated single-fiber SERS sensor. *Sensors and Actuators B-Chemical*, Vol. 69, 1-2, pp. 28-36

Stokes, D. L., Z. H. Chi, et al. (2004). Surface-enhanced-Raman-scattering-inducing nanoprobe for spectrochemical analysis." *Applied Spectroscopy*, 58(3): 292-298

Sutherland, R. M., C. Dahne, et al. (1984). Optical Detection of Antibody-Antigen Reactions at a Glass-Liquid Interface. *Clinical Chemistry*, Vol. 30, 9, pp. 1533-1538

Tai, H., H. Tanaka, et al. (1987). Fiber-Optic Evanescent-Wave Methane-Gas Sensor Using Optical Absorption for the 3.392-micron Line of a He-Ne Laser. *Optics Letters*, Vol. 12, 6, pp. 437-439

Taitt, C. R., G. P. Anderson, et al. (2005). Evanescent Wave Fluorescence Biosensors. *Biosensors and Bioelectronics*, Vol. 20, pp. 2470-2487

Thompson, R. B. (2006). *Fluorescence Sensors and Biosensors* Taylor and Francis, Boca Raton

Thompson, V. S. and C. M. Maragos. (1996). Fiber-Optic Immunosensor for the Detection of Fumonisin $B_1$. *Journal of Agricultural and Food Chemistry*, Vol. 44, pp. 1041-1046

Tien, P. K. (1971). Light Waves in Thin Films and Integrated Optics. *Applied Optics*, Vol. 10, 11, pp. 2395-2413

Tiwari, V. S., R. R. Kalluru, et al. (2007). Fiber optic Raman sensor to monitor the concentration ratio of nitrogen and oxygen in a cryogenic mixture. *Applied Optics*, Vol. 46, 16, pp. 3345-3351

Toppozada, A. R., J. Wright, et al. (1997). Evaluation of a Fiber Optic Immunosensor for Quantitating Cocaine in Coca Leaf Extracts. *Biosensors and Bioelectronics*, Vol. 12, 2, pp. 113-124

VanDyke, D. A. and H.-Y. Cheng. (1988). Fabrication and Characterization of a Fiber-Optic Based Spectroelectrochemical Probe. *Analytical Chemistry*, Vol. 60, pp. 1256-1260

Villatoro, J., D. Luna-Moreno, et al. (2005). Optical Fiber Hydrogen Sensor for Concentrations Below the Lower Explosive Limit. *Sensors and Actuators B*, Vol. 110, pp. 23-27

Vo-Dinh, T., D. L. Stokes, et al. (1999). Surface-enhanced Raman scattering (SERS) method and instrumentation for genomics and biomedical analysis. *Journal of Raman Spectroscopy*, Vol. 30, 9, pp. 785-793

Walczak, I. M., W. F. Love, et al. (1992). The Application of Evanescent Wave Sensing to a High-Sensitivity Fluoroimmunoassay. *Biosensors and Bioelectronics*, Vol. 7, pp. 39-48

Wang, Y., S. Guo, et al. (2008). Facile fabrication of large area of aggregated gold nanorods film for efficient surface-enhanced Raman scattering. *Journal of Colloid and Interface Science*, Vol. 318, 1, pp. 82-87

Wei, Z. Q., A. J. Mieszawska, et al. (2004). Synthesis and manipulation of high aspect ratio gold nanorods grown directly on surfaces. *Langmuir*, Vol. 20, 11, pp. 4322-4326

Wei, Z. Q. and F. P. Zamborini. (2004). Directly monitoring the growth of gold nanoparticle seeds into gold nanorods. *Langmuir*, Vol. 20, 26, pp. 11301-11304

Wiejata, P. J., P. M. Shankar, et al. (2003). Fluorescent Sensing Using Biconical Tapers. *Sensors and Actuators B*, Vol. 96, pp. 315-320

Willets, K. A. and R. P. Van Duyne. (2007). Localized surface plasmon resonance spectroscopy and sensing. *Annual Review of Physical Chemistry*, Vol. 58, pp. 267-297

Winograd, N. and T. Kuwana. (1969). Characteristics of the Electrode-Solution Interface Under Faradaic and Non-Faradaic Conditions as Observed by Internal Reflection Spectroscopy. *Journal of Electroanalytical Chemistry*, Vol. 23, 3, pp. 333-342

Wolfbeis, O. S. (2006). Fiber-Optic Chemical Sensors and Biosensors. *Analytical Chemistry*, Vol. 78, pp. 3859-3874

Xu, H. X., E. J. Bjerneld, et al. (1999). Spectroscopy of single hemoglobin molecules by surface enhanced Raman scattering. *Physical Review Letters*, Vol. 83, 21, pp. 4357-4360

Zhang, Y., C. Gu, et al. (2005). Surface-enhanced Raman scattering sensor based on D-shaped fiber. *Applied Physics Letters*, Vol. 87, 12,

Zhang, Z. Y., Q. Jiang, et al. (2009). A p-type-doped Quantum Dot Superluminescent LED with Broadband and Flat-Topped Emission Spectra Obtained by Post-Growth Intermixing Under a GaAs Proximity Cap. *Nanotechnology*, Vol. 4, pp. 055204-055208

Zhou, C., P. Pivarnik, et al. (1997). A Compact Fiber-Optic Immunosensor for *Salmonella* Based on Evanescent Wave Excitation. *Sensors and Actuators B*, Vol. 42, pp. 169-175

# 19

# Optical Fibers in Phase Space: A Theoretical Framework

Ana Leonor Rivera
*Centro de Fisica Aplicada y Tecnologia Avanzada,*
*Universidad Nacional Autonoma de Mexico*
*Mexico*

## 1. Introduction

Since their discovery, optical fibers have received increasing attention due to its important technological applications (Bottacchi, 2002; Culshaw, 1997; Harmon, 2001; Herrmann, 1973; Keyl, 2002; Lauterborn et al., 1997; Prasad & Williams, 1991; Shimizu et al., 1997; Way, 1998; Young, 2000). A fiber is an optical waveguide in which the propagation of an optical wave is confined to two dimensions (the cross section dimensions). The dimension of the confinement must be comparable to the wavelength of the light which one would like to confine (Kogelnik, 1979).

A fiber is formed by a region with a refractive index larger than that of the surrounding media; this condition assures the total internal reflections at the interfaces required for the beam propagation. If an integer number of wavelengths have traveled between two consecutive reflections, a standing wave pattern will be developed, producing a constructive interference which allows high electromagnetic power to be transmitted along the fiber. A change in the fiber geometry, modifies the reflection angle and consequently the total number of reflections: rays traveling thicker fibers require more reflections to travel along the same length in the propagation direction. This implies that the effective velocity of light in thicker fibers must be slower; this effect also produces a change in the phase of the output signal (Arnaud, 1976). These phenomena are cumbersome to explain by standard optical methods (Kogelnik, 1979; Torchigin & Torchigin, 2003) and are easily predicted by the use of the Wigner Distribution Function (Reyes et al., 1999). One of the main advantages of the phase space approach to solve optical problems is the simplicity of the mathematical calculations compared with the traditional treatment (Bottacchi, 2002). This is due to the fact that in phase space representation, the relevant properties of the system can be obtained by simple matrices products.

Quantum mechanically, the development of nonlinear optics allowed the generation and manipulation of new quantum states of light, going from the simplest and common one, the so-called coherent states (Glauber, 1963), to squeezed states (Walls, 1983; Yuen, 1976), Fock states (Lvovsky et al., 2001) or entangled states (Ou et al., 1992). A full quantum theoretical analysis of the three-photon states is contained in the Wigner function (Leonhardt, 2001) that has proven to be very helpful to visualize in the phase space (the amplitude $\mathbf{q}$ and phase $\mathbf{p}$ quadratures) quantum mechanical system defined by its density matrix. This has already been the case for some quantum states of light such as the coherent state, the squeezed vacuum or the bright squeezed state (Breitenbach et al., 1995; 1997; Smithey et al., 1993), whose Wigner function has been experimentally reconstructed using homodyne quantum tomography, a technique that allows the measurement of the marginal probability distribution that expresses the quadrature amplitude distribution. The use of optical fiber for quantum squeezing has considerable technological advantages, such as generating squeezing directly at the communications wavelength and the use of existing transmission technology (Corney et al., 2008).

More generally, the Wigner function contains the full information about the quantum states (Wigner, 1932) and their moments allow to differentiate between paraxial regime, wave-like regime and chaotic behavior (Rivera et al., 1997). More particularly, it allows us to establish the quantum correlations between the different generated modes in the case of twin photons or photon triplets (Benchekh et al., 2007). The Wigner function is a positive definite function in the phase space only for classical states with Gaussian marginal probability distributions (Rivera & Castano, 2010). However, it can be negative in some circumstances for particular quantum states of light. These negativities are the signature of highly nonclassical behaviour of a quantum state (Rivera & Castano, 2010) as it has been observed for a quantum state of light prepared in a single-photon Fock state (Lvovsky et al., 2001). These quantum negativities are also present in the case of complete degenerate three-photon states obtained by third order optical parametric fluorescence or amplification, and also for aberrated optical systems.

Historically, the so-called *Wigner Distribution Function* (Wigner, 1932) has been of central importance as an alternative description of Quantum Mechanics (Kim & Noz, 1991). However, these phase-space mathematical tool has found exciting applications in a wide range of the physical sciences and even engineering ranging from statistical mechanics (Green, 1951; Mori et al., 1962) to optics (Perinova et al., 1998; Schleich, 2001; Wolf, 2004). Moreover, it has become the basis of an entire discipline: time-frequency representation of wave phenomena (Allen & Mills, 2004; Boashash, 2003; Cohen, 1995; Grochenig, 2000). There exist several reviews of the quantum phase-space distribution functions, in particular of the Wigner distribution function. A concise but authorative review of the quantum distribution functions is that by Wigner (Wigner, 1971). A good mathematical treatment of the quantum distribution functions and related operator algebra is given in the book by (Louisell, 1973). Some extensive reviews of the quantum distribution functions are given by (Balazs & Jennings, 1984; Berry, 1977; Filinov et al., 2008; Groot & Suttorp, 1972; Hillery et al., 1984; Lee, 1995; OConnell, 1983; Takabayasi, 1954). Applications of the Wigner distribution function to Optics are reviewed by (Dragoman, 1997; Dodonov, 2002; Mack & Schleich, 2003; Zalevsky & Mendlovic, 1997), and for the particular case of fibers on the works (Bao & Chen, 2011; Benabid & Roberts, 2011; Benchekh et al., 2007; Corney et al., 2008; Leonhardt, 2001; Rivera & Castano, 2010a).

As Wigner functions, the *Lie Algebra*, due to its mathematical simplicity to solve differential equations by numerical integration, has become an important aid for the solution of different problems in classical and quantum mechanics (Bakhturin, 2003; Frank & van Isacker, 1994; Hamermesh, 1962; Jacobson, 1979). A Lie treatment of geometrical optics and aberrations has been developed by (Dragt & Finn, 1976), and it is a new approach to fiber optics (Reyes et al., 1999; Reyes & Castano, 2000) that simplifies the traditional solution of optical problems (Born & Wolf, 1999) to the determination of the corresponding Symplectic Map associated to the optical system, thus reducing the problem to simple matrices products. The Gaussian Symplectic map helps to find the Wigner distribution function of the probability density of an optical fiber, and from it, it is possible to obtain all the physical information required to analyze the fiber (Rivera & Castano, 2010a).

This chapter presents a brief review of the phase-space analysis applied to fiber Optics, using the Wigner Distribution Function. The hope is that it will show the beauty, elegance and usefulness of this mathematical construction. The rest of the chapter is organized as follows. Section 2 gives a short review of phase space representation using the Wigner Distribution Function which describes some of its important properties and its physical interpretation. Section 3 presents the description of the Maxwell equations under paraxial approach (considering parallel rays close to the optical axis of the system) that describe the light propagation in a fiber by a parabolic type equation that is completely equivalent to the quantum system Schrödinger equation for a bidimensional potential-well time-dependent. Section 4 shows an example, analyzing a gaussian beam propagation through a fiber.

## 2. Phase space representation

The standard formulation of quantum mechanics either in the Schrödinger (Schrodinger, 1946) or in Heisenberg pictures (Heisenberg, 1930) may create an impression that quantum and classical dynamics are completely different (Dirac, 1935). However, there are representations in which quantum dynamics seems to resemble classical statistical mechanics, and where the state of a quantum system is represented by the quasiprobability distribution in phase space of the corresponding classical system (Kim & Noz, 1991). Of course, there are at least two important differences (Hillery et al., 1984):

1. Quasiprobability distributions may take negative values (unlike the true probability distributions).
2. The classical distribution can be localized at a point in phase space, whereas the quantum distribution must always be spread in a finite phase volume, in agreement with uncertainty relations.

Among different quasiprobability distributions Cohen (1995), the Wigner Distribution Function, introduced by Wigner in 1932[1] (Wigner, 1932), is the only one for which the quantum evolution law coincides with the classical one for the case of linear dynamics (Moyal, 1949). The Wigner distribution function is a real valued quasiprobability distribution containing all information available about the system. Its popularity stems from its characteristics (Wigner, 1932):

- It has a close connection to the marginal probability distributions characterizing the probabilities of the outcomes of von Neumann measurements of the system.
- It lends itself to a visualization of quantum states, and some of their properties.
- It is a versatile calculation tool for normally ordered operators.

With the use of this distribution function, it is straightforward to cast quantum mechanics in a form which resembles the classical theory of statistical averages over the classical phase space, with the Wigner distribution function playing a role analogous to a probability function

[1] Wigner's original motivation for introducing it, was to be able to calculate the quantum correction to the second virial coefficient of a gas, which indicates how it deviates from the ideal gas law (Wigner, 1932). Classically, to ca1culate the second virial coefficient one needs a joint distribution of position and momentum. So Wigner devised the simplest joint distribution that gave, as marginals, the quantum mechanical distributions of position and momentum. The quantum mechanics came in the distribution, but the distribution was used in the classical manner. It was a hybrid method. Also, Wigner was motivated in part by the work of Kirkwood (Kirkwood, 1933) who had previously calculated this quantity but Wigner improved it.

(Fairlie, 1964). Consequently, the Wigner distribution function has been used extensively to study the classical limit of quantum mechanical systems (Kim & Noz, 1991; Mayer & Band, 1947; Moyal, 1949).
There is in principle an infinite variety of quantum phase-space distribution functions corresponding to an infinite number of possible ordering rules of two noncommuting operators and their linear combinations (Lee, 1995). The general class of this distributions is given by (Cahill & Glauber, 1969; Cohen, 1966; Kakazu et al., 2007). Distribution functions in general have different properties and are associated with various dynamical equations, so they may be described most conveniently by distribution functions having different characteristics (Hillery et al., 1984). Other distribution functions that have been considered in the past include those of Kirkwood (Kirkwood, 1933), Margenau-Hill (Johansen & Luis, 2004; Margenau & Hill, 1961; Terletsky, 1937)[2], Husimi (Husimi, 1940), Q-functions (Husimi, 1940; Kano, 1965; Smith, 2006), Page (Page, 1952), Glauber-Sudarshan (Glauber, 1963; Sudarshan, 1963), Rihaczek (Rihaczek, 1968), and Choi-Williams (Choi & Williams, 1989). Another very used function is the ambiguity function (Bastiaans, 1980; Marks & Hall, 1979; Woodward, 1963).
Phase space representation, in particular through the so-called *Wigner Distribution Function*, has proven to be a very effective tool applied in many branches of physics (Cohen, 1995; Kim & Wigner, 1987; Kim & Noz, 1991; Mecklenbrauker et al., 1997; Moyal, 1949; Stewart et al., 2002; Wigner, 1932), and more specifically in fiber optics (Dragoman & Meunier, 1998; Kominis & Hizanidis, 2002; Reyes et al., 1999a; Rivera & Castano, 2010a; Sheppard & Larkin, 2000; Voss et al., 1999). The Wigner distribution function was invented by Wigner (Wigner, 1932) to study the quantum corrections to the classical behavior of certain statistical systems described by the Boltzmann formula. For the evaluation of the Wigner function are various implementations (Bala & Prabhu, 1989; Easton et al., 84; Eilouti & Khadra, 1989; Flandrin et al., 1984; Frank et al., 2000; Gupta & Asakura, 1986; Lohmann, 1980; Lopez et al., 2002; Maanen, 1985; Mateeva & Sharlandjiev, 1986; Rivera et al., 1997; Subotic & Saleh, 1984).
In this chapter the notation will be for the optical position coordinates $\mathbf{q}$ (that corresponds to the interaction of the ray with a $z = 0$ reference plane), and for the canonically conjugate momentum $\mathbf{p}$ (which describes the direction of the ray with respect to the normal at the point $\mathbf{q}$ that evolves over the system's optical axis, $z$) (Buchdahl, 1970).
Consider a particle in one dimension. Classically, the particle is described by a *phase space distribution* $P_{cl}(\mathbf{q}, \mathbf{p})$. The average of a function of position and momentum $A(\mathbf{q}, \mathbf{p})$ can then be expressed as

$$\langle A \rangle_{cl} = \int_{-\infty}^{\infty} \mathrm{d}\mathbf{q} \int_{-\infty}^{\infty} \mathrm{d}\mathbf{p}\, A(\mathbf{q}, \mathbf{p})\, P_{cl}(\mathbf{q}, \mathbf{p})\,. \tag{1}$$

A quantum mechanical particle is described by a density matrix $\hat{\rho}$, and the average of a function of the position and momentum operators $\hat{A}(\hat{q}, \hat{p})$ as

$$\langle A \rangle_{quant} = \mathrm{Tr}\,(\hat{A}\,\hat{\rho})\,. \tag{2}$$

It must be admitted that, given the classical expression $A(\mathbf{q}, \mathbf{p})$, the corresponding self adjoin operator $\hat{A}$ is not uniquely defined. The use of a *quasiprobability phase space distribution*

[2] Kirkwood attempted to extend the classical theory to the quantum case and devised the distribution commonly called the Rihaczek or Margenau-Hill distribution to do that. Many years later, Margenau and Hill derived the distribution that bears their name. The importance of the Margenau-Hill work is not the distribution but the derivation. They were also the first to consider joint distributions involving spin.

$P_Q(\mathbf{q},\mathbf{p})$, however, does give such a definition by expressing the quantum mechanical average as

$$\langle A\rangle_{quant} = \int_{-\infty}^{\infty} \mathrm{d}\mathbf{q} \int_{-\infty}^{\infty} \mathrm{d}\mathbf{p}\, A(\mathbf{q},\mathbf{p})\, P_Q(\mathbf{q},\mathbf{p})\,, \tag{3}$$

where the function $A(\mathbf{q},\mathbf{p})$ can be derived from the operator $\hat{A}(\hat{q},\hat{p})$ by a well defined correspondence rule. This allows one to cast quantum mechanical results into a form in which they resemble classical ones. This is a reformulation of Schrödinger quantum mechanics which describes states by functions in configuration space (Kim & Noz, 1991).
In the case where $P_Q$ in (3) is chosen to be the Wigner Distribution function (Wigner, 1932), then the correspondence between $A(\mathbf{q},\mathbf{p})$ and $\hat{A}$ is that proposed by Weyl (Weyl, 1927), as was first demonstrated by Moyal (Moyal, 1949).
The requirement given by Eq. (3) let us to define a function in the $6N$ dimensional $\mathbf{q},\mathbf{p}$ phase space, called the *Wigner Distribution Function* in terms of the density matrix, $\rho$ as:

$$\mathcal{W}_\rho\left(\mathbf{q},\mathbf{p};t\right) \equiv \left(\frac{1}{\pi\hbar}\right)^{3N} \int_{-\infty}^{\infty} \mathrm{d}\mathbf{r}\, \exp\left(\frac{2i}{\hbar}\,\mathbf{p}\cdot\mathbf{r}\right)\, \rho(\mathbf{q}-\mathbf{r},\mathbf{q}+\mathbf{r};t)\,, \tag{4}$$

Because for pure states described by a wavefunction $\Psi$, the density matrix is given by (vonNeumann, 1927)

$$\rho(\mathbf{q},\mathbf{q}') = \Psi^*(\mathbf{q}')\,\Psi(\mathbf{q})\,, \tag{5}$$

the expression (4) for pure states can be rewritten in coordinate representation as:

$$\mathcal{W}_\Psi(\mathbf{q},\mathbf{p};z) \equiv \frac{1}{2\pi\hbar} \int_{-\infty}^{\infty} \mathrm{d}\mathbf{r}\, \Psi^*\left(\mathbf{q}-\frac{1}{2}\mathbf{r};z\right)\, e^{-i[\mathbf{p}\cdot\mathbf{r}]/\hbar}\, \Psi\left(\mathbf{q}+\frac{1}{2}f\mathbf{r};z\right) \tag{6}$$

or taking the Fourier transform (Goodman, 1968) in momentum representation as

$$\mathcal{W}_\Psi(\mathbf{q},\mathbf{p};z) = \frac{1}{2\pi\hbar} \int_{-\infty}^{\infty} \mathrm{d}\mathbf{r}\, \tilde{\Psi}\left(\mathbf{p}+\frac{1}{2}\mathbf{r};z\right)\, e^{-i[\mathbf{q}\cdot\mathbf{r}]/\hbar}\, \tilde{\Psi}^*\left(\mathbf{p}-\frac{1}{2}\mathbf{r};z\right)\,, \tag{7}$$

where $\hbar$ is the Planck constant divided by $2\pi$, $\tilde{\Psi}$ denotes the Fourier transform of $\Psi$ and the asterisk represents the complex conjugate. In geometric optics, $\hbar$ corresponds to the wavelength $\lambda$ of the beam (Wolf, 2004).
In Wigner phase-space representation everything we have said for the coordinate domain holds for the momentum domain because the Wigner distribution is basically identical in form in both domains (compare equations 6 and 7). The complete symmetry between $\mathbf{q}$ and $\mathbf{p}$ in the former definitions of the Wigner function (equations 6 and 7), indicates that space and momentum have equal weight in this description (Moyal, 1949). Due to this, the Wigner distribution function can be thought as the expected value of the parity operator around $(\mathbf{q},\mathbf{p})$ in the phase space (Royer, 1997); i.e. the Wigner function is proportional to the overlap of $\Psi(\mathbf{q},z)$ with its specular image around $(\mathbf{q},\mathbf{p})$, that is a measure of "how much centered" is $\Psi(\mathbf{q},z)$. Note that the Wigner distribution function is a 4-dimensional phase space distribution function, where two dimensions correspond to real space and the other two to momentum space. The Wigner distribution function is a real function that can take either positive and negative values, however, only for a Gaussian the Wigner distribution function is positive everywhere (Hudson, 1974; Soto & Claverie, 1983); therefore, one cannot interpret it as a classical probability function in phase space (Lee, 1995). Tthe value of $\mathcal{W}_\Psi$ mirrors closely the intuitive objects in the model, that in the case of quantum optics may be the coherent states

of the radiation field (Glauber, 1965), and in monochromatic paraxial wave optics, they are often beams with Gaussian position and inclination distributions (Hillery et al., 1984; Rivera & Castano, 2010).

From all phase space representations, the Wigner Distribution Function can be uniquely distinguished (among shift-invariant joint distributions) by imposing a requirement of correct marginals with respect to arbitrary directions in the time-frequency plane, thus connecting the Wigner distribution with the fractional Fourier transform (Atakishiyev et al., 1999). It also contains all the information of the system and it can be proved that it contains the hologram of the signal (Wolf & Rivera, 1997).

For numerical calculations it is very useful to note that the Wigner distribution function is the Fourier transform of the kernel (Wigner, 1932):

$$\mathcal{W}_{\Psi}(\mathbf{q},\mathbf{p};z) = \Psi\left(\mathbf{q}+\frac{1}{2}\mathbf{r}\right)\Psi^{*}\left(\mathbf{q}-\frac{1}{2}\mathbf{r}\right) . \tag{8}$$

Because $\mathcal{W}_{\Psi}(\mathbf{q},\mathbf{p};z)$ is Hermitian $[\mathcal{W}_{\Psi}(\mathbf{q},\mathbf{r}) = \mathcal{W}^{*}_{\Psi}(\mathbf{q},-\mathbf{r})]$, the Wigner distribution function is real (Moyal, 1949).

When we integrate $\mathcal{W}_{\Psi}(\mathbf{q},\mathbf{p})$ over $\mathbf{p}$, we obtain the probability distribution in $\mathbf{q}$, while if we integrate $\mathcal{W}_{\Psi}(\mathbf{q},\mathbf{p})$ over $\mathbf{q}$, we obtain the probability distribution in $\mathbf{p}$, (Moyal, 1949). Then, to recover either the image $|\Psi(\mathbf{q};z)|^2$ (light intensity on the two-dimensional screen of coordinate $\mathbf{q}$ at the optical axis position $z$) or the diffraction pattern $|\tilde{\Psi}(\mathbf{p};z)|^2$, it is necessary to make a simple projection of the Wigner distribution function (Wigner, 1932):

$$|\Psi(\mathbf{q};z)|^2 = \int_{-\infty}^{\infty} \mathrm{d}\mathbf{p}\, \mathcal{W}_{\Psi}(\mathbf{q},\mathbf{p};z) , \tag{9}$$

$$|\tilde{\Psi}(\mathbf{p};z)|^2 = \int_{-\infty}^{\infty} \mathrm{d}\mathbf{q}\, \mathcal{W}_{\Psi}(\mathbf{q},\mathbf{p};z) . \tag{10}$$

If the signal or image of interest is nonstationary, the Wigner distribution function gives the local spectrum centered at $\mathbf{p}$ as a function of location (Bartelt et al., 1980). Thus, the total energy of $\Psi(\mathbf{q},z)$ can be obtained from integration of $\mathcal{W}_{\Psi}(\mathbf{q},\mathbf{p};z)$ over the entire phase space (Hillery et al., 1984).

Moreover $|\mathcal{W}_{\Psi}(\mathbf{q},\mathbf{p};z)| \leq (2\pi\hbar)^{-1}$.

Another interesting property (Schempp, 1986) is that the Wigner distribution function has the same extension and is band-limited as the function $\Psi(\mathbf{q},z)$.

The Wigner distribution function is the expectation value of the parity operator about the phase-space point $\mathbf{q},\mathbf{p}$ (Royer, 1997). To show this, let us first rewrite

$$\mathcal{W}(\mathbf{q},\mathbf{p}) = \left(\frac{1}{\pi\hbar}\right)^{3N} \langle\Psi|\hat{\Pi}_{\mathbf{q},\mathbf{p}}|\Psi\rangle , \tag{11}$$

where the operator $\hat{\Pi}_{\mathbf{q},\mathbf{p}}$ has the following three equivalent expressions:

$$\begin{aligned}
\hat{\Pi}_{\mathbf{q},\mathbf{p}} &= \int_{-\infty}^{\infty} \mathrm{d}r\, e^{2ipr/\hbar}\, |q-r\rangle\langle q+r| , \\
&= \int_{-\infty}^{\infty} \mathrm{d}k\, e^{-2ikq/\hbar}\, |p+k\rangle\langle p-k| , \\
&= \left(\frac{1}{\pi\hbar}\right)^{3N} \int_{-\infty}^{\infty} \mathrm{d}k \int_{-\infty}^{\infty} \mathrm{d}r\, e^{i[k(\hat{R}-q)+r(\hat{P}-p)]/\hbar} .
\end{aligned} \tag{12}$$

Let us now consider the special case $q = 0, p = 0$, and denote $\hat{\Pi}_{q=0,p=0} = \hat{\Pi}$; we have

$$\begin{aligned}
\hat{\Pi} &= \int_{-\infty}^{\infty} \mathrm{d}q \, |-q\rangle \langle q| \,, \\
&= \int_{-\infty}^{\infty} \mathrm{d}p \, |p\rangle \langle -p| \,, \\
&= \left(\frac{1}{\pi\hbar}\right)^{3N} \int_{-\infty}^{\infty} \mathrm{d}k \int_{-\infty}^{\infty} \mathrm{d}y \; e^{i[k\hat{R}+y\hat{P}]/\hbar} \,.
\end{aligned} \tag{13}$$

From (13) it is immediately apparent that $\hat{\Pi}$ is the parity operator (about the origin): it changes $\Psi(\mathbf{q})$ into $\Psi(-\mathbf{q})$ and $\tilde{\Psi}(\mathbf{p})$ into $\tilde{\Psi}(-\mathbf{p})$, or equivalently

$$\hat{\Pi}\hat{R}\hat{\Pi} = -\hat{R} \,, \qquad \hat{\Pi}\hat{P}\hat{\Pi} = -\hat{P} \,, \tag{14}$$

moreover,

$$\hat{\Pi}^{-1} = \hat{\Pi} \,. \tag{15}$$

We now observe that $\hat{\Pi}_{\mathbf{q},\mathbf{p}}$ may be obtained from $\hat{\Pi}$ by a unitary transformation

$$\hat{\Pi}_{\mathbf{q},\mathbf{p}} = \hat{\mathcal{D}}(\mathbf{q},\mathbf{p}) \, \hat{\Pi} \, \hat{\mathcal{D}}(\mathbf{q},\mathbf{p})^{-1} \,; \tag{16}$$

here

$$\hat{\mathcal{D}}(\mathbf{q},\mathbf{p}) = e^{i(p\hat{R}-q\hat{P})/\hbar} \tag{17}$$

is a phase-space displacement operator, introduced by Glauber (Glauber, 1963) in connection with a different, though related, type of phase-space representation of quantum mechanics, the coherent-state representation. We have the actions

$$\hat{\mathcal{D}}(\mathbf{q},\mathbf{p})^{-1} \, \hat{R} \, \hat{\mathcal{D}}(\mathbf{q},\mathbf{p}) = \hat{R} + q \,, \tag{18}$$

$$\hat{\mathcal{D}}(\mathbf{q},\mathbf{p})^{-1} \, \hat{P} \, \hat{\mathcal{D}}(\mathbf{q},\mathbf{p}) = \hat{P} + p \,, \tag{19}$$

$$\hat{\mathcal{D}}(\mathbf{q},\mathbf{p})^{-1} \, F(\hat{R},\hat{P}) \, \hat{\mathcal{D}}(\mathbf{q},\mathbf{p}) = F(\hat{R}+q,\hat{P}+p) \,. \tag{20}$$

From this follows directly

$$\hat{\Pi}_{\mathbf{q},\mathbf{p}}(\hat{R}-q)\hat{\Pi}_{\mathbf{q},\mathbf{p}} = -(\hat{R}-q) \,, \tag{21}$$

$$\hat{\Pi}_{\mathbf{q},\mathbf{p}}(\hat{P}-p)\hat{\Pi}_{\mathbf{q},\mathbf{p}} = -(\hat{P}-p) \,, \tag{22}$$

that is, $\hat{\Pi}_{\mathbf{q},\mathbf{p}}$ reflects about the phase-space point $\mathbf{q},\mathbf{p}$ and is thus the parity operator about that point. Note that

$$(\hat{\Pi}_{\mathbf{q},\mathbf{p}})^2 = 1 \,. \tag{23}$$

The Wigner function, is thus $\left(\frac{1}{\pi\hbar}\right)^{3N}$ times the expectation value of the parity operator about $\mathbf{q},\mathbf{p}$. Alternatively, $\mathcal{W}(\mathbf{q},\mathbf{p})$ is proportional to the overlap of $\Psi$ with its mirror image about $\mathbf{q},\mathbf{p}$, which is clearly a measure of how much $\Psi$ is "centered" about $\mathbf{q},\mathbf{p}$.

### 3. Light propagation on a fiber

Propagation of light in a fiber is governed by Maxwell equations (Born & Wolf, 1999). Consider a monocromatic light beam of frequency $\omega$ propagating through a fiber of refractive index $n$, described by the wavefunction $\Psi(x, y, z)$. It can be shown that this beam obeys the Helmholtz equation (Born & Wolf, 1999):

$$\frac{\partial^2\Psi}{\partial x^2} + \frac{\partial^2\Psi}{\partial y^2} + \frac{\partial^2\Psi}{\partial z^2} + \frac{\omega^2}{c^2}n^2\Psi = 0\,, \tag{24}$$

where $n = n(x, y, z)$ is the refractive index of the fiber.
Under paraxial approach, the beam is almost parallel and close to the optical axis of the system, $z$, then $n = n(0, 0, z)$, and $\Psi$ vary slowly with $z$ allowing to neglect second order derivatives in the $z$ direction. This considerations let to write equation (24) as the parabolic type equation (Leontovich & Fock, 1946)

$$\frac{i}{k}\frac{\partial\Psi}{\partial\tau} = \frac{1}{2k^2}\left(\frac{\partial^2\Psi}{\partial x^2} + \frac{\partial^2\Psi}{\partial y^2}\right) + \left(n_0^2 - n^2\right)\Psi\,, \tag{25}$$

where $n_0$ is the vacuum refractive index, and

$$\tau = -\int_0^z \frac{1}{n_0(z')}dz'\,. \tag{26}$$

Equation (25) shows that the light beam propagation in the paraxial approximation is described by a Schrödinger equation where the wavelenght $\lambda = \frac{1}{k}$ plays the role of the Planck constant and instead of time appears $z$. The potential well is given by the refractive index $n_0^2 - n^2(x, y, z)$. This treatment translate the problem of solving the Helmholtz equation (24) to solve the Schrödinger equation for a system with two degrees of freedom $(x, y)$ in a time-dependent $(z)$ potential well.
This Schrödinger equation (25) is valid for any wave that follows the Helmholtz equation under the paraxial approach (for a detailed description check (Arnaud, 1976; Manko, 1986; Marcuse, 1972). The validity of this approximation can be verified using the moments of the Wigner distribution function of the solution as shown in (Rivera et al., 1997). To solve the problem in fiber optics it can be applied the formalism of symplectic groups through coherent state representation of quantum mechanics (Manko & Wolf, 1985).
In general, the output Wigner function of an optical system is related to the input Wigner through (Castano et al., 1982; Gutierrez & Castano, 1992):

$$\mathcal{W}_{\Psi_{out}}(\mathbf{q}, \mathbf{p}; z) = \mathcal{W}_{\Psi_{in}}(a\,\mathbf{q} + b\,\mathbf{p},\ c\,\mathbf{q} + d\,\mathbf{p};\ z)\,, \tag{27}$$

where $a, b, c$ and $d$ are parameters which depend on the specific system under study.
As an example, the free space propagator is given by

$$\mathcal{W}_{\Psi_{out}}(\mathbf{q}, \mathbf{p}; z) = \mathcal{W}_{\Psi_{in}}\left(\mathbf{q} - \frac{z}{k}\mathbf{p}, \mathbf{p}; z\right)\,, \tag{28}$$

for a lens of focal length $f$, we have

$$\mathcal{W}_{\Psi_{out}}(\mathbf{q}, \mathbf{p}; z) = \mathcal{W}_{\Psi_{in}}\left(\mathbf{q}, \frac{1}{f}\mathbf{q} + \mathbf{p}; z\right)\,, \tag{29}$$

and to obtain a Fourier transform we use:

$$\mathcal{W}_{\Psi_{out}}(\mathbf{q}, \mathbf{p}; z) = \mathcal{W}_{\Psi_{in}}(-\mathbf{p}, \mathbf{q}; z)\,. \tag{30}$$

## 4. Gaussian beam propagation in optical fibers

To model optical fibers it is common to consider gaussian beams that travel freely through space (Rivera & Castano, 2010). Gaussians are also ubiquitous in quantum mechanics, where they are intimately related to the harmonic oscillator (Gitterman, 2003; Moshinsky, 1996; Sako & Diercksen, 2003), to the coherent (Gori et al., 2003; Grewal, 2002; Grosshans et al., 2003; Lauterborn et al., 1993; Lesurf et al., 1993) and squeezed states formalism (Agarwal & Ponomarenko, 2003; Dodonov, 2002; Kim et al., 2002; Sohma & Hirota, 2003). In Quantum Optics, Gaussian beams are fundamental to test and to compare wave optical models and systems (Berry, 1994; Oraevsky, 1998; Rivera et al., 1997).
Using Fermat minimal action principle, it can be proved that the system is governed by the optical Hamiltonian (Rivera et al., 1995):

$$H = -\sqrt{n^2 - p^2}\,. \tag{31}$$

This Hamiltonian generates a ray path, i.e. a unidimensional group of canonical transformations of the points of the optical phase space. In a three-dimensional optical medium we denote the two screen coordinates (perpendicular to the optical axis) by $\mathbf{q} = (x.y)$ and the optical axis coordinate as $z$.
When the canonical transformation has a nonlinear part, it is possible to identify this nonlinearity as the effect of aberrations as is studied in (Rivera et al., 1997; Rivera & Castano, 2010). An alternative approach (called coherent states for Lie groups) uses the continuous representations in quantum mechanics as a particular case of arbitrary Lie groups and can be used in fiber optics for analyzing nonquadratic media under the action of Hamiltonians that are the linear form of the Lie group representation with $z$ dependent coefficients (Klauder, 1964). In geometric optics (paraxial approach), momentum is $|\mathbf{p}| = n\sin\theta$, where $n$ denotes the refractive index and $\theta$ is the angle between the ray and the optical axis (Wolf, 2004).
A Gaussian function $\Gamma$ associated to the one-dimensional real coordinate $x$ is defined as (Simon, 2002)

$$\Gamma(x) = M\,\exp\left[-\frac{(x-x_0)^2}{2w_0} + ip_0 x\right], \tag{32}$$

where $M = \left(\frac{w_1}{\pi|w_0|^2}\right)^{1/4}$, $x_0$, $p_0$ are real numbers, and $w_0 = w_1 + iw_2$, $w_1 > 0$ is a complex number. The dimension of $w_0$ is $[x^2]$, the one of $x_0$ is $[x]$, and that of $p_0$ is $[x^{-1}]$. The pre-exponential factor $M$ guarantees the normalization condition

$$\langle\Gamma \mid \Gamma\rangle = \int_{-\infty}^{\infty} \mathrm{d}x\,\Gamma^*(x)\,\Gamma(x) = 1\,. \tag{33}$$

This Gaussian is centered at $x_0$ and has a complex width $\sqrt{2w_0}$. The value at its maximum is $M$. If $p_0 \neq 0$ or $w_2 \neq 0$, this Gaussian shows oscillations.
The Fourier transform of the Gaussian $\Gamma$ (Equation 32) provides the momentum representation of the beam (Goodman, 1968):

$$\tilde{\Gamma}(p) = \frac{1}{\sqrt{2\pi}}\int_{-\infty}^{\infty} \mathrm{d}x\,e^{-ipx}\,\Gamma(x) = \left(\frac{w_1}{\pi}\right)^{\frac{1}{4}} \exp\left[-\frac{w_0(p-p_0)^2}{2} - ix_0(p-p_0)\right]. \tag{34}$$

Interestingly, it is another Gaussian, centered in $p_0$, with width $\sqrt{2/w_0}$ and it oscillates for $x_0 \neq 0$.

In phase space, the Gaussian $\Gamma$, is represented by its Wigner distribution function (Rivera & Castano, 2010):

$$\mathcal{W}_\Gamma(x,p) = 2\exp\left\{-\frac{(x-x_0)^2}{w_1} - \frac{|w_0|^2}{w_1}(p-p_0)^2 + \frac{2w_2}{w_1}(x-x_0)(p-p_0)\right\}, \quad (35)$$

that is a two-dimensional Gaussian; coordinate centered at $x_0$ with width $\sqrt{w_1}$, momentum center $p_0$ with width $|w_0|/\sqrt{w_1}$, and tilted by $\arctan(2w_2/w_1)$.
A *Vacuum Coherent State* (Dodonov, 2002) is a Gaussian function with $x_0 = p_0 = 0$ and $w_0 = 1$. It has the important property of being the only state described by the same function in both coordinate and momentum representation (Dodonov & Manko, 2000). A generalized *coherent state* is described by a Gaussian function with $w_0 = 1$, but $x_0$ and $p_0$ arbitrary. The state for which $w_0 \neq 1$ is called a *squeezed state* (Dodonov, 2002).
To evaluate the Wigner distribution function of a sectioned fiber we assume that an optical fiber is a cylinder of radius $a$ and infinite length. We will consider a Gaussian traveling through a fiber with constant refractive index $n$ until a break generated by a section with different refractive index $m$ encapsulated by two parabolic surfaces. The Symplectic Map for this system is given by the product of the initial propagator (before the break), first refraction, propagation between refraction surface, second refraction (after the break), and last propagation. It can be shown that the Symplectic Map of this system is (Reyes & Castano, 2000)

$$M_{total} = \ldots e^{:F_4:}\, e^{:F_2:}, \quad (36)$$

where

$$\begin{aligned} e^{:F_2:} &= e^{-\frac{a}{2n}sin(2k\pi z):\vec{\mathbf{p}}^2:} e^{\alpha(n-m):\vec{\mathbf{q}}^2:} e^{-\frac{\gamma}{2m}:\vec{\mathbf{p}}^2:} \\ &\times e^{-\frac{\gamma}{2}:\vec{\mathbf{p}}^2:} e^{-\alpha(n-m):\vec{\mathbf{q}}^2:} e^{-\frac{a}{2n}sin(2k\pi z'):\vec{\mathbf{p}}^2:}, \\ e^{:F_4:} &= e^{\mathcal{A}:(\vec{\mathbf{p}}^2)^2:+\mathcal{B}:\vec{\mathbf{p}}^2(\vec{\mathbf{p}}\cdot\vec{\mathbf{q}}):+\mathcal{C}:(\vec{\mathbf{p}}\cdot\vec{\mathbf{q}})^2:+\mathcal{D}:\vec{\mathbf{p}}^2\vec{\mathbf{q}}^2:} \\ &\times e^{\mathcal{E}:\vec{\mathbf{q}}^2(\vec{\mathbf{p}}\cdot\vec{\mathbf{q}}):+\mathcal{F}:(\vec{\mathbf{q}}^2)^2:}. \end{aligned}$$

Here, $z$ and $z'$ give the propagation before and after the break, respectively.
In Eq. (36), the exponential $e^{:F_2:}$ is the Gaussian term, while $e^{:F_4:}$ corresponds to the aberration term. This method simplifies the optical problem of obtaining the image of an optical system to the determination of the corresponding Symplectic Map associated to the system, thus reducing the problem to simple matrices products.
In order to calculate the Wigner distribution function we need to use the following correspondence

$$\begin{aligned} :\vec{\mathbf{p}}: &\longrightarrow \vec{\mathbf{p}} \\ :\vec{\mathbf{q}}: &\longrightarrow \vec{\mathbf{q}} \end{aligned}$$

to make the Symplectic map, and consider the convolution between a point source and $M_{total}$. The point source is defined by the Dirac Delta

$$F(\vec{\mathbf{p}'},\vec{\mathbf{q}'}) = \delta(\vec{\mathbf{q}'} - \vec{\mathbf{q}}, \vec{\mathbf{p}'} - \vec{\mathbf{p}}). \quad (37)$$

The convolution between $F$ and $M_{total}$ (up to fourth order) is

$$\begin{aligned} F * M_{total} &= F(\vec{\mathbf{p}'},\vec{\mathbf{q}'}) * e^{F_2}e^{F_4}\ldots \\ &\simeq [F(\vec{\mathbf{p}'},\vec{\mathbf{q}'}) * e^{F_2}](1+F_4)F(\vec{\mathbf{p}'},\vec{\mathbf{q}'}) * e^{F_2} \\ &+[F(\vec{\mathbf{p}'},\vec{\mathbf{q}'}) * e^{F_2}]F_4\,. \end{aligned} \quad (38)$$

The aberration of the system respect to the Gaussian ray (second term of the last equation) is

$$F * e^{F_2} = F(\vec{\mathbf{p}}', \vec{\mathbf{q}}') * e^{F_2} = C_1 \frac{\pi}{\sqrt{\alpha_5}} e^{-\left(\frac{\alpha_1 \vec{\mathbf{p}}'^2 - \alpha_2 \vec{\mathbf{p}}' \cdot \vec{\mathbf{p}} + \alpha_3 \vec{\mathbf{p}}^2}{\alpha_4}\right)},$$

where

$$\begin{aligned}
\alpha_1 &= \frac{a}{2n} \sin(2k\pi z')(\beta + \frac{\gamma}{2}), \\
\alpha_2 &= \beta \frac{a}{2n} \sin(2k\pi z'), \\
\alpha_3 &= \beta(\frac{\gamma}{2} + \frac{a}{2n} \sin(2k\pi z')), \\
\alpha_4 &= \beta + \frac{\gamma}{2} + \frac{a}{2n} \sin(2k\pi z'), \\
\alpha_5 &= (\frac{\gamma}{2n} + \frac{a}{2n} \sin(2k\pi z))(\beta + \frac{\gamma}{2n} + \frac{a}{2n} \sin(2k\pi z')).
\end{aligned}$$

Without perturbation ($\gamma \to 0$), the aberration yield

$$F * e^{F_2} = C_1 \frac{\pi}{\sqrt{\frac{a}{2n} \sin(2k\pi z)}},$$

that corresponds to the convolution between $e^{-\frac{a}{2n} sin(2k\pi z)\vec{\mathbf{p}}^2}$ and a point source.
Now we can calculate the Wigner distribution function in the image plane substituting in Equation (6) the Symplectic Map of this system, Eqs. (36) and (38):

$$\begin{aligned}
W(\vec{\mathbf{q}}', \vec{\mathbf{p}}') = &\int_{-\infty}^{\infty} d\vec{\mathbf{r}}\, e^{-i\vec{\mathbf{p}}' \cdot \vec{\mathbf{r}}} \\
&\times \left\{ F * e^{F_2}(\vec{\mathbf{q}}' + \frac{\vec{\mathbf{r}}}{2}) + [F * e^{F_2}] F_4(\vec{\mathbf{q}}' + \frac{\vec{\mathbf{r}}}{2}) \right\} \\
&\times \left\{ F * e^{F_2}(\vec{\mathbf{q}}' - \frac{\vec{\mathbf{r}}}{2}) + [F * e^{F_2}] F_4(\vec{\mathbf{q}}' - \frac{\vec{\mathbf{r}}}{2}) \right\},
\end{aligned}$$

that can be rewritten up to fourth order as

$$\begin{aligned}
&W(\vec{\mathbf{q}}', \vec{\mathbf{p}}') = \\
&C_1^2 \frac{2\pi^3}{\alpha_5} e^{i\pi} e^{-(2/\alpha_4)(\alpha_1 \vec{\mathbf{p}}''^2 - [2\alpha_1 + \alpha_2]\vec{\mathbf{p}}'' \cdot \vec{\mathbf{p}} + [\alpha_1 + \alpha_2 + \alpha_3]\vec{\mathbf{p}}^2)} \\
&\times \left\{ 1 + 2F_4 \delta(\vec{\mathbf{p}}') + \pi^2 \mathcal{F} \left( \frac{d^2}{dp_1'^2} + \frac{d^2}{dp_2'^2} \right)^2 + 8\pi^2 e^{i\pi} \right. \\
&\times \left[ \mathcal{C} \left( p_1' \frac{d}{dp_1'} + p_2' \frac{d}{dp_2'} \right)^2 + 2\mathcal{F} \left( q_1' \frac{d}{dp_1'} + q_2' \frac{d}{dp_2'} \right)^2 \right. \\
&+ \mathcal{E} \left( p_1' q_1' \frac{d^2}{dp_1'^2} + p_2' q_2' \frac{d^2}{dp_2'^2} + (p_2' q_1' + p_1' q_2') \frac{d}{dp_1'} \frac{d}{dp_2'} \right) \\
&\left. \left. + \frac{1}{2} [\mathcal{D}(\vec{\mathbf{p}}') + \mathcal{E}(\vec{\mathbf{p}}') \cdot \vec{\mathbf{q}}'] \left( \frac{d^2}{dp_1'^2} + \frac{d^2}{dp_2'^2} \right) \right] \right\} \delta(\vec{\mathbf{p}}'),
\end{aligned} \tag{39}$$

with $\vec{\mathbf{p}}' = \vec{\mathbf{p}}'' - \vec{\mathbf{p}}$. From this equation is clear that after the break, the Wigner distribution function is a Gaussian with center in $(2\alpha_1 + \alpha_2\vec{\mathbf{p}})/(2\alpha_1)$. The Wigner function found have a general phase of $2\pi$, except in the term $(4\pi^2/3)\mathcal{F}\{\frac{d^2}{dp_1'^2} + \frac{d^2}{dp_2'^2}\}^2\delta(\vec{\mathbf{p}}')$ where the phase is $\pi$. Thus the initial Gaussian is modified by a corrective term, the $F_4$ polynomial (the exponent of the Aberration Lie Operator).
The limiting case without break, $\gamma \to 0$, has a Wigner distribution function given by

$$W(\vec{\mathbf{q}}', \vec{\mathbf{p}}') = \frac{C_1^2 e^{i\pi}}{\frac{a}{2n} sin(2k\pi z)} \delta(\vec{\mathbf{p}}')\,(1 + 2F_4 + \ldots) \tag{40}$$

## 5. Conclusions

As shown in this chapter, phase space approach (through the Wigner distribution function) simplifies the calculation and helps in the description of optical fibers. This is due to the fact that in phase space representation, the relevant properties of the system can be obtained by simple matrices products. Quasiprobability distribution functions are useful not only as calculation tools but can also provide insights into the connections between geometric opticsl and wave optics due to the fact that they allow one to express wave optics averages in a form which is very similar to that for classical averages. In this sense it serves to validate paraxial approximation.

## 6. References

Agarwal, G. S. & Ponomarenko, S.A. (2003). Minimum-correlation mixed quantum states. *Physical Review A*, Vol. 67, article 032103, ISSN: 1050-2947.

Allen, R. L. & Mills, D. (2004). *Signal Analysis: Time, Frequency, Scale, and Structure*. Wiley, Berlin.

Arnaud J. A. (1976). *Beam and Fiber Optics*, Academic Press, New York.

Atakishiyev, N.; Vicent, L. E. & Wolf, K. B. (1999). Continuous vs. discrete fractional Fourier transforms. *J. Comput. Appl. Math.* Vol. 107, pp. 73.

Bakhturin, Y. (2003). *Groups, Rings, Lie and Hopf Algebras*. Kluwer Academic Publishers, New York.

Bala, S. & Prabhu, K. M. M. (1989). New method of computing Wigner-Ville distribution. *Electr. Letter*, Vol. 25, pp. 336.

Bao, X. & Chen, L. (2011). Recent Progress in Brillouin Scattering Based Fiber Sensors *Sensors*, Vol. 11, pp. 4152-4187.

Bastiaans, M. J. (1980). Wigner distribution function display: a supplement to ambiguity function display using a 1-D input. *Applied Optics*. Vol. 19, No. 2, pp. 192-193, ISSN: 1559-128X.

Balazs, N. L. & Jennings, B. K. (1984) Wigner's function and other distribution functions in Mock phase space. *Physics Reports*, Vol. 104, No. 6, pp. 347-391, ISSN: 0370-1573.

Bartelt, H.O.; Brenner, K.-H. & Lohmann, A. H. (1980). The Wigner distribution function and its optical production. *Optics Communications*, Vol. 32, pp. 32-38, ISSN: 0030-4018.

Benabid, F. & Roberts, P. J. (2011). Linear and nonlinear optical properties of hollow core photonic crystal fiber. *Journal of Modern Optics* Vol. 58, No. 2, pp. 87-124.

Benchekh, K.; Gravier, F.; Douady, J.; Levenson, A. & Boulanger, B. (2007). Triple photons: a challenge in nonlinear and quantum optics. *C. R. Physique*, Vol. 8, pp. 206âĂŞ220.

Berry, M. V. (1977). Semiclassical Mechanics in Phase Space: A Study of Wigner's Function. *Philosophical Transations of the Royal Society of London A* Vol. 287, No. 1343, pp. 237-271.

Berry, M. V. (1994). Evanescent and real waves in quantum billiards and Gaussian beams *Journal of Physics A: Mathematical and Theoretical* Vol. 27, No. 11, pp. L391-L398, ISSN 1751-8113.

Boashash, B. (2003). *Time Frequency Analysis*. Elsevier Science, Holland.

Born, M. & Wolf, E. (1999). *Principles of Optics, 7^th^* ed. Cambridge University Press, ISBN 0 521 642221, Cambridge.

Bottacchi, M. (2002). *Optical Fibre Transmission Theory, Technology and Design: Optic Propagation Theory*. John Wiley & Sons, New York.

Breitenbach, G.; Muller, T.; Pereira, S. F.; Poizat, J. P.; Schiller, S. & Mlynek, J. (1995). Squeezed vacuum from a monolithic optical parametric oscillator, *Journal of the Optical Society of America B*, Vol.12, pp. 2304.

Breitenbach, G.; Schiller, S. & Mlynek J. (1997). Measurement of the quantum states of squeezed light. *Nature* Vol. 387, pp. 471, ISSN: 0028-0836.

Buchdahl, H. A. (1970). *An introduction to Hamiltonian Optics*. Cambridge University Press, Cambridge.

Cahill, K. E. & Glauber, R. J. (1969). Density operators and Quasiprobability Distributions. *Physical Review*, Vol. 177, pp. 1882-1896.

Castano, V. M.; Vázquez-Polo, G. & Gutiérrez Castrejón, R. (1982). *Scanning Microscopy Supplement* Vol. 6, pp. 415-418, ISSN: 0892-953X.

Choi, H. I. & Williams, W. J. (1989). Improved time-frequency representation of multicomponent signals using exponential kernels. IEEE Trans. on Acoustics, Speech, and Signal Processing, Vol. 37, pp. 862.

Cohen, L. (1966). Generalized phase-space distribution functions. *Journal of Mathematical Physics*. Vol. 7, pp. 781.

Cohen, L. (1995). *Time-frequency analysis*. Prentice Hall, New Jersey.

Corney, J. F.; Heersink, J.; Dong, R.; Josse, V.; Drummond, P. D.; Leuchs, G. & Andersen, U. L. (2008). Simulations and experiments on polarization squeezing in optical fiber. *Physical Review A*, Vol. 78, article 023831, ISSN: 1050-2947.

Culshaw, B. (1997). *Optical Fiber Sensors: Applications, Analysis and Future Trends*. Artech House, New York.

Dirac, P. A. M. (1935). *The Principles of Quantum Mechanics*. 2^nd^ Ed., Cambridge University Press, Cambridge.

Dragoman, D. (1997). The Wigner Distribution Function in Optics and Optoelectronics. In *Progress in Optics XXXVII*, ed. Wolf, E. Elsevier, Amsterdam.

Dragoman, D. & Meunier, J. P. (1998). Recovery of longitudinally variant refractive index profile from the measurement of the Wigner transform. *Optics Communications*, Vol. 153, pp. 360-367, ISSN: 0030-4018.

Dragt, A. J. & Finn, J. M. (1976). Lie series and invariant functions for analytic symplectic maps. *Journal of Mathematical Physics*, Vol. 17, No. 12, pp. 2215-2227. ISSN: 0022-2488.

Dodonov, V. V. & Manko, O. V. (2000). Universal invariants of quantum-mechanical and optical systems. *Journal of the Optical Society of America A*, Vol.17, No. 12, pp. 2403-2410, ISSN: 1084-7529.

Dodonov, V. V. (2002). Nonclassical states in quantum optics: a squeezed review of the first 75 years, *Journal of Optics B: Quantum and Semiclassical Optics* Vol. 4, No.1, pp. R1, ISSN 1464-4266.

Easton, R. L.; Ticknor, A. J. & Barrett, H. H. (1984). Application of the Radon transform to optical production of the Wigner distribution function. *Optical Engineering*. Vol. 23, pp. 738.

Eilouti, H. H. & Khadra, L. M. (1989). Optimised implementation of real-time discrete Wigner distribution *Electr. Letters*, Vol. 25, pp. 706.

Fairlie, D. B. (1964). The formulation of quantum mechanics in terms of phase space functions. *Proceedings of the Cambridge Philosophical Society*, Vol. 60, No. 3, pp. 581-590, ISSN: 0305-0041.

Filinov, V. S.; Bonitz, M.; Filinov, A. & Golubnychiy, V. O. (2008). *Wigner function quantum molecular dynamics*. Lecture Notes in Physics, Vol. 739, Springer Verlag, Berlin.

Flandrin, P.; Martin, W. & Zakharia, M. (1984). On a hardware implementation of the Wigner-Ville transform, *Proceedings of the International Conference on Digital Signal Processing* 84, pp. 262.

Frank, A.; Rivera, A. L. & Wolf, K. B. (2000). Wigner function of Morse potential eigenstates. *Physical Review A*, Vol. 61, No. 5, article 054102, ISSN: 1050-2947.

Frank, A. & van Isacker, P. (1994). *Algebraic Methods in Molecular and Nuclear Structure*. Wiley Interscience, New York.

Gitterman, M. (2003). Harmonic oscillator with multiplicative noise: Nonmonotonic dependence on the strength and the rate of dichotomous noise. *Physical Review E*, Vol. 67, article 057103, ISSN: 1063-651X.

Glauber, R. J. (1963). Coherent and Incoherent States of the Radiation Field. *Physical Review*, Vol. 131, No. pp. 2766-2788.

Glauber, R. J. (1965). In *Quantum Optics and Electronics*, edited by DeWitt, C.; Blandin A. & Cohen-Tannoudji, C. Gordon and Breach, New York.

Goodman, J. W. (1968). *Introduction to Fourier Optics*. McGraw-Hill, New York.

Gori, F.; Santarsiero, M.; Simon, R.; Piquero, G.; Borghi, R. & Guattari, G. (2003). Coherent-mode decomposition of partially polarized, partially coherent sources. *Journal of the Optical Society of America A* Vol. 20, No.1, pp. 78-84, ISSN: 1084-7529.

Green, H. S. (1951). The Quantum Mechanics of Assemblies of Interacting Particles. *Journal of Chemical Physics*, Vol. 19, pp. 955-956, ISSN 0021-9606.

Grewal, K. S. (2002). Appearance of classical coherent oscillator states under environment-induced decoherence *Physical Review A* Vol. 65, article 052103, ISSN: 1050-2947.

Grochenig, K. (2000). *Foundations of Time-Frequency Analysis (Applied and Numerical Harmonic Analysis)*. Birkhäuser, Boston.

Groot, S. R. & Suttorp, L. G. (1972). *Foundations of Electrodynamics*. North-Holland, Amsterdam.

Grosshans, F.; Van Asschet, G.; Wenger, J.; Brouri, R.; Cerf, N.J. & Grangier, P. (2003). Quantum key distribution using gaussian-modulated coherent states. *Nature* Vol. 421, No. 6920, pp. 238-240, ISSN: 0028-0836.

Gupta, A. K. & Asakura, T. (1986). New optical system for the efficient display of Wigner distribution functions using a single object transparency. *Optics Communications*, Vol. 60, pp. 265-271, ISSN: 0030-4018.

Gutierrez Castrejon, R. & Castano, V. M. (1992). *Optik*, Vol. 91, pp. 24, ISSN: 0030-4026.

Hamermesh, M. (1962). *Group Theory and its application to physical problems*. Addison-Wesley Pub., Massachusetts.

Harmon, J. P. et al. (2001). *Optical Polymers: Fibers and Waveguides*. American Chemical Society, New York.

Heisenberg, W. (1930). *The Physical Principles of the Quantum Theory*. 2nd Ed., Cambridge University Press, Cambridge.

Herrmann, R. (1973). *Optik*, Vol. 37, pp. 91, ISSN: 0030-4026.

Hillery, M.; O'Connell, R. F.; Scully, M.O. & Wigner, E. P. (1984). Distribution Functions in Physics: Fundamentals. *Physics Reports*, Vol. 106, No.3, pp. 121-167, ISSN: 0370-1573.

Hudson, R. L. (1974). When is the wigner quasi-probability density non-negative? *Reports on Mathematical Physics*, Vol. 6, No. 2, pp. 249-252, ISSN 0034-4877.

Husimi, K. (1940). Some formal properties of the density matrix. *Prog. Phys. Math. Soc. Japan*, Vol. 22, pp. 264-266.

Jacobson, N. (1979). *Lie Algebras*. Dover Publications, New York.

Johansen, L. M. & Luis, A. (2004). Nonclassicality in weak measurements. *Physical Review A*, Vol. 70, Article 052115, ISSN: 1050-2947.

Kakazu, K.; Kiyuna, M. & Sakai, E. (2007). Systematic Derivation of the Number-Phase Distribution Functions. *Progress on Theoretical Physics* Vol. 118, pp. 827.

Kano, Y. (1965). A new Phase-Space Distribution Function in the Statistical Theory of the Electromagnetic Field. *Journal of Mathematical Physics*, Vol. 6, No. 2, pp. 1913-1921. ISSN: 0022-2488.

Keyl, M. (2002 ). Fundamentals of quantum information theory. *Physics Reports* Vol. 369, No. 5, pp. 431-548, ISSN: 0370-1573.

Kim, Y. S.& Wigner, E. P. (1987). Covariant phase-space representation for localized light waves. *Physical Review A*, Vol. 36, pp. 1293-1296, ISSN: 1050-2947.

Kim, Y. S.& Noz, M.E. (1991). *Phase space picture of Quantum Mechanics*. World Scientific, Singapore, pp. 37–55.

Kim, M. S.; Son, W. Buzek, V. & Knight, P. L. (2002). Entanglement by a beam splitter: Nonclassicality as a prerequisite for entanglement. *Physical Review A*, Vol. 65, article 032323, ISSN: 1050-2947.

Kirkwood, J. G. (1933). Quantum Statistics of Almost Classical Assemblies. *Physical Review* , Vol. 44, No. 1, pp. 31-37.

Klauder J. R. (1964) Continuous representation theory. II. Generalized relation between quantum and classical dynamics. *Journal of Mathematical Physics*, Vol. 5, No. 2, pp. 177-187. ISSN: 0022-2488.

Kogelnik, H. (1979). In *Integrated Optics*, ed. by Tamir, T. Springer Verlag, Berlin, pp. 15–81.

Kominis, Y. & Hizanidis, K. (2002). The Hamiltonian perturbation approach of two interacting nonlinear waves or solitary pulses in an optical coupler. *Physica D*, Vol. 173, No. 3, pp. 204-225, ISSN: 0167-2789.

Lauterborn, W.; Kurz, T. & Wiesenfeldt, W. T. (1993). *Coherent Optics*. Springer-Verlag, Berlin.

Lauterborn, W.; Kurz, T. & Parlitz, U. (1997). *Journal of the Franklin Institute*, Vol. 334, pp. 865.

Lee, H.W. (1995). Theory and application of the quantum phase-space distribution functions. *Physics Reports*, Vol. 259, No. 3, pp. 147-211, ISSN: 0370-1573.

Leonhardt U. (2001). *Measuring the Quantum State of Light*. Cambridge University Press, Cambridge.

Leontovich, M. & Fock, V. (1946). Solution of the problem of electromagnetic waves along the earth surface by the method of parabolic equations. *Soviet Journal of Physics*, Vol. 10, pp. 13-24.

Lesurf, J.C. et. al., (1993). Selected Papers on Gaussian Beam Mode Optics for Millimeter Wave and Terahertz Systems. S P I E - International Society for Optical Engineering, New York.

Lohmann A. (1980). The Wigner function and its optical production. *Optics Communications*, Vol. 32, pp. 32-38, ISSN: 0030-4018.

Lopez Viera J. C.; Rivera A. L.; Smirnov Yu. F. & Frank, A. (2002) Simple Evaluation of FranckâĂŞCondon Factors and Non-Condon Effects in the Morse Potential. *International Journal of Quantum Chemistry*, Vol. 88, No. 2, pp. 280-295, ISSN: 0020-7608.

Louisell, W. H. (1973). *Quantum Statistical Properties of Radiation*. Wiley, New York.

Lvovsky, A.I.; Hansen, H.; Aichele, T.; Benson, O.; Mlynek, J. & Schiller S. (2001) Quantum state reconstruction of the single-photon Fock state. *Physical Review Letters*, Vol. 87, Article 050402, ISSN 0031-9007.

Maanen, H. R. E. (1985). Duplication of the sampling frequency of periodically sampled signals for the calculation of the discrete Wigner distribution. *Journal of the Audio Engineering Society*. Vol. 33, pp. 892-901.

Marks, R. J. & Hall, M. W. (1979). Ambiguity function display using a single 1-D input. *Applied Optics*, Vol. 18, No. 15, pp. 2539-2540, ISSN: 1559-128X.

Mack, H. & Schleich, W. P. (2003). A photon viewed from Wigner phase space. *Optics and Photonics News* Vol. 14, pp. 28.

Manko, V. I. & Wolf, K.B. (1985). *The influence of aberrations in the optics of gaussian beam propagation*, Reporte de Investigación del departamento de Matemáticas, Universidad Autónoma Metropolitana, México.

Manko, V. I. (1986). Invariants and coherent states in fiber optics, In: *Lie Methods in Optics*, Ed. Sanchez Mondragon, J. & Wolf, K. B., pp. 193-206, Springer-Verlag, ISBN 3-540-16471-5, Berlin.

Marcuse, D. (1972). *Light transmission Optics*, Van Nostrand, New York.

Margenau, H. & Hill, R. N. (1961) Correlation between measurements in quantum theory. *Progress on Theoretical Physics* Vol. 26, pp. 722.

Mateeva Ts. & Sharlandjiev, P. (1986). Generation of a Wigner distribution function of complex signals by spatial filtering. *Optics Communications*, Vol. 57, pp. 153-158, ISSN: 0030-4018.

Mayer, J. E. & Band, W. (1947). On the Quantum Correction for Thermodynamic Equilibrium. *Journal of Chemical Physics*, Vol. 15, pp. 141-153.

Mecklenbrauker, W. et al. (1997). *The Wigner Distribution: Theory and Applications in Signal Processing*. Elsevier, Netherlands.

Mori, H.; Oppenheim, I. & Ross, J. (1962). Some topics in Quantum Statistics. The Wigner function and transport theory, In: *Studies in Statistical Mechanics* Vol. I, Ed. De Boer, J. & Uhlenbeck, G. E. North-Holland Pub., Amsterdam, pp. 213-298.

Moshinsky, M. & Smirnov, Y. F. (1996). *Harmonic Oscillator in Modern Physics: From Atoms to Quarks*, $2^{th}$ ed., Gordon & Breach Publishing Group, New York.

Moyal, J. E. (1949). Quantum Mechanics as a statistical theory. *Proceedings of the Cambridge Philosophical Society*, Vol. 45, No. 1, pp. 99-124, ISSN: 0305-0041.

OConnell, R. F. (1983). The Wigner distribution function 50th birthday. *Foundations of Physics*. Vol.13, pp. 83-92, ISSN 0015-9018.

Oraevsky, A. N. (1998). Gaussian Beams and Optical Resonators. In *Proceedings of the Lebedev Physics Institute Seri3s* Vol. 222, Nova Science Publishers, New York.

Ou, Z. Y.; Pereira, S. F.; Kimble, H. J. & Peng, K.C. (1992). Realization of the EinsteinâĂŞPodolskyâĂŞRosen paradox for continuous variables. *Physical Review Letters*, Vol. 68, pp. 3663, ISSN 0031-9007.

Page, (1952). Instantaneous power spectra. *J ournal of Applied Physics*, Vol.23, pp. 103.
Perinova, V.; Luis A. & Perina, J. (1998). *Phase in Optics*. World Scientific, Singapore.
Prasad, P. N. & Williams, D. J. (1991). *Introduction to Non-Linear Optical effects in Molecules and Polymers*. John Wiley & sons, New York.
Reyes, J.; Rodriguez, R.; Cotorogea, M. & Castano, V. M. (1999). *Optik*, Vol. 110, pp. 305, ISSN: 0030-4026.
Reyes, J.; Saenz, A.; Silva, A. & Castano, V. M. (1999). *Optik*, Vol. 110, pp. 527, ISSN: 0030-4026.
Reyes, J. & Castano, V. M. (2000). *Optik*, Vol. 111, pp. 219, ISSN: 0030-4026.
Rihaczek, A. W. (1968). Signal energy distribution in time and frequency. *IEEE Trans. Information Theory*, Vol. 14, pp. 369-376.
Rivera, A. L.; Chumakov, S. M.; Wolf, K. B. (1995). Hamiltonian foundation of geometrical anisotropic optics. *Journal of the Optical Society of America A*, Vol.12, No. 6, pp. 1380-1389, ISSN: 1084-7529.
Rivera, A. L.; Atakishiyev, N. M.; Chumakov, S. M.; Wolf, K. B. (1997). Evolution under polynomial Hamiltonians in quantum and optical phase spaces. *Physical Review A*, Vol. 55, No. 2, pp. 876-889, ISSN: 1050-2947.
Rivera, A. L. & Castano, V. M. (2010). Linear and non-linear symmetry properties of Gaussian wave packets. *Optik*, Vol. 121, No. 6, pp. 539-552, ISSN: 0030-4026.
Rivera, A. L. & Castano, V. M. (2010). Physical defects in fiber optics: A theoretical framework in phase space. *Optik*, Vol. 121, No. 17, pp. 1563-1569, ISSN: 0030-4026.
Royer, A. (1977). Wigner function as the expectation value of a parity operator. *Physical Review A*, Vol. 15, pp. 449-450, ISSN: 1050-2947.
Sako, T. & Diercksen, G. H. F. (2003). Confined quantum systems: spectral properties of the atoms helium and lithium in a power series potential. *Journal of Physics B*, Vol. 36, No. 7, pp. 1433-1458, ISSN 0953-4075.
Schleich, W. P. (2001). *Quantum Optics in Phase Space*. Wiley, Berlin.
Schempp, W. (1986). Analog radar signal design and digital signal processing: a Heisenberg nilpotent Lie group approach. In: *Lie Methods in Optics*, Ed. Sanchez Mondragon, J. & Wolf, K. B., (pp. 193-206), Springer-Verlag, ISBN 3-540-16471-5, Berlin.
Schrodinger, E. (1946). *Statistical Thermodynamics*. Cambridge University Press, Cambridge.
Sheppard, C. J. R. & Larkin, K.G. (2000). Focal shift, optical transfer function, and phase-space representations. *Journal of the Optical Society of America A*, Vol.17, No. 4, pp. 772-779, ISSN: 1084-7529.
Shimizu, M. et al. (1997). *Optical Fiber Amplifiers: Materials, Devices and Applications*. Artech House, New York.
Simon, M. K. (2002). *Probability Distributions Involving Gaussian Random Variables: A Handbook for Engineers and Scientists*. Kluwer Academic Publishers, New York.
Smith, T. B. (2006). Generalized Q-functions, *Journal of Physics A*, Vol. 39, No. 44, pp. 13747-13756, ISSN 1751-8113.
Smithey, D.T. Beck, M. & Raymer, M. G. (1993). Measurement of theWigner distribution and the density matrix of a light mode using optical homodyne tomography: application to squeezed states and the vacuum. *Physical Review Letters*, Vol. 70, pp. 1244-1256, ISSN 0031-9007.
Sohma, M. & Hirota, O. (2003). Capacity of a channel assisted by two-mode squeezed states. *Physical Review A*, Vol. 68, article 022303, ISSN: 1050-2947.

Soto, F. & Claverie, P. (1983). When is the Wigner function of multidimensional systems nonnegative? *Journal of Mathematical Physics*, Vol. 24, No. 1, pp. 97-100. ISSN: 0022-2488.

Stewart I. et al. (2002). *The Symmetry Perspective: From Equilibrium to Chaos in Phase Space and Physical Space*. Birkhauser, Boston.

Subotic, N. & Saleh, B. E. A. (1984). Generation of the Wigner distribution function of two-dimensional signals by a parallel optical processor. *Optics Letters*. Vol. 9, No. 471.

Sudarshan, E. C. G. (1963). Equivalence of Semiclassical and Quantum Mechanical Descriptions of Statistical Light Beams. *Physical Review Letters*, Vol. 10, pp. 277-279, ISSN 0031-9007.

Takabayasi, T. (1954). The Formulation of Quantum Mechanics in terms of Ensemble in Phase Space. *Progress on Theoretical Physics* Vol. 11, pp. 341.

Terletsky, Ya. P. (1937). *Z. Eksp. Teor. Fiz.* Vol. 7, pp. 1290.

Torchigin, V. P. & Torchigin A. V. (2003). An increase in the wavelength of light pulses propagating through a fiber. *Physics Letters A* , Vol. 311, No. 1, pp. 21-25, ISSN: 0375-9601.

von Neumann, J. (1927). *Gött. Nachr*, pp.273.

Voss, H. U.; Schwache, A.; Kurths, J. & Mitschke F. (1999). Equations of motion from chaotic data: A driven optical fiber ring resonator. *Physics Letters A*, Vol. 256, No. 1, pp. 47-54, ISSN: 0375-9601.

Walls, D. F. (1983). Squeezed states of light *Nature* Vol. 306, pp. 141-142, ISSN: 0028-0836.

Way, W.I. (1998). *Optical Fiber Communication*. SPIE-International Society for Optical Engineering, New York.

Weyl, (1927). *Z. Phys.*, Vol. 46, pp. 1.

Wigner, E. P. (1932). On the Quantum Correction For Thermodynamic Equilibrium. *Physical Review*, Vol. 40, No. 5, pp. 749-759.

Wigner, E. P. (1971). Quantum-mechanical distribution functions revisited. In *Perspectives in Quantum Theory*, Ed. Yourgrau, W. & van der Merwe, A. MIT Press, Cambridge.

Wolf, K. B. & Rivera, A. L. (1997). Holographic information in the Wigner function, *Optics Communications*, Vol. 144, No. 1-3, pp. 36-42, ISSN: 0030-4018.

Wolf, K. B. (2004). *Geometric Optics on Phase Space*. Springer, Berlin, ISBN 3-540-22039-9.

Woodward, P. W. *Probability and Information Theory with Applications to Radar*. Pergamon, Oxford.

Young, M. (2000). *Optics and Lasers: Including Fibers and Optical Waveguides*. Springer-Verlag, New York.

Yuen, H. P. (1976). Two-photon coherent states of the radiation field. *Physical Review A*, Vol. 13, pp. 2226, ISSN: 1050-2947.

Zalevsky, Z. & Mendlovic, D. (1997). Light propagation analysis in graded index fiber-Review and applications. *Fiberand Integrated Optics* Vol. 16, pp. 55-66, ISSN 0146-8030.

# 20

# Effects of Radiation on Optical Fibers

Fuhua Liu[1,2], Yuying An[1], Ping Wang[2], Bibo Shao[2] and Shaowu Chen[2]
[1]*School of Technical Physics, Xidian University*
[2]*Key Laboratory of Laser Interaction with Matter,*
*Northwest Institute of Nuclear Technology*
*China*

## 1. Introduction

Optical fibers have many advantages over metallic lines such as broad bandwidth, low-loss, immunity from interference due to electromagnetic induction, etc. They can be used to implement ultra-fast pulse signal transmission over a long distance under the circumstance with sophisticated electromagnetic radiation. However, while optical fibers are exposed in nuclear radiation environments, changes in their optical properties will occur thus resulting in deterioration of system performance eventually. Optical fibers will be required to withstand exposure to nuclear environments. Since optical fibers were applied in nuclear radiation environments as signal transmission media, people began to study effects of radiation on optical fibers, to measure the changes of optical fiber parameters, e.g. radiation-induced loss, irradiation damage recovery time and to analyze the effecting factors (Mattern et al., 1974; Evans et al., 1974; Golob et al., 1977; Friebele et al., 1978, 1979, 1980). Research results are used to evaluate the variation degree of optical fiber system performance and their working lives under nuclear circumstance, and to search methods for reducing radiation-induced loss (Tsunemi et al., 1986; Akira et al., 1988). As a result, anti-radiation optical fibers are developed subsequently. With the application of anti-radiation optical fibers, the degradation of performance will be reduced and the system life will be extended accordingly. On the other hand, radiation detecting systems based on the parameter changes above-mentioned are established to monitor the ambient radiation doses of underground nuclear exploders, space-aircrafts, radiation reactors and other nuclear facilities (Ramsey et al., 1993; Moss et al., 1994; Tighe et al., 1995; Fernadndez et al., 2002; May, 2006).
When radiation projects to optical fibers, three effects will produce: (1) Increase of optical fibers absorption loss. The additional loss caused by radiation of photons and electrons with lower energy corresponds with the mechanism of color center. The color center spectrum lies usually within the visible and near-infrared wavelength regions, and it is resonant absorption that leads to additional loss. Neutron or alpha particle radiation absorbed by optical fibers can also cause additional loss. It will mainly damage optical fiber matrix structure and produce atomic structure defects and release electrons. (2) Changes of optical fiber refractive index. As a result, boundary conditions will no longer fully meet the optical fiber waveguides, and increase of evanescent field coupling energy will lead to additional loss. (3) Development of optical fiber luminescence. It is usually considered to be fluorescence or Cerenkov effects. It is very difficult to detect the light due to its weak intensity along optical fiber axis.

The purpose of the research is to provide experimental data for reducing radiation-induced loss of optical fibers and to attempt to evaluate system performance degradation of optical fibers under nuclear environments.
This chapter will mainly discuss transient γ-ray effects on commercial optical fibers. Two different dose and dose rate γ-ray pulses are employed to irradiate four types of optical fibers and radiation-induced losses are measured by using five lasers with different wavelengths as carriers.

## 2. Effects of γ-ray radiation on optical fibers

### 2.1 Loss mechanism of optical fibers

Loss is inherent nature of optical fibers. In accordance with different generation mechanisms, loss is usually classified into: absorption loss, scattering loss, waveguide loss and bending loss, etc (Liu et al., 2006). When light-waves propagate in optical fiber media, interactions between photons and atoms occur. Photons will continue to transfer their energy to matrix atoms gradually. This process results in absorption loss. Optical matrix materials and impurities are the main factors influencing the absorption loss. Based on the different absorption subjects, absorption loss is classified into intrinsic absorption loss, impurity absorption loss and atomic defect absorption loss. Collisions of photons with substrate atoms, microscopic changes in optical fiber material density, and uneven composition distribution or structural defects generated during the manufacturing process will produce scattering loss. Rayleigh scattering which is inevitable is the lowest limit of optical fiber loss. Once variation of boundary condition for optical waveguide or waveguide deformation appears, part of light-wave mode energy will leak out, resulting in energy loss, i.e. waveguide loss. Fiber bended to a certain extent, part of the light energy will be lost, resulting in bending loss.

### 2.2 Effects of γ-ray radiation on optical fibers

The photon of γ-ray is the quantum of electromagnetic radiation. Radiation damage of material due to incident photon flux is varied, depending upon the material through which the photon propagate and the photon energy of the radiation. Damage ranges from simple heating, as photons are absorbed, to ionization and even photon-nuclear disintegration if the interacting photon energy is of the order of 10MeV or greater. According to different photon energy, effects of γ-ray on materials include: photoelectric, Compton, electron pair and scattering effects (Mei, 1966). The variation of cross sections for different effects in optical fibers with photon energy is calculated by GEANT4 and shown in Fig. 1. The data exhibited in Fig. 1 serve to point out that Compton Effect is dominant over the behavior with γ-ray radiation exposure on optical fibers. In addition, there is also fluorescence and Cerenkov effects. The penetration of radiation into materials is not only dependent upon the material itself but also upon the type of radiation. The penetration rate of γ-ray into optical fibers is calculated and shown in Fig.2. Atomic and molecular electron energy levels are on the order of a few electron volts, and so an electron bound at an atomic site in a material would not undergo a simple transition to a higher atomic energy level due to interaction with γ-ray. The resulting high energy electron of Compton Effect is the primary source of radiation damage due to γ-ray absorption in optical fibers. Its energy and intensity distribution in the horizontal profile of optical fibers is depicted in Fig.3, 4, and 5 when γ-ray with photon energy of 0.3, 0.8, and 1.0MeV projects along the vertical profile respectively.

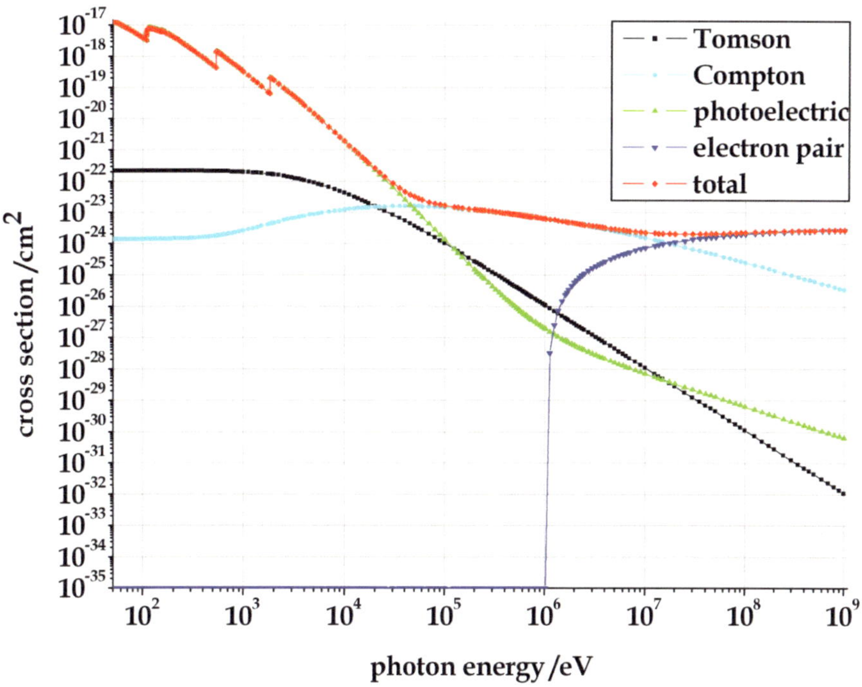


Fig. 1. Curve of different effects cross sections varying with photon energy.

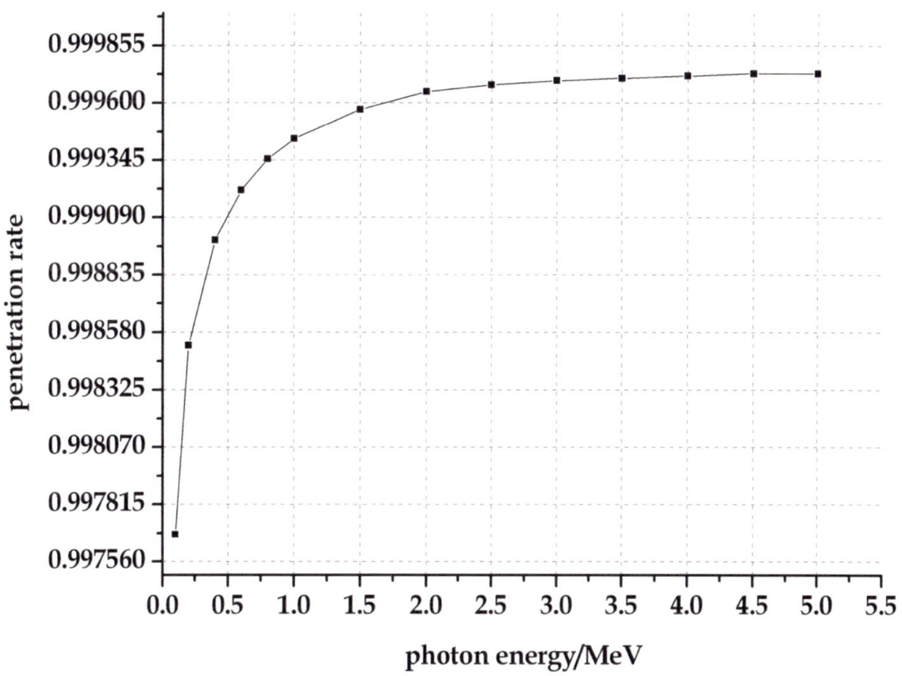


Fig. 2. Graph of penetration rate of γ-ray into optical fibers as a function of photo energy.

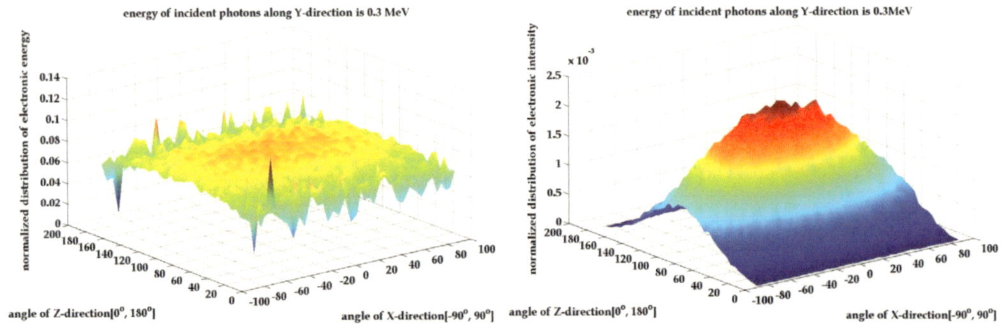


Fig. 3. Diagrams depicting distribution of resulting electronic energy and intensity in horizontal profile of optical fiber while photo energy of γ-ray is 0.3MeV.

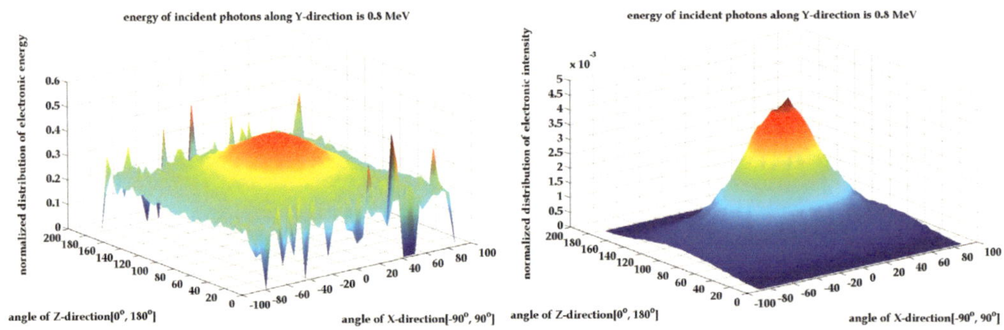


Fig. 4. Diagrams depicting distribution of resulting electronic energy and intensity in horizontal profile of optical fiber while photo energy of γ-ray is 0.8MeV.

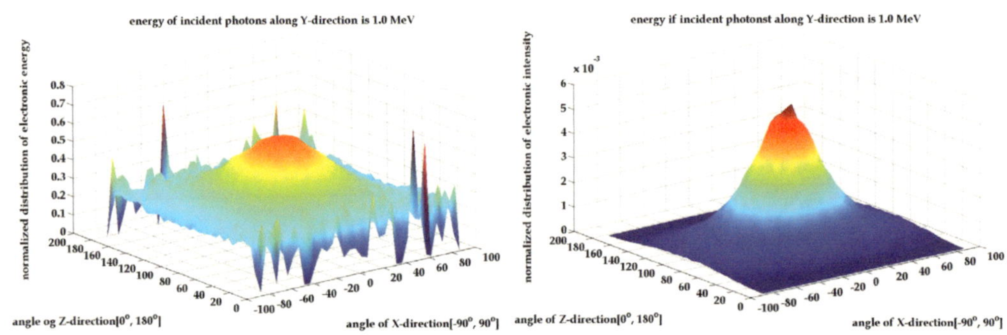


Fig. 5. Diagrams depicting distribution of resulting electronic energy and intensity in horizontal profile of optical fiber while photo energy of γ-ray is 1.0MeV.

If sufficient ionizing radiation of γ-ray with energies from several MeV down into the keV range is absorbed by optical fibers, it causes damages to optical fiber materials. The damages which produce additional radiation-induced loss on light propagation are associated with the energy and intensity distribution of the resulting high energy electrons.

### 2.3 Mode distribution and alteration of refractive index in optical fibers

The index of refraction is attributable to the electromagnetic properties of optical fibers. As in crystalline, similar processes of color center formation by radiation absorption may occur in amorphous. It's reasonable to assume that some changes in the index of refraction may result from radiation exposure. In fact radiation-induced changes in the refractive index distribution of optical fibers will influence distribution of mode field and confinement factor and bring additional waveguide loss. Optical waveguide loss arises from the waveguide imperfections. Confinement factor of waveguide can be described as (1) (Yasuo, 2002)

$$\Gamma = \frac{V + \sqrt{b}}{V + \dfrac{1}{\sqrt{b}}} \tag{1}$$

where $b$ is the normalized propagation constant, $b = \frac{(\beta / k_0)^2 - n_2^2}{n_1^2 - n_2^2}$. $V$ is the normalized frequency, $V = k_0 n_1 a \sqrt{n_1^2 - n_2^2}$. $k_0$ is the propagation constant of plane wave in vacuum, $k_0 = \omega \sqrt{\varepsilon_0 \mu_0}$. $a$ is the radius of the core. $\omega$ is the angle frequency of light-wave.

The computational results of relative mode field distribution in the core and clad of optical fibers, relative distribution of electric field intensity and confinement factors as a function of refractive index changes are shown in Fig.6, 7, 8 respectively. Any alteration in indexes of refraction within optical waveguide will influence the mode distribution and cause waveguide loss eventually.

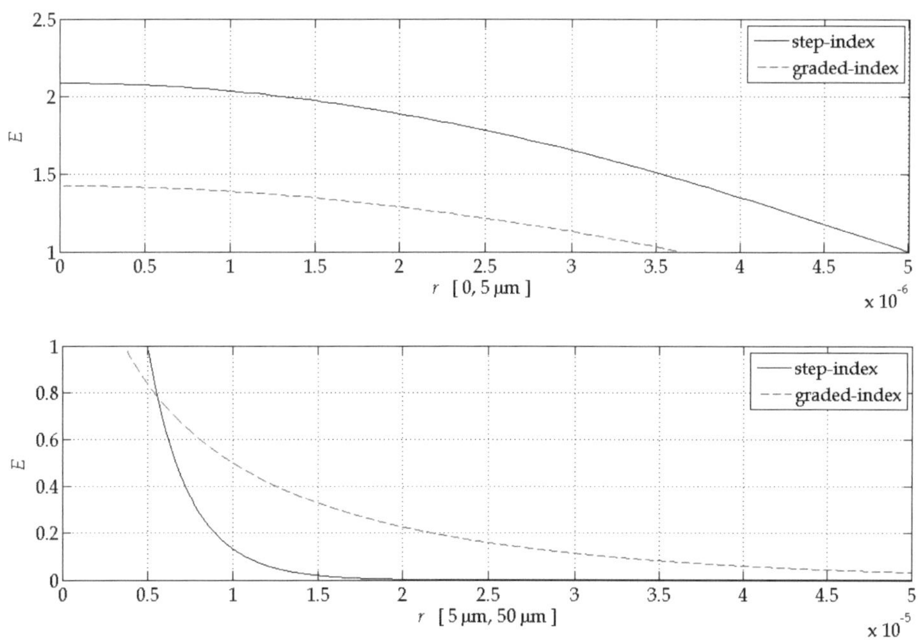


Fig. 6. Relative mode field distribution of electromagnetic wave in core and clad.

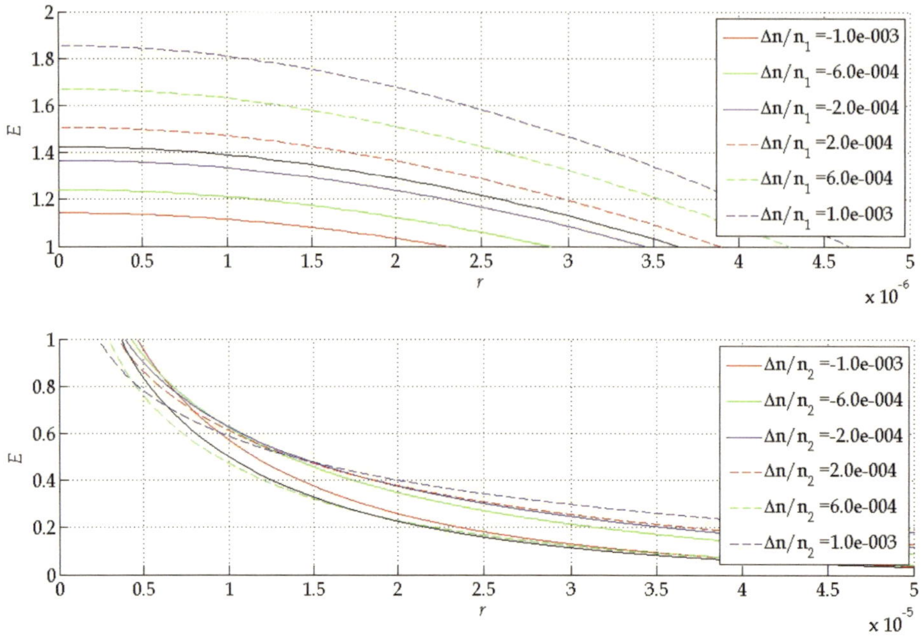


Fig. 7. Relative distribution of electric field intensity in optical fiber as a function of refractive index changes.

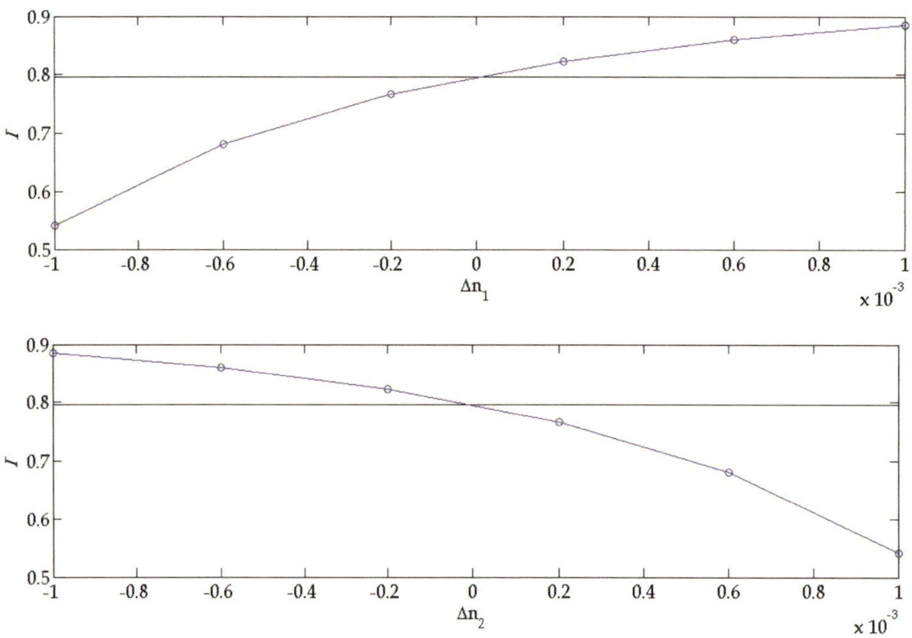


Fig. 8. Confinement factors as a function of refractive index changes.

## 3. Design of experimental measurement system

### 3.1 Measurement system structure

In order to measure pulsed γ-ray radiation-induced loss, a special experimental measurement system will be needed. During the system design, the following factors should be considered comprehensively: (1) Radiation sources. Radiation dose is adjusted by the nature of ray attenuations, with real-time simultaneous multi-point monitoring. (2) Optical fibers. In order to obtain uniform irradiation, optical fibers are coiled into circles with diameters as small as possible e.g. several centimeters, until the bending loss cannot be neglected. Its exposure length can be adjusted conveniently. (3) Response and record time sequence. The system's response time should be controlled within a tenth or less of the time width of the radiation rays in order to reduce influence of the measurement system time characteristics on results. To ensure records of effect signal waveforms, all devices should be set at automatic working states. The measurement system linear dynamic range of amplitude should be as large as possible e.g. 100. (4) Measurement environment. The main radiation source is a large electron accelerator with strong space electromagnetic radiation, so all the electronic equipments should be shielded effectively.

A typical experimental apparatus for measuring the radiation-induced loss in optical fibers is shown in Fig. 9.

### 3.2 Measurement system components

The measurement system consists of three parts: (1) Signal recording section. It contains a trigger, a signal generator and a transient digital oscilloscope. The trigger is used to start the transient oscilloscope and signal generator simultaneously. The signal generator is used to produce pulse signal to drive the analog optical fiber transmission system, thus producing pulsed light signal. The transient oscilloscope is used to record the optical signal while γ-ray impulses on optical fibers. (2) Optical fiber transmission section. It contains a semiconductor laser transmitter and a receiver. The semiconductor laser is used to convert the pulsed electric signal into optical ones, and the semiconductor receiver is used to convert pulsed light signals into electrical ones and send it to the transient oscilloscope. (3) Target section. It contains optical fibers under test and regulating facilities.

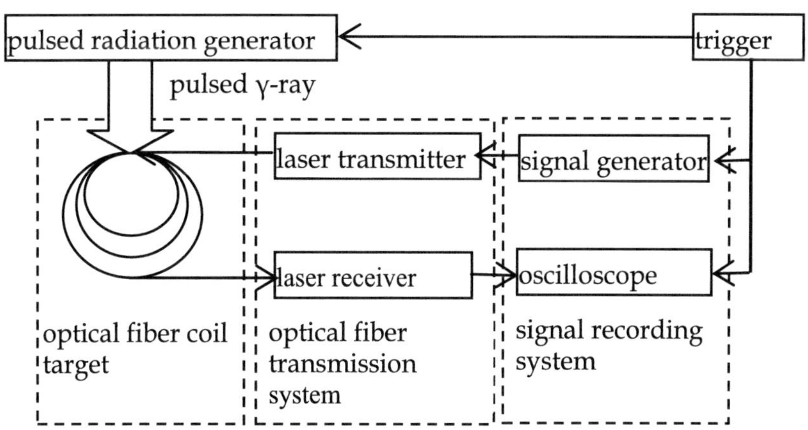


Fig. 9. Schematic diagram of experimental setup for measuring transient radiation-induced loss in optical fibers under pulsed exposure.

### 3.3 Main technical parameters

A comprehensive list of the important parameters in conducting a measurement of the radiation response of optical fibers is as below:

Radiation source I: average photon energy of 0.3 MeV, pulse width of 25ns, dose rate of $2.03\times10^7$ Gy/s.

Radiation source II: average photon energy of 1.0 MeV, pulse width of 25ns, dose rate of $5.32\times10^9$ Gy/s

Trigger: input/output of -10-+10 V adjustable, time interval of 0.001-10μs adjustable.

Signal generator: input and output amplitude of -5-+5 V adjustable, pulse width of 0.0003-10μs adjustable.

Transient oscilloscope: analog bandwidth of 1 GHz, digital sampling rate of 5 GHz.

Optical fiber transmission system: bandwidth of 3 GHz, in-band flatness within ±1dB, linear dynamic range of 100 (non-linear is less than 3%), peak output noise less than 5mV. input / output impedance of 50 Ω, the laser wavelength of 405, 660, 850, 1310 and 1550 nm.

Optical fiber types: ITU G.651(50/125μm, 62.5/125μm), G.652 and G.655 available bare optical fibers.

In considering the effect of radiation, the radiation damage is a dynamic process, i. e. concurrent with the darkening due to the production of color centers by the irradiation is recovery due to emptying of the holes and electrons out of these centers. Thus ,the net optical absorption that is observed is the sum of these two process.

## 4. Development of laser transmitter and receiver

A broad-bandwidth analog optical fiber transmission system is developed for radiation-induced loss measurement under the circumstance with complicated electromagnetic fields. The ultra-fast pulsed electric signal is converted to optical by electro-optic conversion method. With certain kilometers propagation, the optical signal is recovered by photo-electric conversion method. The experimental measurement results of the transmission system indicate that its bandwidth is (0.0003-3)GHz, in-band flatness ±1dB, linear dynamic range 100, output peak-to-peak noise less than 5mV, and input/output standing-wave-ratio less than 2.

### 4.1 Structure and design

#### 4.1.1 Structure

An analog optical fiber transmission system is usually comprised of a transmitter, a certain distance long optical fibers and a receiver. The structure is shown in Fig. 10.

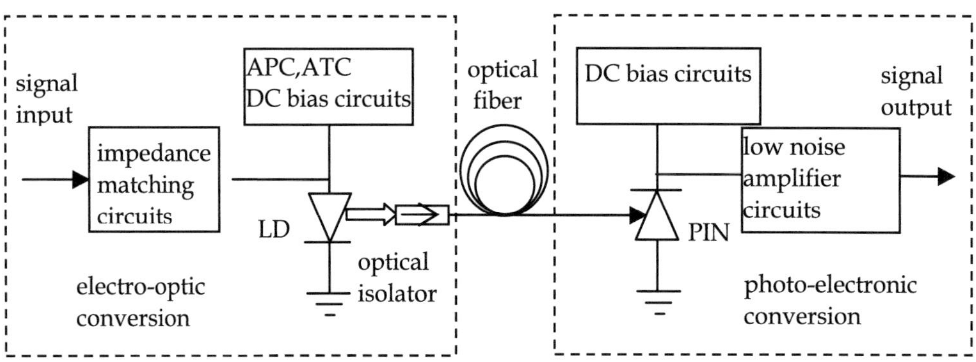


Fig. 10. Schematic diagram of broad bandwidth analog optical fiber transmission system.

Key technologies for designing and fabricating a broad bandwidth analog optical fiber transmission system lie in these aspects: (1) By impedance matching, the response in high frequency band is compensated and the overall system bandwidth is expanded consequently. (2) With peak-to-peak noise of PIN photoelectric diode and preamplifier reduced, the linear dynamic range is extended. (3) The parasitic parameters of components and micro-striplines in the modules reduced, the system can operate stably.

### 4.1.2 Transmitter

The transmitter contains an integrated multi-quantum-well distributed feedback laser diode(MQW-DFB-LD), an optical isolator, automatic power control (APC) circuits (Tanaka et al., 2002; Zivojinovic et al., 2004; Pocha et al., 2007), automatic temperature control (ATC) circuits, and DC bias circuits. The potential improvements of overall optical system performance depend on studying and analyzing LD transient characteristics such as modulation bandwidth, intensity, frequency noise levels and nonlinear distortion. Numerical laser models and sophisticated computation can accurately predict LD transient characteristics of above-mentioned parameters. The time dependent carrier density rate equations for LD in the active region are described as (Huang, 1994; Ghoniemy et al., 2003):

$$\frac{dN}{dt}=\frac{J_e}{ed}-\frac{N}{\tau_e}-GP\left(N-N_{leak}\right) \qquad (2)$$

$$\frac{dP}{dt}=GP\left(N-N_{leak}\right)-\frac{P}{\tau_p}+\beta\frac{N}{\tau_e}$$

where $N$ is the average density of electrons, $P$ is the average density of photons, $N_{leak}$ is the density of leaked carriers, $d$ is the thickness of the active layer, $\tau_e$ is the lifetime of electrons, $\tau_p$ is the lifetime of photons, $G$ is the differential coefficient which expresses light gain, $\beta$ is the simultaneous emission coefficient, $J_e$ is the injected current density, $e$ is the electron charge.

The cutoff modulation frequency of LD is deduced from (2) and given by

$$fc=\frac{1}{2\pi}\sqrt{\frac{\Gamma N_{leak}G\tau_p+1}{\tau_e\tau_p}\left(\frac{J_e}{J_{th}}-1\right)} \qquad (3)$$

where, $\Gamma$ is the light gain confine factor, $J_{th}$ is the threshold current density.

From (3) it can be conclude that the cutoff modulation frequency can be enhanced by increasing the injected current. Considering the carrier transport effects, too large injected current may lead to the variation of the refractive index distribution in the active region, resulting in the deterioration of the modulation performance. The appropriate bias current should be selected by considering the relationship of bandwidth with dynamic range synthetically.

With the help of a light-wave component analyzer, the transient characteristics of LD can be analyzed. The electrical equivalent circuit is shown in Fig. 11. $L_{p1}$, $L_{p2}$ is the lead inductance, $C_p$ the package parasitic capacitance, $C_d$ the PN junction diffusion capacitance, $C_b$ the PN junction barrier capacitance, $R_{diff}$ the PN junction differential resistance, $R_v$ the bulk resistance of semiconductor material, D the equivalent ideal diode. Elemental parameters in the electrical equivalent circuits provide P-Spice simulation software with original data.

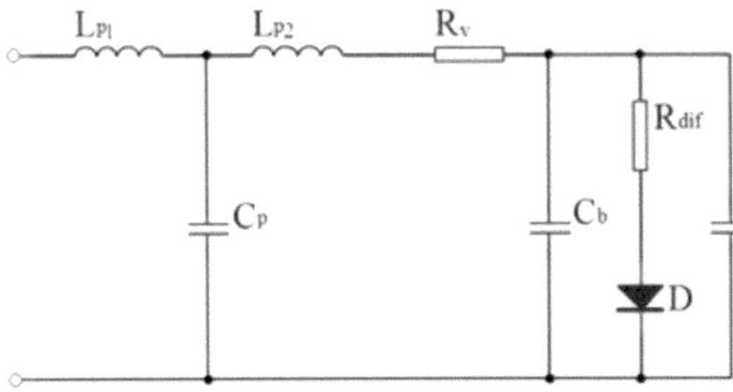


Fig. 11. Equivalent AC circuit diagram of LD.

A basic driving circuit is designed to meet LD analog amplitude modulation, and shown in Fig. 12. In the figure $D_1$ is designed to prevent LD from reverse breakdown. $R_1$-$R_4$, $C_1$-$C_3$ are mainly used for impedance matching and high frequency response compensating. $L_1$, $C_4$, and $C_5$ provide the bias decoupling, reducing transient current impact on LD, $V_{in}$ is input modulation signal, and $I_{bias}$ DC bias current.

APC circuits are used to provide LD with static operating point current. Precise temperature control is needed for LD to operate stably, since its threshold current, output power and peak wavelength will vary with junction temperature fluctuation. ATC circuit will meet the demand of junction temperature control. With APC and ATC circuits, the LD's variation of output power keeps within 2%, and the variation of operating temperature keeps within 0.1°C.

There may be some parasitic effects imposed on modulation characteristics to a certain extent. Simulation results show that when a LD is working in high-frequency modulation, the parasitic lead inductance will be impacted obviously on the amplitude-frequency characteristics. Efforts should be made to reduce the distribution parameters to a level as low as possible in different package forms in order to make the LD work stably.

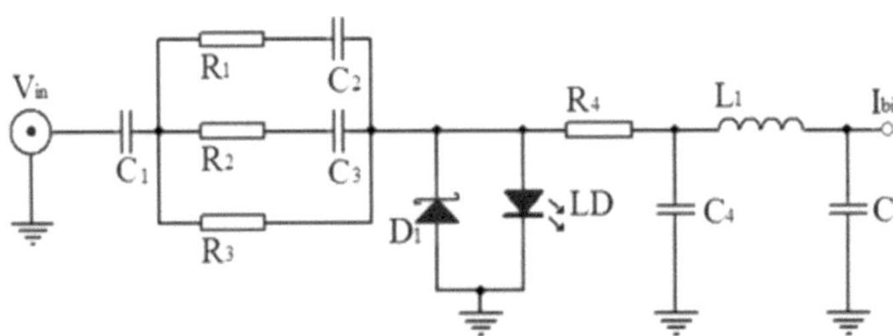


Fig. 12. Principle circuit diagram of impedance matching for electro-optic conversion module.

#### 4.1.3 Receiver

The receiver is comprised of a PIN detector, bias circuit and broad band-width low-noise amplifier. The PIN converts optical signals into electrical ones. The broad bandwidth low-noise amplifier enlarges the signal to an advisable level suitable for recording. Since the PIN output signal is usually weak, such techniques as low-noise, high-gain amplification and impedance matching are needed to design the receiver.

The equivalent PIN AC circuit is shown in Fig. 13 (An, 2002). In the figure $I_i$ is an ideal current source, $C_j$ the junction capacitance, $R_j$ the junction resistance, $R_p$ and $C_p$ equivalent parasitic values, and $R_{load}$ equivalent load for the amplifier. Amplifier bandwidth, noise characteristics, and impedance matching should be taken into account in designing photo-electric conversion module. Its principle circuit is shown in Fig. 14.

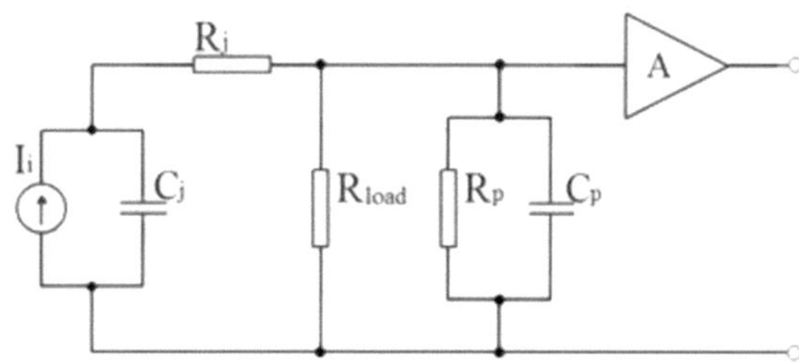


Fig. 13. Equivalent AC circuit diagram of PIN detector.

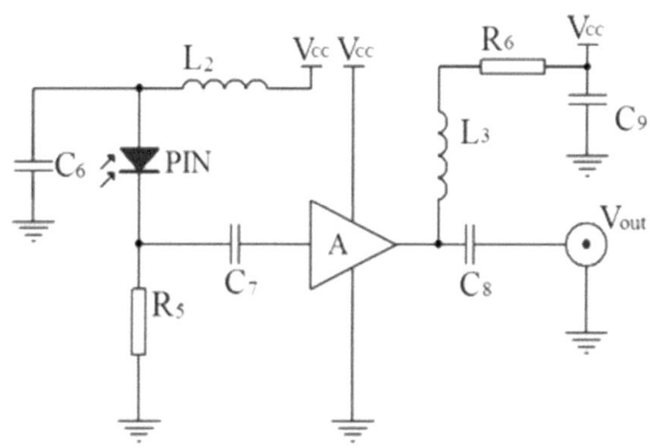


Fig. 14. Principle circuit diagram of photo-electric conversion module.

### 4.2 Performance
#### 4.2.1 Bandwidth

Bandwidth indicates the response of the input electrical signal frequency components (Hinojosa et al., 2001). Both frequency and time domain measuring methods are employed in the experiments. In frequency-domain measurement a light-wave component analyzer is used. The results show that bandwidth is (0.0003-3) GHz, band flatness ±1dB. The measured amplitude-frequency characteristic curve is shown in Fig. 15 (frequency sweep range (0.0003-3) GHz, amplitude coordinate scale 2 dB/div). With a electrical sub-nanosecond pulse signal generator and a broad bandwidth digital oscilloscope, response in time domain is measured. The typical pulse waveform recorded is shown in Fig. 16. In the figure R1 is input pulse signal wave, and R2 is output waveform. The wave front of R1 and R2 is 153.5ps and 169.1ps respectively. According to the Gaussian approximation formula estimation, its bandwidth is approximately 3 GHz.

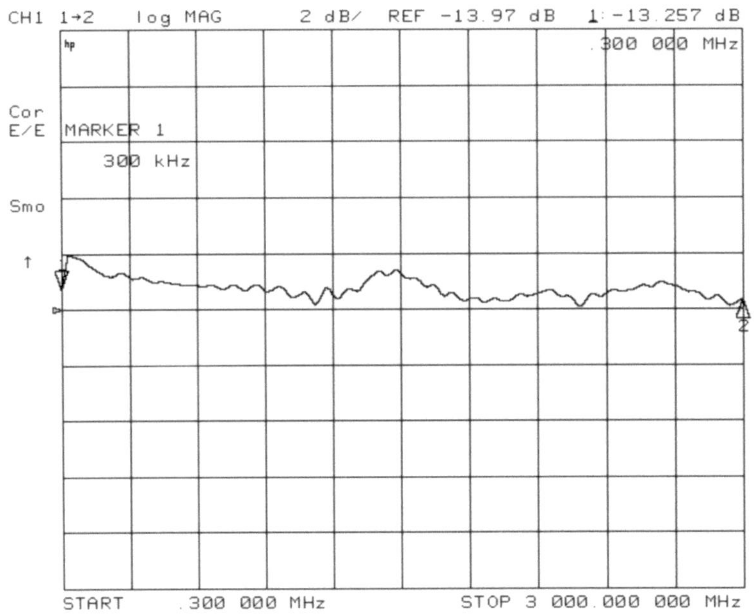


Fig. 15. Curve of amplitude-frequency characteristics.

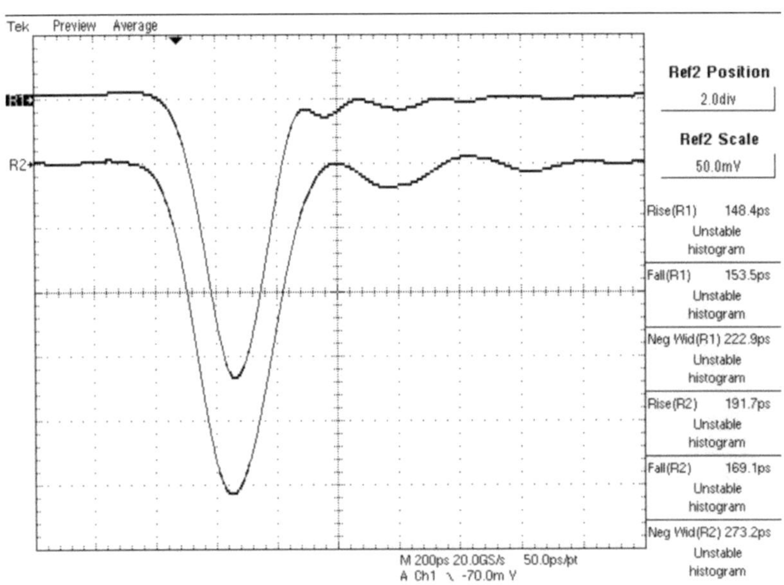


Fig. 16. Waveform of pulse response.

### 4.2.2 Linear dynamic range

The linear dynamic range of the optical fiber transmission system is measured by point-to-point scanning method. The input/output data and fitting curve are shown in Fig. 8. The result indicates that its dynamic range is greater than 100 with non-linear error less than 3%.

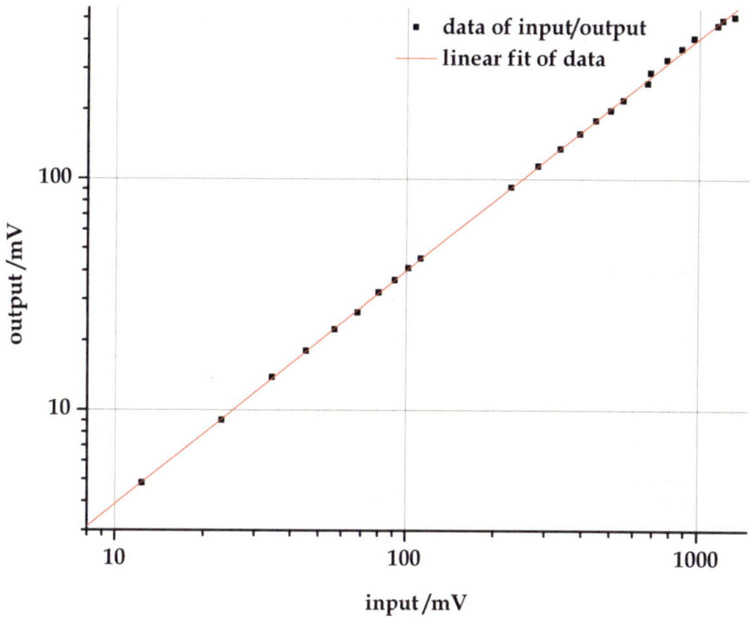


Fig. 17. Data and linear fitting curve of response.

### 4.2.3 Output noise

A broad bandwidth digital oscilloscope is used to record the output noise while the input is zero. A typical waveform recorded is shown in Fig. 18(time scale of 50ns/div, amplitude scale of 1mV/div). It can be seen from the waveform that the peak output noise $V_{p-p}$ is less than 5mV (3.34mV recorded).

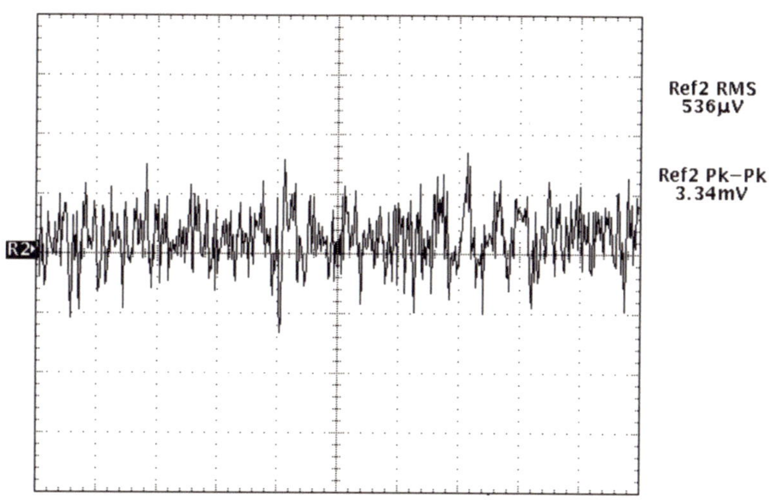


Fig. 18. Waveform of output noise.

### 4.2.4 Standing wave ratio

By a light-wave component analyzer, the measured curve is shown in Fig. 19. (frequency sweep range of (0.0003-3) GHz, amplitude coordinate scale of 1dB/div). The result shows that SWR is less than 2(1.59 recorded).

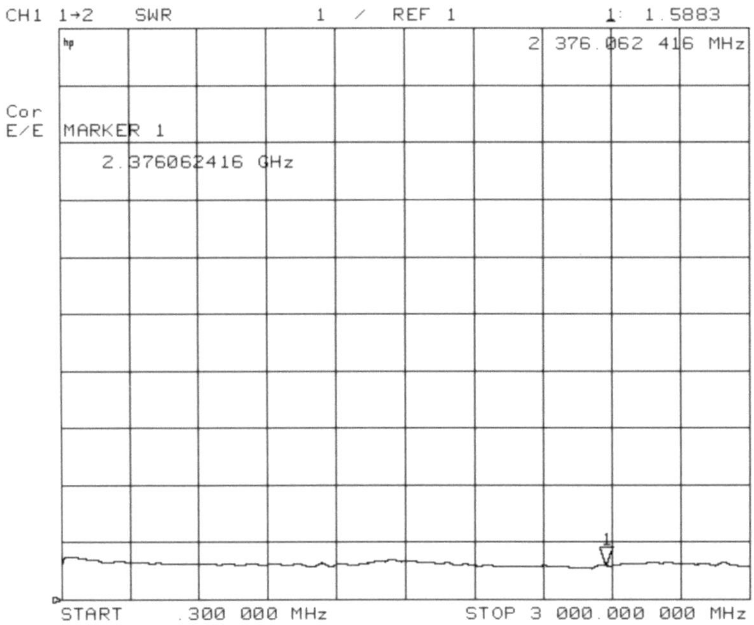


Fig. 19. Curve of standing-wave-ratio in all bandwidth.

## 5. Results and analysis

Two different kinds of pulsed γ-ray devices with average photon energy of 0.3MeV, pulse width of 25ns, dose rate of $2.03\times10^7$Gy/s and average photon energy of 1.0MeV, pulse width of 25ns, dose rate of $5.32\times10^9$Gy are employed as irradiation sources in the experiment. The transient radiation-induced loss of pulsed γ-ray effecting on single-mode and multi-mode optical fibers have been measured. Optical fiber transmission systems with several different wavelength such as 405, 660, 850, 1310 and 1550nm are involved in the experimental measurement system.

### 5.1 Amplitude performance of transient radiation-induced loss

The experiments have been accomplished on two devices with high/low ray flux respectively. The high flux device makes detection system saturate, which has shorter detection wavelengths (405, 660 and 850nm). The low flux device exerts very low response on detection system which has longer detection wavelength (1310 and 1550nm). The signal level is too low to detect. The average radiation-induced loss of optical fibers under relative low and high flux pulsed γ-rays are shown in Table 1. and 2. respectively. It has become evident that the radiation-induced loss experienced by optical fibers is extremely larger than the intrinsic loss and dependent on the fiber type. It is appear that the single mode fiber may be influenced to a lesser degree than multi mode fibers. It is likely from the difference in fractions of optical power that propagates within the cladding of the two types of fiber. To the response of same type fiber, the radiation-induced loss relies upon the wavelength of laser carrier. It is obvious that the shorter the wavelength of laser carrier, the larger the radiation-induced loss of optical fiber.

| optical fiber types | G.651/50/125 | | G.651/62.5/125 | | G.652 | | G.655 | |
|---|---|---|---|---|---|---|---|---|
| detection wavelength /nm | 1310 | 1550 | 1310 | 1550 | 1310 | 1550 | 1310 | 1550 |
| radiation-induced loss /dB/m/Gy | 0.027 | 0.016 | 0.037 | 0.017 | 0.023 | 0.015 | 0.031 | 0.011 |

Table 1. Average transient irradiation-induced loss under low pulsed γ-ray.

| optical fiber types | G.651/50/125 | | | G.651/62.5/125 | | |
|---|---|---|---|---|---|---|
| detection wavelength /nm | 405 | 660 | 850 | 405 | 660 | 850 |
| radiation-induced loss /dB/m/Gy | 6.94 | 6.34 | 1.87 | 7.55 | 6.92 | 2.15 |

Table 2. Average transient radiation-induced loss under high pulsed γ-ray.

### 5.2 Time performance of transient radiation-induced loss

Typical light signal waveform is shown in Fig. 20. The calculation result shows that the response time of transient radiation-induced loss is approximately 5ns. In order to measure the recovery of the tested optical fibers, the optical spectrum loss at the range of 700-1600 nm is measured and compared with non-irradiated ones. The results are shown in Fig. 21. It can be seen from Fig. 21 that the radiation-induced loss of pulsed γ-ray effecting on optical fibers still remains, especially evident in the range of 700-1000nm and around 1390nm. The permanent radiation-induced loss increases with the decrease of wavelength in this range. To verify the heat recovery of radiation effects, the optical fibers have been heated to 80°C for 2h and the above-mentioned loss measurements are repeated then. It can be found that the loss have no changes virtually indicating the existence of permanent radiation-induced loss under the irradiation conditions in this experiment.

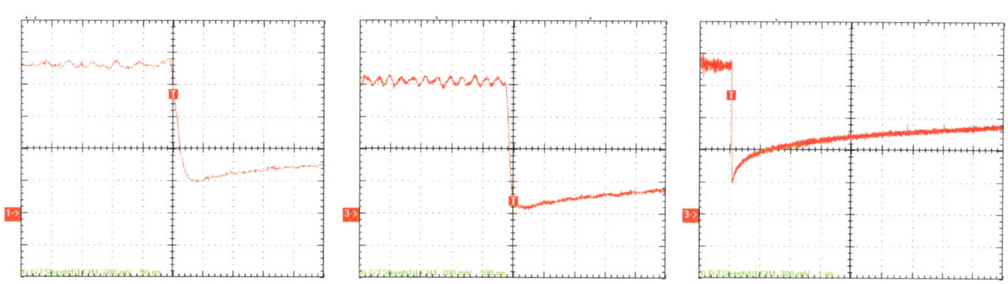

(a)200mV/div 50ns/div (b)200mV/div 100ns/div (c)200mV/div 1000ns/div

Fig. 20. Typical signal waveform recorded in experiments(200mV/div, 50, 100,200ns/div).

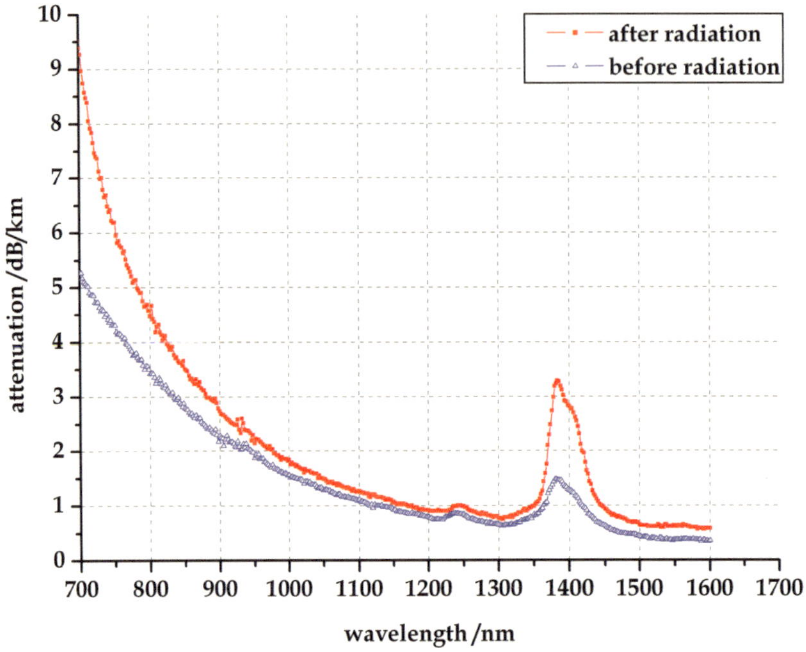


Fig. 21. Optical spectrum loss curve comparison before and after radiation exposure.

### 5.3 Relationship between transient radiation-induced loss and radiation dose

By 850nm laser measurement system, the measured data of low flux γ-ray transient radiation-induced loss on optical fiber G. 651 (62.5/125μm) and fitted curve are shown in Fig. 22. It can be drawn that the transient radiation-induced loss has an approximate linear relationship with total dose in the range of 0.1-3.5 Gy. It is observed that the radiation-induced loss tends to saturation with increasing dose. The saturation is associated with the total radiation-induced color centers in a given length of fiber under test. In the experiments, it is not observed that the decreasing in radiation-induced loss with increasing dose, so called radiation annealing.

Luminescence and Cherenkov lights are not observed in the experiments due to their weak intensities along fiber axis. The sensitivity of the measurement system is to be improved.

### 5.4 Effect analysis

The essential difference between the crystalline and amorphous solid is that there is not long-range order in the latter. Instead there are localized regions of ordered atomic arrangements in amorphous solids that exist only over a few atomic diameters. Therefore there will be localized electronic states within glasses which account for the optical properties in such materials. A color center is an impurity or imperfection within an otherwise well-ordered system. Generally there will be a set of energy levels available for electronic transitions. These energy levels then represent an absorption spectrum while light not absorbed gives the material its characteristic color. Impurities, atomic defects, irregular arrangements of atoms or trapped charge carriers can cause color centers to form with sets of specific absorption spectrum.

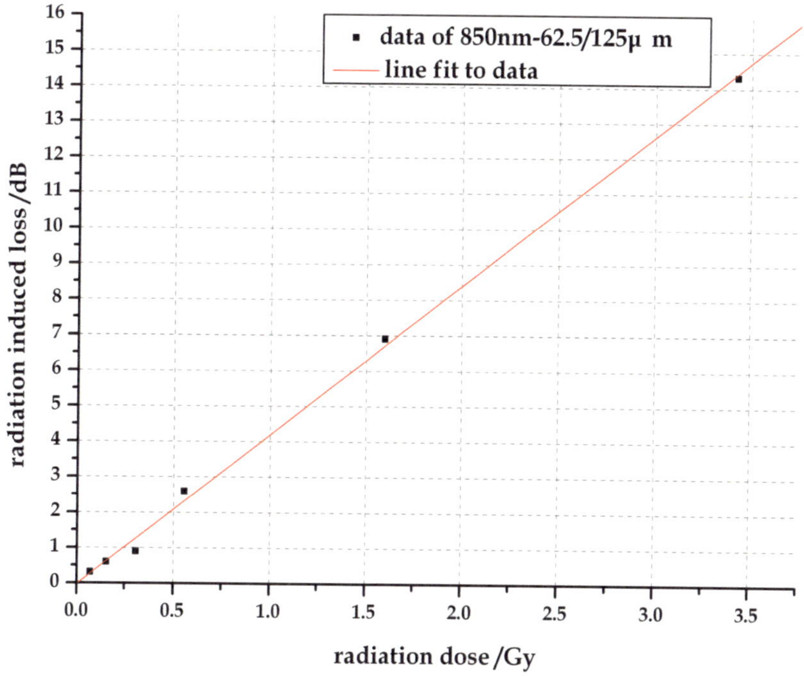


Fig. 22. Experimental data and fitting curve of transient radiation-induced loss.

When γ-ray irradiates on optical fibers, Compton effect will occur usually and the resulting high energy electrons causing the primary radiation damage due to γ-ray absorption in optical fibers. High energy electrons will increase the concentration of color centers, which lead to additional absorption of incident light. Short-lived color center will continue to prevent the formation of new color centers. Therefore in a certain range of radiation dose, radiation-induced loss presents an approximately linear relationship with doses, but as the dose increases, radiation-induced loss tend to get saturated. γ-ray radiation also results in overall optical waveguide deterioration due to changes in the indexes of refraction of core, cladding, or both. It is likely that we could not be able to clearly separate the absorption effect and the index of refraction effect. But in this case, it may be manifest that the absorption effect is dominant.

The role of external heat may accelerate the transition of excitation level, the relaxation of carriers from trapping, or diffusion of color centers, for many forms of color centers are unstable and thermal processes are sufficient to restore the material to its original state. So heat may be beneficial for recovery of radiation-induced loss. But external heat appears invalidation to the permanent radiation-induced loss, which arises from stable color centers.

Radiation-induced loss is somewhat dependent on the fiber types. It appears that the single-mode fibers may be less affected than multi-mode fibers. Single-mode fibers have more concentrated electromagnetic energy distribution than multi-mode fibers. In the same irradiation conditions, the absorption of single-mode fiber radiation dose is relatively small in the mode distribution region, so the radiation-induced loss of single-mode fiber is lower than that of multimode fiber.

## 6. Conclusion

In this chapter, two different dose rates of pulsed γ-ray devices are used to irradiate four kinds of optical fibers. By using near infrared and visible wavelength measurement system, the radiation-induced loss is measured. It can be drawn from the experimental results: (1) Under the same experimental condition, the radiation-induced loss of multimode fibers is slightly larger than single-mode fibers. (2) Radiation-induced loss will increase as the detection laser wavelength shifts from near-infrared to visible regions of optical spectrum. Within a certain dose range transient multi-mode fiber radiation-induced loss displays a nearly linear dependence upon the total dose. (3) Two models are invoked to explain radiation-induced loss. One is that the generation of new color centers in fiber materials will increase the absorption loss in the near infrared and visible region. The other is that the changes of refractive index will lead to additional waveguide loss. Both radiation-induced loss mechanisms exist simultaneously; therefore, radiation-induced loss is the result of joint action of the two. (4) Radiation-induced fluorescence density along the optical fiber axis is so low that measurement system with higher sensitivity is needed (e.g. photoelectric multiple tube). Taking the advantage of effects of radiation, on the one hand scientists can seek methods for decreasing additional loss and develop anti-radiation optical fibers suitable for transmission systems under radiation environments, and on the other hand they can also manufacture radiation dose meters based on this effect.

## 7. Acknowledgment

The work is sponsored by Northwest Institute of Nuclear Technology. The authors would like to express their thanks to Honggang Xie and Weiping Liu in Northwest Institute of Nuclear Technology for helping with valuable discussions and computations.

## 8. References

Akira I. and Junich T.(1988). Radiation resistivity in silica optical fibers[J]. *IEEE Journal of Lightwave Technology*. vol. 6(2), pp. 145-149, ISSN 0733-8724

An Y. Y., Liu J. F., Li Q. H., et al.(2002). *Optoelectronic Technology*[M]. Beijing: Electronic Industry Press, pp.117-120, ISBN 7-5053-7565-2 (in Chinese)

Evans B. D., Sigel G. H. & Jr.(1974). Permanent and transient radiation-induced loss in optical fibers[J]. *IEEE Transactions on Nuclear Science*. vol. NS-21, pp. 113-118, ISSN 0018 9499

Friebele E. J., Sigel G. H. & Jr.(1978). Radiation response of fiber optic waveguides in the 0.4 to 1.7μm region[J]. *IEEE Transactions on Nuclear science*. vol. NS-25(6), pp. 1261-1266, ISSN 0018-9499

Friebele E. J.(1979). Optical fiber waveguide in radiation environments[J]. *Optical Engineering*. vol. 18(6), pp. 552-561, ISSN 0091-3286

Friebele E. J., Lyon P. B., Blackburn J., et al.(1990). Interlaboratory comparison of radiation-induced attenuation in optical fibers. Part III: Transient exposures[J]. *IEEE Journal of Lightwave Technology.* vol. 8(6), pp. 977-989, ISSN 0733-8724

Fernadndez A., Berghmans F., Brichard B., et al.(2002). Toward the development of radiation- tolerant instrumentation data links for thermonuclear fusion experiments[J]. *IEEE Transactions on Nuclear Science,* vol. 49(6), pp.2879-2887, ISSN 0018-9499

Golob J. E., Lyon P. B. & Looney L. D.(1977). Transient radiation effects in low-loss optical waveguides[J]. *IEEE Transactions on Nuclear science.* vol. NS-24(6), pp. 2164-2168, ISSN 0018-9499

Ghoniemy S., Maceachern L., Mahmond S.(2003). Extended robust semiconductor laser modeling for analog optical link simulations[J]. *IEEE Journal of Selected Topics in Quantum Electronics,* vol. 9(3), pp.872-878, ISSN 1077-260X

Huang D. X.(1994). *Semiconductor optoelectronics*[M]. Chengdu: Electronics Science and Technology University Press, pp.144-153, ISBN 7-81016-151-2 (in Chinese)

Hinojosa J.(2001). S-parameter broadband measurements on-coplanar and fast extraction of the substrate intrinsic properties[J]. *IEEE Microwave and Wireless Components Letters,* vol. 11(2), pp.80-82, ISSN 1531-1309

Liu S. H.& Li C. F.(2006). Optoelectronic technology and application [M] Guangzhou, Hefei: Guangdong Science and Technology Press, Anhui Science and Technology Press. pp.800-801, ISBN 7-5359-4186-9 (in Chinese)

Mei Z. Y.(1966). *Nuclear Physics*[M].Beijing: Science Press. pp. 1-36 (in Chinese)

Mattern P. L., Watkins L. M., Skoog C. D., et al. (1974). The effects of radiation on the absorption and luminescence of fiber optic waveguides and material[J]. *IEEE Transactions on Nuclear Science.* vol. NS-21, pp. 81-95, ISSN 0018-9499

Moss C. E., Casperson D. E., Echave M. A., et al.(1994). A space fiber-optic X-ray burst detector[J]. *IEEE Transactions on Nuclear Science.* vol. 41(4), pp. 1328-1332, ISSN 0018-9499

May M. J., Clancy T., Fittinghoff D., et al.(2006). High bandwidth data recording systems for pulsed power and laser produced plasma experiments[J]. *Review of Scientific Instruments,* vol. 77(10), pp.1032-1035, ISSN 0034-6748

Pocha M. P., Golddard L. L., Bond T. C., et al.(2007). Electrical and optical gain level effects in InGaAs double quantum-well diode lasers[J]. *IEEE Journal of Quantum Electronics,* vol. 43(10), pp.860-868, ISSN 0018-9197

Ramsey A. T., Adler H. G. & Hill K. W.(1993). Reduced optical transmission of SiO2 fibers used in controlled fusion diagnostics[R]. *DE93008516*

Tsunemi K., Naoki W., Kazuo S., et al.(1986). Radiation resistance characteristics of optical fibers[J]. *IEEE Journal of Lightwave Technology.* vol.4(8), pp. 1139-1143, ISSN 0733-8724

Tighe W., Adler H., Cylinder D., et al.(1995). Proposed experiment to investigate use of heated optical fibers for Tokamak diagnostics during D-T discharges[R]. *DE95007355*

Tanaka T, Hibino Y., Hashimoto T., et al.(2002). Hybrid-integrated external-cavity laser without temperature-dependent mode hopping[J]. *IEEE Journal of Lightwave Technology*, vol. 20(9), pp.1730-1739, ISSN 0733-8724

Yasuo K. *Light-wave Engineering*[M]. Kyoritsu Shuppan Co., Ltd. and Science Press, pp.131-137, ISBN 7-03-010186-3, Beijing (in Chinese)

Zivojinovic P., Lescure M. & Tap-Béteille H.(2004). Design and stability analysis of a CMOS feedback laser driver[J]. *IEEE Transactions on Instrumentation and Measurement*, vol. 53(1), pp.102-108, ISSN 0018-9456